ACCESS TO KNOWLEDGE
IN THE AGE OF INTELLECTUAL PROPERTY

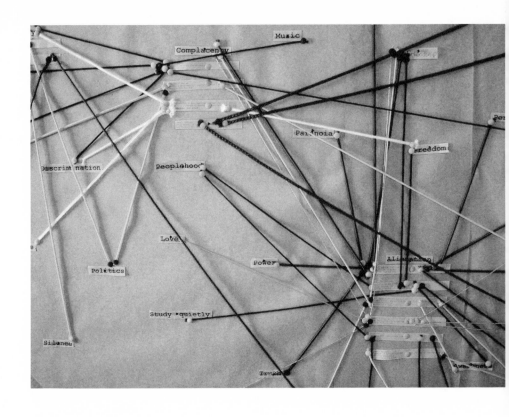

ACCESS TO KNOWLEDGE
IN THE AGE OF INTELLECTUAL PROPERTY

edited by

Gaëlle Krikorian and Amy Kapczynski

ZONE BOOKS · NEW YORK

2010

The publisher wishes to acknowledge the generous support of
the Open Society Institute.

Printed in the United States of America.

Distributed by The MIT Press,
Cambridge, Massachusetts, and London, England

Frontispiece: Graeme Arendse, *Chimurenga Library*
(photo Stacy Hardy).

Library of Congress Cataloging-in-Publication Data

Access to knowledge in the age of intellectual property /
edited by Gaëlle Krikorian and Amy Kapczynski.
    p.   cm.
  Includes bibliographical references.
  ISBN 978-1-890951-96-2
     1. Intellectual property. I. Krikorian, Gaëlle, 1972–
II. Kapczynski, Amy.

K1401.A929 2010
346.04'8—dc22
               2009054048

Contents

EPILOGUE: A2K IN THE FUTURE—VISIONS AND SCENARIOS

Preface

In a hospital in South Korea, leukemia patients are expelled as untreatable because a multinational drug company refuses to lower the price of a life-saving drug. Thousands of miles away, a U.S. group called the Rational Response Squad is forced by the threat of a copyright lawsuit to take down a YouTube video criticizing the paranormalist Uri Geller. Could we—should we—see these two events, so seemingly remote from one another, as related? Yes—or such is the premise of a new political formation on the global stage, one that goes under the name of the "access to knowledge movement"—or more simply, A2K.

A2K is an emerging mobilization that includes software programmers who took to the streets to defeat software patents in Europe, AIDS activists who forced multinational pharmaceutical companies to permit copies of their medicines to be sold in South Africa, and college students who have created a new "free culture" movement to "defend the digital commons"—to select just a few. A2K can also be seen as an emerging set of theoretical commitments that both respond to and reject the key justifications for "intellectual property" law and that seek to develop an alternative account of the operation and importance of information and knowledge, creativity and innovation in the contemporary world. (The quotes reflect the fact that A2K calls the concept of "intellectual property" into question, because of its tendency to reify the form of legal regulation that it represents. Some argue that the term itself should be banished; we nonetheless use it here because most A2K advocates have found it indispensable, as a term that designates the broad and diverse restrictions on the exchange of information and knowledge against which they have emerged and mobilized.) *Access to Knowledge in the Age of Intellectual Property* takes as its subject this new field of activism and advocacy and the new political and conceptual conflicts occurring in the domain of intellectual property.

Why is intellectual property becoming the object of a new global politics today? Can file sharers, software programmers, subsistence farmers, and HIV-positive people find useful common cause in their joint opposition to existing regimes of intellectual property? What concepts might unite the emerging A2K coalition, and what issues might fracture it? What is at stake with the use of the term "access" as a fulcrum of this mobilization? Is A2K more than an agenda for those opposed to restrictions on intellectual property—and should it be?

This volume takes such questions as its object. It aims to make this new field of political contention accessible to those unfamiliar with it and to provide a place for those generating it to analyze its evolution, goals, tensions, and future. The contributions come from a varied mix of activists and academics and from different parts of the world. This makes for an eclectic and sometimes even uncomfortable mix, one true to the emerging dynamics of the A2K movement itself. Their subjects are also diverse, part of our own editorial attempt to avoid narrowly prescribing the contours of A2K even as we inevitably, through these same selections, construct them.

The book itself is divided into four parts and an epilogue. The first section offers two introductions to the field of A2K. It should serve to orient readers entirely new to debates over intellectual property, but also to provide fodder for debate among those who consider themselves peripheral or central actors in the movement itself. The first introduction, by Amy Kapczynski, offers a conceptual genealogy of the A2K movement—an account of the concepts and arguments that its participants are generating in order to theorize their common condition and to undermine the narrative about intellectual property that has justified the expansion of this form of law and governance over the past few decades. The second introduction, by Gaëlle Krikorian, examines A2K as a field of activism. It describes how the mobilization has emerged and organized itself using the issue of "access," the technological and political context to which the movement corresponds, the representations and practices it engages, and its political stakes both as a form of social mobilization and as an alternative to intellectual property rights extremism.

The second section of the book provides a geography of the new field of activism and advocacy that constitutes A2K. With no pretense to being comprehensive, it illuminates a series of historical moments that have decisively marked the emergence of the politics of A2K. It thus identifies a series of fronts along which intellectual property conflicts are crystallizing and sketches A2K mobilizations across a spectrum of political space and time.

In this section, Ahmed Abdel Latif describes how A2K has been framed as a concept and the genesis of the A2K name, thereby locating A2K as a field of forces gathering together under a common banner. Thereafter, several historical moments in A2K illustrate how, where, and when certain key issues surfaced and were rendered the subject of politics. Ellen 't Hoen describes how health activists work-ing on pharmaceutical policy came to conceptualize intellectual property as central to their struggles. Sangeeta Shashikant narrates the behind-the-scenes forces that led to one of the most salient moments of success for A2K, the Doha Declaration of the World Trade Organization, which declared that intellectual property rights do not trump public health. Moving from medicines to the emerging politics of hackers, Philippe Aigrain analyzes the successful mobilization against the codifica-tion of software patents at the European Parliament. The last contribution in this section comes from Viviana Muñoz Tellez and Sisule F. Musungu, who describe two recent and dramatic defeats for intellectual property absolutism at the World Intellectual Property Organization (WIPO). In the first, A2K activists working with developing-country governments outflanked their opponents, proposing a new "development agenda" that seeks to reorient the work of WIPO to respond to the needs of those living in the Global South. In the second, A2K activists and their allies mobilized to defeat a new WIPO Broadcasting Treaty that had been heav-ily promoted by forces in the old media seeking to extend their control over the domain of new media.

The third section of the book offers varying visions—perhaps complementary, perhaps at odds with one another—of the conceptual terrain of the A2K move-ment. It charts the evolution of ideas and the surfacing of arguments within the movement and thereby explores how the issue of intellectual property has been politicized and how our collective understandings of what is at stake in these debates have been tentatively transformed by A2K activists.

The section begins with Peter Drahos's account of the global mobilization of intellectual property owners that preceded and helped to shape A2K. That mobi-lization was exceptionally successful—in a matter of years, it secured a dramatic reordering of the global governance of intellectual property, most importantly by inserting intellectual property obligations into the new World Trade Organization. These efforts were sustained by the ideological interventions that Drahos describes.

In response to these interventions, A2K advocates have attempted to reframe public understandings regarding the just and efficient conditions for the use, creation, and re-creation of knowledge. Many use the issue of access as a lens, possibly theoretical and certainly strategic, to refocus traditional political con-figurations around intellectual property and to set out their claims. Yochai Benkler articulates the "information commons" as the central concept of A2K and describes the historical and political forces that converged to create the conditions for this striking new field of political coalition. Interventions by Carlos M. Correa, Roberto Verzola, Gaëlle Krikorian, Jeffrey Atteberry, and Lawrence Liang explore paradoxes and tensions in the emerging discourse of A2K along vectors ranging from indigenous knowledge, in the essay by Carlos Correa, to the notion of the commons, in Jeffrey Atteberry's contribution, and the figure of the pirate, in one of the essays by Lawrence Liang. Robert Verzola and Lawrence Liang, in another essay, each offer us new paradigms for the relationship between knowledge and the production and control of knowledge-embedded goods, thus offering us new ways in which to think about the struggle between A2K and intellectual property. Verzola theorizes the commonalities between technological measures used to disrupt the reproducibility of information in the digital and agricultural realms and challenges us to rethink the domain of information production as one of abundance and fertility, rather than scarcity. Liang explores etymological links between identity and property and considers the implications of thinking about intellectual and cultural production through the dynamics of relationality, rather than possession. Gaëlle Krikorian, focusing on free-trade agreements, offers an analysis of the political environment and the political rationales of the maximalization of intellectual property protection and examines some of the perspectives and experiences of the resistances to it.

The section closes with an opening, reproducing questions that we distributed to a group of A2K actors who have different approaches to and involvements in the movement—Onno Purbo, Jo Walsh, Anil Gupta, and Rick Falkvinge. The questions invited them to elaborate on the concepts and ideology central to A2K, and their responses illustrate the diversity of views on these matters that exist within the movement.

A2K activists have proven remarkably creative and successful in recent years, not only in contesting the contours of intellectual property law, but also in identifying weaknesses and failures in the regime of intellectual property, spaces where new regimes for generating and managing knowledge and knowledge goods might evolve. The third section of the book describes A2K by exploring its strategies and tactics. It thereby seeks to illuminate how the mobilization has politicized this previously "technical" area of law and policy and at times has successfully combated very well-resourced and politically powerful opponents.

By comparing different strands within A2K, Susan K. Sell articulates the various grammars of claims-making of movements within the movement. A series of detailed case studies of strategies deployed in specific contexts then permits us to mark and critically assess the choices and stances being made in the name of A2K: in India, the choice NGOs made to master and rework the discourse of patent law in order to oppose drug patents (Chan Park and Leena Menghaney); in Thailand, the efforts made to reduce medicine prices by pressing the government legally to override patents (Jiraporn Limpananont and Kannikar Kijtiwatchakul); in South Africa and elsewhere, the deployment of the rhetoric and law of competition to attack exclusive rights in information (Sean Flynn); in an NGO in the United States, the creation of an open-access journal that sought to develop knowledge-governance principles and practices consistent with the commitments of the movement (Manon A. Ress); at technological standard-setting organizations, debates over the nature and terms of open standards (Laura DeNardis); at WIPO, attempts to introduce new multilateral agreements to defend the rights of the visually impaired and rebalance the current copyright regime (Vera Franz); and finally, in the domain of global health law, the promotion of alternative models for medical research and development that would better combine the twin goals of access and innovation (Spring Gombe and James Love).

This section next reproduces a series of questions and responses solicited from advocates (Harini Amarasuyiya, Vera Franz, Heeseob Nam, Carolina Rossini, and Dileepa Witharana) regarding contemporary strategic and tactical opportunities and dilemmas in A2K. Participants were invited to reflect upon how the movements and groups with which they are associated have articulated their principles and campaigns, defined their goals and translated these into practice, and related to law, the state, private interests, and others in the A2K coalition.

The section closes with two interviews that provide practical as well as theoretical dialogues on the transformations associated with A2K as they affect society and the economy. Yann Moulier Boutang and Gaëlle Krikorian engage the implications of the emergence of "cognitive capitalism" for knowledge industries as well as for governments and individuals. Charles Igwe and Achal Prabhala discuss the knowledge-governance and dissemination strategies that characterize the Indian and Nigerian film industries and how these might inform debates about A2K.

To end the volume in a mode that invites continuing reflection, an epilogue offers a series of visions of the future by authors—Sarah Deutsch, Gaëlle Krikorian, Eloan dos Santos Pinheiro, Hala Essalmawi, and Roberto Verzola—who were asked to imagine best-case and worst-case scenarios of the regulation and production of knowledge in their field of interest. Unconstrained by the imperative to describe "likely" scenarios, they offer us alternative visions that illuminate the stakes of the choices that we make today and how these choices could portend radically different futures for access to knowledge.

As the diversity of the volume demonstrates, the conceptual and political dynamics of the A2K movement reveal it as a mobilization that is very much still in motion. Neither in the introductions that follow nor in this collection as a whole do we purport to describe fully, account for, or locate the movement for access to knowledge. The name itself is contestable and may not be the one that represents this new politics over time. Nor is it clear what shape this new politics will take— how much it will tend toward conceptions of information and how much toward issues of knowledge, how much it will attend to or be driven by the concerns of the Global South as opposed to those of the North, what modes of engagement with law and with activism will characterize the mobilization over time, or who will constitute the center and who the periphery when historians write the story of A2K.

But despite this still-provisional nature, the A2K movement has already begun to reveal an important reality: Today, freedom and justice are increasingly mediated by decisions that were until recently considered supremely technical—decisions about the scope of patent law, about exceptions and limitations to copyright for the blind, about the differential virtues of prizes and patents for stimulating government investment in neglected diseases. By politicizing a discourse that was once highly technocratic, the A2K movement is rendering visible once-obscure vectors of the transmission of wealth and of power over life and death. It demands that the concepts and terms central to intellectual property be introduced into everyday discourse and become legible in their political implications around the world. This volume, we hope, will assist in that project.

PART ONE

INTRODUCTION

Access to Knowledge: A Conceptual Genealogy

Amy Kapczynski

A decade or two ago, the words "intellectual property" were rarely heard in polite company, much less in street demonstrations or on college campuses. Today, this once technical concept has become a conceptual battlefield. A Google search for the term, for example, first turns up a ferociously contested Wikipedia definition.[1] When I did the search, after two links to the World Intellectual Property Organization (WIPO) Web site, the next most important page according to Google's ranking algorithm was an article called "Did You Say 'Intellectual Property'? It's a Seductive Mirage," by free-software guru Richard Stallman.[2]

Criticisms of the existing state of intellectual property law have gone viral, turning up around the world in domains as diverse as software, agriculture, medicine, and music. Activist efforts to challenge the contours of intellectual property law are increasingly interconnected and gathered (especially globally) under the call for "access to knowledge" or "A2K."[3] A2K is a mobilization very much in process — it hasn't yet been subject to the kind of histories or hagiographies that would render one description or account of it authoritative. Rather than provide such an account, this introductory essay seeks to locate A2K in two ways: as a reaction to structural trends in technologies of information processing and in law, and as an emerging conceptual critique of the narrative that legitimates the dramatic expansion in intellectual property rights that we have witnessed over the past several decades.

As the following pages describe, new information-processing technologies have made certain kinds of knowledge and information increasingly critical to the accumulation and distribution of global wealth, as well as to the terms of our bodily and social existence. Information-processing industries responded to these shifts by pressing for — and achieving — unprecedented extensions of intellectual property rights in order to gain more control over the use and exchange of information across the globe.

This move was not just a naked expression of lobbying power, although it was that, too. Importantly, a conceptual narrative legitimated this shift. As we'll see, this narrative is not a single theory, but an amalgam of theories drawn from different domains and spun together to appear as one coherent account. The A2K movement is challenging the coherence of this account by formulating a series of critical concepts, metaphors, and imaginaries of its own—concepts such as the "public domain" and the "commons" and ideals such as "sharing," "openness," and "access." These concepts are sometimes self-consciously cultivated by activists and at other times can more accurately be said to be immanent in their claims.

One way to map the A2K movement, then, is to explicate the most important of these concepts by analyzing the work that they do to challenge the prevailing justifications for intellectual property law. A conceptual genealogy of this sort can help us not only better understand the political conflicts that are emerging around issues of intellectual property rights, but also determine who is or may become part of the A2K mobilization. Finally, it can also help us map key conceptual tensions in the field of A2K, ideational vectors that pull this new discourse in one direction or another along the spectrum of political vision and action where the A2K movement is being assembled. This introduction thus closes by articulating a series of questions that confront A2K as it looks to the future.

HOW KNOWLEDGE MATTERS

To understand why and how a new politics of intellectual property is arising today, we must first understand something about why and how knowledge matters in the world today—both how it makes a difference in our world and how it is implicated in the materialization, the making into matter, of that world.

Although knowledge has always mattered to the organization of human societies, in recent years, prominent economists and social theorists have sought to demonstrate that knowledge has come to matter in a new way. When the purported shift happened and what it means depends upon how the change is characterized.

In the economic perspective, knowledge matters in its technological capacity, for its effect on productivity and growth. Karl Marx and Joseph Schumpeter early on posited that capitalism relies on technological dynamism,[4] but the role of knowledge was not recognized in the neoclassical paradigm until the work of Robert Solow in the 1950s. Solow posited a connection between knowledge and economic growth, arguing that the vast proportion of gains in productivity in early twentieth-century America could be attributed not to factors related to the use of labor or capital, but to a "residual" that he described as technical change.[5] Solow's residual came to be understood as a range of advances in knowledge—from new machines (such as tractors) to new management techniques (such as Fordism)—that made processes of production more efficient.[6]

Mainstream economists soon began to contend that knowledge is not only important, but *increasingly* important to economic growth, positing that the world's most developed economies have been becoming more knowledge intensive. Fritz Machlup took note of the way the U.S. economy was changing in the 1960s, a change that was first marked by "an increase in the share of 'knowledge-producing' labor in total employment."[7] At the turn of the twentieth century, for example, one-third of U.S. workers were employed in the service industries. By 1980, close to seven in ten were.[8] The trend that Machlup and his colleagues were identifying in the United States was in fact occurring across so-called developed economies as agricultural and to a lesser extent industrial jobs steadily lost ground to jobs in sectors such as education, finance, information technology, and the culture industry.[9] The most productive component of these economies shifted from industrial sectors to "information-processing" sectors such as financial services, marketing, biotechnology, and software.[10]

Perhaps the most prominent theorist of this shift, Manuel Castells, refers to this as a transition to the "informational" mode of development. Informationalism is not identified by the importance of knowledge to the economy, for knowledge was essential to the industrial mode of development too. Rather, it derives from the fact that "the action of knowledge upon itself [is] the main source of productivity." New information and communications technologies permit accelerating feedback loops of innovation and information processing, making the human mind "the direct productive force, not just a decisive element of the production system."[11] Manufacturing and agriculture of course do not disappear, but information processing—for example, in computing, genetic engineering, or management techniques—decisively determines their productivity.

Can the shift truly be characterized as global, given that it is centered in a few of the world's wealthiest countries? Castells says yes, because the economy today can work "as a unit in real time ... on a planetary scale" and because local economies everywhere depend "on the performance of their globalized core," which includes "financial markets, international trade, transnational production, and, to some extent, science and technology, and specialty labor."[12] Also, developing countries that have long labored under a trade imbalance with regard to manufactured goods and raw materials and the unequal distributions of wealth generated by these now labor under a "new form of imbalance" regarding "the trade between high-technology and low-technology goods, and between high-knowledge services and low-knowledge services, characterized by a pattern of uneven distribution of knowledge and technology between countries and regions around the world."[13]

The discourse about the rising centrality of knowledge to economic growth seems to imply a claim that human society—and more specifically, certain societies— are becoming more knowledgeable, leaving others behind. (Note how Castells refers to the "uneven distribution of knowledge... between countries and regions around the world.") In fact, the claim should be understood to be narrower because of the circumscribed form of "knowledge" implicated here. For Castells, for example, knowledge is defined as "a set of organized statements of facts or ideas, presenting a reasoned judgment or an experimental result, which is transmitted to others through some communication medium in some systematic form."[14] The focus here is thus on those forms of knowledge that are central to economic productivity and efficiency—namely, technical and scientific knowledge. There are, of course, many other kinds of knowledge, such as ethical knowledge or knowledge of a person. As I will describe later, in its broadest sense, knowledge can be described as a competence that only sometimes relates to a technical effect.

The claim that knowledge is increasingly central to the global economy—or that the global economy is today "informational," rather than industrial—thus must be understood as a more specific claim: that advances in the ability of humans to codify, organize, exchange, and test certain kinds of scientific and technical knowledge have created revolutionary changes in modes of economic productivity. These changes can be traced back many centuries, for example, to the advent of the printing press—a technology that made copying much more reliable and written texts much more widely available and that enabled feedback loops that allowed information to be collected and corrected over time.[15] Newer information and communications technologies have intensified this process by increasing the speed of information transfer and processing, earlier through technologies such as the railroad and telegraph and more recently through the pervasive networking of digital technologies that we associate, for example, with the Internet.[16] This increased capacity to codify, store, process, and exchange information has been a precondition for the development of information-intensive sectors from biotechnology to financial engineering. It is also a precondition of the shift toward more flexible, networked, information-intensive business systems such as just-in-time production.[17]

Of course, such shifts have implications far beyond the realm of economics. The same transformations that have made scientific and technical knowledge more central to the global economy, for example, have also made such knowledge more central to human health. Globally, life expectancy has increased by almost twenty years since the 1950s.[18] This can be attributed in substantial part to advances in scientific knowledge about disease and to increased access to such knowledge, for example, as embodied in better sanitation and vaccines.[19] The rise of new forms of knowledge management and the application of sophisticated information-processing schemes to fields such as health and agriculture means that our relationships to our very bodies—how we eat, whether we live—are more intimately governed by scientific and technical knowledge and information than ever before.

For Castells, as well as for earlier theorists such as Daniel Bell, not just our economies, but our societies thus have become increasingly knowledge intensive or informational. In this sociological conception, changes in our ability to codify, communicate, and process knowledge have inaugurated a new relationship between knowledge and society. This shift is reflected, for example, in a new ordering of occupations, one in which professional and technical classes gain preeminence.[20] It is also reflected in governance, because policy formation is newly focused around knowledge and expertise "for the purpose of social control and the directing of innovation and change."[21]

For example, the rise of statistics and the field of "political arithmetic" led to the development of the modern census, which made possible the use of population data in government for the first time.[22] New fields of social knowledge such as psychoanalysis, penology, and pedagogy also came into being, subjecting the human to new forms of technological production and surveillance.[23] Knowledge thus has become central to the "activities of government and to the very formation of its objects, for government is a domain of cognition, calculation, experimentation and evaluation."[24] From philosophy to medicine, accounting to education, and town planning to social insurance, "know-how" and technology make modern governance possible.[25]

New systems of knowledge and information technologies also inaugurate shifts in the relationship between individuals and these processes of economic production, social control, and governance. The digital network revolution, for example, places the technologies of information production and exchange in the hands of (at least some) "average" citizens in a way that was not true in the era of the industrial assembly line and the printing press. As Yochai Benkler argues, the contemporary processing power of computers ubiquitously linked together creates a platform for new kinds of collaborative human action and production, exemplified by projects such as Wikipedia and free software. This shift creates the potential for "an increasing role for nonmarket production in the information and cultural production sector, organized in a radically more decentralized pattern than was true of this sector in the twentieth century." It also creates the possibility of new forms of political activism and new relationships between those who govern and those who are governed.[26] One new arena where this activism has developed and where the relationship between those who govern and those who are governed has played out is the realm of intellectual property law, which has expanded globally to an unprecedented extent in the past few decades.

In 2006, the "ex-gay" group Exodus International sought to force blogger Justin Watt of justinsomnia.org to remove the parody (bottom) of its billboard (top) from the Internet, accusing the blogger of copyright infringement. See Lia Miller, "Both Sides in Parody Dispute Agree on a Term: Unhappy," *New York Times*, March 27, 2006.

Intellectual property rights are legal entitlements that give their holders the ability to prevent others from copying or deploying the covered information in specific ways. Patents, copyrights, and trademarks are the most familiar forms of intellectual property.[27] Each regulates information in a different way. Patents typically cover forms of technological invention—once things such as machines and mousetraps and today things such as new molecules, plant varieties, and software. By describing his invention and showing that it is new, useful, and "nonobvious," an inventor can obtain a patent that gives him the right to prevent others from making, using, or selling the invention for a period of 20 years. Copyrights typically cover expressive or literary works—classically, maps, charts, and books, but today also things such as sound recordings and software. The holder of a copyright can prevent others from copying or performing the protected expression or creating "derivatives" of that expression (for example, creating a screenplay out of a novel) for upward of 100 years.[28] Trademarks protect the use of a distinctive trade name in commerce, permitting the holder of the mark (for example, Rolex™) to restrict its use, most centrally to ensure that consumers are not confused about the origin of a good.[29]

The grouping of these different modes of regulation under the rubric of intellectual property is not uncontroversial.[30] Nonetheless, the rubric usefully helps us to identify a mode of legal regulation that applies to different areas of technology and commerce. In an alchemy that turns immaterial expressions and ideas into tradable commodities, intellectual property rights effectively give creators the ability to market information while also preventing it from being imitated and reproduced by others. These rights can, of course, lead to substantial revenues for those who hold them (and also to substantial economic costs for society, as I'll describe in a moment). Less obviously, but no less importantly, intellectual property doctrines that govern the ownership of creations made in the course of employment structure the distribution of benefits between corporations and employees. The so-called "work for hire" doctrine, for example, regulates whether the inventions or creations that a person makes at work belong to her or to her employer, and over the course of the nineteenth and early twentieth centuries, this doctrine became far more favorable to employers.[31]

But shifts in intellectual property law, like shifts in the way that knowledge and information matter, have effects beyond the domain of the economy. They also directly mediate human experience, well-being, and freedom. The rules of copyright, for example, regulate who can speak and read. Examples of copyright owners seeking to censor speech with which they disagree emerge with relentless regularity. Copyright also endemically shapes how we learn and think, because, for example, it affects the prices of textbooks and the viability of online archives.

Intellectual property law is perhaps at its most controversial in public debates where it regulates life itself—that is, in the domain of medicine. Because patents limit competition, they tend to raise the price of pharmaceuticals. That can put life-saving treatments out of reach, especially for the world's poor. Patents also shape the priorities of our medical research and development (R&D) system. Our existing system, which relies heavily on patents—and thus on high prices—to incentivize R&D has directed enormous sums into treating the ailments of the very rich and almost nothing into treating those of the very poor.

Because intellectual property law regulates strategies of information production and the appropriation of value from information in the marketplace, it has become a central battleground in the struggles over the structure and spoils of the contemporary economy. Because intellectual property law also regulates much more—from how we are able to learn, think, and create together to how and whether we have access to the medicines and food that we need to live—it has become a central site of political struggle, not just locally, but globally.

Both trends have been accelerated by the explosive expansion of intellectual property rights that has occurred in recent years. In countries such as the United States, for example, intellectual property rights have become broader (covering more kinds of information), deeper (giving rights holders greater powers), and more punitive (imposing greater penalties on infringers).[32] Supplemental measures have also been introduced to increase the technological control of rights holders and to counter the way that digital technologies facilitate copying. Anticircumvention laws have been introduced, for example, that prohibit the cracking of technological locks, such as forms of encryption that a copyright holder might place on a song or DVD to control how it is played.

This shift has been called a "second enclosure movement," a metaphorical move that casts it as a modern-day analogue of the privatization of common lands that occurred in stages in England from the fifteenth through the nineteenth centuries.[33] Metaphors of enclosure and its antipode, the commons, have been central to the attempt to mobilize against the encroachments of exclusive rights in the digital age. But they are also problematic.[34] Drawing as it does on the post-feudal history of England, for example, the concept of enclosure domesticates what is better understood as a global phenomenon. The most dramatic expansions of intellectual property rights in recent years have occurred across, rather than within national borders.

In many ways the most striking aspect of the expansion of intellectual property law is the shift inaugurated by the TRIPS (Trade-Related Aspects of Intellectual Property Rights) Agreement.[35] Adopted in 1995, TRIPS was the brainchild of key players from the multinational information industries, that is, companies whose primary business is the production and processing of information and informational goods. CEOs from companies such as Pfizer, Merck, Monsanto, DuPont, General Motors, IBM, and Warner Communications, through a high-powered lobbying group known as the Intellectual Property Committee, persuaded the United States, Europe, and Japan that the agreement was needed to protect their national interests in strong intellectual property protection.[36]

The TRIPS Agreement represented a radical shift in at least three ways. Although treaties on intellectual property were not new (and indeed are remarkably old), before TRIPS, such treaties were generally overseen by the WIPO. WIPO had no enforcement capability, and countries could choose to join treaties in "à la carte" fashion. TRIPS was instead to be part of the new World Trade Organization (WTO). Under the WTO's "single undertaking" rule, countries would not be able to join the WTO without also adhering to the TRIPS Agreement. Because the WTO carried with it a new dispute-resolution system, violations of TRIPS would now be punishable with trade sanctions. Finally, the intellectual property standards incorporated into the agreement were far more expansive than those that were in force in many countries at the time, particularly for developing countries. For example, TRIPS required members to offer patent protection for medicines, to create property rights in new varieties of plants, and to impose criminal penalties for those who "pirate" copyrighted movies or trademarked handbags.

The negotiations that produced TRIPS were a terrain of open struggle between countries of the Global North and those of the South. Developing countries generally opposed the suturing of intellectual property laws into the new regime of world trade, arguing that intellectual property law restricts, rather than promotes free trade, that Northern countries had developed under conditions of low intellectual property protection, and that TRIPS is simply a mechanism to transfer wealth from the South (overwhelmingly an importer of informational goods subject to intellectual property rights) to the North (whose corporations own the vast majority of what constitutes intellectual property today).

Northern countries, led by the United States and pushed by multinational companies, were unyielding: Regime change in the area of intellectual property was to be a condition for membership in the WTO. The United States was eventually able to prevail through "a sophisticated process of trade threats and retaliation" that forced key countries to yield.[37] As Peter Drahos analyzes it:

For the U.S. state there [was] also a payoff. By helping its multinational clientele to achieve *dominium* over the abstract objects of intellectual property, the U.S. goes a long way towards maintaining its *imperium*.... A global property regime offers the possibility that abstract objects come to be owned and controlled by a hegemonic state. Algorithms implemented in software, the genetic information of plants and humans, chemical compounds and structures are all examples of abstract objects that form an important kind of capital.[38]

TRIPS was an exceptionally audacious attempt to extract value from and exert control over informational domains in virtually all of the countries of the world. As such, it has less in common with localized enclosure movements than with colonial strategies of conquest.

In the words of the great chronicler of empire Joseph Conrad, "The conquest of the earth...is not a pretty thing when you look into it too much. What redeems it is the idea only. An idea at the back of it...and an unselfish belief in the idea — something you can set up, and bow down before, and offer a sacrifice to."[39] Here, that idea is one that is not propounded by any particular theorist, but rather that is mobilized in political discourse, occupies the realm of popular political culture, and is used to justify the dramatic expansion of intellectual property that we have seen in recent decades.

LEGITIMATING INTELLECTUAL PROPERTY IN THE INFORMATION AGE

The legitimation narrative of intellectual property today is not a coherent theory, but a thaumatrope — two different images on a card or disk, recto and verso, that when spun on an axis give the appearance of a single, unified image. One image is derived from the field of information economics, but omits the skepticism about intellectual property present in that field. The other screen is derived from the theories of the Chicago School of economics about the superiority of private-property rights in material resources, but suppresses the many significant differences between the economics of land and the economics of information.[40]

We can call the result the "despotic dominion" account of intellectual property law — the notion that the right to intellectual property is, or should be, as William Blackstone described the right to material property, "that sole and despotic dominion which one man claims and exercises over the external things of the world, in total exclusion of the right of any other individual in the universe."[41] Property here is defined as the right of a single individual to be the gatekeeper with respect to a resource and to act autocratically with respect to decisions about its use. This vision of property is sustained by the notion that only the individual owner, and not the state, community, or nonowners, may make decisions about the price or terms of transactions around that property.

This account should not be confused with actual existing intellectual property law (or actually existing property law, for that matter).[42] Rather, the despotic dominion account is a narrative that has been used to justify the aggressive expansion of intellectual property rights in recent years, and it is thus this narrative that A2K confronts as it seeks to change the politics of intellectual property law today.

The first image in the despotic dominion account draws selectively on the field of information economics, arguing that intellectual property is needed to promote investment in informational goods. Information, we are told, is typically expensive to produce, but cheap to reproduce. For example, it is relatively expensive to synthesize and test a new pharmaceutical compound or to produce a major motion picture. Under today's technological conditions, it is also relatively cheap to reverse engineer a drug or to copy a DVD. In an unregulated market, second-comers could reproduce the drug or movie, paying only the cost of copying and without paying the full costs of the producing of the drug or movie in the first place. These "free riders" would be able to drive the innovator from the marketplace, because they would be able to sell the drug or movie more cheaply. The result: Rational actors will not develop drugs or make major motion pictures, because they will be unable to turn a profit, and indeed may suffer a loss, being unable to recoup their original investment.

Enter the deus ex machina of intellectual property rights. Patents and copyrights give individuals (or more likely, firms) the right to prevent others from copying their creations for a period of time. This lets them recoup their investments and make a profit. Exclusion rights thus generate markets in information, solving the free-rider problem and aligning individual incentives with social good.

Consider the suppositions of this first image: Creative and scientific works are best generated by rational, self-interested market actors who are motivated by profit. Intellectual property law provides the control needed to "incentivize" this creativity, because it permits individuals to profit through the sale of informational goods. Individual legal entitlements such as these are necessary because rational creators will not create if they cannot profit and/or if others can ride free. When they can profit, creators will create in accordance with social welfare, as expressed by demand for commodities in the marketplace. In this model, if we want creativity and the benefits associated with it, we must pay for it. The best, most efficient way to pay is with a system of private, individual rights.

This account is not to be confused with theories of intellectual property as articulated in the field of information economics. That field tends to be much more ambivalent about the effects of intellectual property rights because of the inefficiencies that accompany them.

In economic terminology, information is a "nonrival" good: One person may "consume" it without limiting the amount available to another. Another way of putting this is that information—inherently—is not consumable. If I have an apple, either you can eat it, or I can eat it. (We can share it, but we can't each have the whole apple.) But if I make up a catchy tune, we can both sing it. I won't have any less of it because you have more of it. All information—from cooking recipes, to scientific formulas, to MP3 files—has this infinitely shareable quality. In economic terms, the marginal cost of production of information is zero.[43] Once a scientist divines a new scientific theory, she can share it freely without spending any more energy or time to produce it again.[44] Because the marginal cost of information is zero, the ideal price of information in a competitive market is also zero. As a result, intellectual property rights create "static" (short-run) inefficiencies. They tend to raise the prices of informational goods above their marginal cost of production, meaning that fewer people have access to these goods than should.[45]

Where there are no adequate substitutes for a good, as may be the case with a patented medicine, intellectual property rights can also generate monopolies. Under conventional economic models, a monopolist will raise prices and reduce output, generating more profits for itself, but also generating deadweight social loss—a further static inefficiency.[46] Intellectual property also has ambiguous effects on dynamic (long-run) efficiency. Because information is both an input and an output of its own production process, intellectual property gives previous creators the power to tax new creators, thus raising the cost of producing the next generation of innovations and pricing out some potential creators.

Other mechanisms to promote investment in new informational goods are widely discussed in the field of information economics. The government can pay, as it often does, for example, with direct grants to scientists or artists or by the creation of financial or reputational prizes that can induce innovation. When innovation and creativity are paid for in this way, the results can be made freely available, as they are, for example, when the U.S. government funds certain basic scientific research or the creation of weather or mapping data. This eliminates the inefficiencies associated with intellectual property rights, leading eminent economists to conclude that government provisioning is superior to intellectual property rights as a strategy to solve the provisioning problem of information.[47]

So why, then, should we conclude that private intellectual property rights are superior to other systems of promoting creativity and innovation, such as direct government funding? Here, the image drawn from information economics is spun together with a narrative drawn from theories of the economics of private property rights in material resources (such as land) popularized in the 1980s and 1990s. Such theories, often associated with the Chicago School of economics, have their roots in the famous account that Harold Demsetz developed of the ability of private property to solve the "tragedy of the commons."[48] When property is held in common, Demsetz argued, individuals will fail to invest in its maintenance or improvement, because they cannot keep others from reaping the benefits of their efforts. Common pastures will be overgrazed, because each individual farmer has an incentive to graze his livestock beyond the point of sustainability. If his sheep don't eat the grass, another farmer's sheep will. A system of private property rights aligns farmers' incentives with social welfare, because it permits them to "internalize" or capture the benefits of their investment in their land, as well as suffer the harms of their failures to invest.

But why is private property superior, say, to community-negotiated rules limiting the hours that a farmer could graze, or a government tax-and-spend regime that organizes investments in land? Here the antiregulatory theories of the Chicago School come in. Individuals are characterized as generally having information superior to that of the government (or a collective) in making investment decisions, as well as in valuing uses in land. If they are free to transact, on this theory, "private" property is more efficient than communal or state-based regulation of property (or, more accurately, private property is the most efficient form of state-based regulation of property, since of course, a private-property regime itself is a form of regulation). Individual farmers will know best, for example, whether land can most profitably be used for sheep grazing or for peach farming. If a peach farmer is able to offer to buy a sheep farmer's property for more than the sheep farmer could make from farming it himself, the property will change hands and be turned into an orchard. Since the latter use is more profitable, it is associated with higher social welfare. Society is thus benefitted by the mutually selfish behavior of the farmers, if they are given the tools of private property rights. Antiregulatory theorists are also skeptical of government intervention in markets because of the concern that state regulations or programs provide a soft target for lobbyists seeking to capture benefits for themselves.[49]

Even as applied to material property such as property in land, there are many difficulties with this account, some of which I'll discuss below. More importantly for our purposes at this point, the sketch drawn from such Demsetzian theories suppresses many of the important distinctions between information and material goods—distinctions that are treated as essentially important in the construction of the first image. But explaining precisely why this is so should await a discussion of the development of the concept of the "commons" in the access to knowledge movement—for it is that discussion that has made this point clear.

INVENTING A POLITICS FOR THE INFORMATION AGE

Against this backdrop of enclosure and conquest has emerged a field of activism that here goes under the name of the access to knowledge movement. One mark of this new mobilization is the attempt to articulate a common language in which to contest the contours of existing intellectual property rules. That language has become centered on a few key terms such as the "public domain," the "commons," "sharing" or "openness," and "access" that are mobilized both to destabilize the despotic dominion account of intellectual property and to conjure forth an alternative ethic of the conditions of creativity and freedom in the information age.

THE PUBLIC DOMAIN

The concept of the public domain is central to the new politics of A2K, although not, as we will later see, always uncontroversially so. It is drawn from judicial and legal discourse, where it has long been used to refer to informational works that are not covered by intellectual property law, for example, because the copyright or patent term has expired.[50] In the 1980s and 1990s, scholars critical of the expansion of intellectual property rights seized upon the term to carve out a positive identity for the "outside" of intellectual property.[51] As James Boyle put it, "The environmentalists helped us to see the world differently, to see that there was such a thing as 'the environment' rather than just my pond, your forest, his canal. We need to do the same thing in the information environment. We have to 'invent' the public domain before we can save it."[52] Key here was early work of David Lange, who argued that no intellectual property right "should ever have affirmative recognition unless its conceptual opposite is also recognized. Each right ought to be marked off clearly against the public domain."[53]

Lange's early articulation of the term marks the abiding influence of intellectual property law on the concept of the public domain. The public domain here is defined as the "conceptual opposite" of the domain of exclusion rights protected by intellectual property. The same relationship is emphasized in James Boyle's definition of the public domain as "material that is not covered by intellectual property rights" as well as "reserved spaces of freedom inside intellectual property."[54]

In the simplest sense, then, A2K advocates use the term positively, to bring into focus the negative space of intellectual property law and to articulate its importance for innovation and creativity. The public domain thus becomes not just the opposite of intellectual property, but also an essential—and endangered—component of our creative and informational ecology. Included herein are not just older works in the literary or technical arts, but also resources such as language and scientific theories that are free of intellectual property rights and to which we have a common right. Many of these resources were never protected as intellectual property at all, thus demonstrating that private rights are not necessary to the production of all informational goods. Such goods and the ability to use them freely are also clearly central to our ability to think and create. The emphasis on the public domain thus is used to counter "the romantic idea of creativity that needs no raw material from which to build" that characterizes the despotic dominion theory of intellectual property and to call attention to the need of every creator to have access to the scientific or cultural domain that precedes and surrounds her. Boyle, for example, contends that the "public domain is the place we quarry the building blocks of our culture. It is, in fact, the *majority* of our culture."[55]

The A2K movement calls upon the public domain in this way to make the case that the account offered by the despotic dominion theory of intellectual property is radically incomplete as a description of both the world as it is and the world as it should be. Even now, in the most absolutist period of intellectual property law we have known, our creative world remains largely beyond the reach of intellectual property rights. And intellectual property rights as we know them bear little resemblance to property rights over material resources, with far greater freedoms reserved for nonowners. If so-called "real property" rights worked like copyrights, for example, the home you built would be turned over to the public some fifty to seventy years after your death. In the meantime, if others wanted to use your front porch to criticize you, you would have to permit it.[56] It turns out that ideas are different from material goods and are treated as such by the law. The concept of the public domain calls attention to this fact—a fact that the despotic dominion account papers over.

The concept of the public domain calls the despotic dominion account of intellectual property into question in yet another way, by emphasizing the "public" values that a public domain serves—and that the privatization of intellectual creations threatens. This is the public domain as opposed to the *private* domain—the domain that the despotic dominion conception of intellectual property equates with the public good.

We can begin by asking what is "public" about the public domain. Is it public like a public park? Like public assistance? Like the public good? Like a public figure? A2K narratives about the public domain treat what is public as synonymous with what is "open to all," but in two different dimensions: that of permission and that of price.

Public-domain material is presented as important to our creative ecology, on the one hand, because one need not ask permission to use it—which is to say, no one has the legal privilege to deny another the ability to use it. If you want to rewrite a Jane Austen novel, retaining most of her words, but inserting zombies, no representative of Austen's estate can deny you permission, because the work is now in the public domain.[57] A2K advocates thus celebrate the public domain as a place free of the political control or personal caprice of others. This is contrasted with the world of intellectual property, where owners of works may stop others from using their creations in ways of which they disapprove.[58] When DJ Danger Mouse became an overnight sensation for an album remixing the Beatles and Jay-Z, for example, he also earned the attention of lawyers for the Beatles's label, who forced him to stop distributing the album. Copyright facilitates consolidated control and disrupts semiotic recoding. The need to obtain permission, A2K advocates argue, is thus in tension with the desire for an open and democratic culture.[59]

The public domain is "public" in another sense. Like a public street, it may be traversed and used by all comers without individualized permission. But also like a public street (if not necessarily a public highway), it may be traversed without payment. (In the phraseology of Richard Stallman of free software fame, it is both "free as in speech" and "free as in beer.") No one pays for what they take from the public domain (there is no licensing fee), so works available in the public domain are available, in theory, at or close to their marginal cost of distribution—the cost of printing and selling a book, for example, without an additional fee for the author who wrote it. And of course, in a world of pervasive digital networks, the cost of distribution indeed moves toward zero, meaning that works out of copyright may be available for no cost at all. The public domain thus has a differential value for those who have limited financial means. In this sense, it is public in the way that public assistance is public—it represents a kind of state subsidy for those who cannot afford the licensing fees and lawyering costs associated with private markets in information.[60]

THE COMMONS

The commons is another concept critical to the attempt by A2K theorists to construct a collective object for their politics. It draws upon the history of property in land and more particularly upon the enclosure of communally managed field and forest resources in Europe. Unlike the public domain, the commons as conceived of by the A2K movement is governed,[61] but unlike private property, the commons is governed collectively.[62] It is not free of the requirement of permission (or, necessarily, of price), but demands permission from a collective, rather than an individual.

Free software is often cited as the paradigmatic example of an informational commons.[63] It is written by legions of volunteers who are not hierarchically organized or governed in the way that employees within a firm are organized and governed. This is not to say that there is no governance of open-source projects—on the contrary, such projects may be highly organized and closely managed. Such projects are also not entirely without either hierarchy or stratification.[64] But they are more modular, participatory, collaborative, and open than equivalent projects organized in proprietary firms.[65]

Free software depends upon a "copyleft" licensing scheme developed by programmers. The best-known such license, the GNU General Public License or GPL, turns copyright on its head by mandating sharing, rather than exclusivity—it permits users to modify, copy, and share the covered work as long as they pass along to others these same freedoms.[66] This is a commons of enforced cooperation, where those who participate are assured that their efforts will manifest themselves in a collective product that they may all access in the future with the added benefit of one another's contributions. Programmers do not have the ability to determine unilaterally the terms of the licensing of free software, but decisions about free software are subject to community comment and deliberation and to the collective ability of communities of programmers to vote with their labor hours.[67] They also have certain rights that those working in a proprietary context as a rule would not—primary among them, the assurance that they will continue to have access to the software they help produce on equal terms with all others, to exploit for profit or otherwise.

The commons as invoked by A2K advocates works in two ways to undermine the despotic dominion conception of intellectual property. At times, A2K theorists call upon the term to distinguish a material commons (for example, a grazing commons or a collectively managed fishery) from a commons of the mind. The despotic dominion justification for private property, recall, is based on the presumption that individuals will overuse a resource if not disciplined by private property rights.

But as Boyle explains it:

> Unlike the earthy commons, the commons of the mind is generally "non-rival." Many uses of land are mutually exclusive. If I am using the field for grazing, it may interfere with your plan to use it for growing crops. By contrast, a gene sequence, an MP3 file, or an image may be used by multiple parties; my use does not interfere with yours. To simplify a complicated analysis, this means that the threat of overuse of fields and fisheries is generally not a problem with the informational or innovational commons.[68]

In other words, we are more likely to see in the informational domain what property scholar Carol Rose has called a "comedy of the commons" than a tragedy of the commons, because more use tends to produce social gains, rather than social losses.[69]

But A2K advocates also use the concept of the commons to invoke the successful history of common property schemes in material goods and thus to undermine the contention that individual management of resources is superior to collective management. Elinor Ostrom recently won a Nobel Prize in economics, in part for the work she did to document and analyze prosperous and stable commons regimes governing rival resources such as land and fisheries, demonstrating that communities can organize both investment in and extraction of resources to ensure sustainability.[70] As Roberto Verzola points out in this volume, for example, a herder with a long-term and cooperative viewpoint would see the potential for the collapse predicted by theorists of "the tragedy of the commons" and work with others to avoid that result.[71] With a presumption of cooperation and foresight, the narrative of the tragedy of the commons can thus be inverted, resulting in "a system of insurance or social security, a type of commons that reduces individual risk by pooling resources."[72]

The concept of the commons is thus intended to do important work to delegitimate the despotic dominion conception of intellectual property. On the one hand, it calls upon the differences between the immaterial and the material to demonstrate that tragedy is far less likely in the former case. On the other, it rejects the view that tragedy necessarily follows common management of material resources, insisting instead that collective management can work. It insists upon the viability of an alternative governance regime for intellectual property—one characterized by relatively flat hierarchies and where the rights of individuals to participate in decision making as well as to participate on equal terms as creators and beneficiaries are central. To call upon the image of the commons is to insist that communities, without the imposition of market or governmental ordering systems, have the power and perhaps the right to set the terms of their collective endeavors.

Here the discourse of the commons meets up with that of the public domain, suggesting that more communal strategies of governance do better than a despotic dominion model at facilitating broadly distributed collaboration, soliciting forms of effort and motivation that may be crowded out in a corporate and proprietary (which is to say, profit-motivated, more hierarchical) context, and facilitating participatory decision-making processes.

SHARING AND OPENNESS

Sharing and openness are prominent memes in the A2K movement, deployed, to name just a few examples, for "share and share alike" copyright licenses, "open-source software," "open standards," and "open-access publishing."[73]

Sharing and openness are here posited against the ethic of exclusion embodied in the despotic dominion conception of intellectual property. A "share and share alike" license in the context of copyright, for example, uses the exclusive right permitted by copyright against itself, requiring those who modify or build upon the work to share their work with others. Copyleft licenses are premised on the same move. In open standards and open-access publishing, "openness" refers to different practices. The former insists that technical standards not be dominated by the rights of certain intellectual property owners and the latter that certain publications (for example, those that are the product of research funded by the government) be made available in databases that are available generally to the public without a fee.[74]

What work does an insistence on sharing and openness do when measured against the despotic dominion account of intellectual property? For one thing, it raises a challenge to the neoclassical model of the rational, self-interested actor upon which that account is based. As Yochai Benkler has noted, the very existence of free software, which is developed largely by unpaid volunteers who participate on the condition that their work will be shared freely with others, demonstrates that a model based on profit-driven self-interest is radically incomplete.[75] There is room for debate over the volunteers' motivations, but as Boyle puts it:

> Assume a random distribution of incentive structures in different people, a global network: transmission, information sharing and copying costs that approach zero, and a modular creation process. With these assumptions, it just does not matter why they do it. In lots of cases, they *will* do it. One person works for love of the species, another in the hope of a better job, a third for the joy of solving puzzles, and so on.... Under these conditions... we *will* get distributed production without having to rely on the proprietary/exclusion model. The whole enterprise will be much, much, much greater than the sum of the parts.[76]

The notion that the "whole is greater than the sum of its parts" is central to understanding the ideal of sharing and openness. If the whole is greater than the sum of its parts, the parts cannot be adequately described or divided from one another. In other words, we cannot isolate and locate credit, labor, or value for creative endeavors in any individual or set of individuals. The maxim can also be understood as an insistence that the thing being "summed"—here, the creative endeavor—happens not within individuals, but among a group. This is an insistence on the generativity of the crowd, on the notion that there is a creative and productive force that resides *between*, rather than within individuals—or more radically, in the infrastructure of their connection, in the network itself. As free-software theorist Eben Moglen memorably puts it, "if you wrap the Internet around every person on the planet and spin the planet, software flows in the network. It's an emergent property of connected human minds that they create things for one another's pleasure and to conquer their uneasy sense of being too alone." Intellectual property law is then "the resistance in the network," disrupting, rather than generating creativity.[77]

We can detect here a certain commitment to the unknowability and unquantifiability of the creative endeavor. We cannot, A2K advocates suggest, fully catalogue and locate human motivations and capacities, nor can we individualize them, as if they are established prior to and apart from exchanges between people. "Knowledge" and "information" are also cast as highly complex phenomena that inevitably elude strict control or management. (How do you survey the limits of an idea?) The domain of access to knowledge is thus pictured as a domain of unbounded, unboundable exchange. This vision is of course opposed to the despotic dominion notion of private property in ideas and to neoliberal theories that put their faith in "privatization, and the creation and defense of secure property rights as the cure for all ills."[78]

Ideals of openness and sharing, like those of the commons and the public domain, also align the A2K movement with the political values of self-determination and autonomy, as well as those of collective governance. As one open-source proponent puts it:

> Proprietary software increases the dependence of individuals, organizations, and communities on external forces—typically large corporations with poor track records on acting in the public interest. There are dependencies for support, installation and problem fixing, sometimes in critical systems. There are dependencies for upgrades and compatibility. There are dependencies when modification or extended functionality is required. And there are ongoing financial dependencies if licensing is recurrent. Political dependencies can result from the use of proprietary software, too. . . . Nearly exact parallels to this exist in agriculture, where the patenting of seed varieties and genome sequences and the creation of non-seeding varieties are used to impose long-term dependencies on farmers. . . . Proprietary software not only creates new dependencies: it actively hinders self-help, mutual aid, and community development.[79]

Others declare more grandly that "access to software determines who may participate in a digital society" and conclude that "only the Free Software model grants equal rights and freedoms to all Member States, their corporations and citizens."[80] Or as the founder of the Linux operating system, Linus Torvalds, puts it, open-source software is like "democracy in the sense that you don't surrender control."[81]

The demand for sharing and openness is thus also a demand that the ability to access and manipulate knowledge and information be democratized.[82] What is being shared and opened is not just a set of commodities, but also the processes by which we communicate with one another and create together and the processes by which we act as citizens of our increasingly informational societies.

A2K also invests with great significance the concept of "access." First associated with the access-to-medicines campaign, the importance of the term to the broader coalition is perhaps best marked by its presence in the name "access to knowledge" itself.[83] The demand for access is an inherently relational one—a claim from those excluded that they be included, that they be given something that others already enjoy. In this sense, it marks perhaps the only—or at least the most prominent— demand for *distributive* justice emanating from the A2K movement, which otherwise borrows more from discourses of freedom.[84]

How, then, are we to understand this demand? We can begin by considering the development of the campaign for access to medicines. Although the claim might seem to be very simply a demand that medicines available to the rich also be made available to the poor, from its inception, the movement has been intimately bound up with claims about intellectual property. It emerged from the crucible of the global HIV/AIDS pandemic and specifically from the recognition that treatment would never be available to the vast majority of those who needed it unless the prices of medicines could be reduced. At the time that the campaign began, AIDS medicines sold for about $10,000 per patient per year. Activists versed in intellectual property law such as James Love teamed up with groups such as Médicins Sans Frontièrs to demonstrate that this price is not a fact of nature or a reflection of the sophistication of antiretroviral medicines, but rather an artifact of patent law. Generic copies of the medicines cost as little as $350 per year (and even less today), but patents—and the aggressively propatent trade policies of countries such as the United States—stood in the way.

The demand for access to AIDS medicines has thus been, from the beginning, a demand for access to *copies* of AIDS medicines. Or, as the logo of the AIDS activist group Act Up–Paris puts it:

The emblem illustrates two important elements in the demand for access. First, claims to access are framed squarely against the backdrop of intellectual property. Second, they are rooted in claims of right that supersede the claims of right made by owners of intellectual property. The right *to* the copy claimed by activists is written over the right *of* the copy claimed by rights holders.

The demand for access thus appears first as a refusal. It emanates not from the discourse of intellectual property, but from the language of human rights.[85] It seeks to elevate the latter over the former, as through the demand, commonly seen at access-to-medicines demonstrations, for "patients' rights not patent rights."

At the level of the slogan, the concept of access seems to embody an outright rejection of the logic of intellectual property and of the type of cost-benefit analyses and arguments about innovation upon which it is based. In fact, however, the discourse of access-to-medicines campaigners has become intimately bound up with the logic of intellectual property, because their attempt to contest the legitimacy narrative of intellectual property law has drawn them into the economic discourse that dominates the field.

As activists sought to challenge the existing law of intellectual property, they found themselves up against the despotic dominion account of intellectual property. Calling upon this account, pharmaceutical companies insisted that they, too, are in the "access" business and that patents are the only way to ensure the development of new medicines. The conditions of access are contested, that is, precisely in the terms of the discourse underlying the concept of intellectual property, requiring A2K advocates to do more than simply argue that they are for access because they are opposed to exclusive rights in medicines. The demand for access is by necessity constructed on a deeper theory of what it means to make medicines accessible—one that is built upon the values of freedom and openness that are evolving within the discourse of the A2K movement, but anchored in the demands for distributive justice that motivate the call for access.

Access-to-medicines campaigners argue, for example, not only that patents artificially raise prices and thus hurt patients, but also that they do not provide the innovation benefits that the despotic dominion account claims for them, particularly for the poor. They point out, for example, that patent-based innovation systems link innovation to high prices. Because the poor cannot pay these high prices, patent-based companies ignore the needs of the poor and instead cater to the needs of the rich. Thus, we have a pharmaceutical R&D system that prioritizes drugs for baldness and erectile dysfunction over lifesaving treatments for ailments such as tuberculosis and malaria.

They also point out that patents can create barriers to research and thus interfere with innovation—and argue that they are particularly likely to do so where poor patients are concerned. They point out, for example, that multinational companies that make AIDS drugs were unwilling to undertake the negotiations that would have been needed to combine the multiple drugs needed for the HIV cocktail into a single pill that would be easier for patients to take. The work was first done not by patent-holding firms, but by Indian generic companies that were unconstrained by patents. Like the discourses of the public domain and openness, the discourse of access here attacks the despotic dominion account's claim that intellectual property invariably promotes innovation. Unlike the other concepts, this one makes central a distributive justice claim—that freedom from intellectual property restrictions is especially important to the poor.

The access-to-medicines campaign also takes aim at the model of private control that is central to the despotic dominion account. Notably, access-to-medicines campaigners have consistently opposed drug company donation programs, staking a claim for a form of access that is defined by nonexclusive sharing of the informational component of drugs, rather than their price per se.[87]

Why? Why would it matter where the drugs come from, as long as they come? For access-to-medicines campaigners, the issue is one of accountability and control. They argue that drug company donation programs are unacceptable because they leave power over life in the hands of private actors, who retain the privilege of charity, the privilege to make good on their promises or not. Overriding patents is cast as a way to insist instead on values of participation and accountability. The demand for access to medicines, like the call for free software, thus places the concept of democracy at the center of the A2K movement and opposes it to the despotic dominion conception of intellectual property.

FOUR QUESTIONS FOR A2K

The concepts that A2K activists are developing and articulating and around which they are mobilizing create a set of political commitments and the contours of a movement through a process of accretion. These concepts often coincide, but they are also at times in tension with one another. The same can be said of some of the values and discourses that A2K activists draw upon when making their arguments. Having mapped the central concepts of the discourse of A2K allows us to pose a series of questions about the conceptual and political commitments being invoked. The answers will help determine the future shape and implications of this new field of politics. What is the nature of the freedom that A2K demands? Is A2K committed more to the model of the public domain or of the commons, and can it be committed to both? Is information really different *enough* from material goods? And finally, can the A2K movement in fact make good on its attempt to create a politics not just of information, but also of knowledge? Or to put it another way, what are the proper limits of the politics of A2K?

WHAT IS THE NATURE OF THE FREEDOM THAT A2K DEMANDS?

Often, A2K thinkers speak of freedom (such as the freedom of the public domain) as a place free of permission. Lawrence Lessig states it most plainly: "The opposite of a free culture is a 'permission culture.'"[88] But are A2K advocates really committed to a vision that posits freedom as a space where one never needs permission—as a space beyond control? If so, what of the very substantial controls that some groups, from free-software programmers to proponents of traditional knowledge, seek to impose upon certain forms of knowledge? Creative Commons, a high-profile organization that Lessig himself founded, offers individuals a set of copyright licenses that they can use to give others more freedoms than copyright law otherwise would. But some of these licenses—not uncontroversially within A2K circles—preclude others from creating derivative works, making use of precisely the power of permission in the service of authorial control.

In fact, no such simple principle of opposition to control can be derived from the thought of A2K. If it could, it would commit A2K also to a series of what are likely to be untenable positions with respect to nonproperty forms of control that can be described as demands for "permission," such as those related to privacy and network security. Is it in fact possible to assume a simple opposition between freedom and control, or are the two instead intimately interconnected and interdependent in the age of digital networks?[89] A2K advocates must envision a particular *mode* of control or demand for permission that they oppose. How, though, should this be characterized?

The A2K movement's conception of freedom also contains within it a certain fractured relationship to markets. The public domain, for example, is sometimes figured as a space free *from* markets, a space where noncommercially motivated creators have the resources and room to play.[90] At other times—and perhaps more often—it is figured as a space free *for* markets where not only amateurs can forage, but where corporations can compete without monopolies, to the benefit of the public as consumers.[91] Can the same domain be both the space of freedom from commerce and the space of freedom for commerce?

When A2K advocates articulate the public domain as a space that is equally—and properly—open to the exploitation of capital and communities alike, it suggests that this competition is itself a free and equal one. But is the public domain in fact universally "free" in a substantive fashion, when those who create from its resources may enclose the results? Does leaving the public domain free in this sense simply mean that those with resources will be able to make use of this (publicly renewed and subsidized) resource and then enclose the results, to the systematic disadvantage of those who continue to operate outside of the confines of property? Is this freedom a structurally unequal freedom, one that can be remedied only by a positive concept of public property (or of a commons) that cannot be the subject of such extraction?

This question is raised most acutely by groups focused on the Global South, such as the farmers' rights group GRAIN, which expresses skepticism about "the merits of concepts such as the 'public domain'... if putting seeds in the public domain means Monsanto can inject them with Terminator genes to destroy peasant agriculture."[92] The muted (or repressed) debate within the A2K movement over the proper status of traditional knowledge (is it rightfully the property of local communities, or part of the public domain open to all?) also evinces the strains of this tension.[93]

Finally, can the freedom imagined by A2K be produced by merely formal lack of (the wrong kind of) constraint, for example, by the lack of the constraints imposed by intellectual property law? Or does it require something more substantive, an affirmative ability, for example, to access works in the public domain, or the tools of the new "remix culture"?[94] Is the freedom of the public domain or the commons really worthy of the name if the majority of the world has no access to the means needed to participate in it—for example, education, computers, and affordable access to digital networks? At the close of 2007, only one-fifth of the world's population was using the Internet, and this use was highly skewed geographically: Only 4 percent of people in sub-Saharan Africa had such access.[95] Although A2K thinkers invoke a robust conception of freedom that would require the ability in *fact* to access the goods of which they speak, in practice, they devote little attention to the profound inequalities in access to digital networks.[96] Can A2K advocates really claim to have a vision of freedom in the digital age if they do not do more to theorize and demand affirmative access to the tools to create and exchange information and knowledge?

IS A2K COMMITTED MORE TO THE MODEL OF THE PUBLIC DOMAIN OR OF THE COMMONS? CAN IT BE COMMITTED TO BOTH?

The A2K movement valorizes the space of both the public domain and the commons, and yet as we've seen, these two spaces are governed in importantly different ways. The commons is controlled, often through the use of intellectual property law itself. The public domain is instead a space beyond intellectual property law, where no one has the right to extract permission or price.

Can the A2K movement be committed to both? If so, this would require restructuring how the commons and public domain are each understood. A2K rhetoric today arguably pastoralizes the commons, eliding the degree to which communal decision making may be characterized by hierarchy and exclusion, rather than by equality and open participation. To put it differently, why should we view a collective despot as an improvement over an individual despot?

In fact, A2K advocates cannot and most of the time do not envision the commons as just any kind of collectivity. Some systems of collective management are, after all, fully compatible with expansive conceptions of intellectual property rights, such as the collective rights organizations that enforce the rights of copyright holders in music.[97] Corporations that mobilize intellectual property norms in the service of exclusivity and maximal profit are of course in some sense "collective" entities, governed by groups of corporate officers and answerable to shareholders. The A2K commons thus cannot be understood simply as a preference for collective over individual governance. Some content must be given to the concept of the collective and its terms of engagement. Like the concept of freedom, the concept of the commons (if it is to lay claim to an ethic that differs substantially from that of intellectual property) must be more substantively defined.

As the example of free software discussed above suggests, when A2K advocates invoke the commons, they conjure forth a community that labors *cooperatively* and that labors under *shared* norms. Those norms differ not just in their recognition rule—the metarule that determines what counts as valid law—but also in their substance from the rules of intellectual property.[98] The commons of software in fact has much in common with the public domain, because its rules of engagement are similar to those that characterize the public domain. Still, they are not identical. Individuals can take from the public domain and not replenish it with their creations. Moreover, its contours and rules are not established by a community of creators, but rather by a community of citizens who authorize the law of intellectual property—which in turn defines the limits of the public domain. Which is the appropriate community of lawmakers, and which the appropriate relation to what came before?

IS INFORMATION REALLY DIFFERENT *ENOUGH*?

Within the emerging ideology of the A2K movement is a strand that envisions it as postideology, even, perhaps, postpolitical. This is evident particularly in the self-styled political agnosticism that characterizes the free and open-source software movement and in the writings of A2K thinkers who are most immersed in the discourses of open source and the revolutionary potential of the networked digital age.[99] In this volume, Benkler, for example, argues that the ideas of A2K, and in particular of "the information commons and the rise of networked cooperation" can "subvert the traditional left-right divide ... and provide the platform on which political and economic interests meet around a common institutional and organizational agenda." A2K can appeal, he argues, to "libertarians, liberals, the postsocialist left, and anarchists," unifying forces on the left and right that usually understand themselves to be at odds with one another.[100]

Such ideological catholicism, even pragmatism, is perhaps one of the most appealing aspects of the A2K movement, particularly at a time when some on the left are calling for a more serious reckoning with the benefits of well-regulated markets and the dangers of ideological rigidity.[101] But the notion that the A2K movement can exceed the traditional divide between classical free-market liberals and the progressive left, that A2K can embrace both the market and the nonmarket, and that A2K advocates need not decide between frames of freedom, justice, or efficiency is surely contestable.

At its core, the sense that the A2K movement can exceed these divides rests crucially on the claim that information is subject to different dynamics than the world of material goods, particularly in the networked digital age. For Benkler, for example, it is "the rise of the networked information economy [that] has created the material conditions for the confluence of freedom, justice, and efficacy understood as effective learning and innovation." That is because in this new environment, productivity and efficiency can be achieved through increasingly open dynamics of sharing and cooperation, both within and outside of markets. "Freedom and efficacy, then, will be the interface with both liberalisms, market and social. Justice and freedom in the sense of the dissipation of structured, stable hierarchical power will be the interface between liberalism and the left."[102]

But the question is, is information different *enough*? As noted above, some within the A2K movement doubt that the poor can compete in a realm of "free" information if that freedom is granted equally to the powerful and the powerless. To paraphrase Anatole France, is this just a kind of majestic equality that leaves the rich and poor equally free to exploit the potential of biotechnology and software engineering? Will resources determine, ultimately, who is heard in the space of "free and open" networks? Can true democratization emerge from spaces of creation and meaning making that are not themselves first radically democratized?

Or is the point of A2K thinkers instead that in the realm of information, we are *relatively* more free and can do more than ever before—if not everything—to reconcile our commitments to freedom, justice, and efficiency? There is a difference, after all, in a competition between the subsistence fisherman and the commercial fishing fleet and between the unknown garage band and the corporately manufactured pop star. There are only so many fish to go around, but there is no limit, theoretically, to the number of songs that can be written. As importantly, according to A2K advocates, garage bands can increasingly compete with studio-driven stars because of the power of digital networks to give creators access to a public and the power of these same networks to lower dramatically the costs of production of informational goods. In the information realm, in a sense, there are always more fish, because the fish there are subject to the rules of immaterial, rather than material goods. And the advent of ubiquitous digital networks means a less unequal competition in the struggle to create new information and to gain access to new publics.

The claim that the A2K movement can move beyond the traditional ideological battles between formal and substantive conceptions of freedom, between the freedom of the market and freedom from the market, is thus intimately bound up with the idea that we can move beyond scarcity in the information age. As Verzola puts it, material abundance is limited because "it must eventually express itself in terms of biomass," but information abundance "is of the nonmaterial variety. Thus, information goods offer the promise of practically unlimited abundance."[103]

In what sense is it useful to conceptualize information as having a kind of abundance that exceeds the material or that is "practically unlimited"? Verzola allows that the realm of information is in fact constrained, in his view "mainly by the limits of human creativity, the storage capacity of media, and the availability of electricity to power servers on the Internet twenty-four hours a day."[104] But there is a utopian strand in A2K thinking that tends to minimize such constraints of mind and environment, suggesting that they need not stand in the way of our ability to think and compute our way to a more just and equal world.

The most enthusiastic proponents of the biotech and open-source software revolutions imagine an era when biology and informatics merge to move us beyond the limits of the physical. But today, half a million women each year still die in childbirth, almost all in developing countries and more than fifty years after the technologies to avert almost all such deaths were developed.[105] We already have the technologies and resources to feed and care for many more people than we currently do, suggesting that there is a primary and prominent set of problems that are not technological, but political.[106] The dynamics of networked informationalism might help overcome political problems where those problems are rooted in struggles over scarce resources. They could also facilitate more transparency and political participation, addressing failures of political accountability more directly.[107]

But critical to the postscarcity aspirations of the A2K movement are questions of degree, distribution, and velocity: Will the informational component of our world advance rapidly or evenly enough to overwhelm the persistent inequalities in the material? Will such advances be distributed evenly enough to make the promise of living beyond scarcity a reality for any but the world's richest? Can we expect a leveling of the pervasive material inequalities in the world if the poor lack access to the labs, computers, and textbooks that would allow them to do more for themselves and if they also lack access to the kind of political power and voice that would allow them to change the terms on which resources and informational goods are currently distributed? Can A2K advocates build a theory of freedom that is based upon the radical political possibilities of the immaterial while also accounting for the crucial moment when the informational intersects with the material in the places that we create and communicate, that we live and die?

The A2K movement was deliberately structured around a demand for access to "knowledge." And yet this introduction and the pages that follow make it clear that A2K actors operate routinely in the idiom of "information," for example, extolling the importance of the information commons or the lessons of information economics. What difference might this difference make? There are at least two ways to approach the question—by asking what A2K activists invest in their own choice of terms and by investigating the etymological implications of the distinction between information and knowledge.

If A2K theorists talk often about information, why isn't the A2K movement instead the A2I movement—a mobilization for "access to information"? Ahmed Abdel Latif, in his account of how the term "A2K" was chosen, explains that "at the conceptual level, knowledge, rather than information, is at the heart of the empowerment of individuals and societies. While information is certainly a prerequisite in the generation of knowledge, acquisition of knowledge remains the ultimate goal. Knowledge processes information to produce ideas, analysis, and skills that ideally should contribute to human progress and civilization."[108]

The decision to articulate the movement's demands in relation to knowledge was in part a response to perceived conceptual differences between knowledge and information. Knowledge is a capacity that is central to empowerment—one that relies upon, but is not reducible to information.

How precisely, though, should we understand the difference between knowledge and information? A2K theorists such as Benkler define the distinction in this way: Information is "raw data, scientific reports of the output of scientific discovery, news, and factual reports," while knowledge is "the set of cultural practices and capacities necessary for processing the information into either new statements into the information exchange, or more important in our context, for practical use of the information in appropriate ways to produce more desirable actions or outcomes from action."[109] Thus, information is objective and external, while knowledge is the capacity to use information to create new information or to use information to generate technical effects in the world (knowledge as "know-how").

This is narrower than the definition of knowledge that we might derive from etymology or contemporary usage. According to the dictionary, we can "know" anything that we understand through "experience or association."[110] The English word "knowledge" corresponds to the German *kennen* and French *connaître*, designating a kind of understanding that comes from the senses. But "knowledge" also incorporates the concepts of *wissen* and *savoir*, designating a kind of understanding that is derived from the mind. It thus designates basic acts of human cognition:

recognition, acquaintance, intimacy, consciousness, or, "the fact, state, or condition of understanding."[111]

In its broadest sense, then, knowledge is more than the ability to process information into more information and more know-how. As Jean-François Lyotard writes, knowledge is

> a competence that goes beyond the simple determination and application of the criterion of truth, extending to the determination and application of criteria of efficiency (technical qualification), of justice and/or happiness (ethical wisdom), of the beauty of a sound or color (auditory or visual sensibility), etc. Understood in this way, knowledge is what makes someone capable of forming "good" denotative utterances, but also "good" prescriptive and "good" evaluative utterances.... It is not a competence relative to a particular class of statements (for example, cognitive ones) to the exclusion of all others.[112]

Knowledge is here a capacity more than it is an object or a possession—a power immanent to intellectual, social, cultural, and technological relations between humans.[113] Information, in turn, is the externalized object of this capacity, the part of knowledge that can be systematized and communicated or transmitted to others.[114]

What would it mean for the A2K movement to take the distinction between knowledge and information seriously and to theorize itself as a movement for access not just to information, but to knowledge? At a minimum, using the narrower definition of knowledge proposed by Benkler, it would require a focus not only on extending access to information, but also on extending individual capacities to produce information and to make use of information to produce practical effects in the material world.

As Benkler points out, there is "a genuine limit on the capacity of the networked information economy to improve access to knowledge." Knowledge cannot be fully externalized into information—it is a capacity, rather than an object. As such, it does not partake of the same dynamics of plenty that is said to characterize the informational domain. While better access to learning materials can enhance education, learning by doing requires local practice, and the practice of education generally "does not scale across participants, time, and distance."[115]

The A2K movement might focus on forms of information regulation that affect the development of knowledge, as it has done to date in work on access to learning materials, open courseware, and lowering intellectual property barriers to distance learning. These moves are more efforts to increase access to information than access to knowledge. If the A2K movement is to embrace its initial identification with the concept of access to knowledge, it must recognize that while access to some information is clearly a prerequisite of building knowledge in Benkler's sense, more ubiquitous access to information is not the same thing as more ubiquitous access to knowledge.

Can the A2K movement—as invested as its logic has become in the model of information technologies and the economics of the copy—build a politics of knowledge as a competence? The dream of perfect (and zero-cost) transmissibility cannot survive an encounter with this concept of knowledge, because a competence that cannot be fully externalized and traded, and thus that is embedded in the material, cannot be nonrival. And if knowledge cannot be accessed through a simple download, then a politics of A2K must reach far beyond a politics of enclosure and intellectual property.

Does this mean broadening the A2K mandate to include work on, for example, the financing of primary schools or the effects of austerity budgets on universities around the world? That is one possible outcome. More modestly, it might instead mean that A2K groups recognize their focus is on improving access to information, acknowledge that knowledge is not an object that can simply be downloaded from North to South, and engage openly with those who worry that more information could in some cases not improve, but rather threaten access to knowledge.

What if the A2K movement were instead to embrace the definition of knowledge that corresponds not just to technical or intellectual knowledge, but also, for example, to artistic or ethical knowledge? This would fit well with its attempt to embrace the literary arts, as well as science and technology, but it would also unmoor the movement from the conception of knowledge present in Benkler's definition. Lyotard's broader definition requires us to recognize that the criteria for successful knowledge are created, rather than given.

For the A2K movement, such a recognition would imply the need for a politics not just of access to knowledge, but of what *counts* as knowledge and of who gets to decide what counts. Would this work a fundamental harm to the universalizing aspirations of the A2K movement? Or would it instead make room for A2K advocates to begin to reckon with existing tensions in the movement, for example, surrounding issues of traditional knowledge and the concept of the commons versus the public domain?

CONCLUSION

A critical genealogy of the concept of access to knowledge allows us to map the sometimes contradictory and often complex interventions that are coming to constitute A2K's theoretical commitments. The first and foremost effect of these interventions is to destabilize the dominant legitimation narrative of intellectual property today, the despotic dominion account that treats the privatization of information as the necessary condition for its efficient production and exploitation.

But the images and values that this new lexicon draws upon should also be examined critically as a place to think about the dilemmas that the A2K movement faces as it seeks to consolidate its critiques of intellectual property and constitute an affirmative vision of its aims. That is the purpose of the questions raised above: What does A2K mean by "freedom"? How can it mediate between its commitments to the public domain and to the commons? Is information different enough to justify the postpolitical and postscarcity elements of A2K thought? And is A2K a movement about knowledge, or about information?

These questions are offered in the spirit of committed criticism: What are those of us engaged in A2K building? Can it be what we claim for it in our most righteous and universalizing moments? Who, ultimately, will decide? What might it mean for us to win what we seek, and how might some of the paths that we have chosen lead us further away from or closer to realizing that aim? My aim here is to articulate these questions. If they are to be resolved, it will be through the iterative and networked process of debate and action that constitutes the A2K movement itself, to which the volume that follows aims to contribute.

NOTES

The author is grateful to Talha Syed, Cori Hayden, Pam Samuelson, and Molly Van Houweling for their insightful comments.

1 Wikipedia, s.v. "Intellectual Property": http://en.wikipedia.org/wiki/Intellectual_property (last accessed February 24, 2010). You can get a sense of the debate by reading the archived "talk" pages, where editors argue about the problems with the definition. See http://en.wikipedia.org/wiki/Talk:Intellectual_property (last accessed February 24, 2010).

2 The Stallman essay is available at www.gnu.org/philosophy/not-ipr.html (last accessed February 1, 2010). Google's algorithm (itself subject to intellectual property protection) answers search queries recursively, using the link structure of the Web to rate the relevance and popularity of a particular Web site.

3 See Gaëlle Krikorian, "Access to Knowledge as a Field of Activism" in this volume and Amy Kapczynski, "The Access to Knowledge Mobilization and the New Politics of Intellectual Property," *Yale Law Journal* 117 no. 5 (March 2008), available on-line at http://papers.ssrn.com/sol3/papers.cfm?abstract_id=1323525.

4 F. M. Scherer, *New Perspectives on Economic Growth and Technological Innovation* (Washington, D.C.: Brookings Institution Press, 1999), pp. 25–28.

5 *Ibid.*, p. 24.

6   Manuel Castells, *The Rise of the Network Society*, 2nd ed. (Oxford: Blackwell, 2000), pp. 169–72.

7   Fritz Machlup, *The Production and Distribution of Knowledge in the United States* (Princeton, NJ: Princeton University Press, 1962), p. 9.

8   Daniel Bell, *The Coming of Post-Industrial Society* (New York: Basic Books, 1973), p. 129.

9   See James R. Beniger, *The Control Revolution: Technological and Economic Origins of the Information Society* (Cambridge, MA: Harvard University Press, 1986), p. 23, fig.1.1; Castells, *The Rise of the Network Society*, pp. 212–31.

10 Castells, *The Rise of the Network Society*, pp. 225–26.

11 *Ibid.*, pp. 13–21, 30, 17, 31.

12 *Ibid.*, p. 101. The globalization of informationalism has been spectacularly illustrated by the recent global economic crisis. The implosion of one exquisitely informational domain in the United States—that of structured finance—cascaded around the world, generating an unprecedently rapid contraction in global trade and production. The World Bank estimated a global contraction of GNP of 1.7 percent in 2009. World Bank, "Global Economic Prospects 2009, Forecast Update," (World Bank, March 2009), available on-line at http://siteresources. worldbank.org/INTGEP2009/Resources/5530448-1238466339289/GEP-Update-March30.pdf (last accessed February 1, 2010), p. 1. The WTO predicts a concomitant contraction of world trade by 9 percent, the biggest since World War II. WTO Secretariat, "WTO sees 9% Global Trade Decline in 2009 as Recession Strikes," press release, March 23, 2009, available on-line at http://www.wto.org/english/news_e/pres09_e/pr554_e.pdf (last accessed February 1, 2010). This is not to say that all regions have been equally affected by the crisis or that all economies are equally dependent upon informationalism—they demonstrably are not. It is to say, rather, that distance from the centers of world finance did not insulate even the poorest from the effects of the recent crisis. Its effects, mediated by communication technologies and techniques such as just-in-time production, are felt through channels such as decreased remittances and increased volatility in commodity markets. The interconnection of markets—itself dependent upon (though not determined by) advances in information technology—is thus one vector for the globalization of the implications of informationalism.

13 Castells, *The Rise of the Network Society*, pp. 108–109.

14 *Ibid.*, p. 17 n.25, citing Daniel Bell. Information, in turn, is defined as "data that have been organized and communicated." *Ibid.*, citing Marc Porat.

15 Elisabeth L. Eisenstein, *The Printing Press as an Agent of Change: Communications and Cultural Transformations in Early-Modern Europe* (Cambridge: Cambridge University Press, 1979).

16 Digital systems—think here of the binary language of zeros and ones that computers use—deploy discrete variables and thus allow information to be transmitted more faithfully. This is in contrast to continuous variables, which are used by analog systems. On the impact of digitalization, see Beniger, *The Control Revolution*, pp. 25–26: "Digitalization promises to transform currently diverse forms of information into a generalized medium for processing and exchange by the social system, much as, centuries ago, the institution of common currencies and exchange rates began to transform local markets into a single world economy." See also Eben Moglen, "Anarchism Triumphant: Free Software and the Death of Copyright," *First Monday* 4, no. 6 (August 2, 1999), available on-line at http://firstmonday.org/htbin/ cgiwrap/bin/ojs/index.php/fm/article/viewArticle/684/594 (last accessed February 1, 2010). Moglen writes, "the movement from analog to digital is more important for the structure of social and legal relations than the more famous if less certain movement from status to

contract." By the same token, nondigitalizable forms of knowledge are lost. See Jean-François Lyotard, *The Postmodern Condition: A Report on Knowledge*, trans. Geoff Bennington and Brian Massumi (Minneapolis: University of Minnesota Press, 1984).

17   Castells, *The Rise of the Network Society*, p. 185.

18   World Health Organization, *World Health Report 2003: Shaping the Future* (Geneva: World Health Organization, 2003), p. xii, available on-line at http://www.who.int/whr/2003/en/whr03_en.pdf (last accessed February 1, 2010).

19   World Health Organization, "World Report on Knowledge for Better Health: Strengthening Health Systems," (Geneva: World Health Organization, 2004), p. 1, available on-line at http://www.who.int/rpc/meetings/en/world_report_on_knowledge_for_better_health2.pdf (last accessed February 1, 2010).

20   Bell, *The Coming of Post-Industrial Society*, pp. 15–18.

21   *Ibid.*, p. 20.

22   Paul Starr, *The Creation of the Media: Political Origins of Modern Communications* (New York: Basic Books, 2004), pp. 97–98.

23   See Michel Foucault, *Discipline and Punish: The Birth of the Prison*, trans. Alan Sheridan (New York: Vintage Books, 1995); Michel Foucault, *The Birth of the Clinic: An Archaeology of Medical Perception*, trans. A. M. Sheridan Smith (New York: Vintage Books, 1994).

24   Nikolas Rose and Peter Miller, "Political Power beyond the State: Problematics of Government," *British Journal of Sociology* 43, no. 2 (June 1992): pp. 173 and 175. See also Max Weber, "Characteristics of Bureaucracy," in *From Max Weber: Essays in Sociology*, ed. and trans. H. H. Gerth and C. Wright Mills (New York: Oxford University Press, 1946), pp. 196–244.

25   Rose and Miller, "Political Power beyond the State," p. 178.

26   Yochai Benkler, *The Wealth of Networks: How Social Production Transforms Markets and Freedom* (New Haven, CT: Yale University Press, 2007), pp. 3, 212–72. See also Krikorian, "Access to Knowledge as a Field of Activism," in this volume.

27   New and more exotic forms of intellectual property rights have also emerged in recent years, such as the "geographical indications" that reserve the use of terms such as "Champagne" to sparkling wine made in certain geographical regions or the exclusive protections for databases that have been implemented in Europe.

28   Copyright terms are required by the TRIPS Agreement to last at least fifty years plus the life of the author, but many countries have longer terms. For example, the current term in the United States is seventy years plus the life of the author.

29   Today, trademark owners may take advantage of more expansive and controversial rights in countries such as the United States, for example, preventing the "dilution" of the value of a mark, even where the use in question would not confuse consumers about the origin of a product.

30   Stallman, "Did You Say 'Intellectual Property'? It's a Seductive Mirage."

31   Catherine L. Fisk, "Removing the 'Fuel of Interest' from the 'Fire of Genius': Law and the Employee-Inventor, 1830–1930," *University of Chicago Law Review* 65, no, 4 (Fall 1998); Catherine L. Fisk, "Authors at Work: The Origins of the Work-for-Hire Doctrine," *Yale Journal of Law and the Humanities* 15, no. 1 (2003).

32   Kapczynski, "The Access to Knowledge Mobilization," pp. 821–22.

33   James Boyle, *The Public Domain* (New Haven, CT: Yale University Press, 2008), pp. 43–45.

34   See the discussion below for more on the potentially problematic nature of the metaphor of the historical "commons" in land.

35 The TRIPS Agreement is available on-line at http://www.wto.org/english/tratop_e/trips_e/ t_agm0_e.htm (last accessed February 3, 2010).

36 The best descriptions of this process are found in Susan K. Sell, *Private Power, Public Law: The Globalization of Intellectual Property Rights* (Cambridge: Cambridge University Press, 2003) and Peter Drahos with John Braithwaite, *Information Feudalism: Who Owns the Knowledge Economy?* (New York: New Press, 2002).

37 Peter Drahos, "Global Property Rights in Information: The Story of the TRIPS at the GATT," *Prometheus* 13, no. 1 (1995): p. 16.

38 *Ibid.* See also Gaëlle Krikorian, "Interview with Yann Moulier Boutang," in this volume.

39 Joseph Conrad, *Heart of Darkness* (London: Penguin 1983), pp. 31–32.

40 We might add a third image (at the risk of disrupting the metaphor), derived from natural rights arguments that creators have an inherent entitlement to control or profit from their creations. This argument is less common than the more economically oriented claims, but turns up not infrequently in debates over copyrightable works. Again, the justification for strong copyright as it appears in these debates suppresses substantial disagreement among theorists about the nature and extent of any such natural rights. See, for example, Jeanne L. Schroeder, "Unnatural Rights: Hegel and Intellectual Property," *University of Miami Law Review* 60, no. 6 (July 2006); Seana Valentine Shiffrin, "Lockean Arguments for Private Intellectual Property," in Stephen R. Munzer (ed.), *New Essays in the Legal and Political Theory of Property* (Cambridge: Cambridge University Press, 2001), pp. 138–67; Jeremy Waldron, "From Authors to Copiers: Individual Rights and Social Values in Intellectual Property," *Chicago-Kent Law Review* 68 (1993): pp. 842–87.

41 William Blackstone, *Commentaries on the Laws of England, Volume 2*, ed. Wayne Morrison (London: Cavendish, 2001), p. 3.

42 Contrary to the despotic dominion account, intellectual property rights are bounded in numerous ways. For example, they typically expire after a period of time (for example, twenty years for a patent), are subject to affirmative rights of users (for example, research rights in patent law and fair use rights in copyright), and may be overridden by governments in certain circumstances (for example, through the mechanism of compulsory licensing). Rights to material property are limited by many different doctrines of property law, for example, the law of eminent domain (which permits the government to take land for public use in exchange for just compensation) and the doctrine of necessity (which permits individuals to trespass on the land of another to prevent something such as a threat to life).

43 The marginal cost of production is the cost required to produce one additional unit of the good. For example, the marginal cost of producing the millionth watch for the next customer is the cost to a watchmaker of producing one additional watch. The concept of marginal cost is important to economists, because competitive-market theory indicates that in a competitive market, price should equal marginal cost. The watchmaker will make and sell another watch if the next customer can pay the marginal cost.

44 It may cost something to distribute the theory to others (the marginal cost of distribution may not be zero), but that is a separate matter.

45 Intellectual property rights may help solve the dynamic (long-run) problem of provisioning, but they do so at this short-run cost—a cost that not all strategies of information production generate.

46 Deadweight loss of this sort can be eliminated with perfect price discrimination, but that

is not expected in practice. The price-discrimination solution also generates distributional effects, transferring wealth from consumers to producers.

47 See, for example, Kenneth J. Arrow, "Economic Welfare and the Allocation of Resources for Invention," in Universities–National Bureau (ed.), *The Rate and Direction of Inventive Activity: Economic and Social Factors, National Bureau of Economic Research.* Special conference series 13 (Princeton, NJ: Princeton University Press, 1962), p. 623, available on-line at http://www.nber.org/chapters/c2144.pdf?new_window=1 (last accessed March 8, 2010).

48 Harold Demsetz, "Toward a Theory of Property Rights," *American Economic Review* 57, no. 2 (May 1967): pp. 347–59. The "tragedy" term comes from Garrett Hardin, "The Tragedy of the Commons," *Science* 162, no. 3859 (December 13, 1968), who famously argued that collective ownership of resources under conditions of scarcity would lead to their destruction.

49 For an early work in this field, see Mancur Olson, *The Logic of Collective Action: Public Goods and the Theory of Groups* (Cambridge, MA: Harvard University Press, 1971).

50 Pamela Samuelson, "Enriching Discourse on Public Domains," *Duke Law Journal* 55, no. 4 (2006): pp. 783 and 786.

51 Boyle, *The Public Domain*, p. xiv.

52 *Ibid.*, p. xv.

53 David Lange, "Recognizing the Public Domain," *Law and Contemporary Problems* 44, no. 4 (1981): pp. 147 and 150–51. For other important early work on the public domain, see the sources cited in Kapczynski, "The Access to Knowledge Mobilization," p. 856 n.232. For a detailed discussion of the ways that scholars have used the term "public domain," see Samuelson, "Enriching Discourse on Public Domains."

54 Boyle, *The Public Domain*, p. 38. In the first definition, Boyle means to include, for example, resources such as language that have never been subject to intellectual property rights, as well as works that have fallen out of protection, for example, because the terms expired. In the second, he means to include domains of freedom that are preserved with respect to protected works, for example, the aspect of a patented invention that may be used because of the "experimental use" exception to patent law.

55 *Ibid.*, p. 41.

56 Copyrights, that is, expire after a fixed term and are limited by fair use rights that protect direct appropriations in certain circumstances, such as for the purpose of parody.

57 See Jane Austen and Seth Grahame-Smith, *Pride and Prejudice and Zombies* (Philadelphia: Quirk Books, 2009); Michael Weinberg, "What Do Ebooks, Zombies, and Copyright Terms Have in Common Besides this Headline?" *Public Knowledge*, December 14, 2009, available on-line at http://www.publicknowledge.org/node/2815 (last accessed February 5, 2010).

58 Fair-use and fair-dealing exceptions place limits on this power, but are widely criticized as too vague and narrow to provide sufficient protection.

59 Lawrence Lessig, *Free Culture: How Big Media Uses Technology and the Law to Lock Down Culture and Control Creativity* (New York: Penguin, 2004), pp. 8–11; William W. Fisher III, *Promises to Keep: Technology, Law, and the Future of Entertainment* (Stanford, CA: Stanford University Press, 2004), pp. 28–31.

60 The public domain can be seen as a "subsidy" in the sense that works within the public domain are either funded directly by the state (for example, the collection of weather data or basic scientific research) or facilitated by the intellectual property rules and other rules administered by the state. But the public domain is also unlike a subsidy for the poor,

because it is neither preferentially available to the poor or populated through a process that seeks to provide the poor with the informational goods that they particularly need.

61 Boyle, *The Public Domain*, p. 39. It is worth noting, however, that there is a genealogical (and as far as I know, unnoticed) link between the commons and the public domain. The term "public domain," according to the *Oxford English Dictionary*, was to designate "land belonging to the public; common land." One could thus speak in the eighteenth century of "cattle that fed on the public domain." *Oxford English Dictionary*, s.v. "Public Domain," available on-line at http://www.oed.com (subscription required).

62 The commons is "generally used to denote a resource over which some group has access and use rights." Boyle, *The Public Domain*, p. 39.

63 *Ibid.*, p. 184.

64 While small open-source projects may be quite informal in their governance, larger projects tend to have more elaborate decision-making procedures. Steven Weber, *The Success of Open Source* (Cambridge, MA: Harvard University Press, 2004), p. 64. For example, Linux is governed by debates on e-mail lists, procedures for reviewing code, and a "hierarchy of gate-keepers" who decide whether a piece of code is included. *Ibid.*, pp. 63–64. Apache is governed by "a formal de facto constitution that is built around a committee with explicit voting rules for approval of new code." *Ibid.*

65 Yochai Benkler, "Coase's Penguin, or, Linux and the Nature of the Firm," *Yale Law Journal* 112, no. 3 (2002), pp. 369–446. See also Steven Weber, *The Success of Open Source*, p. 62: "The key element of the open source process, as an ideal type, is voluntary participation and voluntary selection of tasks" (italics omitted).

66 Free Software Foundation, "The GNU General Public License," available on-line at http://www.gnu.org/copyleft/gpl.html (last accessed February 5, 2010).

67 The latest version of the GPL, for example, was subject to extensive comment and debate, facilitated by the Free Software Foundation among others. See http://gplv3.fsf.org (last accessed February 5, 2010).

68 James Boyle, "The Second Enclosure Movement and the Construction of the Public Domain," *Law and Contemporary Problems* 66, nos. 1–2 (Winter–Spring 2003): p. 41.

69 Carol M. Rose, "The Comedy of the Commons: Commerce, Custom, and Inherently Public Property," in *Property and Persuasion: Essays on the History, Theory, and Rhetoric of Ownership* (Boulder, CO: Westview Press, 1994).

70 Elinor Ostrom, *Governing the Commons: The Evolution of Institutions for Collective Action* (Cambridge: Cambridge University Press, 1990).

71 See Roberto Verzola's essay "Undermining Abundance: Counterproductive Uses of Technology and Law in Nature, Agriculture, and the Information Sector" in this volume.

72 *Ibid.* Ostrom argued that communities can successfully organize the necessary cooperation only under certain conditions, for example, where the community is sufficiently bounded and the members sufficiently proximate. The concept of a digital commons challenges some of these assumptions and raises important questions about the conditions of successful common management in the digital age.

73 See, respectively Creative Commons, "Creative Commons—Attribution—Share Alike 3.0 Unported" license, available on-line at http://creativecommons.org/licenses/by-sa/3.0 (last accessed February 7, 2010) and Philippe Aigrain, "An Uncertain Victory: The 2005 Rejection of Software Patents by the European Parliament," Laura DeNardis, "The Global Politics of

Interoperability," and Manon A. Ress, "Open-Access Publishing: From Principles to Practice," in this volume.

74 See DeNardis, "The Global Politics of Interoperability," and Ress, "Open-Access Publishing."

75 Benkler, "Coase's Penguin," pp. 371–72.

76 Boyle, "The Second Enclosure Movement," p. 46.

77 Moglen, "Anarchism Triumphant."

78 Boyle, "The Second Enclosure Movement," pp. 51 and 41 n.33.

79 Danny Yee, "Development, Ethical Trade and Free Software," *First Monday* 4 (1999), available on-line at http://firstmonday.org/htbin/cgiwrap/bin/ojs/index.php/fm/article/viewArticle/709/619 (last accessed February 7, 2010).

80 Georg C. F. Greve, "Statement by Free Software Foundation Europe (FSFE) at the Inter-Sessional, Inter-Governmental Meeting on a Development Agenda for WIPO," (April 2005), available on-line at http://fsfe.org/projects/wipo/statement-20050413.en.html (last accessed February 7, 2010).

81 Linus Torvalds, quoted in Yee, "Development, Ethical Trading and Free Software."

82 See the remarks by Onno Purbo in "Virtual Roundtable on A2K Politics," in this volume.

83 See Ahmed Abdel Latif's essay, "The Emergence of the A2K Movement: Reminiscences and Reflections of a Developing-Country Delegate," in this volume.

84 Yochai Benkler, in "The Idea of Access to Knowledge and the Information Commons: Long-Term Trends and Basic Elements" in this volume, notes that in the campaign for access to medicines, the "language of justice is most easily available and has been dominant."

85 Thus, the manifesto of the first global march for HIV/AIDS treatment access begins not with a discussion of patents or prices, but rather by asserting the "fundamental rights of health-care and access to life-sustaining medicines." "Global Manifesto, Treatment for All...Now!" Durban AIDS Conference Reports (July 2000), available on-line at http://www.actupny.org/reports/durban-access.html (last accessed February 7, 2010).

86 See the essay by Spring Gombe and James Love, "New Medicines and Vaccines: Access, Incentives to Investment, and Freedom to Innovate," in this volume.

87 To be sure, activists demand free HIV medicines as well as generic HIV medicines, because most of the people in the world living with HIV/AIDS cannot pay even the cost of generics. But the demand is not that drug companies give the drugs away for free, but that governments buy generic medicines for those in need.

88 Lessig, *Free Culture*, p. xiv.

89 Wendy Hui Kyong Chun, *Control and Freedom: Power and Paranoia in the Age of Fiber Optics* (Cambridge, MA: The MIT Press, 2006).

90 Benkler writes most eloquently of this, aligning the public domain with freedom from "hier-archical relations of production" and "tightly scripted possibilities." Benkler, *The Wealth of Networks*, p. 138.

91 Boyle, "The Second Enclosure Movement and the Construction of the Public Domain," p. 62. Lessig also overtly aligns the freedom of "free culture" with "'free markets,' 'free trade,' [and] 'free enterprise.'" Lessig, *Free Culture*, p. xiv.

92 GRAIN, "Freedom from IPR: Towards a Convergence of Movements," editorial, *Seedling* (October 2004), p. 3, available on-line at http://www.grain.org/seedling/?id=301 (last accessed February 7, 2010).

93 See Carlos M. Correa, "Access to Knowledge: The Case of Indigenous and Traditional

Knowledge" and Jeffrey Atteberry, "Information/Knowledge in the Global Society of Control: A2K Theory and the Postcolonial Commons" in this volume.

94   Lawrence Lessig, *Remix: Making Art and Commerce Thrive in the Hybrid Economy* (New York: Penguin Press, 2008).

95   United Nations, *The Millennium Development Goals Report 2009* (New York, 2009), p. 52, available on-line at http://www.un.org/millenniumgoals/pdf/MDG_Report_2009_ENG.pdf (last accessed February 7, 2010).

96   For the first position, see, for example, Lessig, *Free Culture*, p. 123; Benkler, *The Wealth of Networks*, pp. 139–40 and 240; and Benkler, "The Idea of Access to Knowledge and the Information Commons" in this volume, where he argues that "freedom, justice, and innovation all require *effective* agency, not merely formal permission to act." Benkler does briefly address the digital divide, but refers to it as a "transitional problem." Benkler, *Wealth of Networks*, pp. 236–37.

97   Collection societies like ASCAP, for example, act as clearinghouses to manage and enforce copyright owners' public-performance rights. Collective management here is a strategy to extract value from copyrighted works. Such societies typically engage in blanket licensing schemes, charging fixed royalties in order to save copyright holders the cost of enforcement. For a discussion of the operation and implications of collective rights management organizations, see Robert P. Merges, "Contracting into Liability Rules: Intellectual Property Rights and Collective Rights Organizations," *California Law Review* 84, no. 5 (1996): pp. 1293–1385.

98   See H. L. A. Hart, *The Concept of Law* (Oxford: Clarendon Press, 1961). In the commons, the recognition rule would be the rule that determines the validity of governing norms.

99   On the former, see Gabriella Coleman, "The Political Agnosticism of Free and Open Source Software and the Inadvertent Politics of Contrast," *Anthropological Quarterly* 77, no. 3 (2004): pp. 507–19.

100 Benkler, "The Idea of Access to Knowledge and the Information Commons." Boyle adopts the same spirit when he calls the politics of open source a "curious mix of Kropotkin and Adam Smith." Boyle, "The Second Enclosure Movement," p. 46.

101 Roberto Unger, *What Should the Left Propose?* (London: Verso, 2005).

102 Benkler, "The Idea of Access to Knowledge and the Information Commons."

103 Verzola, "Undermining Abundance: Counterproductive Uses of Technology and Law in Nature, Agriculture, and the Information Sector."

104 *Ibid.*

105 Lynn P. Freedman et al., "Who's Got the Power: Transforming Health Systems for Women and Children, Summary Version," United Nations Millennium Project, 2005, pp. 1–2, available on-line at http://www.unmillenniumproject.org/documents/TF4Childandmaternalhealth.pdf (last accessed February 7, 2010).

106 On the political nature of famines, see, for example, Amartya Sen, *Poverty and Famines: An Essay on Entitlement and Deprivation* (Oxford: Clarendon Press, 1982) and Alex de Waal, *Famine Crimes: Politics and the Disaster Relief Industry in Africa* (London: African Rights and the International African Institute, 1997).

107 Benkler, in particular, contends that the networked information economy can facilitate greater political freedom. See *The Wealth of Networks*, pp. 176–272.

108 Latif, "The Emergence of the A2K Movement: Reminiscences and Reflections of a Developing-Country Delegate."

109 Benkler, *The Wealth of Networks*, p. 313.

110 *Merriam-Webster Online Dictionary* (2009), s.v. "Knowledge," http://www.merriam-webster.com/dictionary/knowledge (last accessed February 7, 2010), defining knowledge as gained "experience or association."

111 *Oxford English Dictionary* (2009), s.v. "Knowledge," available on-line at http://www.oed.com (subscription required).

112 Lyotard, *The Postmodern Condition*, p. 18.

113 This definition of knowledge differs substantially from the one offered by theorists such as Castells, who mean by it a "set of organized statements of fact." Castells, *The Rise of the Network Society*, p. 17 n.25.

114 *Ibid.*, defining information as "data that have been organized and communicated."

115 Benkler, *The Wealth of Networks*, pp. 314–15.

Access to Knowledge as a Field of Activism

Gaëlle Krikorian

Mobilizations around issues involving access to knowledge (A2K) can be seen as a phenomenon highly symptomatic of political as well as technological changes in our society. The neoliberal revolution, beginning at the end of the 1970s,[1] and the emergence of digital media and the Internet, a central phenomenon of the past two decades, are prime examples of such shifts. Both have played a role in the contemporary trend toward the development of new and/or increasingly exclusive intellectual property rights. Since the late 1990s, this evolution triggered the mobilization of groups and individuals around the world that are now brought together under the banner of A2K—or are perceived as belonging to a general movement.

This book aims at investigating the forms that this phenomenon is taking, as well as the changes it calls for and the transformations that it might effect in our society. In this introduction, I intend to discuss the technical and political settings that have provoked or sustained the existence of this movement and to explore some of the social tensions involved. The A2K movement raises fundamental questions about the conception and production of ideas, goods, and services created in the current knowledge-based economy and about access to such ideas, goods, and services. In doing so, and in order to be in a position to challenge effectively the prevailing practices in these areas, it also questions more broadly the representations and actions that legitimize, organize, and ensure the functioning and sustainability of the existing system based on intellectual property rights. It discusses the place and role of the various actors involved in this system (the state, the corporations, the individual, the market), as well as the relations and interactions between them.

As the A2K movement structures itself, it develops and offers its own readings of the world—readings that invite us to explore new possibilities in apprehending and organizing our societies—and as such could gain from the spirit of the gleaner and from Michel de Certeau's insight that "everyday life invents itself by *poaching* in countless ways on the property of others."[2]

THE DIGITAL ERA AND IMMATERIAL WALLS

Intellectual property rights protection is the main framework for the control and regulation of the production and of the use of knowledge and information. Standards of protection of intellectual property rights are established and governed in various ways: in national laws and regulations, but also via international agreements, including multilateral ones such as the TRIPS (Trade-Related Aspects of Intellectual Property Rights) Agreement of the World Trade Organization (WTO), in numerous treaties of the World Intellectual Property Organization (WIPO) and the World Customs Organization, and finally by bilateral or regional agreements, treaties, or conventions. Over the years, the variety of institutions establishing the norms, rules, and procedures involved in the governing of intellectual property rights has stretched and expanded, thereby implicating an increasing number of actors and an increasing variety of aspects of social life. These developments have built on the evolution of the conception of what intellectual property rights are and what precisely can be subjected to intellectual property laws. Thus, the invention of new means of creating exclusive rights has grown all the more important in recent decades.

Though the ways in which intellectual property rights protections have been extended may sometimes seem minor—a few words added in a law, a few concepts reinterpreted, a chapter concerning intellectual property rights added to a free-trade agreement—their effects are often significant. The way intellectual property rights are handled also reflects changes in the strategies employed by intellectual property owners in the face of technological as well as political developments. They successfully have changed the goals of intellectual property law, goals ranging from authorizing private property while limiting access to materials held in common to prioritizing property and its defense.[3] Rather than creating physical walls to protect material property, they have sought to create immaterial legal walls to enclose information and knowledge, the immaterial property of the digital age.

Information and knowledge are the raw material of which immaterial goods, ideas, and inventions are made, and as such, they are key to individual as well as collective human development and welfare.[4] On the scale of the global economy, what is at stake in exclusive intellectual property regimes is nothing less than the control of existing stocks of information and knowledge and of their flows, along with the management and harnessing of the innovations that such information and knowledge can allow to produce. As in the current international economy growth and competitiveness have become increasingly dependent on the production, processing, and circulation of information and knowledge, the empire of intellectual property rights was expanded. The ramifications of the intellectual property system thus extend to techniques, technologies, know-how, and skills in all sectors, whether they concern financial speculation, aerospace engineering, medical or military research, agronomics, textiles, shipbuilding, cooking, or music composition.

The Internet and the digital era have changed the relationship of users of technologies to production and creation, opening up new possibilities that quickly have translated into the emergence of new practices. On the individual level, this new technological context has contributed to the blurring of the line between consumers and creators and to the characteristic status usually allocated to each, such as passivity versus productivity, inertia versus efficiency. New technologies and new formats (VCRs, VHS, CDs, DVDs, and so on) have made it possible for anybody equipped with the proper equipment—and in the capitalist economy of wealthy countries, access to such equipment has been rapidly democratized— to copy, adapt, mix, or perform sounds, images, or motion pictures. Because the Internet facilitates large-scale and nearly instant exchanges—features that many would recognize as being specific to contemporary "globalization"—creation by means of these technologies is characterized by the marginal costs of production and the high speed and low geographical concentration of distribution. The ways in which one creates have not fundamentally changed. Creation has always been inspired and made possible by what already exists, and it continues to be, but the space and time in which the act of creating can be performed by ordinary people has been significantly transformed with the unfolding of an immaterial world in which new possibilities of creation have become increasingly accessible to many.

As Lawrence Lessig has pointed out, these technological changes have introduced the "potential to expand the reach of this creativity to an extraordinary range of culture and commerce."[5] First, new types of goods and products enriching the economy of the immaterial keep emerging, and their importance keep growing.[6] Second, changes in production due to the fact that digital technologies "create and replicate reality much more efficiently than non-digital technology does" have affected not only what people can do at their own, individual level, in their private spheres, but also at the level of the economy itself.[7] For instance, what inspires and provides incentives for economic actors tends to change. Because innovations and goods are easier to copy and more difficult to protect, providing services often becomes more economically rewarding than selling physical products. The economy of immaterial goods develops according to specific ways and via specific means that in return bear the potential to transform the functioning of the economy as a whole. As Yann Moulier Boutang explains in an interview in this volume, digital technologies, owing to their dramatically reduced production and distribution costs, offer opportunities to revise the sharing of those costs in many different sectors.[8] In the book-publishing industry, for instance, such changes could benefit the creator's ability to work by allowing us to revise the way in which they are paid and thus improve greatly a manner of compensation that is largely inadequate for many of them in the current intellectual property system. Creators in general can also benefit from easier access to the works of others that facilitates potentially new forms of work and cooperation that favor research and creativity. Entire industries and economic activities, not only those specialized in immaterial goods, changed with the integration of digital tools and began to transform even more substantially with the move toward the concept of open innovation and network-based peer production.[9] Some companies have invited consumers into the innovation process—in some cases even through challenges directly posted on the Internet to encourage people to come up with new ideas and to share them.[10] They have used consumers as a source of inspiration for new products or designs that could attract and interest new customers and create new markets. In recent years, groups such as IBM, Proctor and Gamble, Lego, and Unilever have adopted such strategies to develop new products. These new ways of doing business have called for important transformations in traditional business models, company cultures, and management strategies. For instance, innovative strategies may shift the focus of many companies from keeping formulas, recipes, or components secret to extracting the value of ideas, increasing the speed at which products can be brought to market, reducing the cost of research and development, and improving the fit between their products and consumers' desires or modes of consumption.[11]

However, despite the real or potential changes that information technologies have introduced and the substantial new prospects they have opened up in the economy, widespread transformations of corporate practices still remain rare. Even companies that have incorporated a certain amount of open innovation have mostly remained intent on maintaining control of ideas as soon as they are generated. New modes of creation and consumption have emerged, but the immaterial has become essentially and only a new field in which capitalist logic can operate, and the principles on which capitalism is based have remained unchanged.

However, insofar as the interaction allowed by digital technologies offers opportunities for intense exchanges and production, as well as for new ways of commodifying goods and services, the transformation of the knowledge economy has certainly affected capitalist ways of functioning. Industrial capitalism now coexists with a new form of capitalism, called by some "cognitive capitalism," or "knowledge capitalism," that is both a new type of accumulation (of intellectual capital) and a new mode of capitalist production.[12] It corresponds to the development of an economy based on the distribution of knowledge goods in which the production of knowledge is the central element in the valorization of capital.[13]

Changes in the means by which capital is produced and accumulated have brought about changes in the position that social groups occupy in relation to the rules of production. In knowledge capitalism, the laboring class no longer holds a central role, and capital is produced mainly by a new class composed of technocrats and people working predominantly for the service sector. This development contributes to the disruption of the preexisting social order as new tensions and power relations between social groups began to arise. With knowledge capitalism comes a reconfiguration of class interests and of the relations between classes and thus a reconfiguration of what defines them. As we will discuss later, in this context and under the rules of intellectual property protection that regulate the production as well as the use of goods, it is the issue of *access* that draws new dividing lines between people and groups, dividing lines that are superimposed on former divisions.

Some people have the means to exist and to thrive in the digital world, while others do not. What is necessary for participation in the immaterial world is not only a computer, the right software, and an Internet connection, which already excludes a large portion of the world population, but also the codes and filters normally acquired through education. These are indispensable for navigating this environment, because only they enable participants to locate and to make use of the resources available in ever-expanding proliferation and to take an active role in the production thereof.[14] But the education necessary for the acquisition of such codes and filters remains a near monopoly of the privileged classes. At the same time, with the unfolding of the knowledge-based economy, the strengthening of intellectual property protections, and the central place that the market occupies in the neoliberal context, potential inequalities in access increase: Knowledge appropriation plays an increasingly important role in the economy and in peoples' lives in general, but is also more than ever subject to market rules. Individuals with no economic and/or cultural capital generally cannot compete on an equal footing with others, and their access to knowledge is easily compromised. Thus, inequalities in access to knowledge reinforce and perpetuate social and class inequalities, while the current knowledge economy and the intellectual property regime overlay an old class structure with new tensions.

As in every capitalist model, in knowledge capitalism, the issue of the transfer of property is a key issue. One of the most salient characteristics of knowledge goods is their electronic transferability. Consequently, in the field of knowledge capitalism, to ensure ownership and control of knowledge goods and thus benefit from the profit derived from their production and subsequent commercial exchange, one must find ways to prevent or slow down a transmission made so simple and easy by digital networks.[15]

This is where intellectual property rights come into play. One might think that it is the same old game, appropriating the means of production, only taking place in a new environment. And to some extent, it is. Some argue that this is one of the problems with the term "intellectual property;" that is, the fact that it determines the granting of legal rights through the establishment of property, what James Boyle describes as the "second enclosure movement."[16] Historically, at the time of the first property enclosures, land ownership was at stake. To enforce it, apart from the use of legal acts, walls and barriers were used to delineate the property. In the knowledge-based economy, intellectual property rights holders, through their efforts to establish property rights over knowledge, are building other kinds of walls to channel access and to regulate who can benefit from what is produced in the immaterial realm. Though not as visibly obvious as physical walls erected throughout the world, in the era of globalization, they are just as determinative in the establishment of national and international social orders.[17]

This enclosure effort is all the more evident as enforcement and repression become increasingly important pieces of the intellectual property right owners' agendas. The purpose of intellectual property walls is not to demarcate space, to differentiate an inside and an outside, each having different characteristics and status, but they are far from serving a merely symbolic function. When, for example, the Chinese government sends tapes showing police raids and the destruction of unauthorized copies of DVDs to the U.S. trade representative, this theatricalization of police efforts is both a performative action intended to prove the goodwill of the Chinese government to its U.S. counterpart and a publicly demonstrated materialization of the existence of intellectual property rights and of the consequences that the act of trespassing in the immaterial world can generate. Such performances—from raids in Moroccan souks to the arrest of teenagers and other Internet users in Hong Kong, France, the United Kingdom, and the United States—are becoming more numerous and increasingly visible in public space.[18]

However intellectual property walls do not always efficiently prevent access. If people are determined to find breaches, they often can. But in doing so, they will knowingly commit an illegal act and thereby run the risk of sanctions. In our societies, most of those who commit an illegal act to access a territory or a good are those who do not have the means to play by the rules or those who consider they have little to lose in comparison with what they hope to gain. Those with the resources and capital, on the other hand, are rarely refused access to a territory or a good. As a result, walls work not so much as real barriers, but as socially polarized filters. They selectively hinder certain people and filter societies as they regulate access to information and knowledge—a mechanism that inevitably makes the issue of access political.

The second most salient characteristic of knowledge goods is what economists call their "nonrival" and "nonexcludable" nature. Nonrival goods are goods whose consumption by one person does not prevent its consumption by another. This book is a nonrival good: After you have read it, another still can. On-line, even *while* you read it, another can. Nonexcludable goods are goods whose consumption can't be prevented once they have become publicly available. Architecture, such as the Cathedral of Notre Dame, is a nonexcludable good: Anybody who can appreciate it can do so for free. Because these goods can be used by multiple individuals simultaneously, it is harder, sometimes impossible, to expropriate them. Both attributes operate as constraints on capitalist exchanges and make rights holders fearful that the technology of digital copying will render their legal rights and sources of profit ineffective. The advent of digital technologies and the popularization of the Internet brought the prospect of huge financial benefits, but at the same time, uncontrolled consumption and production, which is materially limited in the physical world, also took on new proportions, given that the spaces in which these take place are numerous and ubiquitous and that the cost of enforcement is high. Who or what entity could indeed possibly observe everything taking place in every potential offender's living room or bedroom or monitor every Saturday-night party in every small town in every country?[19]

Consequently, a race began between the "cops and the robbers." As Yann Moulier Boutang describes it, "the cops never get a head start. There is a delay, and their route is full of pitfalls."[20] The enforcers have acquired new ways to locate the infringers, but the technical possibilities have intensified their interactions and exchanges. What makes the Internet a public space—a space that if not all individuals, at least many can access and inhabit and where freedom can appear—also makes it a space difficult to control.[21] But along with the freedom that the Internet provides to users, it also gave rights holders a cheaper way to watch individual activities on a global scale and to monitor and locate infringements, if not when people are enjoying the use of illegally acquired material, then when they are merely acquiring it. As such, as Cory Doctorow notes, the Internet and the personal computer represent "a perfect storm for bringing ordinary peoples' ordinary activity into the realm of copyright."[22] If activities that infringe intellectual property rights cannot be eradicated, they can be criminalized. And if the act in itself cannot be prevented, social condemnation can affect people's behavior, repression leading to suppression.

One consequence of such condemnations is to marginalize appropriative and sharing practices and to make them disappear from public spaces. As Lessig observes, if "we can't stop our kids from using these tools to create, or make them passive," we can "drive it underground, or make them 'pirates.'" Examining developments in the field of copyright, he adds: "We are in the middle of... what some call 'the copyright wars.'"[23] If there is war over copyrights, we should ask ourselves who the fighting camps are. On one side stand the owners of intellectual property rights. They are rather easy to identify: They sponsor laws in public forums and pay for advertisements on TV; they promote a moral position as well as an understanding of economics for the public to adopt; they argue that their camp is the righteous side of the debate, the side of struggling artists who need to be protected from dishonest plagiarists, of quality magazines endangered by blogs and free publications. Who are their enemies? Many different profiles fit into this category, including kids "stealing" songs or movies with their computers, unknown artists copying and transforming very well-known ones, and individuals using peer-to-peer platforms to share files and software.

A wide range of individuals thus started to be targeted as "pirates" for the improper use, sharing, and production of materials using copyrighted matter, and the more copyrights expand, the more favorable are the conditions of the production of "pirates." They easily fit into a political environment that is predominant in many Western countries in which security and repression had become routine. Increasingly, public, social, and legal resources have been encouraged to be or actually have been mobilized in an effort to enforce intellectual property protections and to limit exchanges of protected material, resources that often seem particularly unreasonable to deploy in many developing countries when one compares them with the national budget of such countries and when one considers essential, but unmet local needs. Meanwhile, the motors of innovation and creativity are jeopardized by an ever more restrictive judicial and legal environment—despite the fact that this, by definition, is contradictory to capitalist interests, which rest on the continuous delivery and marketing of new products to generate accumulation.[24]

Various tensions and conflicts about the effects of and the justifications for intellectual property rights have emerged and crystallized in the past decade. They have taken the form of negotiations and contentious relations between states within international organizations such WIPO, the WTO, the World Health Organization (WHO), and UNESCO, the UN Educational, Scientific and Cultural Organization. They emerged noticeably in an election context recently with the constitution and election to the European Parliament of a Swedish pirate party.[25] They have resulted in lawsuits brought by people with AIDS against pharmaceutical companies or the conviction of farmers who have campaigned against genetically modified organisms. And they have inspired demonstrations and lobbying campaigns for access to medicines, against software patents, against biopiracy, and for the mobilization of students, librarians, and researchers. They have provoked the organization of university meetings for open sources, for open publishing, and for access to knowledge. They have triggered conflicts between major corporations, lawsuits between competitors, and debates in many parliaments, senates, and congresses. Each of these contentions can be seen as expressions and elements of the formation of access to knowledge as a field of activism.

These mobilizations and the common framing of their claims manifest a willingness to give the issue of access a central position in the contestation of fundamental political and social issues today. As the A2K movement seeks to promote the visibility of challenges to access and uses the issue of access to structure its discourse, it encompasses social and political contests specific to the inequalities created by the rules governing the appropriation of value and property under the current regime of neoliberal capitalism. As such, as we will see later, A2K can be seen as the development of a response to continued efforts to extend intellectual property rights, efforts that themselves can be seen as both a political and a social mobilization.

POLITICAL CONTEXT AND POLITICAL PRACTICES: FROM INTELLECTUAL PROPERTY TO ACCESS TO KNOWLEDGE

To understand how the interests of intellectual property rights owners became state policy, we need to look more closely at the way those advocating increased intellectual property protections have been organized and mobilized, because their strategy rests as much on their harnessing of an ideological/political context and their manufacturing of conceptual tools as on the details of the ways in which they have organized their mobilization.

The fact that the market is in the foreground of most contemporary political theories, or more exactly, that no other views than those putting it in the foreground could establish themselves successfully in the past forty years, illustrates the spread of neoliberal rationality, which "extend[ed] and disseminat[ed] market values to all institutions and social actions" across the globe. Far from being only an economic doctrine, this ensemble of political practices and institutions has enveloped the state, which has incorporated its economic logic and redefined itself according to the search for profitability, progressively linking its legitimacy to its capacity to sustain and fuel the market. The market, while remaining a "distinctive player," has become the organizational principle that is applied to the state, as well as to individuals and society.[26]

Contrary to the assumption that associates neoliberalism with a weak and quasi-absent state, in this regime, the state, while it must be kept subject to the logic of the market, has a key role to play to guarantee that the market operates properly: It needs to provide laws, regulations, and institutions that establish optimal conditions for its development.[27] Thus, far from being noninterventionist, the state actively participates in an array of domains through policy arbitration, dismantling welfare programs, and deregulating entire sections of socioeconomic life while controlling, encouraging, or criminalizing social activities and behaviors.

In this context, those representing the state and those representing private interests have started to act as partners. This rapprochement has been facilitated by the phenomenon of the "revolving door," according to which the members of the dominant classes, including many who have been advocates for the establishment of neoliberalism, hold positions, often in tandem or succession, in both public and private institutions.

The action of advocates for an increase in intellectual property protection benefited, both literally and symbolically, from neoliberalism's successful establishment at the end of the 1970s and 80s as the dominant political system internationally. Neoliberal rationality indeed constitutes a favorable environment for strengthening intellectual property rights. On the one hand, neoliberalism promotes individual entrepreneurship and private property, while on the other, it encourages free trade and the multiplication of agreements that have proven to be a key vehicle for the expansion of intellectual property restrictions across the world.

Of course, the maximization of intellectual property rights can also be seen as paradoxical, if not contradictory, with regard to the neoliberal doctrines, since these, in theory at least, promote competition and do not encourage institutionalized monopolies. But such is the beauty and the efficiency of neoliberal rationality that it is malleable enough to allow it to incorporate paradoxes without losing its apparent cohesion and strength. Thus, the state, while lauding free-market theory and spreading it both in discourses and through international agreements, in effect undermines it, allowing monopolies and limiting competition in order to ensure and expand the rights of intellectual property owners.[28]

Besides providing intellectual property rights owners with a practical vehicle for the realization of their agendas, neoliberalism has also has helped them establish their sociocultural position and direct their political actions. Ideologies are a powerful instrument in the production of legitimacy, and the laissez-faire ideology of classic economic liberalism has provided neoliberalism with the benefits of legitimacy and historical weight. Although those who advocate for increased intellectual property restrictions generally publicly condemn what they call "ideology," which they usually associate with left-leaning political utopias, Communism, socialism and other such "evils," the ideological domination of neoliberalism that is now (and still) understood as inevitable, having imposed itself with no alternative and as having coincided with Francis Fukuyama's "end of history," has proven largely instrumental to the successes of the movement to increase intellectual property protections and the lack of questioning of the vision that it promotes. At the same time, the contributions of intellectual property advocates to the neoliberal revolution through the success with which they have promoted their own goals has helped power the rise of this broader political movement.

But if the strategy of intellectual property rights owners has benefitted from their harnessing of an ideological/political context, it also has rested on the way they have manufactured conceptual tools and organized their mobilization. At first blush, it may seem like a misconception to treat advocacy for intellectual property rights as an organized social movement. After all, those who advocate for increased restrictions on intellectual property rights tend to belong to the dominant class, and what is usually considered as an alliance consisting of a group of property owners, a network of industries, or a cartel of multinationals has a priori no need for mass mobilization and the confrontational tactics that most organized social movements rely on in order to persuade states to act in their interests.

However, it worth noting that, like other social mobilizations, intellectual property right owners do not employ the traditional political means of representative politics to further their agenda. Of course, they soon succeeded in making the state a partner in realizing that agenda, rather than a source of opposition to it, and while confrontations between intellectual property interests and the state do occur, for instance when intellectual property advocates lobby and even threaten governments, these confrontations take a very different form from those expected from typical social movements. Nevertheless, a number of features central to the concept of a contemporary social movement do seem to describe the mobilization for increased intellectual property protection.

A social movement is a product of its time and reflects actions and reactions to a particular political and economic condition, or what some social scientists refer to as a political opportunity structure. The emergence of the movement in favor of increased protection of intellectual property rights, taken as such, offers insights into evolution of the information society and into the constraints and opportunities that such an evolution has presented to intellectual property owners as the basis for their mobilization.

Strategically, as is the case in numerous social movements, success has rested on building a common identity that goes beyond the interests of one group and on mobilizing collective action that encompasses a range of tactics that goes well beyond merely lobbying those who govern.[29] Using an emphasis on the concept of property as the basis of this shared identity, promoters of increased intellectual property protection successfully incorporated the project of a handful of corporate executives into international trade negotiations, U.S. government policy, and even international trade rules.[30] An initiative launched by a specific segment of society, but couched in terms of social progress (that is, of increased social welfare and development) thus successfully achieved changes in basic social and political norms.

The A2K mobilization integrates and responds to the specific ways of doing politics that the movement for greater intellectual property protections has used. This is undoubtedly inevitable, because A2K advocates are engaged in discussing and criticizing the effects of intellectual property rights, and consequently, they incorporate the legal language that articulates those rights and engage with the institutional frame that produces them.[31] Participants in the A2K movement are keen to monitor their counterparts' moves, and they take inspiration from the manner in which the movement for intellectual property restrictions has successfully incorporated its agenda into the state's agenda. Consequently, they regularly employ the technical and legal language of the various institutional contexts where they try to counterbalance the effect of their opponents or where they try to intervene before their opponents do so, whether at the WHO, WTO, or WIPO, during the negotiations of bilateral trade agreements, during negotiations over the United Nations Millennium Development Goals, or in negotiations over national regulations and laws. Furthermore, A2K advocates of necessity employ the dominant economic logic when they seek to promote a balance between public and private rights based on criticisms of the way the market functions, for example, or when discussing the need for competition, the effect of monopolies, or the exclusive impact of prices. The two opposing movements can thus be seen as adverse forces at one moment in history, which also implies that they to some extent share a common culture and experience.

Both movements likewise participate in and exploit the effects of globalization. Globalization in the neoliberal context both results from and in turn provokes transformations of the existing power structure and the practices of power. Internalizing the context of globalization, both movements elaborate their actions inside and outside national borders. Both contribute to the ways that politicization occurs outside the framework of representative politics and to what results from it. Both compete to influence the state's performance and what its role should be. As much as neoliberalism rests on state intervention and control to strengthen and facilitate market logic, claims for an open and protected public domain hold the necessity of an active power structure that bears responsibilities for public interests and that enforces rules.[32]

All of these factors affect the movement for access to knowledge as a field of activism, determining its concrete strategies and tactics. For many A2K advocates, opposing intellectual property extremism in public debates and concretely establishing optimal conditions for the creation of new things and for socioeconomic prosperity requires reviving awareness of the social value of spaces where all people can share and make use of knowledge produced there and of the need to secure such spaces.[33] This process involves questioning the arbitration performed by those who govern and who mediate between the public and private spheres and between public and private interests. Facing the alliance between the state and multinational corporations, A2K advocates are confronted with a particular power structure and the form of governance established thereby. In this context, the issue to debate is not so much the intrinsic capacity of the state to control resources or whether the state or the market works better at doing so, largely because these two entities are no longer in opposition with one another. Rather, the discussion concerns the objectives that the state pursues, its priorities, and the ways it operates within neoliberal rationality, all of which call for A2K advocates to develop a critical understanding of the role of neoliberalism in the development of the global intellectual property regime. More broadly, A2K advocates interrogate not only the role and place given to the governing powers, but also the role and place given to the individual and the relation between the two of them, as well as the relations between individuals themselves.[34] They thus address fundamental and age-old issues regarding the governing of societies, as well as current transformations of power and the legitimate expectations that individuals can experience as a consequence.

WHAT IS A2K?

In 2004, the term "access to knowledge" emerged as a common umbrella under which individuals and organizations could denounce inequalities and injustices related to intellectual property. But it remains a fair question to ask whether this gathering is more than the pooling of problems and demands, more than a juxtaposition of identities that have provisionally focused their energies on a common hindrance.

A consideration of the trajectory followed in terms of commitment by the individuals who today take part in A2K mobilizations provides a useful understanding of the emergence of the phenomenon. Some describe their participation as a logical evolution of their involvement in other political issues. Indeed, many A2K advocates have been or are currently active on other fronts. Analyzing the mechanisms at the heart of the problems they focused on is often what led them, at the end of the day, to describe those problems in terms of inequalities in access to knowledge, giving rise to the recognition of an underlying cause and the formulation of a common framework around which others could be rallied. In a typical example of this process, one activist may have joined mobilizations to end discrimination against HIV-positive people. This initial effort may logically have turned into a campaign for access to drugs in developing countries. And such a campaign might, as a matter of course, lead him or her to denounce the negative effects of intellectual property protections, which bar access to such medicines. As the activist intensified his or her critique of the dampening effect such protections have on the generation of medical innovation, key critiques surrounding access-to-knowledge issues would begin to be formulated. The activist would soon find that, perhaps even almost unwittingly, he or she had joined the A2K mobilization.

A2K does not look like a mass movement. It does not rely on massive street demonstrations as a constitutive means to confront the power structures that it challenges. Perhaps a more massive form of mobilization and a more cut-and-dried political stand would do a better job of advancing the purposes of A2K. Or perhaps the A2K movement could better serve itself by drawing on what it already possesses, which is a composite form of mobilization that provides the potential to cement together a multitude of actions.

As in a Venn diagram, movements fit into one another and overlap, each one bringing its own obsessions, tactics, networks, and savoir-faire. A2K can be seen as movement of movements, resting on the capacity of its participants to hear and share their various messages or, more specifically, the common denominators within their messages, without allowing their differences to develop into obstacles. The A2K umbrella is large enough to allow for an intense variety of participants, issues, and actions, as well as to allow creativity to express itself through various modes without being limited by the hierarchical structures that often hamper conventional organizations. This is not unlike the form of political activity witnessed throughout the course of Barack Obama's presidential campaign: a general rhetoric not only coexisted with, but was actually energized and shaped in the eyes of the public (or in the eyes of enemies) by a variety of actions from individuals and networks stepping forward from multiple places. The very qualities that gave this movement its blurry outlines also enhanced its force by enlarging its federative power. Within the A2K movement, individuals bring whatever they have to the table, be it their handicaps or their positions of privilege, and draw upon them as resources for collective action in order to formulate political questions. When an HIV-positive person asks for access to life-saving medication or a visually impaired person asks for access to educational material, their point cannot be easily dismissed publicly or ignored by political leaders. Meanwhile, when an academic from a prominent U.S. university presents an analysis of knowledge gaps or business models, his opinion is likely to be heeded and to be echoed in political spheres and media circles.

Although each element within the collective may have its priorities or its particular raison d'être, the movement as a whole does not impose a hierarchy of matters of concern. It sustains a plurality of claims and actions without undermining the particularities or the autonomy of individual groups or national coalitions. A campaign for the local production of generic drugs in Brazil or South Korea takes place simultaneously alongside a movement at WIPO for access to reading material for visually impaired persons. At the same time, an international mobilization takes place to defend a professor of philosophy taken to court in Argentina for putting Spanish translation of texts by Jacques Derrida on the Internet, and a backlash is organized against the European Commission for allowing its customs arm to block the transit through its harbors of life-saving drugs from India to other developing countries.

At some levels, a dampening out of particularities, singularities, and diverse priorities or choices normally occurs in favor of the vision of those within a movement who enjoy certain advantages, whether by speaking a dominant language, having a higher level of education, associating with upper-class social networks, or possessing greater financial resources. But although the consequences of such power relations are by no means absent within the A2K, so far, this has not seemed to affect the apparent cohesion of the movement.[35]

So far, the A2K model of activism allows for such heterogeneity without sacrificing the capacity to function as a common entity advancing a common cause. Each protagonist may be focused on one particular issue or may be involved in several different fights at the same time. However, as a member of the A2K mobilization, he or she agrees to represent a collective identity—whether it is in a very active or less committed way—thus becoming part of an entity that transcends the elements that constitute it, a movement that contributes to the emergence of a common imaginary.

Intellectual property rights affect and encompass a variety of issues that are diverse and separate in nature. Paradoxically, the reification of intellectual property rights as one coherent concept that embraces copyrights, patents, and trademarks has enabled the emergence of an extremely diverse A2K front. Opposition to institutions with so wide a footprint as the WTO or to policies with such far-reaching effects as the TRIPS Agreement or free-trade agreements favors the coalescence of groups or movements originally focused on specific and discrete concerns because such groups oppose entities that structure interconnections between domains. Each A2K actor not only addresses a specific effect of the strengthening of intellectual property rights, but, as a member of a collective, embraces multiple issues, becoming sensitive to the echo and similarities between their causes and taking into consideration the broader logic and structure manifested in the specific legal provisions they oppose. The nature of the enemy determines the organization of the resistance mobilized against it, forcing it to become more systemic. In order to make the best of collective action, A2K advocates therefore cannot limit themselves to a juxtaposition of diverse demands or criticisms, but must instead develop an integrated common agenda, or at least try to do so. If this is not yet where the A2K movement is, it is definitively a trend along which it tends to evolve. A2K advocates are pushed to formulate a global vision for society, rather than simply denounce legal dispositions or policies, and the time frame in which they plan their actions to serve longer-term objectives expands accordingly.

This dynamic usually carries benefits for movements: an increased presence in forums and political spaces, the capitalization of resources, networks, and the benefits from specific actions, the ability to move back and forth between the grassroots and the political spheres, and so on. At the same time, the trend toward integration and consolidation can weigh down an organization and raise problems within it, disrupting the specific culture of action that characterizes it, potentially blurring their initial objectives, and creating internal tensions. Contesting the specific effects of the intellectual property system without abandoning larger A2K claims in terms of creativity, innovation, and access certainly has the potential to take activists further than they first decided to go, to get them involved in politics in a more totalizing way than they intended when they first demanded a right or denounced what they identified as an injustice.

By identifying themselves as A2K constituencies, individuals and groups show an interest in defining themselves and in being perceived not simply as contradictors or opponents of the intellectual property system, but as promoters of a positive and cohesive agenda—something bigger than mere opposition.[36] Though they were originally brought together by objections to a common enemy,[37] a conscious strategic move was made by many A2K advocates for the purpose of allowing them to reframe the issues outside of the logic of the intellectual property rights system. Whether they invoke notions of the public domain or knowledge goods and knowledge spaces as commons, A2K actors are trying to formulate a debate outside of the dialectic of opposition, in a discursive space in which they can set at least part of the terms and in which intellectual property represents only one among several options.[38]

Not only does A2K not look like a mass movement, but many of its advocates are not very radical, and, as a whole, the movement is rather utilitarian. Finally, A2K is not as confrontational as many other social mobilizations. Most A2K advocates so far seem interested in withdrawing from the dialectal logic of direct power struggles, either with the state or with industry.[39] Members of the A2K constellation are not constituted as activists through a confrontational relation with "the" public, the way mobilizations of minorities can be, for instance. The aim of people who recognize themselves in the A2K discourse is mostly not to mark themselves off from the general public. They do not wish to materialize "subaltern counterpublics" dedicated to the formulation of "oppositional interpretations of their identities, interests, and needs," but instead seek to alter the cultural horizon represented by the dominant vision and in doing so transform the main discursive arena and constitute *an* alternative general public.[40]

The A2K movement is not so much based on a claim of the "unity" of all people and their struggles, but rather on the effort to convince others that they are affected, should be concerned, and should act accordingly.[41] A2K constituencies are mobilized against a peculiar enemy, but they are organized in the name of a "shared" interpretation of their interests and needs, which is understood to extend to the interests and needs of *the* public. Individual particularities (disabilities, privilege, or status) are used by the collective to establish or legitimate its political power, but its dynamic is not based on the affirmation of particularities the way identity politics are. It seeks to increase awareness of the various problems and various needs of specific groups, but it also attempts to have a structural effect on the system as a whole by promoting equality of access. It thus also participates in the articulation of political claims for redistribution and recognition within a politics of justice such as the one Nancy Fraser advocates.[42]

Herein lies an interesting tension, however. Even as there are minorities among its ranks speaking in the name of their own individual experiences (the visually impaired, AIDS patients, and so on), that is, from their "situated knowledge," to use Donna Haraway's formula, the claims that A2K advances (the defense of a public domain or of the commons, for example) are presented as possessing a universal range. The A2K movement does not try to construct a "universal" subject, as "unity" movements sometimes do, but it does succeed in allowing the translation of knowledge between communities and the realization of alliances between multiplicities.[43] Without prejudging the A2K movement's success or its future efficacy, we can observe that, for now, it manages to compose a collective interest that can be seen and presented as universal. Considered in this light, the A2K movement resembles the "multitude" that Michael Hardt and Antonio Negri describe as "singularities acting in common," a heterogeneous collective inheriting its intentionality in the progression from the singular to the common in which the concept of access becomes itself a dispositive of the organization of singularities.[44]

The A2K movement presents itself and is being seen as a nongovernmental force. It offers a critique of the standard decision-making processes—of pressures exerted on legal and executive institutions by the private sector, of the willingness of these institutions to surrender to these pressures—showing how they exclude, dissimulate, or neglect. But instead of emphasizing a position of exteriority, it incorporates the logic of governments, institutions, and industries and actually even also includes, in a very open way, individuals working for governments, as well as for private multinationals, in hopes of integrating these into something larger that it will have itself contributed to designing. The A2K movement occupies well-established institutional political spaces such as the WTO or WIPO and, at the same time, also seeks to create new political spaces and to legitimize them by co-opting institutional representatives from traditional political and economic power structures. It contributes to a blurring of the lines between genres through its ubiquitous and rather flexible ways of being and of engaging in debates. At the same time, it demonstrates the clear willingness of its constituencies to acquire a say in debates and to make use of what Hannah Arendt called their "power of speech" to establish themselves as political actors.[45]

GOVERNANCE AND RESISTANCE SEEN THROUGH THE PRISM OF ACCESS

The field of A2K inequalities is a composite: It includes new types of inequalities, significantly increased preexisting inequalities, and those brought to light by the structuring role of the market in knowledge capitalism. A2K targets issues specifically posed in the new digital society, but fundamentally, it raises classical problems, such as inequalities in the distribution of resources, or social justice. What is novel is the prism used to analyze the problems, as well as the manner in which they are addressed—the modes of organization of political action that the A2K movement employs.

As I've noted, the A2K movement comprises a diversity of references, political traditions, and forms of mobilization. Consequently, the phenomena and actions that A2K activists find intolerable and the reasons that spark their reactions are very diverse. As such, the movement illustrates and fits very well within the larger and more general movement of civil-society actors engaging in nongovernmental politics.[46] This new conceptual field of political mobilization emerged in relation to two different trends in politics. On the one hand, beginning in the early 1970s, there was an increasing public demand for government accountability. On the other, attacks led by Margaret Thatcher and Ronald Reagan on the welfare state and on anticapitalist institutions such as labor unions and class-based political parties moved people away from traditional representative politics in the 1980s. Like other types of nongovernmental mobilizations, the A2K movement can be seen as the extension of politics "beyond the realm of representation" in reaction to "dysfunctions of the political realm."[47] Individuals and groups involved in the

A2K movement share a common concern with the way intellectual property governance is exercised and more generally question the way in which they and we are governed.

Those who hold the reins of the dominant power structure try to convince people and societies that the established rules are made to guarantee equality of opportunity between individuals, as well as to ensure progress and wealth within the society. In the face of these attempts, organized denunciations put forward by social movements expose inequalities, hierarchies, and power relations that then cannot always easily be justified publicly, even if they are rationalized and validated within privileged circles. A2K advocates denounce the divergence between the theoretical promotion of innovation, a value commonly accepted by all and invoked by those who govern to legitimate their policies, and the actual effect of the rules of intellectual property protection. They seek to "question the social norms that enable governing bodies to call upon unimpeachable principles in order to justify objectionable policies."[48] From the alliance between claims to rights of access and utilitarian criticisms of the intellectual property system thus emerges a movement generating its own particular politics: a politics of access. As such, the A2K movement illustrates an evolution in the culture of mobilization and collective action.

Using the prism of access, the A2K movement attempts to reveal the dissonances in what James C. Scott calls the "public transcript," the "open interaction between subordinates and those who dominate," of the dominant powers regarding matters of equality and democracy. It analyzes the asperities lying beneath supposedly calm political surfaces, thereby rendering possible the perception of injustice, as well as the moral questioning of it.[49] Refusing to accept the normalization of restrictions on access, the A2K movement makes needs visible and imposes upon political leaders the duty to meet them. Its existence tends to prove that despite—or perhaps because of—the willingness of the dominant powers to see inflexible intellectual property rules adopted and implemented, they fail in their attempt to naturalize social inequities regarding access. On the contrary, the issue of access is particularly useful and efficient as a way to catalyze questioning of the dominant powers and their regimes.

However, one could argue that the act of voicing criticism and complaints against a situation that is perceived as having been imposed upon individuals by stronger forces (be it their leaders or their gods) is simply one aspect of a technique adopted in order to withstand a situation perceived as inevitable. So when movements or individuals denounce the intellectual property system and the inequalities that result from it, they contribute to a background noise of criticism whose existence does little more than reveal a power relation within the established order, following Foucault's idea that every power goes hand in hand with a form of resistance to it. Yet the fact that the dominant power structure is continuously trying to legitimize itself and its political decisions does not mean that these efforts are effective. That is, efforts to persuade the dominated do not necessarily lead to their consent. And as James C. Scott points out, the level of effort invested in maintaining a given power structure also provides one element that allows us to estimate the level of instability of this regime.[50] Thus, when it is possible to elude the hypnosis that the dominant power structure aims to induce, decoding the means that it uses to maintain its hegemony and to make acts of resistance invisible informs the observer of both the weaknesses and the limits of the hegemonic power.

One of the strengths of the A2K movement thus is the way in which a large number of different issues lead directly to questioning of the claims made by dominant powers and their regimes in order to legitimize themselves and their actions. When merely a small percentage of the people with HIV/AIDS across the world had access to the only drugs that could keep them alive, questions about access were raised and political tensions resulted. This phenomenon can be explained by the existence of a crisis situation (an uncontrolled and deadly pandemic) and the fact that the issue is a matter of life or death.[51] But where access is in question, contention forms and gains legitimacy in situations and on issues that do not necessarily correspond to what are generally viewed to be people's most vital needs, such as the enjoyment of the arts or access to educational material, aspects that prove to be indispensable to the well-being, the wealth, and the stability of individuals and societies.

As Lawrence Liang notes, some needs, as they are understood according to common representations, do not necessarily conform to the "essential character" of what Gayatri Spivak calls the "subaltern subject."[52] Subalterns are usually seen as "the poor" in wealthy countries, "people from developing countries," and anybody who is discriminated against and essentialized as being inferior and consequently thought of as having mainly rudimentary needs. For these populations, needs that coincide with their desire to improve their well-being or their position in society— and hence needs that possibly go with tendencies to question or transgress the established order—tend to be easily disqualified. Similarly, regardless of what is stated by international declarations or conventions, what constitutes human rights often varies depending on whose rights are being discussed. A2K advocates question what qualifies as "primary" or "essential" needs or rights. They hold a variety of conceptions of needs and rights, from the need to save lives to Amartya Sen's notion of positive freedom and the necessity of taking into account a person's concrete ability to be or do something beyond the mere existence of theoretical "rights."[53]

Endeavors to politicize otherwise accepted situations and to promote what A2K advocates see as the legitimate expectations of individuals or of societies are all the more necessary because in a taken-for-granted hierarchy of needs, vital, but unmet needs systematically operate to obscure others that are perceived as less important. For example, the need for life-saving medicines obscures the need for cultural goods. By the same token, obvious price barriers that prevent people from meeting their basic needs mask the effect of less noticeable and less tangible obstacles: The effect of the high prices of books trumps the effect of copyright regimes. Moreover, most affected people do not necessarily have a clear understanding of their own exclusion, for the principle underlined by Pierre Bourdieu operates fully: As cultural deprivation increases, the awareness thereof actually decreases.[54] Interviews investigating barriers to access to knowledge in Thailand revealed that "ordinary people" (by this I mean people who have no special interest in or knowledge of intellectual property) often have a difficult time identifying the concrete effect of intellectual property in their own daily lives.[55] The study found that a mother was quicker to blame changes in the curriculum at her child's school that required her to buy brand new books for her second child, instead of using the ones her first child used, than she was to identify the many barriers instituted by the enforcement of intellectual property rights.

As Bourdieu has noted, for such people in such situations, "the problem is that, for the most part, the established order is not a problem."[56] It is the characteristic of power structures to dissemble the problems that they author and/or condone. Therefore, an important goal for the A2K mobilization is to educate people and encourage them to insist upon their needs and rights and the redefinition of them, rather than relinquish them in the face of situations that tend to obfuscate the true sources of their frustration.[57]

The A2K mobilization bears the political forms of an organized and formal movement confronting institutionalized powers, and it has been increasingly considered as such. It also includes concealed individual acts of resistance and everyday-life actions that occur under the radar of the usual observers of social movements. Many such acts would be considered utilitarian, rather than as the result of a concerted or deliberate effort of resistance, but still, they contradict the rules of the dominant power. Music exchanges between friends, the occasional purchase of copies of DVDs or CDs in street markets in New York or Casablanca, the sharing of software among students and colleagues—each of these, albeit mostly unwittingly, constitutes grains of sand in the cogs of the system of intellectual property rights.

New potentialities derived from changes in the production of wealth provide tools that allow ordinary people to resist domination discretely through simple daily actions, even while appearing to be mere passive and accepting subjects. These are the unstructured, hidden acts of resistance that belong to what Scott calls "infrapolitics."[58] Most individuals throughout history and still today have not

enjoyed the luxury of becoming part of an organized and public political movement. However, a vast array of sporadic and often almost invisible political actions
is commonly undertaken by subalterns in order to embellish their everyday life
within the system or in order to weaken the forces of oppression, injustice, and
repression.[59] Their actions are the subversive ways of being that are the province
of the dominated: poaching, escaping, finagling, pirating, getting around the law
when they are outcasts, or trying to use loopholes in the law, leaking documents
when they are government agents.

One of the challenges for the A2K movement is to help "ordinary people" realize that they have an interest in engaging in these battles and that they have the
means to do so. This is how the conditions of the reproduction of domination—
including the fact that they appear legitimate to the dominated themselves—can
be disrupted. In encouraging resistance by ordinary people, the A2K movement
can take advantage of the possibilities resulting from the alliance of new technologies with the new aspirations that have emerged with them.

PERSPECTIVES AND REPRESENTATIONS SEEN THROUGH THE A2K PRISM

The potential success of the A2K movement thus depends not only on its ability
to make visible and emphasize the social dimensions and effects of the protection
of intellectual property, but on its ability to help individuals to perceive the world
differently.

A2K scholars and activists try to give a higher profile to concepts such as the
"commons" or the "public domain," concepts that offer resources and alternatives
in the organization of society on a pragmatic level and that also provide individuals with ways to question and rethink their relations to their economic, social, and
political environment. This means undoing, to some extent, prevailing conceptions
and beliefs by digging breaches in the imaginary established by the advocates
for increased intellectual property restrictions and by summoning references and
knowledge ignored or disregarded in the current system.

Many of the questions that are discussed by A2K advocates, such as the commons or the role of the public domain, have been raised in the past. However,
the memory of these previous discussions seems to have been lost, and the terms
of the debates have been reversed. Contrary to discussions in the eighteenth and
nineteenth centuries, today, the ultimate fear is not seeing the public domain
eviscerated, but rather seeing intellectual property contravened. Reflection and
analysis are massively focusing on how to create new exclusive rights and how
to enforce them, while needed debates on how the commons can be organized
and managed, for instance, occupy a small minority of people. For things to be

different, an inversion of perspectives is required, something that would shake up the realm of representations. Doing so requires the movement to be able to question the social values inculcated in the system in which we live—values at the origin of the production of representations and of behaviors.

I won't discuss in detail here the arguments and concepts employed by defenders of access to knowledge and their critiques of the claims advanced by advocates of intellectual property extremism. They appear throughout the book and are discussed in depth in Amy Kapczynski's introductory essay, "Toward a Conceptual Genealogy of Access to Knowledge." Rather, I will conclude here by exploring what is at stake in the maintenance and reproduction of the representations and social values that underwrite the effort to preserve and extend intellectual property restrictions and in the A2K movement's efforts to disrupt those representations and appeal to other values.

The conversion into privately owned goods of goods and ideas that once constituted property held in common or that were not included in the realm of property diminishes the sense of the role commons or public goods play in society, because valorizing private property depreciates the value of goods accessible by all. Thus, the emergence of the concept of intellectual property and the exponential growth of the protection of intellectual property rights have accompanied the relinquishment of the belief in the "productive power of the commons" and has inhibited recognition of the possibility that types of economy based on something other than the possession and exploitation of private property are viable.[60] Meanwhile, the logic of privatization, together with the rationality claimed for markets, both of which are fostered and publicly praised by neoliberals, largely have helped undermine the values attributed to all that is public. These developments slowly, yet profoundly change the way that societies are organized, as well as their dynamics of consumption, the relationship between individuals, and even individuals' understanding of themselves. At the time of the first enclosure movement, the privatization of the land was claimed to be a way to limit "strategies of overuse and underinvestment," and the transformations of economies that it generated were often viewed as progressive and beneficial to society.[61] Although these assertions are debatable, the notion that "this innovation in property systems allowed an unparalleled expansion of productive possibilities" still prevails without being subject to direct discussion in most cases.[62] Those are beliefs that accompanied the establishment of the current intellectual property system and on which claims for the apparent superiority of the model since then have rested.

In today's society, some A2K advocates argue that a second enclosure movement, in the form of increasing restrictions on intellectual property, immaterial property, and other forms of information and knowledge mostly favors strategies that "undermine processes of abundance intrinsic to nature" and thus organize "artificial scarcity" while ultimately harming the innovative potential of individuals and societies.[63] Some believe that the technological changes that took place in the past several decades should lead to radical change in the way the knowledge economy is thought about and organized, including the ways that intellectual property rights are conceived and used.[64] As critics point out, for instance, nonrivalrous goods are increasingly involved in human activities, and there can be "no tragedy for nonrivalrous goods left in the commons," because by definition, "a nonrivalrous resource can't be exhausted."[65] Such evolutions require a general reassessment of attitudes and laws regarding property, because what may have been seen as predominantly beneficial to societies in the past and in the context of mercantile or industrial capitalism leads in the context of today's knowledge economy to dead ends and dangerous imbalances in economic distribution. Developments such as these are in themselves favorable to the emergence or reemergence of alternative visions to the escalation of the protection of intellectual property rights.

For the A2K movement, ideology is perceived as a trap, either because it is a label placed upon the movement to undermine its credibility (when it is called "Communist" by its enemies) or because it is seen as a dogmatism that does not correspond to the movement's flexibilities and aspirations to inclusiveness. The A2K movement therefore bases its legitimacy on other things in order to promote the desired transition from the intellectual property regime to the A2K paradigm. However, in this time of global financial, economic, and ecological crisis, which may contribute to dispelling the illusion of the efficiency and cohesion of the neoliberal state and undermine its credibility, alternative propositions for society might find a better environment for their development and reception.[66] Under present conditions, it may become possible to question the dominant discourses and make other fictions intelligible, other possibilities comprehensible.

From the struggle over intellectual property emerges the common understanding that creativity, whether it is used as a justification for intellectual property rights or sought simply for its own good, represents an ultimate goal and a shared value in our contemporary world.[67] Of course, A2K advocates and the defenders of copyrights and other exclusive rights remain opposed regarding the issue of what makes creation possible and of how creation takes place. Here, the conflict is between those who deem that the best way to meet this goal is through more control over intellectual property and those who, on the contrary, think it is through increased freedom.

One of the major criticisms of the proprietarian approach is that it in fact fails to fulfill what it promises: the promotion and guarantee of innovation. Instead, while shrinking the public domain, it actually jeopardizes or "cripples" creativity, to use Lessig's eloquent expression.[68] The first thing that property rules limit is indeed that freedom to assimilate and transform knowledge, a freedom that is critical to creation, whether it be achieved through direct copying or mere quoting. Extremist intellectual property positions tend to banish totally the possibilities of quotation and borrowing. This is the case with audiovisual technologies, for example, for which the right to quotation is almost nonexistent, in spite of the fact that it has always existed for other media and disciplines. To take a commonplace example, imagine what science would be like if scientists couldn't quote and use each other's work and thereby expose and criticize or improve preexisting assumptions or demonstrations. At the other end of the spectrum of quoting practices, the collection of quotations assembled by philosopher and critic Walter Benjamin, which was at the center of his work, provides a good example of how the montage and rearrangement of existing pieces of text can generate original creation, reveal hidden aspects of what has been taken for granted, and forge novel understanding of reality: "Benjamin's ideal" was to produce "a work consisting entirely of quotations, one that was mounted so masterfully that it could dispense with any accompanying text.... The main work consisted in tearing fragments out of their context and arranging them afresh in such a way that they illustrated one another and were able to prove their *raison d'être* in a free-floating state, as it were."[69]

Similarly, the creative function of quoting and borrowing has always been essential to music composition. Citations, manipulations "à la manière de," and blatant references to preceding masters have figured throughout the classical music composition that marked the eighteenth and nineteenth centuries and the movie soundtracks of the twentieth century, to mention only two examples. Today, artistic and intellectual production using digital technologies continues to be dependant on the ability to copy and to borrow. Perhaps in an even more obvious manner than before, previous works are the raw material of future creation. As Cory Doctorow notes: "if copying on the Internet were ended tomorrow, it would be the end of culture on the Internet too. YouTube would vanish without its storehouse of infringing clips; LiveJournal would be dead without all those interesting little user-icons and those fascinating pastebombs from books, news-stories and blogs; Flickr would dry up and blow away without all those photos of copyrighted, trademarked and otherwise protected objects, works, and scenes."[70]

But the freedom to use existing things affects not only our ability to produce, determining the number of tools and quantity of raw material that we access, but also the ways we comprehend things and relate to them. The act of quoting allows us to place ourselves and our discourses within a heritage, a continuity. Thus, the sphere of immaterial public goods or immaterial commons involves more than just sources of inspiration or mere material resources of creativity. It offers individuals shared landmarks that inscribe them in a common temporality that is a present time, but also a common past. One who subtracts material from the public field disrupts and impoverishes the collective memory and in so doing affects the capacity of individuals to think for themselves, collectively as well as individually. As Hannah Arendt put it, what Walter Benjamin called the "collector's attitude" results in the withdrawal of things from the public and with them "all kinds of things that were once public property" as a way to establish himself in the past.[71] Consequently, others are deprived of the opportunity to make the same kind of connection, because the act of withdrawing goods from the public domain can be compared to the act of removing elements of a common past, thereby foreclosing the possibility of common relationships to them that otherwise would be allowed to the rest of society. Public goods are made of a common past shared in the present, and access to partial or truncated material limits not only individuals' ability to act and to create in that present, but also their relationship to the past, to their culture, and to their history. It is through material traces left by others that the past is made accessible, allowing history to play its proper social role and positioning individuals as cultural mediators.

The cultural losses, impoverishment, and amnesia that occur in the name of progress and innovation also occur in a geopolitical dimension. In developing countries, and that means many countries of the world, portions of their cultural heritage are lost at the hands of the macroprocesses associated with capitalist pressure that favors the spread of Western culture, as well as by virtue of the limited means (financial, technological, or legal) available in these countries to store and share their cultural resources. This is not to say that there is necessarily an extinction of the cultures that fail to qualify as dominant, or that these cultures do not disseminate ideas and works on a global scale, as well, or even that globalization necessarily transforms people in non-Western countries into simple consumers of imported cultural goods. Indeed, appropriation and transformation processes take place everywhere. They happen when Indians watch, enjoy, and reinterpret American TV shows, when Americans shoot movies that aim at portraying the reality of Indian cities in a more authentic way than Bollywood does, and when the booty bass music from U.S. ghettos becomes the *baile* funk of the Brazilian *favelas* before being adapted by Italian musicians and distributed by German producers at Parisian parties. But the control by intellectual property rules of access to technologies and resources surely favors certain flows and certain directions for these flows. It limits or harms the preservation and transmission of certain cultural productions. At the same time, knowledge that is privatized and removed from the public domain in wealthy countries or anywhere else on the planet is thus made inaccessible to most people in these countries, since the only way to access resources is then to pay for access. "As in the relation between colonized and colonizer, knowledge is anything but symmetrical."[72]

This is what is called cultural domination. And if it has implications for the production of and access to knowledge, it can also interfere with the conceptual work of the A2K movement itself. For example, despite the best of intentions, use of a notion such as the public domain might reinforce existing cultural domination. As Carlos M. Correa and Jeffrey Atteberry mention in this volume, advocating the defense and expansion of the public domain applied to traditional knowledge or biodiversity can also allow corporations to seize whatever knowledge or resources they want, while indigenous people remain once more without control over the means needed to protect their own resources and the ways in which they are valued and employed.[73]

This is where the A2K constellation might want to consider its geographical positions and contours with caution, for tensions over the issues of intellectual property are produced not only in the context of exploiting or opposing the operations of capitalist logic, but also in the context of exercising or resisting the exercise of imperialist logic. To serve the equality or universality sought by A2K advocates, their own practices of conceptualizing what they consider to be desirable ends must constantly be reflected on and questioned, taking such contexts into account. A dominant culture "tends to produce the representations through which it is perceived (whether rejected or imitated) by others,"[74] which naturally affects the criticisms that are formulated against it and the solutions called upon to facilitate emancipation therefrom. In the process of elaborating "its own representations of the dominant culture," any struggle against this domination needs to question the values that the critics of the dominant culture themselves promote and the way they define and use them, whether the topic be property, private ownership, or the concept of the commons or the public domain.[75] It is particularly what constitutes their common share, the metal of the two faces of the coin that unites proponents and opponents of the dominant culture, that requires meticulous questioning.

Cultural domination has many ways to influence people's behaviors and their comprehension of what is possible and what is not. Fear is an essential element employed in the arguments for intellectual property rights protection and in the discourses of neoliberal rationality as a means of influencing people's choices and behavior. We live in an era when, at least in Western countries, the fear of getting in trouble for having encroached upon others' property (even when the boundaries that define it are invisible) is almost omnipresent. Penalties are involved: Minor encroachments can make you liable for damages, subject you to social opprobrium, and even lead to legal punishments. But in a less draconian manner, we've been taught to fear that unless you've paid well for what you acquire, you can end up less well off than you were before, due to the possibility that what you are getting is "bad" by nature: It is of poor quality, won't last, and/or it is harmful. This follows as a matter of course from the "wisdom" that holds that if you want quality, you have to pay for it. As Lessig has noted "lurking in the background of our collective thought is a hunch that free resources are somehow inferior."[76] Bad quality for instance, will have consequences on your health if what you are getting is medicines, and it can damage your computer or make you lose all your data if it is software.

As Wendy Brown has put it, in the neoliberal context, the state "attempts to construct prudent subjects through policies that organize such prudence."[77] And people are forced to choose sides. Antipiracy advertisements remind us in blunt terms of the choice that is offered: If you buy copies of DVDs or software, "you are either for terrorism or support the war on terror."[78] Under the current dominant political rationality, what is considered "property infringement," "piracy," and "terrorism" seem easier and easier to equate with each other and assimilate to each other. Professionals who work in the copyright field can testify that this conflation of infringements of intellectual property rights with terrorism is not only propaganda intended for the general public, but a message that shapes policy making. It is indeed becoming common to hear people in national policy-making or international meetings equating piracy with terrorism without hesitation or any sense of restraint. In the end, the result is a vicious circle: Even if enforcement policies cannot pin down each act of infringement, the scarecrow effect of criminalization fuels fears and suspicions that deter people's infringing behaviors.

In Argentina in 2002, workers took control of the Zanon plant in an occupation that continued for several years. The plant, a ceramics factory that had been closed by an owner who no longer considered it profitable, had previously received millions of dollars in public subsidies as part of the corporate welfare program of the Carlos Menem government. The workers decided to reject their fate and called for the expropriation of the plant in the name of the public investment made, declaring Zanon "of the people."[79] They exercised the "right to reappropriation" of the multitude that, according to Hardt and Negri "is first of all the right of reappropriation of the means of production."[80] They decided to seize control of the engine of production and render it the common property of those who actually operated it. In the face of this attempt to impose an alternative way of being, Menem declared that "we will impose order . . . we will impose respect of the state of law. Among other things, the right to private property."[81]

In this episode, two ethics conflicted. In the end, as force was deployed by the state, the right to private property prevailed in most of the plants. But it is certainly easier to enforce property rights and hunt down pirates in the physical world than in the digital one. In the immaterial realm, reappropriation does not necessarily require expropriation. It is use that is the key.[82] Thus, the terms of the confrontation may change.

Use per se can be subversive. If beliefs and fears do not curb it, the use that people make of protected material can open breaches in immaterial barriers and render them useless as filters, taking away from them the power to confer profit and social position. Massive use by individuals of data covered by intellectual property rights can easily become uncontrollable, and organized digital networks can impose a commons by simple virtue of producing and making available their production. To refuse the manufacturing of illegality and criminalization appears in itself as a legitimate and useful horizon for the A2K movement, starting with questioning the definition of what is illegal by performing threatened acts and challenging the hold of fear.[83]

But what the issue of use determines is also the way questions are politicized. With the advent of a world in which the production of immaterial goods is increasing can also come new ways to conceptualize and formulate politics as well as to elaborate political action, because in such a world, possession is not exactly the same thing as ownership. The conflict over the Zanon ceramics plant was staged over the material means of production. Where immaterial goods are involved, benefiting from capitalist appropriation does not necessarily require possessing the means of production. Instead, it involves having control over what is to be given value—goods and elements that in fact cannot anyway be materially owned. In this context, it is access that needs to be controlled, rather than the strict ownership of goods. That is why access is such a central issue, why it is at the center of so many conflicts today.

The issue of access is a central issue because it is a product of its time.[84] Yet emphasizing it is also a strategic choice: The machine that we are ourselves is obviously more complex than a machine in a plant. So are its relationships and interactions with the material and immaterial environments that surround it. Thus, the agency that is involved in its use challenges control in many ways and can more easily evade it, as the ever more repressive attempts to enforce intellectual property rights demonstrate. Thus, seen through the prism of access, the terms that crystallize political mobilization differ. They no longer begin with the distinction between public and private property. When attention is focused on access, what is invoked is justice, equality, or freedom, without directly confronting the issue of property.

But of course, the issue of property cannot so easily be liquidated. On the contrary, the current struggles over access to knowledge reopen a debate over property. Through efforts to build social rights based on the new, emerging possibilities for sharing and an ideal of access, the movement questions property and the role it should play in our societies. It is a whole field of contests that open along new lines of conflict. Meanwhile, as in other contexts, to exercise freedom—here, the freedom of use—requires being in the position to do so, and thus, there is no escape from the issue of the conditions necessary to exert freedom.

CONCLUSION

The global increase and strengthening of norms of intellectual property protection in the past decades is symptomatic of the emergence of information technologies and attempts by capitalist entrepreneurs to benefit from these changes. Because we are in a phase of intense globalization of economic exchanges and communication in which the materiality of property dissolves, capitalists, seeking accumulation, have tried to define, categorize, and make the most of immaterial property. But in the knowledge economy, old models have become obsolete on many levels. Intellectual property rights owners do not seem to realize it, but cognitive capitalism relies on forms of creation and innovation derived from new modes of production. The production process is increasingly based on "a new relationship between production and consumption," and as Maurizio Lazzarato points out, "what is 'productive' is the whole of the social relation."[85] In the end, this new relationship may advance the legitimacy of the A2K movement more than anything else, and rights owners may simply prove unable to adapt: To date, their responses to the major technological changes that have altered the functioning of capitalism have prioritized the pursuit and the fortification of old strategies. Attempts to maintain these models by criminalizing and forbidding human behavior, although they fit well the ways of neoliberal governance, appear obsolete, poorly adapted to the current reality, and redolent of the "putrid and tyrannical obsolescence" that Hardt and Negri evoke.[86] As Immanuel Wallerstein says about the future of the United States, it may be that factors that originally worked to establish the intellectual property system's hegemony will ultimately bring about its end.[87]

New forms of communication and production also influence the way people interact with each other and engage in the production of culture and information.[88] The free and open-source software movement shows the efficacy of these social practices of production, sharing, and distribution. Such practices are not in themselves new, but exercised in the current technological and political context, they bring new potentialities in terms of individual capacities and collective action and may contribute to the production of the conditions for the emergence of a new political subjectivity. They represent a power in the capitalist world on which the A2K movement can rely to promote a new politics and new ways of doing politics. The relational aspects of creation and knowledge production imply a form of equality in the process through which individuals contribute and exchange. This equality is both a practice of creation (the "horizontal" communication and collaboration allowed by new technologies and formats, such as blogs and social-networking sites) and a political value promoted by A2K advocates who take into account the various needs and particularities of specific groups of individuals. Because of the nature of digital goods, the exercise of the freedom of use represents a possible course of action for movements and a key value to them. The A2K movement thus corresponds to a form of mobilization that can conciliate an inheritance from identity politics and more traditional claims for social justice. In that respect, the A2K movement represents an evolution in the modes of social mobilization.

The alliance between new forms of production based on new approaches to cooperation, on the one hand, and ways to conceptualize politics developed and brought to the forefront by the A2K movement, on the other, may provide an opportunity to oppose intellectual property extremism. More broadly, such a development also offers the chance to act outside of the immaterial walls erected by intellectual property protection and to exit at least intermittently from the constraints of neoliberal rationality, even perhaps causing the state to act in a different manner, as well.

The neoliberal form of colonization is a cultural and political attempt at subjection affecting every individual. A2K advocates are looking for a way out, promoting a morality based on values they wish to see emerge in politics. In the eyes of those in the movement, the current system is not good for two main reasons: It does not fulfill its own objectives in terms of innovation, and it is not fair in terms of access. As a field of activism, the A2K movement fights on both of these fronts and articulates one to the other, one preceding the other according to the context of the discussion within political spheres or in public debate. While the utilitarian concerns of the A2K movement anchor it in the capitalist system, its social justice claims reinject a moral dimension into the discussion, because the issue of access offers a new way to consider the issue of property. A2K advocates make use of the issue of access to rally many different constituencies, taking advantage of the flexibility of the notion to elude a number of pitfalls, whether these be representations they want to evade (dominant ideologies, political labels) or political traditions they do not want to identify with or be identified with because they seem outdated and do not appear to offer successful avenues for change.

The use of the issue of access makes it possible to avoid such pitfalls, but ultimately, it also leads in its own way to a critique of property. If everybody can access a good, the benefit of property instantly crumbles. The other side of the coin is that in the immaterial world, the control of access de facto replaces ownership. Thus, the political result of the A2K strategy may lead to greater disruptions than what many actors were anticipating when they first became involved in this battle. This is where a new politics of access to knowledge starts.

NOTES

The author is grateful to Harriet Hirshorn, Karyn Kaplan, and Tunde Oyewole for their assistance in translating this piece into English.

1   David Harvey, *A Brief History of Neoliberalism* (New York: Oxford University Press, 2005).

2   For a cinematic illustration of the spirit of the gleaner, see *Les glaneurs et la glaneuse* (The Gleaners and I, 2000), written and directed by Agnès Varda, where gleaning can be seen both as a possible choice for individual existence in our consumer societies as a way to create in the context of the digital technologies. Michel de Certeau, *The Practice of Everyday Life*, trans. Steven Rendall (Berkeley: University of California Press, 1984), p. xii.

3   James Boyle, "The Second Enclosure Movement and the Construction of the Public Domain," *Law and Contemporary Problems* 66, nos. 1–2 (Winter–Spring 2003): pp. 39–40.

4   For a detailed discussion of information and knowledge and the implications of the differences between the two terms for the A2K movement, see Amy Kapczynski's essay "Access to Knowledge: A Conceptual Genealogy" in this volume.

5   Lawrence Lessig, *The Future of Ideas: The Fate of the Commons in a Connected World* (New York: Random House, 2001), p. 9. The book is available under a Creative Commons Attribution-Noncommercial License at: http://www.lessig.org/blog/2008/01/the_future_of_ideas_ is_now_fre_1.html (last accessed January 26, 2010).

6   On the global scale, for example, the volume of financial exchanges is fifty times greater than the volume of exchanges of concrete products in the real economy.

7   Lessig, *The Future of Ideas*, p. 7.

8   See Gaëlle Krikorian, "Interview with Yann Moulier Boutang," in this volume.

9   Yochai Benkler shows how open-network, open-source innovation developed from the computer world and the organizational superiority of this model. See Yochai Benkler, *The Wealth of Networks: How Social Production Transforms Markets and Freedom* (New Haven, CT: Yale University Press). The book is available under a Creative Commons Attribution Noncommercial Sharealike License at http://cyber.law.harvard.edu/wealth_of_networks/Download_ PDFs_of_the_book (last accessed January 26, 2010).

10 "The Love-In: The Move Toward Open Innovation Is Beginning to Transform Entire Industries," *The Economist*, October 11, 2007, available on-line at http://w4.stern.nyu.edu/news/ docs/love.pdf (last accessed January 26, 2010).

11 In cases where the control of creation through exclusive rights remains part of the business strategy, changes in business models and practices nevertheless call into question and revise the exact role and place given to intellectual property protection and the types of protections that are more adapted for a given product or context—patents or brands, for example.

12 Yann Moulier Boutang, *Le capitalisme cognitif: La nouvelle grande transformation* (Paris: Éditions Amsterdam, 2007), pp. 82–116.

13 Carlo Vercellone, "Les politiques de développement à l'heure du capitalisme cognitive," *Multitudes*, no. 10 (October 2002): p. 14, available on-line at http://multitudes.samizdat.net/Les-politiques-de-developpement-a (last accessed January 26, 2010).

14 Sorting information is a key issue for many in the virtual world. You may be surrounded by information and still not have access to knowledge or be in a position to enjoy it or use it if you cannot sort and select the information you need or the appropriate knowledge. It is not for nothing that strategies to counter censorship of the Internet are in part based on the camouflaging of pertinent information in an accumulation of useless data, forcing would-be censors to lose as much time as possible in the search for "subversive" content. For example, these tactics were used by individuals trying to communicate inside and outside Iran during the presidential election in June 2009. See Blondeau Olivier and Allard Laurence, "L'Iran ou la guerre des proxies: Vers une culture publique de la sécurité informatique?" Fondation pour l'Innovation Politique, available on-line at http://www.fondapol.org/les-travaux/toutes-les-publications/publication/titre/liran-ou-la-guerre-des-proxies-vers-une-culture-publique-de-la-securite-informatique.html (last accessed January 26, 2010).

15 Moulier Boutang, *Le capitalisme cognitif*, p. 157.

16 Another problem is that behind the use of the term "intellectual" as an umbrella adjective to characterize legal rights that concern disparate intangible assets in forms and natures lies an ideological manipulation to legitimize a political move dedicated to serve simultaneously the interest of a heterogeneous set of industries. James Boyle, "The Second Enclosure Movement and the Construction of the Public Domain," *Law and Contemporary Problems* 66, nos. 1–2 (Winter–Spring 2003).

17 Electrified fences, concrete blocks, thermal, metric, and biometric detection systems, sensors, and other devices, such as those built in Palestine, Cyprus, around Ceuta and Melilla in the western Sahara, on the border between the United States and Mexico, in Kashmir, or on the border of Botswana, are only a few contexts in which these physical walls were built. For an analysis of the multiple functions of contemporary walls in neoliberal globalization during the contemporary decline of sovereignty, see Wendy Brown, *Walled States, Waning Sovereignty* (New York: Zone Books, 2010).

18 The strategy of the Recording Industry Association of America (RIAA) for example, consists of picking a sample of individuals (around forty thousand) out of a pool of millions who have been identified as sharing music, then claiming copyright infringement and threatening to take them to court. In August 2007, four major music labels, the Universal Music Group, Warner Music Group, EMI, and Sony Music, filed a suit against Joel Tenenbaum, a graduate student, for file downloading and illegal sharing. Tenenbaum, along with Jammie Thomas-Rasset, were then the only persons in the United States accused of illegal file sharing to have their cases taken before a jury. The vast majority of people when contacted by copyright owners claiming copyright infringement agree to pay the requested financial settlement. In June 2009, Thomas-Rasset was found liable for copyright infringement and ordered to pay nearly $2 million. In July 2009, Tenenbaum was condemned to pay $675,000 for illegally downloading and distributing thirty copyrighted songs. He decided to appeal. See Joel Tenenbaum, "How it Feels to be Sued for $4.5m," guardian.co.uk Music Blog, available on-line at http://www.guardian.co.uk/music/musicblog/2009/jul/27/filesharing-music-industry (last

accessed January 26, 2010). Meanwhile, file sharing and illegal file downloading are common practices. A research into the music consumption of fourteen-to-twenty-four-year-olds commissioned by the British Music Rights (BMR) organization and carried out by the University of Hertfordshire showed that, 95 percent of those surveyed had copied music illegally and that on average, half of the content on their MP3 players was downloaded illegally. A study also indicated that uploading music for others to share was seen by them as "altruistic" behavior. See Rosie Swash, "Half the Music on Average Teenager's MP3 Player is Illegal," guardian.co.uk, June 16, 2008, available on-line at: http://www.guardian.co.uk/music/2008/jun/16/news.rosieswash3 (last accessed January 26, 2010).

19 Cory Doctorow, "Why I Copyfight," *Locus Magazine*, November 2008, available on-line at http://www.locusmag.com/Features/2008/11/cory-doctorow-why-i-copyfight.html (last accessed January 26, 2010).

20 Moulier Boutang, in "Interview with Yann Moulier Boutang" in this volume.

21 It would be worth discussing at length whether the Internet can be seen as a public space in Hannah Arendt's sense and whether the public domain can be understood as being equivalent to the political domain. The political implications of this question are important for the issue of the preservation of public space and as a context for the ability of individuals to realize political action. According to Arendt, a public space is a space that allows active citizenship, a space in which individuals can gather and encounter one another to deliberate about matters of collective concern. In such political spaces, each citizen is in a position to exercise the powers of agency, to develop the capacities for judgment, and to develop concerted action aiming at political efficacy. Initiatives such as the Web site www.nosdeputes.fr, dedicated to the observation by citizens of the activities of the members of the French Parliament, the sharing of political information, and the pooling of citizens' comments, resemble an attempt to create such a space.

22 Doctorow, "Why I Copyfight."

23 Lawrence Lessig, "In Defense of Piracy," *Wall Street Journal*, October 12, 2008, available on-line at http://online.wsj.com/article/SB122367645363324303.html (last accessed January 27, 2010).

24 Moulier Boutang, *Le capitalisme cognitif.*

25 With 7.1 percent of the votes in its country, the Swedish Piratpartiet won two seats in the European Parliament during the European election on June 7, 2009. The party, which was created three years earlier, claimed around fifty thousand members and emerged as a new political force—it was ranked third during the election in Sweden. Philippe Rivière, "Après les élections européennes: Emergence du pouvoir pirate," *Le Monde Diplomatique*, June 12, 2009. This mobilization can in part be explained as a reaction to the condemnation by a Swedish court of The Pirate Bay in April 2009. The Pirate Bay offered a search engine making copyright-protected files accessible for illegal file sharing via the Piratebay.org Web site, one of the most visited BitTorrent destinations in the world. The Stockholm District Court sentenced four persons who contributed to the development of the Web site, Gottfrid Svartholm Warg, Peter Sunde, Fredrik Neij, and Carl Lundstrom, to one year each in prison each and assessed $3.6 million in penalties to pay to the record and film companies whose rights were infringed.

26 Wendy Brown, "Neoliberalism and the End of Liberal Democracy," in *Edgework: Critical Essays on Knowledge and Politics* (Princeton, NJ: Princeton University Press, 2005), pp. 40, 46–53.

27 Michel Foucault, *The Birth of Biopolitics: Lectures at the Collège de France, 1978–1979*, ed. Michael Senellart, trans. Graham Burchell (New York: Palgrave Macmillan, 2008); Brown, "Neoliberalism and the End of Liberal Democracy," p. 52.

28 Peter Drahos discusses the moves of the industry and its collaboration with governments in "'IP World'—Made by TNC Inc." in this volume. And in "Free-Trade Agreements and Neoliberalism: How to Derail the Political Rationales that Impose Strong Intellectual Property Protection" in this volume, I develop the relationship between neoliberalism and the transformations triggered by the industry regarding intellectual property rights. As Michel Feher notes, however, the motivation for the neoliberal state to support industry is not systematical. It is mostly fueled by the political gain that the state can foresee in the crisis of sovereignty that it is facing in its desire to optimize its "residual power." For a discussion of the functioning of neoliberalism see the analysis developed by Michel Feher in *Vacarme*, starting in *Vacarme* 51 (Spring 2010) and continuing in the following issues. If that proves true, despite the social determinism that influences actors and other class interests at stake, an interesting avenue could open to A2K advocates, as long as they succeed in framing what they would like the state to do in a way that convinces it that such actions will serve to shore up its political power. This is something that has begun to happen to some extent with environmental issues: The adoption of policies in favor of the environment has become politically rewarding. Former vice president and presidential candidate Al Gore's campaign against global warming is probably both a trigger and an illustration of this trend. See, for example, the movie *An Inconvenient Truth*, which revolves around Al Gore's travels and his efforts to educate the public about the climate crisis.

29 Comparing the range of modes of action employed by the A2K movement and the advocates of increased intellectual property protections, one notices that most of them are the same. In addition to lobbying and trying to infiltrate political arenas, both use media, public actions, and the denunciation of state policies. Advocates of increased intellectual property protections benefit from public actions by activist groups, as was the case when an unknown group that started to be called "the Red Shirts" by AIDS activists because they wore red shirts during their actions, took the floor in support of the big pharmaceutical companies on several occasions during an international AIDS conference in Bangkok in 2004. They also occasionally resort to aggressive campaigns in newspapers. USA for Innovation, an NGO advocating for the pharmaceutical company Abbott, published harsh attacks against the Thai government when it issued compulsory licenses. Although intellectual property advocates do not organize public marches, the difference between the registers of intervention that they employ and those of the A2K movement is mostly a matter of proportion, not of kind.

30 On the advocacy and rhetorical strategies of the proponents of strong intellectual property rights, see Peter Drahos, with John Braithwaite, *Information Feudalism: Who Owns the Knowledge Economy* (New York: The New Press, 2002), pp. 61–62, 68–71, 189–90; Peter Drahos, "Global Property Rights in Information: The Story of TRIPS at the GATT," *Prometheus* 13, no. 1 (1995): pp. 6–19; Susan K. Sell, *Private Power, Public Law: The Globalization of Intellectual Property Rights* (Cambridge: Cambridge University Press, 2003), pp. 81–97 and 97–100; Amy Kapczynski, "The Access to Knowledge Mobilization and the New Politics of Intellectual Property," *Yale Law Journal* 117, no. 5 (March 2008): pp. 839–51, available on-line at http://papers.ssrn.com/sol3/papers.cfm?abstract_id=1323525 (last accessed January 28, 2010).

31 See Amy Kapczynski's discussion on the gravitational pull and discursive effects that law

exerts on framing processes in "The Access to Knowledge Mobilization and the New Politics of Intellectual Property," pp. 809–10 and 861–73.

32 Just as, according to neoliberal views, the market cannot exist "by itself," in nature, there cannot be a commons or public goods left free, untouched by rules or regulations. For them to exist and be accessible implies that some agent or agency establishes them through an act of sovereignty and keeps them free.

33 The status and the conditions of definition, occupancy, and use of these spaces are grounds for discussion and vary greatly depending on the model proposed, whether it is patent pools, Creative Commons licenses, open source collaboration, or something else.

34 Irrespective of debates about the condition or the future of nation-state or of state sovereignty, note that our discussion concerns the state because, even if its mode of operation and its relations with other social actors change over time, it currently represents the dominant model of governing and of power. Such a focus by no means denies the possibility that competing or alternative governing forms may emerge.

35 However, the existence of these power relations raises the question of what it is that the A2K movement excludes or truncates, besides its capacity to welcome issues and individuals. The essays by Carlos Correa, "Access to Knowledge: The Case of Indigenous and Traditional Knowledge," and Jeff Atteberry, "Information/Knowledge in the Global Society of Control: A2K Theory and the Postcolonial Commons," in this volume, provide elements to reflect on this issue, notably on the issue of traditional knowledge.

36 See Gaëlle Krikorian, "Fabrication non gouvernementale de traités internationaux: Entretien avec James Love," *Vacarme* 34 (Winter 2006): p. 102, available on-line at http://www.vacarme.org/article537.html (last accessed January 29, 2010). To refuse to define oneself with regard to one's enemy bears in itself an attempt to question the use of the term "intellectual property." Scholars and activists have discussed the conceptual effect of imposing this term to designate, under one appellation, copyright, patents, trademarks, industrial design and undisclosed information. The term was manufactured and imposed as an ideological vehicle by the movement of private interests promoting exclusive rights. See James Boyle, "The Second Enclosure Movement and the Construction of the Public Domain." The common use of the term "intellectual property" disseminated the notion of embedding forms of knowledge in the realm of physical property. See Richard M. Stallman, "Did You Say 'Intellectual Property'? It's a Seductive Mirage," available on-line at http://www.gnu.org/philosophy/not-ipr.html (last accessed January 29, 2010). Stallman argues because of the fact that copyright, patent, and trademark laws are very different from property laws, the use of the term "intellectual property" creates an artificially coherent category. This is why some groups are campaigning for the term to be abandoned.

37 Kapczynski, "The Access to Knowledge Mobilization and the New Politics of Intellectual Property," pp. 804–85.

38 In doing so, their approach tends to differ from, for example, controversies over patents or criticisms that were made during the eighteenth and nineteenth centuries by Adam Smith, Thomas Jefferson, or Thomas Babington Macaulay. These focused on the effects of exclusive rights and monopolies and as such contemplated possibilities mostly from within this system. See Boyle, "The Second Enclosure Movement and the Construction of the Public Domain," p. 57. Meanwhile, contrary to what was assumed about property rights in the eighteenth or nineteenth centuries, the A2K movement cannot count on the then commonplace view that

intellectual property rights, while they are needed, also have to be controlled, limited, and sometimes abrogated.

39 This does not mean that A2K advocates are not keen to organize the theatrical and unequivocal public condemnation of corporate industries and to capitalize on these campaigns to increase their own political power, the same way that groups from the environment movement do. To take only one recent example among many, consider the AIDS activist protest campaign outside the headquarters of Roche in September 2008, "Roche: Feasting on Our Dead Bodies," depicting the pharmaceutical company as a hyena and stating in French, "If you have a monopoly, you don't need to be human." See http://www.flickr.com/photos/hughes_leglise/2909698636 (last accessed January 29, 2010).

40 Nancy Fraser, "Rethinking the Public Sphere: A Contribution to the Critique of Actually Existing Democracy," in Craig Calhoun (ed.), *Habermas and the Public Sphere* (Cambridge, MA: The MIT Press, 1992), p. 123.

41 In this, too, it distinguishes itself from the identity-based and minoritarian movements that have been described as being "new social movements."

42 Nancy Fraser and Axel Honneth, *Redistribution or Recognition?: A Political-Philosophical Exchange* (New York: Verso, 2003).

43 Antonella Corsani, "Knowledge Production and New Forms of Political Action: The Experience of the Intermittent Workers in France," trans. Timothy S. Murphy, available on-line at http://transform.eipcp.net/transversal/0406/corsani/en/#sdfootnote13anc (last accessed January 29, 2010).

44 Michael Hardt and Antonio Negri, *Multitude: War and Democracy in the Age of Empire* (New York: Penguin, 2004), p. 105.

45 Hannah Arendt, *On Revolution* (New York: Penguin Books, 1985), p. 19.

46 Michel Feher, Gaëlle Krikorian, and Yates McKee (eds.), *Nonovernmental Politics* (New York: Zone Books, 2007), p. 14.

47 *Ibid.*, pp. 14 and 26.

48 *Ibid.*, p. 17.

49 James C. Scott, *Domination and the Arts of Resistance: Hidden Transcripts* (New Haven, CT: Yale University Press), p. 2.

50 Gilles Chantraine and Olivier Ruchet, "Dans le dos du pouvoir: Entretien avec James C. Scott" *Vacarme* 42 (Winter 2008), available on-line at http://www.vacarme.org/article1491.html (last accessed January 30, 2010). See also Scott, *Domination and the Arts of Resistance*.

51 A crisis situation certainly favors the questioning of the legitimacy of the state and its policies. See Pierre Bourdieu, *Pascalian Meditations*, trans. Richard Nice (Cambridge: Polity, 2000), p. 178.

52 See Lawrence Liang, "Beyond Representation: The Figure of the Pirate," in this volume.

53 See for instance Amartya Sen, "Equality of What," in Robert E. Goodin and Philip Pettit (eds.), *Contemporary Political Philosophy: An Anthology* (Oxford: Blackwell, 1997), pp. 476–86. Of course, the choice of the terms used to frame the discussions is not neutral or meaningless, and to use a rhetoric of rights, of needs, or of positive freedom corresponds to different political conceptions or strategies. Within the A2K movement, such claims coexist without one dominating the others—yet.

54 Pierre Bourdieu and Alain Darbel, *L'Amour de l'art: Les musées d'art européens et leur public* (Paris: Éditions de Minuit, 1969), p. 157; available in English as *The Love of Art: European Art*

Museums and Their Public, trans. Caroline Beattie and Nick Merriman (Cambridge: Polity, 1991).

55 These interviews were conducted by Consumers International in the context of a study carried on in Thailand and finalized in 2006 in the Project on Copyright and Access to Knowledge. See http://www.idrc.ca/en/ev-67263-201-1-DO_TOPIC.html (last accessed January 31, 2010).

56 Bourdieu, *Pascalian Meditations*, p. 178.

57 This is also a common tactic of political leaders, lately excessively used by the right wing in France, for example: putting the blame for people's social and economic difficulties on scapegoats (immigrants, poor) and/or obstacles or enemies (the poor who are unionized, for instance) whom the dominant power wants to eliminate politically. See Mona Chollet, *Rêves de droite: Défaire l'imaginaire sarkozyste* (Paris: Zones, 2008), pp. 41 and 93.

58 Scott, *Domination and the Arts of Resistance*, p. 183.

59 Chantraine and Ruchet, "Dans le dos du pouvoir: Entretien avec James C. Scott."

60 Boyle, "The Second Enclosure Movement and the Construction of the Public Domain," p. 8. We can see an evolution in the functioning of capitalism: for instance, enrichment is not solely based on the physical appropriation of resources and the means of production and capitalism is increasingly based on the allocation and transfer of credit, not on the production and sale of goods. (For a discussion of entrepreneurial neoliberalism and the functioning of capitalism, see the analysis developed by Michel Feher in *Vacarme*, starting in *Vacarme* 51 [Spring 2010] and continuing in the following issues). However, at the base of this system and in the distribution of social positions that it produces still remains the ownership of property, be it only the potential to own: Some people are powerful in this system without materially possessing capital in the form of property, but for many reasons, they are believed by others to retain the capacity to do so. Of course, at the end of the day, "rich" people, even if they make use of huge volumes of immaterial assets that they don't in fact possess, in most of the cases also own a lot of both tangible and intangible resources. Nevertheless, ownership is socially constructed and based on others' perceptions.

61 Boyle, "The Second Enclosure Movement and the Construction of the Public Domain," p. 3. See also Kapczynski, "Access to Knowledge: A Conceptual Genealogy."

62 Boyle, "The Second Enclosure Movement and the Construction of the Public Domain," p. 3.

63 See Roberto Verzola's essay "Undermining Abundance: Counterproductive Uses of Technology and Law in Nature, Agriculture, and the Information Sector" in this volume.

64 See Gaëlle Krikorian, "Interview with Yann Moulier Boutang," in this volume.

65 Lessig, *The Future of Ideas*, p. 22.

66 Bourdieu, *Pascalian Meditations*, p. 178. Such alternatives came into play in what happened when the HIV/AIDS crisis contributed to making intolerable the inequalities in access to medicines between rich and the poor countries.

67 See Liang, "Beyond Representation": "Underlying much of copyright's mythology are the modernist ideas of creativity, innovation, and progress. The narrative conjunction of these ideas is represented as universal, and indeed, it is shared by both advocates of stronger copyright and advocates of the public domain." The differences between creativity, innovation, and progress and the different political ranges with which they can be associated could be discussed. The French pirates, for instance, while they call for the liberation of creation, reject the term "innovation," which they assimilate to the notion of progress, as an acceptable horizon for their mobilization.

68 Lessig, *The Future of Ideas*, p. 14.

69 Hannah Arendt, "Introduction. Walter Benjamin: 1992–1940," in Walter Benjamin, *Illumina-tions*, ed. Hannah Arendt, trans. Harry Zohn (New York: Schocken, 1968), p. 47.

70 Doctorow, "Why I Copyfight."

71 Arendt, "Introduction. Walter Benjamin: 1992–1940," p. 43.

72 Eric Fassin, "Same Sex, Different Politics: 'Gay Marriage' Debates in France and the United States," *Public Culture* 13, no. 2 (Spring 2001): p. 215.

73 See Correa, "Access to Knowledge: The Case of Indigenous and Traditional Knowledge" and Atteberry, "Information/Knowledge in the Global Society of Control: A2K Theory and the Postcolonial Commons," in this volume.

74 Fassin, "Same Sex, Different Politics," p. 215.

75 *Ibid.*

76 Lessig, *The Future of Ideas*, p. 13.

77 Brown, "Neoliberalism and the End of Liberal Democracy," p. 43.

78 See Liang, "Beyond Representation," in this volume.

79 See *The Take* (2004), a documentary directed by Avi Lewis and Naomi Klein.

80 Hardt and Negri, *Empire* (Cambridge, MA: Harvard University Press, 2000), p. 406.

81 Carlos Menem, quoted in *The Take*.

82 Jeremy Rifkin, *The Age of Access: The New Culture of Hypercapitalism, Where All of Life is a Paid-for Experience* (New York: J. P. Tarcher/Putnam, 2000). For a discussion of the condition of neoliberalism and the role of the issue of access, see the analysis developed by Michel Feher in *Vacarme*, starting in *Vacarme* 51 (Spring 2010) and continuing in the following issues.

83 Pirates, for example, challenge the merging of economic and security issues as established by neoliberalism. See Wendy Brown, "Souveraineté poreuse, démocratie murée," *Revue inter-nationale des livres et des idées*, no. 12 (July 2009). They also undertake practices of freedom that may help erode the culture of fear and prudence, and when these practices rest on trans-formative production, they can offer fruitful routes to opening breaches in dominant repre-sentations.

84 See Rifkin, *The Age of Access*.

85 Maurizio Lazzarato, "Immaterial Labor," trans. Paul Colilli and Ed Emory, in Michael Hardt and Paolo Virno (eds.), *Radical Thought in Italy: A Potential Politics* (Minneapolis: University of Minnesota Press, 1996), p. 147, available online at: http://www.generation-online.org/c/fcimmateriallabour3.htm.

86 Hardt and Negri, *Empire*, p. 410.

87 Immanuel Wallerstein, *The Decline of American Power: The U.S. in a Chaotic World* (New York: The New Press, 2003).

88 In the course of the conversations that unfold through the act of sharing in the knowledge society, even one's relationship to the self and the other is revisited. See Lawrence Liang's essay "The Man Who Mistook His Wife for a Book" in this volume.

PART TWO

THE EMERGENCE OF THE POLITICS OF A2K

How do **you** say A2K?

Access to Knowledge • Acesso ao Conhecimento • Dostep do wiedzy • Acceso a los Conocimientos • Acceso al Conocimiento • Accès au savoir • 知识获取 • Sifuna ulwazi • Πρόσβαση στη Γνώση • Adgang til kundskab • Adgang til viden • Ona Ogbon ati Oye • Zugang zu Wissen • Доступ до знань • الوصول إلى المعرفة • Доступ к знаниям • Toegang tot Kennis • Klíč k vědění • Tilgang til kunnskap • Waniko yeruzivo • Достъп до Знания • Pristup znanju • Приступ знању • Пристап до знаење • Accès à la connaissance • U swikelela ndivho • ცოდნის ხელმისაწვდომობა • Teacht ar eolas • Rok no and • Accesso alla conoscenza • БИЛИМ АЛУУГА ЖЕТИШӨӨ • Arivukkaana Vaaipu • Komunikimi me dijen • জ্ঞানের রাজ্যে প্রবেশ • Denumata Praveshaya • Katamelletso ho Tsebo • Dostop do Znanja • Ukufinyelela ulwazi • ការទទួលព័ត៌មានឯកសារ • Gye nyansa • Kuva nemukana wekuwana ruzivo • Kufinyelela Elwatini • Teisė žinoti • Toegang tot inligting • ການເຂົ້າເຖິງຂໍ້ມູນຂ່າວສານ • Njira ya tsopano yopezera chidziwitso • Hebbaade gàndal • Hunes e mand • Accès aux connaissances • Dugg buntu xamxam • Puleleho tsebong • Monyetla wa ho ithuta • Ukufikeleleka Kulwazi • Ukufinyelela Olwazini • Kufikelela kutiva • Okufuna okumanya

How do you make it happen? WIPO/IIM, June 20–22, 2005 Be There!
www.tacd.org • www.cptech.org/a2k

CPTech (now known as Knowledge Ecology International) and eIFL organized a collaborative effort to translate "How do *you* say A2K?" into many languages on the A2K listserve. The design was used for T-shirts and posters by several groups involved in the A2K campaign.

The Emergence of the A2K Movement: Reminiscences and Reflections of a Developing-Country Delegate

Ahmed Abdel Latif

> As long as lions do not have their historian, hunting stories and tales will always be to the glory of the hunter.
> —African proverb from Bernard Njonga, *Le poulet de la discorde*

Since their emergence on the international scene, developing countries have sought to reform and adapt global rules regulating the generation and dissemination of knowledge to take into consideration their specific socioeconomic circumstances and levels of development. Their participation in what is now known as the access to knowledge (A2K) movement is part of their effort to achieve this objective.

In this context, my assignment to the Permanent Mission of Egypt in Geneva (2000–2004), which was to follow intellectual property (IP) issues, first at the World Intellectual Property Organization (WIPO) and then also at the World Trade Organization (WTO), led to my involvement in the formation of the A2K movement. The following is thus an account of this process from the viewpoint of a Geneva-based delegate of a developing country. It aims to be a contribution to the narrative of the genesis of the A2K movement, rather than a definitive account of a process in which many different actors, in particular academics and civil-society activists, were also actively involved. This account focuses on developments and initiatives that took place in Geneva-based international forums and organizations, particularly at WIPO, that played an important structuring role in the emergence of A2K as a movement and in the framing of A2K as a concept.

GLOBAL INTELLECTUAL PROPERTY DEBATES: A2K IN HISTORICAL CONTEXT FROM THE PERSPECTIVE OF DEVELOPING COUNTRIES

Intellectual property rights have become the predominant framework for regulating the generation, dissemination, and use of knowledge. With the globalization of intellectual property rights and the expansion of the scope in intellectual property protection, the main institutions involved in international deliberations and rule making on intellectual property issues, particularly the WTO and WIPO, have acquired unprecedented importance. It is thus not surprising that recent efforts by developing countries aiming at adapting and reforming rules regulating knowledge have been centered on these two organizations.

In this regard, it is important to recall that, already in the 1960s and 1970s, developing countries had sought to reform the main international conventions in the area of intellectual property, such as the Berne Convention for the Protection of Literary and Artistic Works (1886) on the protection of copyright and the Paris Convention on the Protection of Industrial Property (1883), with a view toward making these instruments more responsive to developing countries' socioeconomic needs in terms of access to educational material, scientific knowledge, and technology. These attempts did not result in the expected reforms pursued by developing countries and have progressively fallen into oblivion.[1]

The conclusion of the 1994 WTO Trade-Related Aspects of Intellectual Property Rights Agreement (the TRIPS Agreement) brought many of these concerns back to the surface, because for developing countries, it represented a landmark development in the process of strengthening intellectual property rights at the global level. TRIPS globalized new rules with an important bearing on the dissemination of knowledge, such as the extension of patent protection to pharmaceutical products and the protection of computer programs (software) by copyright. TRIPS also laid down minimum standards for the enforcement of intellectual property rights, and it came under the aegis of the WTO dispute settlement system, which could be used in cases of noncompliance, features that were lacking in existing intellectual property agreements under WIPO.[2] A powerful discourse accompanied the conclusion of TRIPS, arguing that strengthened intellectual property protection in developing countries would promote innovation and lead to increased flows of investment and technology transfers.[3]

Furthermore, after the adoption of TRIPS, developed countries quickly signaled their determination to pursue the establishment of new intellectual property standards further, beyond the minimum standards contained in the TRIPS Agreement ("TRIPS-plus" standards).

These TRIPS-plus standards promoted by developed countries resulted either from norm-setting activities in WIPO or from intellectual property chapters in bilateral and regional free-trade agreements,[4] which often required adherence to WIPO instruments such as the 1996 Internet Treaties. These treaties strengthened copyright protection in the digital environment, establishing new obligations in an area that was not specifically addressed by the TRIPS Agreement. The 1999 WIPO

Digital Agenda promoted adherence to these treaties in the context of efforts to grapple with the challenges to traditional copyright protection brought by the Internet and information and communication technologies.[5] The European Union, which had adopted a sui generis regime for the protection of nonoriginal databases, was pressing for the adoption of a similar regime of protection in the context of WIPO's Standing Committee on Copyright and Related Rights.

Alongside this evolving landscape, a campaign for access to medicines emerged and gained significant momentum with the defeat of a lawsuit brought by thirty-nine pharmaceutical companies against the South African government in 1998, culminating with the adoption of the 2001 Doha Declaration on TRIPS and Public Health.[6]

For many of the actors involved in this mobilization, including developing countries, this campaign was extremely effective in addressing the impact of the newly globalized intellectual property standards on public health and in firmly putting the issue of patents and access to medicines on the global agenda. It was often cited as exemplary in the way it framed the issue, attracted public attention, and forged a coalition made of developing countries (including Brazil, India, and the African Group) and of civil-society and nongovernmental organizations (NGOs) such as Médecins Sans Frontières (MSF), the Consumer Project on Technology (CPTech, now Knowledge Ecology International), and the Third World Network, in addition to public-health grassroots organizations in developing countries such as South Africa, Thailand, and Brazil.[7]

During the access-to-medicines campaign, a collaboration developed between negotiators from developing countries, particularly Geneva-based ones, and several of the most active NGOs, which often provided these negotiators with information, legal analysis, and technical support.[8] Developing countries, on the other hand, articulated positions that coincided with the public-policy concerns advanced by many of these NGOs. The achievement of a more development-oriented intellectual property system that would be supportive of public health was a common objective of both the developing countries and the NGOs representing civil-society consumers and patients. This convergence of interests and strategies was most effective in the deliberations leading to the adoption of the Doha Declaration.

In terms of its wording, the Doha Declaration on TRIPS and Public Health was a significant development in global deliberations on intellectual property because its formulations embodied a balanced approach to intellectual property protection that contrasted with the maximalist intellectual property discourse that was prevalent until that time. For many developing countries, this balanced and powerful message had a wider significance beyond the WTO, because it signaled the importance of implementing intellectual property protection in a manner that is supportive of public-policy objectives.

Soon after the adoption of the Doha Declaration, the influential report of the UK Commission on Intellectual Property Rights (CIPR) was released in September 2002. The report underlined the need to achieve a more balanced international intellectual property system that would not be based on a "one size fits all" approach and that would take into consideration the different needs of countries, as well as their different levels of development. It emphasized that "access to books and learning materials is still a real problem in many developing countries."[10] The report invited developing countries to improve access to copyrighted works and to achieve their goals for education and knowledge transfer by adopting measures fostering competition under copyright laws, as well as by maintaining or adopting broad exemptions for educational, research, and library uses in their national copyright laws.[11]

The CIPR report had a significant impact in intellectual property and development circles for several reasons:[12] the creation of the CIPR came at the initiative of a developed country, the United Kingdom; the commission's membership included a number of prominent experts from both developing and developed countries, as well as representatives of industry and academia, and the commission's report contained many recommendations that addressed pressing policy issues with which most countries, particularly developing countries, were confronted in international forums and processes.[13]

More importantly, in terms of its content, the CIPR report captured very accurately a growing trend of opinion that distanced itself from both a maximalist discourse that promoted the absolute benefits of intellectual property and a discourse that was unequivocally critical of intellectual property as a matter of principle. It thus recognized both the benefits and costs of intellectual property protection, emphasizing the need to ensure that the costs do not outweigh the benefits, particularly for developing countries.

In many instances, the report echoed several of the criticisms lodged by developing countries against the international intellectual property system and the TRIPS Agreement. Furthermore, the report contained the first direct criticism of WIPO's orientations to be advanced in international policy debates beyond specialized circles of intellectual property scholars and NGOs.[14] In this regard, the report underlined that WIPO "should give explicit recognition to both the benefits and costs of IP protection" and "should act to integrate development objectives into its approach to the promotion of IP protection in developing countries."[15]

The publication of the CIPR report coincided with the launch by UNCTAD (the United Nations Conference on Trade and Development) and the ICTSD

(International Centre for Trade and Sustainable Development) of the Bellagio Dialogues on Development and Intellectual Property Policy with the support of the Rockefeller Foundation. These dialogues also pointed to the need to integrate the development dimension in the setting of global intellectual property standards. The first of these dialogues, in 2002, identified various areas of concern for efforts directed toward achieving a more balanced and development-oriented intellectual property system, such as dealing with the danger of the further harmonization of intellectual property rights laws calibrated on the high standards being promoted by developed countries,[16] the dangers posed by the promotion of TRIPS-plus standards, and the importance of building capacity for self-development in developing countries.[17] From 2002 to 2005, the Bellagio Dialogues brought together key intellectual property negotiators, experts, and representatives of civil society. Several of their recommendations, which converged with those of the CIPR report, influenced developments in global policy forums and deliberations on intellectual property.[18]

More generally, the backlash against the "roaring" nineties and against the categorical assumptions and assertions about the absolute benefits of economic globalization was in full swing. A more nuanced discourse on globalization from the Global North and the Global South was emerging and gaining ground on the international scene.[19]

FROM ACCESS TO MEDICINES TO A2K

All these developments induced a number of like-minded Geneva-based representatives of developing countries (from Argentina, Brazil, Egypt, and India) to believe that the debate should move beyond TRIPS and public health to address other substantive areas where global intellectual property rules had a significant impact on public-policy objectives of importance to developing countries, such as access to educational material and scientific knowledge. The copyright issue, for instance, had been the sleeping giant in the debate on intellectual property in the 1990s. Concerns had appeared regarding the impact of the WIPO Internet Treaties and the legislation implementing them—such as the U.S. Digital Millennium Copyright Act—on access to information and fair use in the digital environment.

A shared conviction emerged among them that the most effective way to mobilize on these issues was to replicate the elements that had proven successful in the access-to-medicines campaign, especially the focus on careful framing of the issue and on building a coalition that would include developing countries, as well as NGOs.

However, the prospects for such a mobilization seemed uncertain in 2002. The clusters of issues around knowledge and information lacked the emotional impact and sense of urgency that characterized the patents-and-medicines debate. Access to medicines, particularly life-saving HIV drugs, is a matter of immediate human survival. It has a compelling humanitarian dimension that is more difficult to establish in issues relating to knowledge and information.

At the same time, many developing countries were wary of opening a new front that could be construed as a more general contestation of the international intellectual property system, in contrast with the more limited and pointed mobilization on the patents-and-medicines issue. Issue-based actions and mobilizations were more likely to succeed, in their experience, than a frontal and systematic opposition. Furthermore, deliberations were still taking place at the WTO on paragraph 6 of the Doha Declaration on TRIPS and Public Health concerning the situation of countries that lacked manufacturing capacities in the pharmaceutical sector.

With their limited capacities and expertise, many developing countries relied on their Geneva-based representatives to articulate their positions in many technical discussions relating to trade and intellectual property discussed in Geneva, and thus it was difficult for these countries to be engaged simultaneously in in-depth negotiations on intellectual property matters in different forums.[20] In addition, many developing countries faced coordination problems, because they often had separate representations to the United Nations and the WTO, or even within the same mission, different persons were assigned to follow WTO TRIPS issues and UN agencies such as WIPO.[21] With regard to NGOs, many of those involved in the access-to-medicines campaigns continued to focus mostly on the TRIPS and public-health issue and were still devoting much of their resources and organizational capabilities to it.[22] Finally, while the WTO was clearly the forum in which to act on the issue of patents and medicines because the TRIPS Agreement had extended patent protection to pharmaceutical products, WIPO appeared to be the appropriate forum where the more general debate on the regulation of global knowledge was to be raised, particularly in view of the new intellectual property standards being advanced by developed countries at WIPO.

Indeed, discussions were taking place at WIPO on a new treaty to protect broadcasting organizations in the digital environment, as well as on a new treaty to harmonize substantive patent law. WIPO's centrality in shaping the global intellectual property discourse, particularly in developing countries, was becoming manifest, as well as its role in the implementation of the TRIPS Agreement through its technical-assistance programs and legislative advice in the context of the 1995 WIPO-WTO agreement on technical cooperation,[23] as well as the 1998 WIPO-WTO joint initiative on technical cooperation for developing countries and the 2001 joint initiative for least-developed countries.[24]

At that time, WIPO was still perceived as a technical organization and was relatively unknown to the larger public and to many activists, academics, and

NGOs involved in intellectual property debates, which focused mostly on the TRIPS Agreement. The majority of the recent literature on intellectual property and development also had concentrated nearly entirely on the TRIPS Agreement and its implications. Furthermore, there was some skepticism, including among some experts and negotiators who had been involved in the TRIPS and public-health negotiations, regarding the possibility of bringing any significant change or reforms to WIPO. The organization was perceived as predominantly influenced by developed countries and owners of intellectual property-rights, particularly in the area of norm setting, a perception reinforced by the fact that WIPO derived nearly 90 percent of its revenues from the use by the private sector of its registration systems, most notably the Patent Cooperation Treaty.[25] *structure of patent treaties?*

As for developing countries, while they had built expertise in engaging with WTO and TRIPS issues, their knowledge of WIPO processes remain limited, as did their participation in the organization's standard-setting activities.[26] The linkages between the discussions on TRIPS at the WTO and the deliberations at WIPO were not evident for most of them. Few developing countries were actively engaged in both forums.[27]

THE FORMATION OF THE A2K MOVEMENT

Some NGOs shared the views on the need to engage more actively in WIPO processes.[28] CPTech was one of them. In fall of 2002 and early 2003, discussions took place between James Love, the director of CPTech, and a number of like-minded Geneva based delegates from developing countries on the means to pursue such a greater engagement. Love subsequently played an important role in providing much-needed links between developing countries, civil-society groups, and academia based in the North, particularly in the United States, which had been mobilized for a number of years on domestic issues relating to copyright, knowledge, and information, in particular in the context of the implementation of the Digital Millennium Copyright Act. CPTech started increasingly to focus its advocacy on WIPO's mission and role in order to raise public interest in its activities and its approach to intellectual property protection. An important stage was set for the coming together of the forces behind the A2K movement.

THE WORLD SUMMIT ON THE INFORMATION SOCIETY

However, there was still the need for opportunities that would act as a catalyst in forging a new coalition on the regulation of knowledge. One of them came inadvertently in the form of the World Summit on the Information Society (WSIS).

In 2001, the United Nations General Assembly approved a proposal by the

International Telecommunications Union to hold the WSIS in two parts: the first part in Geneva in December 2003, and the second part in Tunisia in November 2005. The objective of the summit was to discuss the new challenges and opportunities created by the digital revolution and the role of information and communication technologies in improving living standards, achieving the UN Millennium Development Goals,[29] and bridging the digital divide between countries and within societies.[30]

The Geneva phase of the summit aimed at formulating the political vision to build the information society and the practical steps to achieve this objective. The Preparatory Committee convened on three occasions in preparation of this first phase (in July 2002, in February 2003, and in September, November, and December 2003).

A number of controversial issues started to appear in the deliberations of the Preparatory Committee, such as the role of the media, freedom of expression, Internet governance, and financing. After the February 2003 meeting, intellectual property also emerged as one of the divisive issues in the deliberations. Delegates with expertise in information and communication technologies who had been mostly representing developing countries in these meetings were ill-equipped to handle such controversial matters. This prompted the more active involvement of a number of Geneva-based missions from developing countries in the negotiations, particularly on intellectual property issues.

During the negotiations, developed countries and the private sector insisted that intellectual property protection is "essential in the Information Society" and that "existing IP regimes and international agreements should continuously provide this protection . . . thus promoting the necessary balance between owners and users of IP."[31] On the contrary, developing countries and many NGOs stressed that the continuous expansion in intellectual property protection could negatively affect creativity and the dissemination of information. In addition, they opposed the inclusion of language that claimed that international intellectual property agreements were "balanced" or "promoting the necessary balance," particularly in view of the numerous criticisms made at the TRIPS Agreement discussions in this respect. After long and tortuous negotiations, a paragraph on intellectual property that represented a compromise formulation was included in the WSIS Geneva Declaration of Principles, the political declaration adopted by the summit. It states that: "IP protection is important to encourage innovation and creativity in the Information Society; similarly, the wide dissemination, diffusion, and sharing of knowledge is important to encourage innovation and creativity. Facilitating meaningful participation by all in intellectual property issues and knowledge sharing through full awareness and capacity building is a fundamental part of an inclusive Information Society."[32]

Intellectual property protection thus is described in this paragraph only as "important" in the information society, and not as "essential," as first advocated by the developed countries and the private sector. In addition, placing intellectual property protection and the dissemination of knowledge on an equal footing implied, from the view point of the developing countries and the NGOs, that intellectual property protection might not always necessarily achieve the dissemination of knowledge, particularly if it is not balanced and supportive of public-policy objectives.

As it is usually the case on intellectual property matters, Brazil was the most active developing country in these negotiations, particularly regarding the wording of the first sentence of the paragraph, where intellectual property and the dissemination of knowledge were placed on an equal standing. African countries were insistent on the references to participation and capacity building in the second sentence of the paragraph.

Apart from the issue of intellectual property, developing countries and many NGOs were also keen to raise the larger issue of access to information and knowledge in the context of the world summit.[33] Ultimately, this issue was included in the WSIS Geneva Declaration in the section on principles governing the information society under the title "Access to information and knowledge." This was the first time, to my knowledge, that the terms "access to information and knowledge" appeared in official UN documents as the result of negotiations between governments.

In this regard, the WSIS Geneva Declaration states that the "sharing and strengthening of global knowledge for development can be enhanced by removing barriers to equitable access to information . . . and by facilitating access to public domain information" (paragraph 25). The declaration further highlights that "a rich public domain is an essential element for the growth of the Information Society" (paragraph 26). It mentions that "access to information and knowledge can be promoted by increasing awareness among all stakeholders of the possibilities offered by different software models, including proprietary, open-source and free software" (paragraph 27). It also aspires to "promote universal access with equal opportunities for all to scientific knowledge and the creation and dissemination of scientific and technical information, including open access initiatives for scientific publishing" (paragraph 28).

Retrospectively, the WSIS appeared as a landmark development for the emerging A2K movement, because the movement succeeded, for the first time, in including A2K concerns in a major policy document that was endorsed by heads of state and governments. The references to the role of the public domain as a necessary element for the growth of the information society, to the importance of raising awareness about the possibilities offered by different models of software, including free and open-source software, and to open-access initiatives in the area of scientific publication were groundbreaking from this perspective. These same elements would later be raised by developing countries and NGOs in WIPO.

The WSIS discussions reinforced the conviction of those in the nascent A2K movement that multilateral deliberations represented the most appropriate venue for them to reach relatively balanced formulations and views on intellectual property protection. Indeed, the multilateral setting provides developing countries with an equal opportunity to put forward their positions and points of view and to shape outcomes, in contrast, for instance, with bilateral or plurilateral processes such as the negotiation of free-trade agreements, where they are often faced with the overwhelming weight of developed countries, particularly in the economic and trade areas.

ENGAGING WIPO

In this regard, another event occurred in 2003 that was important in the formative stage of the A2K movement. A group of prominent public figures, scientists, and academics, including Nobel Prize winners Joseph Stiglitz, Sir John Sulston, and Harold Varmus, addressed an open letter to the director general of WIPO in July 2003 requesting him to convene a meeting in 2004 to examine new open, collaborative development models such as the Humane Genome Project and open academic and scientific journals. The letter stated that "these models provided evidence that one can achieve a high level of innovation in some areas of the modern economy without intellectual property protection, and indeed [that] excessive, unbalanced, or poorly designed intellectual property protections may be counter-productive."[34]

Commenting on the matter, a U.S. official was quoted in the media affirming that "open-source software runs counter to the mission of WIPO, which is to promote intellectual property rights," adding that "to hold a meeting which has as its purpose to disclaim or waive such rights seems to us to be contrary to the goals of WIPO."[35] This comment triggered a strong reaction in the United States, particularly among academics and civil-society groups, because many open-source collaborative models use copyright. However, the incident equally shed light on the narrow manner in which WIPO's mandate was construed by its single most influential member and the implications this carried for the organization's activities and its approach to intellectual property, which it seemed to consider an end in itself, rather than a means to achieve the public-policy goals of the generation and dissemination of knowledge.

Shortly after, a meeting with the title "WIPO's Work Programme and How to Involve Consumers" was organized by the Trans Atlantic Consumer Dialogue

Special Group on Intellectual Property, in Lisbon in October 2003.[36] I was the only Geneva-based developing-country delegate participating in this meeting. The meeting was the first of its kind in recent years to address WIPO's mandate and activities from the perspective of NGOs and consumers. Although this meeting did not attract the same media coverage as the letter previously mentioned, it signaled that WIPO's activities and its narrow approach to intellectual property were beginning to become the subject of increased attention by civil-society groups and NGOs.

As a further reflection of this evolution, the first policy paper focusing exclusively on WIPO was published in 2003 by the Quaker United Nations Office in Geneva and the Quakers International Affairs Programme, based in Canada.[37] The paper developed a number of views, building on the CIPR report and arguing that WIPO's mandate should not be narrowly limited to the "promotion of IP," as stated in the 1967 Stockholm Convention establishing it, but should be properly construed in the context of the 1974 agreement with the UN by virtue of which WIPO became a UN specialized agency. Under Article 1 of that agreement, the "UN recognized WIPO as its specialized agency with the responsibility for taking appropriate action in accordance with its basic instrument . . . to promote creative intellectual activity"—not intellectual property. The paper also emphasized that WIPO, as a UN agency, should fully integrate and mainstream the development dimension into its activities, as was done in the rest of the UN system.

In September 2003, the second UNCTAD-ICTSD Bellagio Dialogue on Development and Intellectual Property Policy identified a number of priority areas for the reformist intellectual property agenda, with a specific mention of one of WIPO's initiatives, the WIPO Patent Agenda.[38] It thus referred to "challenging the institutional framework in which intellectual property policy is developed . . . including opposition to moves to harmonize the patent regime, such as through WIPO's Patent Agenda." The meeting also addressed important priorities of the A2K movement, such as "supporting the consideration and development of complementary open models for promoting innovation and affordable access to technologies in developing countries, including open source and other collaborative approaches."[39]

By the end of 2003, NGOs and civil-society groups had begun to participate more actively in WIPO meetings. Until then, NGO representation in these meetings was almost exclusively limited to representatives of rights holders' organizations that, in general, favored an increase in the levels of intellectual property protection standards. Public-interest NGOs and civil-society groups had been virtually absent from WIPO deliberations.

This period also witnessed an increased participation of developing countries in WIPO's norm-setting discussions. The South Centre contributed to this process when, beginning in 2002, it started to support developing countries in enhancing their participation in WIPO's activities and in coordinating their positions.[40] It did so by convening meetings where developing countries could prepare in advance for WIPO meetings and by providing analytical notes that highlighted the development and public-policy implications of WIPO's deliberations, particularly in the area of standard setting, such as on patent harmonization and copyright.

After the Lisbon meeting in 2003, the Trans Atlantic Consumer Dialogue (TACD) convened another meeting, "Global Access to Essential Learning Tools," in April 2004 in New York.[41] The meeting included panels on access to textbooks, academic journals, and distance education and software.

I was invited to participate in this meeting, where I moderated the panel on access to textbooks. At that time, apart from being the Egyptian delegate to WIPO, I was coordinating the work of the African Group at WIPO, given that since January 2004, Egypt had assumed this responsibility, which rotates between the members of the group. This task involved presenting the positions of the African Group regarding issues discussed at WIPO. It also entailed assisting the group in building common positions, taking into consideration differences of views that might arise between countries in the group.

THE GENESIS OF THE TERM "A2K" AND THE FRAMING OF A2K AS A CONCEPT

At the New York meeting, a side meeting on strategy took place that brought together key actors that had been active since 2003 in the WSIS and WIPO processes. While it was clear that a coalition was emerging on a number of issues relating to access to knowledge and information, the coalition still lacked a clear and distinctive identity. It had a fragmented constituency that was made up of a number of disparate groups with a focus on very specific issues that at first glance appeared to be not very much related to each other. These groups included opponents of greater protection for databases, advocacy groups promoting free and open-source software, groups advocating open-access initiatives in the area of scientific research journals, plus librarians, consumer organizations, and the visually impaired promoting greater use of exceptions and limitations to copyright, as well as groups promoting the public domain. In addition, a number of developing countries such as Brazil, India, Egypt, and South Africa were sympathetic to all or some of the demands advanced by these groups.

A conceptual framework was lacking to bring all these groups and issues together under a single banner. There was agreement among these actors that the issue of access was central and common in all their efforts and activities. Then the question arose: Access to what?

Several recurrent terms were discussed at the New York meeting, such as access to "information." On my part, I made a strong plea for the exclusive use of the term "access to knowledge" instead, for several reasons.

First, at the conceptual level, knowledge, rather than information, is at the heart of the empowerment of individuals and societies. While information is certainly a prerequisite in the generation of knowledge, acquisition of knowledge remains the ultimate goal. Knowledge processes information to produce ideas, analysis, and skills that ideally should contribute to human progress and civilization.

Second, "access to knowledge" appeared as the appropriate response to the term "knowledge economy" that had been increasingly used, since the end of the 1990s, to describe the new, prevailing paradigm that reflected the changes in the global economy brought about by globalization and new technologies. Often this term was used to promote an expansion in the scope of intellectual property rights and to increase the levels of intellectual property protection. Thus, if the "knowledge economy" was the new paradigm in the global allocation of wealth and resources, then "access to knowledge" became the indispensable other side of the coin in order to make the economic globalization process underpinning the knowledge economy inclusive and equitable.

Third, for tactical considerations, I was concerned that the use of the term "information" would be strictly associated with the deliberations of the WSIS process and potentially could engulf the emerging coalition in the myriad controversial issues that had plagued the WSIS process, such as human rights (freedom of expression), media regulation, and privacy. While these issues are close to the concerns of the emerging A2K movement, they were not to be, in my opinion, the main focus of the advocacy efforts of the movement, because there were many other groups and communities, in particular human rights groups, that were mobilized around them. This doesn't mean that the human rights dimension is not important in the framing of A2K. On the contrary, it is imperative to root the A2K concept deeper in the human rights regime and discourse, particularly in relation to economic and social rights such as the right to health,[42] the right to education,[43] and the "right to take part in cultural life and to enjoy the benefits of scientific progress and its applications."[44]

Finally, the term "access to knowledge" possesses a universal appeal and legitimacy that is powerful. While there might be differences about how to achieve access to knowledge, the goal would be difficult to oppose in itself. It embodies a positive agenda and is not only a "reaction" to trends in expanding intellectual property protection that the world had witnessed since the 1990s. This was a central consideration.

Indeed, previous efforts by developing countries to reform the intellectual property system had confronted it in a frontal manner, tried to act mainly from within the intellectual property system itself, and were ultimately overtaken by the ability of the system to maintain the status quo. This had to be avoided. On the contrary, it was vital that the emerging A2K movement did not define itself exclusively in relation—and even less in opposition—to the intellectual property system, but rather that it would work at building a public-policy objective, such as had occurred in the case of public health, that the intellectual property system could be made to support. The rationale for this was reinforced by the fact that access to knowledge is a cross-sectoral issue by its very nature and affects many areas, such as science, education, research, and many other public-policy areas. Thus, it was important for the A2K movement to expand into the policy debates in relation to knowledge, education, science, and research, rather than become immediately and exclusively engulfed in the technical discussions of the intellectual property system.

These considerations in relation to intellectual property were also central for the inclusiveness of the movement and its capacity to reach different stakeholders, including, for instance, the private sector. It was clear from issues such as increased protection for databases, infringement liability for Internet service providers, and the digitalization of books that there was an important part of the technology industry, particularly in the United States, that shared some of the concerns of the A2K movement and that supported some of its proposals—on open standards, for instance.

However, beyond this initial framing of the A2K concept, its vitality was to be reflected in the extent to which other actors from academia, civil-society groups, and governments would participate in its further elaboration and development, thus ultimately participating in its wider ownership and diffusion, as well.[45] The conceptualization of A2K would remain a work in progress, and the contours of its agenda should continue to evolve and adapt to the challenges raised by the issues in the globalization of knowledge.

Shortly after my plea for the term "access to knowledge" at the New York TACD meeting, the term became increasingly used as the single rallying cry of the movement. James Love then came up with the term "A2K" as a short brand name, which became central in the movement's advocacy efforts.

THE WIPO DEVELOPMENT AGENDA
By mid-2004, it was clear that the momentum for seeking change in WIPO was gaining strength. NGOs were becoming increasingly active, particularly in the context of WIPO discussions on the broadcasting treaty. Developing-country participation in WIPO's substantive debates had also increased significantly, particularly in relation to the proposed Substantive Patent Law Treaty, which raised a number

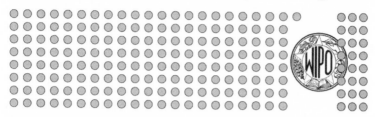

Balance in the WIPO Development Agenda?

Hardly. Of the 193 non-governmental organizations (NGOs) eligible to attend WIPO's Development Agenda summit, only 24 work explicitly on improving conditions in developing countries. So when WIPO holds a meeting about intellectual property in the developing world, the groups that actually work there will be outnumbered 7 to 1.

This image is in
the public domain

Balance in the WIPO Development Agenda? (Ren Bucholz/Electronic Frontier Foundation).

of concerns for developing countries in terms of its impact on the flexibilities they enjoyed under the TRIPS Agreement.[46]

However, it became increasingly apparent for a number of like-minded representatives of developing countries active in WIPO discussions that all these efforts remained fragmented. They started to realize that only a major policy initiative could bring change to WIPO and advance all these dispersed efforts. Such an initiative would go beyond these efforts to address specific standard-setting proposals for increased intellectual property protection being advanced at the organization. It would address in a systematic and comprehensive manner the organization's culture in promoting intellectual property, particularly in the areas of norm setting and technical assistance.

On a substantive level, the initiative would seek to include many of the proposals and recommendations that had been put forward on WIPO since 2002, in particular, the CIPR report and the outcomes of the UNCTAD-ICTSD Bellagio Dialogues, this in addition to bringing to WIPO the global debate over ideas about intellectual property that was taking place outside of it and in which the A2K coalition had become an important actor.

On a procedural level, the initiative was also motivated by the desire to establish, for the first time in a long time, a grouping of like-minded representatives of developing countries at WIPO who shared the same perspective on intellectual property, rather than simply just belonging to the same geographical group.

Indeed, the initiative relied on geographical groups to steer the work of the organization in procedural and substantive matters. However, while these groups might be useful in a number of procedural matters, such as allocating the membership of bodies or electing officials or committee chairs, reliance on them in substantive matters often made less sense, because in many cases, countries within the same group had significantly diverging views on intellectual property.[47] Indeed, it could be problematic for countries to reach common positions on substantive intellectual property negotiations at the international level if their national intellectual property laws differed significantly and did not grant the same level of intellectual property protection. This is the case, for example, in Latin and Central America, where the gap is particularly acute, for instance between countries that have adopted TRIPS-plus standards as a result of free-trade agreements concluded with developed countries such as the United States and other countries that have refrained from entering in such agreements. While the practice of having representatives of like-minded groups of developing countries coordinating on specific issues was current practice in the WTO TRIPS context, it was not the case at WIPO.

On their part, civil-society groups had done significant work since 2002 in preparing the ground through increased advocacy aimed at bringing WIPO into the wider public-policy debate.

The time was thus ripe for bringing such a new initiative forward. At the end of August 2004, Brazil and Argentina circulated the proposal for a development agenda for WIPO. Egypt joined them, with a group of other countries, to present this proposal at the session of the WIPO Assemblies in late September.[48] Egypt's adherence to this initiative came in continuity with its important contribution to efforts by developing countries to achieve a more development-friendly international trade and intellectual property system.

Indeed, in the 1980s, Egypt was among the developing countries that resisted the inclusion of intellectual property issues in the Uruguay Round of the General Agreement on Tariffs and Trade. During the round itself, in 1990, it joined a number of developing countries such as Bolivia, Colombia, Peru, Venezuela, Argentina, Brazil, Chile, China, Colombia, Cuba, Egypt, Nigeria, Peru, Tanzania, and Uruguay in submitting a developing-countries draft text for the intellectual property agreement that was under negotiation.[49]

Egypt was also among the developing countries that availed themselves of the possibility to use the Appendix of the Berne Convention for the Protection of Literary and Artistic Works (1971 Paris version), which provided—subject to just compensation to the rightful owner—"for the possibility of granting non-exclusive and non-transferable compulsory licensing in respect of (i) translation for the purpose of teaching, scholarship or research, and (ii) reproduction for use in

(Myth of the free-rider "problem")

connection with systematic instructional activities, of works protected under the Convention."

In its national legislation implementing the TRIPS Agreement (Law 83 of 2002), Egypt had incorporated many of the public-health-related flexibilities of the agreement. And while it has concluded a number of free-trade agreements, such as the EU-Egypt Association Agreement and the EU-EFTA Agreement, it has succeeded to a great extent in avoiding taking on new, extensive TRIPS-plus obligations with a bearing on public-policy objectives such as public health.

A2K AND THE WIPO DEVELOPMENT AGENDA

Although the WIPO Development Agenda initiative was not only about A2K, A2K-related issues were clearly an important component of the proposals and ideas that the initiative was seeking to advance. This was reflected in the original document containing the development agenda proposal, which included key elements and concerns of the A2K movement such as:

- An indication that adding new layers of intellectual property protection to the digital environment would obstruct the free flow of information and scuttle efforts to set up new arrangements for promoting innovation and creativity through initiatives such as the "Creative Commons."
- An expression of concern at the ongoing controversy surrounding the use of technological protection measures in the digital environment.
- A reference to the importance of safeguarding the exceptions and limitations existing in the domestic laws of member states.
- A mention of the need to bear in mind the relevance of open-access models for the promotion of innovation and creativity in order to tap into the development potential offered by the digital environment and an invitation to WIPO to consider undertaking activities with a view toward exploring the promise held by open, collaborative projects to develop public goods, as exemplified by the Human Genome Project and open-source software.
- A reference to the need to examine the potential development implications of several of the provisions of the proposed Treaty on the Protection of Broadcasting Organizations that the Standing Committee on Copyright and Related Rights was currently discussing, taking into consideration the interests of consumers and of the public at large.[50]

Before the WIPO Assemblies started, the A2K movement mobilized support for this initiative. The South Centre published an analytical note emphasizing the need to integrate development into WIPO activities and processes, thus supporting the rationale for such an initiative.[51]

The TACD convened a "Future of WIPO" meeting in mid-September of 2004 in Geneva. The meeting included a number of prominent figures who had been at the forefront of efforts for a more balanced intellectual property system, particularly in the United States and the United Kingdom, such as Larry Lessig, Yochai Benkler, James Boyle, John Sulston, and Tim Hubbard. It also included leading activists such as Martin Khor and Richard Stallman. A Geneva-based delegate from Argentina also spoke at the meeting. A "Geneva Declaration on the Future of WIPO" was shortly launched, signed by leading figures from academia, NGOs, and civil society.

Of course, it was no coincidence that this meeting was held a few days before the WIPO Assemblies, which would examine the proposal for a development agenda. It reflected once more the close collaboration between developing countries and a number of NGOs that had become active in WIPO processes since 2003.

The few developing-country Geneva delegates who had been actively involved in efforts leading to the launch of the WIPO Development Agenda,[52] including this author, were convinced that such an initiative on the part of developing countries in WIPO would have limited chances of success if it was not supported by civil-society groups from the North, which were capable of mobilizing public opinion and the media through their networks and advocacy in a manner that would have an impact in developed countries. At the same time, NGOs that had identified the reform of WIPO as a central part of their advocacy to achieve a more balanced and consumer-friendly intellectual property system saw in the WIPO Development Agenda a vehicle for moving forward their efforts in this area.

Consequently, in the same way that the original proposal for a development agenda incorporated key elements of the A2K agenda that were important to many NGOs and developing countries, the Geneva Declaration on the Future of the World Intellectual Property Organization lent its support to the WIPO Development Agenda proposal, stating that it "pointed in the right direction" and "created the first real opportunity to debate the future of WIPO."[53] Furthermore, it also addressed issues that primarily concerned developing countries, such as the call for a fundamental reform of WIPO's technical assistance programs so as to enable developing countries to use to the full the flexibilities in the TRIPS Agreement to promote access to medicines for all.[54]

The presence of NGOs from the South in these developments was weaker in comparison with that of the NGOs and civil-society groups from the North.[55] However, this came as no surprise. Many groups in the South that are mobilized on issues of the environment, human rights, public health, and poverty alleviation were not that engaged or familiar with global debates on access to and ownership of knowledge and even less so had means to mobilize around them. The notable exception was the Third World Network which was actively engaged in support of the WIPO Development Agenda through its advocacy efforts.[56]

It was also only natural that groups from the North had more experience in dealing with some of the issues of concern to the A2K movement, such as technological protection measures in the digital environment in view of the more advanced digital economies and legislations in developed countries. However, their experiences in this area were also useful for developing countries that were faced with obligations in this area through free-trade agreements or as a result of adherence to the WIPO Internet Treaties.

Thus, both the WIPO Development Agenda process and the A2K movement brought together developing countries, consumer-based civil-society groups, and NGOs, particularly from the North, in a mutually beneficial collaboration. This, of course, doesn't necessarily mean that the importance for each one of them of their shared priorities is the same.

In this regard, the A2K concept could be compared to a coin. On one face, we find the "A2K for development" concerns of developing countries that seek flexibilities in intellectual property rules, while on the other, there are the "A2K for innovation and creativity" priorities lying beyond the traditional intellectual property system, in which consumers and NGOs are faced with high intellectual property standards in developed countries. Both are mutually relevant and important for each other, but each has a different emphasis. For instance, exceptions and limitations for educational purposes might be of greater importance to developing countries, given their vast educational needs. On the other hand, alternative innovation models and open collaborative projects, whose role and impact are still limited in developing countries, are more likely to be an immediate priority for consumer organizations in developed countries that seek alternatives to proprietary models of knowledge generation. There is thus a constant balance to be maintained within the A2K movement between these two faces of the coin. There is no doubt however, that both developing countries and NGOs, respectively as predominantly importers and consumers of knowledge goods, share the goal of achieving more balanced intellectual property and information and communication technology regimes that enable greater access to knowledge.

After the launch of the WIPO Development Agenda, CPTech organized two meetings in 2005 on the A2K treaty proposal, the first with the Third World Network and the International Federation of Library Associations, and the second with the TACD. By May 2005, the current draft A2K treaty was completed.[57] Thus, by 2005, the A2K movement was fully formed and had come forward with a major norm-setting proposal, the draft A2K treaty.

By that time, I had left Geneva and had returned to Cairo to assume new professional obligations. I was struck upon my return to Egypt by the extent to which A2K and the WIPO Development Agenda, as well as the debates underpinning them, had remained confined to a number of specialized circles following intellectual property issues and to the Geneva multilateral setting. Important efforts needed to be made, particularly in the area of raising awareness, to bring a more extensive awareness of these debates to developing countries.

In this regard, a Regional Arab Dialogue on intellectual property and sustainable development was organized by the Bibliotheca Alexandrina, ICTSD and UNCTAD in June 2005.[58] As a follow-up to one of the recommendations of this dialogue, the Bibliotheca Alexandrina organized a regional seminar entitled "New Tools for the Dissemination of Knowledge and the Promotion of Creativity and Innovation: Global Developments and Regional Challenges" on September 7 and 8, 2006.[59] This regional seminar adopted a number of recommendations to promote A2K in the Arab world, such as raising awareness about open collaboration and new tools for the dissemination of knowledge (the Creative Commons, open academic and scientific journals, and so on) and establishing a regional research agenda.[60]

FUTURE DIRECTIONS FOR THE A2K MOVEMENT

While the emergence of the A2K movement involved a number of groups engaged in intellectual property debates, mostly in the North, as well as developing countries at the multilateral level in Geneva, the future growth and consolidation of the movement lies in its capacity to mobilize interest and support in the South, particularly among governments, civil-society groups, and academia. This is a long-term strategic priority for the A2K movement. It is thus imperative to continue raising awareness about the importance of A2K issues for developing countries and in developing countries.

In this regard, it is important to clarify that access to knowledge is not the antithesis of intellectual property. Developing countries can combine balanced intellectual property policies where they use creatively both intellectual property and open business models in some areas (such as in the creative industries) while also pursuing overall A2K policies and measures to address their vast educational needs and requirements for building scientific and technological capabilities. Many policies relevant to knowledge pursued by developing countries at the national level, such as in the areas of education, culture, intellectual property and information and communication technologies, and science and technology in general, could be framed and conceptualized in the context of the overall A2K paradigm and its objectives.

"ratchet" mechanism...
in IP, pension benefits, etc.
have to take back what
has been given.

have for
my self
paper...

At the international level, there is still much that can be done in terms of mainstreaming A2K in policy positions adopted by developing countries in international forums, as well as in the diverse groupings to which they belong.[61] Another pressing priority for the A2K movement at this level is to have a concrete impact on policy processes and deliberations in international processes and forums or face the risk of falling into irrelevance or of becoming a purely academic exercise.

In this regard, the proposal for work on exceptions and limitations made by a number of developing countries at the March 2008 session of WIPO's Standing Committee on Copyright and Related Rights paves the way for a process of deliberations of significant importance, one in which the A2K movement should be fully engaged and which it should support.[62] The proposal stems from the premise that often, developing countries have comparatively fewer exceptions and limitations for research, education, the visually impaired, and so in their national copyright laws, compared with developed countries, and make less use of them, although they might be more in need of them, given their vast needs for access to educational material. The proposal raises the possibility of elaborating an international instrument on exceptions and limitations that would provide normative guidance in this area and include a mandatory set of exceptions and limitations that would be common to all WIPO member states.[63]

created/
possibility
possibilities
arise.

Another opportunity for the A2K movement lies in the implementation of the WIPO Development Agenda recommendations relating to access to knowledge. In effect, after two years of intense debates, in September and October 2007, the WIPO Assemblies adopted by consensus forty-five recommendations aiming at the establishment of a development agenda for WIPO.[64] The assemblies established the Committee on Development and Intellectual Property to monitor the implementation of these recommendations. At the committee meetings held in March and July 2008, member states discussed the implementation of a number of recommendations, particularly those relating to intellectual property technical assistance.

The implementation of the recommendations relating to the strengthening of the public domain or to the consideration of open, collaborative models requires the identification of specific activities and concrete proposals to promote these issues in WIPO's activities, such as through awareness raising (seminars, workshops, publications, and so on) legislative advice, or norm setting. The A2K movement should contribute to the identification of these activities and proposals and remain active in this implementation phase, because it might be even more critical than the phase of deliberations that took place from 2004 to 2007. It is the actual implementation of these recommendations that will determine the extent to which the WIPO Development Agenda will have been able to orient WIPO's activities toward a more balanced and development-oriented approach to intellectual property and toward promoting creativity and innovation beyond traditional intellectual property categories in a manner that effectively contributes to A2K.

Beyond WIPO, the A2K movement should carefully examine influencing other relevant policy processes and forums. In this regard, UNESCO, the United Nations Educational, Scientific, and Cultural Organization, can be an organization of significant importance for promoting access to knowledge. Until now, it has not been central in the preoccupations and advocacy of the A2K movement. In addition, the experience of UNESCO in concluding the Convention on Cultural Diversity in a relatively short period of time (2003 to 2005) might bear valuable lessons for the A2K movement.

The Internet Governance Forum (IGF) that resulted from the World Summit on the Information Society is another venue where the A2K movement has pursued its objectives and should continue to do so.

Beyond WIPO, UNESCO, and the IGF, another interesting possibility would be launching an A2K initiative at the UN General Assembly, the competent body for discussing all political, economic, and social issues at the UN, particularly issues that have a cross-sectoral nature and touch upon many areas of the UN's work, such as development, education, culture, science, intellectual property, and so on. This step would be critical in the further diffusion and adoption of the A2K paradigm by the entire UN system, as has occurred in the past with other concepts, such as human development or sustainable development. Of course, this is a long-term process, and it requires an active role on the part of governments, as well as engagement and leadership from developing countries. The trajectory of the environmental movement from the first environmental summit in Stockholm in 1972 (where only a handful of heads of state and governments were present) to the Rio summit in 1992 (in which more than a hundred heads of state and governments participated) represents a valuable experience for the A2K movement to study for extracting the appropriate lessons.

Finally, it is of fundamental importance to generate empirical studies and academic work that further contribute to the advancement of the A2K paradigm and that underpin it. In this regard, the Yale Law School Information Society Project A2K conferences, starting in 2006, have played a valuable role in strengthening the links between the A2K movement and academia. Similarly, the A2K country studies (Brazil, China, Egypt, India, South Africa) commissioned by the Information Society Project will enrich A2K advocacy as well as raise awareness on A2K in developing countries.[65]

CONCLUSION

In its further development and growth, the A2K movement continues to confront the need to achieve a delicate balance. On the one hand, while the A2K movement emerged from an interaction with globalized intellectual property rules and processes, particularly at WIPO, it should go beyond these and engage with other national and international rules and processes in the areas of education, culture, human rights, the environment, and so on. This interaction could further enrich A2K conceptually, while at the same time, issues in these areas could benefit from being framed in A2K terms.

However, it is also equally important for the A2K movement to remain focused in its advocacy on the original set of issues that led to its emergence—the generation, dissemination, and use of knowledge and their regulation—and not to disperse itself in areas close to its concerns, but where other social actors and movements are already doing valuable and effective advocacy work. This consideration is essential, particularly for developing countries, because new challenges loom in the horizon in the form of proposals for higher enforcement standards or new enforcement agreements that could possibly be detrimental to the objectives of the A2K movement.[66]

NOTES

The views expressed in this paper do not necessarily reflect the views or opinions of any institution with which the author is affiliated. The author is grateful to Pedro Roffe, Edna Ramírez Robles, Amy Kapczynski, and Gaëlle Krikorian for their valuable comments. He is particularly indebted to Gaëlle Krikorian for her perseverance and patience.

1 See Susan K. Sell, *Power and Ideas: North-South Politics of Intellectual Property and Antitrust* (Albany: State University of New York Press, 1998) and Christopher May and Susan K. Sell, *Intellectual Property Rights: A Critical History* (Boulder: Lynne Rienner Publishers, 2006).

2 For an analysis of the TRIPS Agreement and its implications for developing countries, see Carlos M. Correa, *Intellectual Property Rights, the WTO, and Developing Countries: The TRIPS Agreement and Policy Options* (London: Zed Books, 2001) and United Nations Conference on Trade and Development (UNCTAD) and International Centre for Trade and Sustainable Development (ICTSD), *Resource Book on TRIPS and Development* (Cambridge: Cambridge University Press, 2005).

3 See Susan K. Sell, *Private Power, Public Law: The Globalization of Intellectual Property Rights* (Cambridge: Cambridge University Press, 2003).

4 See, for instance, David Vivas-Eugui, "Regional and Bilateral Agreements and a TRIPS-plus World: the Free Trade Area of the Americas (FTAA)," *TRIPS Issues Papers* 1 (2003), Quaker United Nations Office, Quaker International Affairs Programme, and the International Centre

for Trade and Sustainable Development and Sisule Musungu and Graham Dutfield, "Multilateral Agreements and a TRIPS-plus World: The World Intellectual Property Organisation," *TRIPS Issues Papers* 3 (2003), Quaker United Nations Office and Quaker International Affairs Programme.

5   For the WIPO Digital Agenda, see http://www.wipo.int/copyright/en/digital_agenda.htm (last accessed March 13, 2009).

6   *Declaration on the TRIPS Agreement and Public Health, WTO Doha Ministerial Conference, 4th Sess.*, WTO Doc. WT/MIN (01)/DEC/W/2 (November 14, 2001). See Sangeeta Shashikant's essay, "The Doha Declaration on TRIPS and Public Health: An Impetus for Access to Medicines," in the present volume.

7   See Susan K. Sell, "TRIPS and the Access to Medicines Campaign," *Wisconsin International Law Journal* 20, no. 2 (2002): pp. 481–522.

8   The Quaker United Nations Office in Geneva played a particularly important role in providing technical support to developing-country delegates, with the assistance of a number of legal experts.

9   Commission on Intellectual Property Rights (CIPR), *Integrating Intellectual Property Rights and Development Policy* (London: Commission on Intellectual Property Rights, 2002).

10   *Ibid.*, p. 102.

11   *Ibid.*, p. 104.

12   For views on the report in the media, see http://www.iprcommission.org/graphic/views.htm (last accessed March 13, 2009).

13   Some developing countries began to cite the report in deliberations on intellectual property issues. As an illustration, see para. 152 of the *General Report of the 37th Series of Meetings, Assemblies of Member States of WIPO, September 23 to October 1, 2002*, contained in document A/37/14, available on-line at http://www.wipo.int/edocs/mdocs/govbody/en/a_37/a_37_14.pdf (last accessed March 13, 2009).

14   See, for instance, Peter Drahos, "Developing Countries and International Intellectual Property Standard-Setting," *The Journal of World Intellectual Property* 5, no.5 (September 2002).

15   CIPR, *Integrating Intellectual Property Rights and Development Policy*, p. 157.

16   Such as the Substantive Patent Law Treaty (SPLT) at WIPO.

17   See "Towards Development-Oriented Intellectual Property Policy: Setting an Agenda for the Next Five Years," ICTSD-UNCTAD Dialogue, The Rockefeller Foundation's Bellagio Conference Center, October 30–November 2, 2002, available on-line at http://www.iprsonline.org/unctadictsd/bellagio/docs/BellagioOutcome_Report.pdf (last accessed March 13, 2009).

18   For more on the impact of these meetings, see Joe Karaganis, *The Bellagio Global Dialogues on Intellectual Property* (New York: Social Science Research Council, 2006).

19   Joseph Stiglitz was one of the most prominent advocates of this discourse. See Joseph Stiglitz, *Globalization and Its Discontents* (New York: Norton, 2002). For a view from the South, see Martin Khor, *Rethinking Globalization: Critical Issues and Policy Choices* (London: Zed Books, 2001).

20   On the role of Geneva representations of developing countries, see CIPR, *Integrating Intellectual Property Rights and Development Policy*, p. 164.

21   For these coordination problems, see Ahmed Abdel Latif, "Developing Country Coordination in International Intellectual Property Standard-Setting," *Working Paper* 24 (T.R.A.D.E), South Centre, Geneva (2005).

22 It is important to note, however, that a number of NGOs active in the access-to-medicines campaign had already started to raise questions about the extent to which WIPO's technical assistance integrated the public-health flexibilities contained in the Doha Declaration on TRIPS and Public Health. See Médecins Sans Frontières, Consumer Project on Technology, Health Action International, and Oxfam, *Implementation of the Doha Declaration on the TRIPS Agreement and Public Health: Technical Assistance—How to Get It Right* (March 2002), available online at http://www.haiweb.org/campaign/access/ReportPostDoha.pdf (last accessed March 14, 2009).

23 The text of the agreement is available on-line at http://www.wto.org/english/tratop_e/trips_e/wtowip_e.htm (last accessed March 14, 2009).

24 See http://www.wto.org/English/news_e/pres98_e/pr108_e.htm and http://www.wtocenter.org.tw/SmartKMS/fileviewer?id=7973 (both last accessed March 14, 2009).

25 See Drahos, "Developing Countries and International Intellectual Property Standard-Setting," and Musungu and Dutfield, "Multilateral Agreements and a TRIPS-plus World."

26 See Abdel Latif, "Developing Country Coordination in International Intellectual Property Standard-Setting," and Musungu and Dutfield, "Multilateral Agreements and a TRIPS-plus World."

27 Countries that were involved included Argentina, Brazil, Egypt, and India.

28 These included the Center for International Environmental Law, the International Centre for Trade and Sustainable Development, the Quakers United Nations Office, and Third World Network.

29 The eight Millennium Development Goals, which range from halving extreme poverty to halting the spread of HIV/AIDS and providing universal primary education, were adopted in September 2000 at a meeting of world leaders in New York that endorsed the United Nations Millennium Declaration. This Declaration set out a series of time-bound targets— with a deadline of 2015—that have become known as the Millennium Development Goals. For more information, see http://www.un.org/millenniumgoals/bkgd.shtml (last accessed March 14, 2009).

30 See Laura DeNardis's essay, "The Global Politics of Interoperability," in this volume.

31 Text of intercessional work between the second and third convening of the Preparatory Committee, new para. 40 in the document WSIS03/PCIP/DT/4(Rev.3)-E, available on-line at http://www.itu.int/wsis/documents/doc_single.asp?lang=en&id=699 (last accessed March 14, 2009).

32 See para. 42 of the Geneva Declaration of Principles contained in document WSIS-03/GENEVA/DOC/4-E, available on-line at http://www.itu.int/wsis/documents/doc_multi.asp?lang=en&id=1161|1160 (last accessed March 14, 2009).

33 The NGOs included CPTech, IP Justice, the Electronic Frontier Foundation, and others.

34 Letter available on-line at http://www.cptech.org/ip/wipo/kamil-idris-7july2003.pdf (last accessed March 14, 2009).

35 William New, "U.S. Official Opposes Open Source Talks at WIPO," *Technology Daily*, August 19, 2003, available on-line at http://lists.essential.org/pipermail/random-bits/2003-August/001090.html (last accessed March 14, 2009).

36 "Workshop on the WIPO Work Program and How to Involve Consumers," Trans Atlantic Consumer Dialogue, Lisbon, October 17, 2003.

37 See Musungu and Dutfield, "Multilateral Agreements and a TRIPS-plus World."

38   See Carlos M. Correa and Sisule Musungu, "The WIPO Patent Agenda: The Risks for Developing Countries," T.R.A.D.E *Working Papers* 12, South Centre (Geneva, 2002).

39   See the report of the 2nd UNCTAD-ICTSD Bellagio Dialogue on Development and Intellectual Property Policy, *Towards Development-Oriented Intellectual Property Policy: Advancing the Reform Agenda* (September 18–21, 2003), available on-line at http://ictsd.net/i/events/dialogues/35259 (last accessed March 14, 2009).

40   The South Centre is an intergovernmental organization of developing countries with headquarters in Geneva established by an intergovernmental agreement that came into force on July 31, 1995. The organization is intended to meet the need for the analysis of development problems and experience, as well as to provide the intellectual and policy support required by developing countries for collective and individual action, particularly in the international arena. For more information, see http://www.southcentre.org/index.php?option=com_content&task=view&id=1&Itemid=1 (last accessed March 14, 2009).

41   "Global Access to Essential Learning Tools," New York, April 5, 2004. See Manon A. Ress's essay, "Open-Access Publishing: From Principles to Practice," in this volume.

42   Article 12 of the International Covenant on Economic, Social and Cultural Rights.

43   Article 13 of the International Covenant on Economic, Social and Cultural Rights.

44   Article 15 (a) and (b) of the International Covenant on Economic, Social and Cultural Rights.

45   This indeed happened with the series of A2K conferences at Yale University starting in 2006.

46   See Correa and Musungu, "The WIPO Patent Agenda: The Risks for Developing Countries." See also Viviana Muñoz Tellez and Sisule F. Musungu's essay, "A2K at WIPO: The Development Agenda and the Debate on the Proposed Broadcasting Treaty," in this volume.

47   For an elaboration on this issue, see Abdel Latif, "Developing Country Coordination in International Intellectual Property Standard-Setting," p. 32.

48   These countries were Bolivia, Cuba, the Dominican Republic, Ecuador, South Africa, Egypt, Kenya, Iran, Peru, Sierra Leone, Tanzania, and Venezuela.

49   See Peter Drahos, "Developing Countries and International Intellectual Property Standard-Setting," *Study Paper 8 for the CIPR Commission* (2002), available on-line at http://www.iprcommission.org/papers/pdfs/study_papers/sp8_drahos_study.pdf (last accessed March 15, 2009).

50   The proposal is contained in document WO/GA/31/11, available on-line at http://www.wipo.int/edocs/mdocs/govbody/en/wo_ga_31/wo_ga_31_11.pdf (last accessed March 15, 2009).

51   *Integrating Development into WIPO Activities and Processes: Strategies for the 2004 Assemblies*, South Centre, Analytical Note, SC/TADP/AN/IP/2 (August 2004), available on-line at http://www.southcentre.org/index.php?option=com_content&task=view&id=80&Itemid=67t (last accessed March 15, 2009).

52   See para. 284 in the General Report of the 43rd Series of Meetings, Assemblies of Member States of WIPO, contained in document A/43/16, available on-line at http://www.wipo.int/meetings/en/doc_details.jsp?doc_id=88952 (last accessed March 15, 2009).

53   Geneva Declaration on the Future of the World Intellectual Property Organization, available on-line at http://www.cptech.org/ip/wipo/genevadeclaration.html (last accessed March 15, 2009).

54   *Ibid.*

55   This has evolved since 2004 with the involvement of institutions from the South in A2K advocacy, such as the Fundação Getulio Vargas in Brazil and the Bibliotheca Alexandrina in Egypt.

56　See, for instance, *Third World Resurgence Magazine*, no. 171–72 (November–December 2004), available on-line at http://www.twnside.org.sg/focus.htm (last accessed March 15, 2009).

57　The draft treaty is available on-line at http://www.cptech.org/a2k/a2k_treaty_may9.pdf (last accessed March 15, 2009).

58　For more information, see http://www.iprsonline.org/unctadictsd/dialogue/2005-06-26/2005-06-26_desc.htm (last accessed March 15, 2009).

59　For more information, see http://www.bibalex.org/a2k/Event/BAEventList.aspx (last accessed March 15, 2009).

60　See http://www.bibalex.org/a2k/attachments/BAEvents/Recomendation.pdf (last accessed March 15, 2009).

61　Such as the G77, the Nonaligned Movement, the African Union, the New Economic Partnership for Africa (NEPAD), the League of Arab States, the Organization of the Islamic Conference (OIC), as well as the India-Brazil-South Africa forum (IBSA).

62　See the proposal by Brazil, Chile, Nicaragua, and Uruguay for work related to exceptions and limitations contained in document SCCR/16/2, available on-line at http://www.wipo.int/meetings/en/doc_details.jsp?doc_id=107712 (last accessed March 15, 2009).

63　For the rationale and policy options to elaborate such an instrument, see P. Bernt Hugenholtz and Ruth L. Okediji, *Conceiving an International Instrument on Limitations and Exceptions to Copyright* (May 6, 2008), available on-line at http://www.soros.org/initiatives/information/articles_publications/publications/copyright_20080506/copyright_20080506.pdf (last accessed March 15, 2009).

64　Available at http://www.wipo.int/ip-development/en/agenda/recommendations.html (last accessed March 15, 2009). The forty-five recommendations are divided into six clusters: Cluster A—technical assistance and capacity building; Cluster B—norm setting, flexibilities, public policy, and the public domain; Cluster C—technology transfers, information and communication technology, and access to knowledge; Cluster D—assessments, evaluation, and impact studies; Cluster E—institutional matters, including mandates and governance; and Cluster F—other issues.

65　The first study to be available is on Brazil. Lea Shaver (ed.), *Access to Knowledge in Brazil: New Research on Intellectual Property, Innovation and Development*, Information Society Project, 2008, available on-line at http://www.law.yale.edu/intellectuallife/6620.htm (last accessed March 15, 2009).

66　Proposals for higher enforcement standards include the SECURE standards advanced in the context of the World Customs Organization. See Viviana Muñoz Tellez, "World Customs Organisation: Setting New Standards of Intellectual Property Enforcement through the Back Door?" *South Bulletin*, no. 13 (April 16, 2008), p. 6, available on-line at http://www.southcentre.org/index.php?option=com_content&task=view&id=592&Itemid=105_ (last accessed March 15, 2009). Proposals for new enforcement agreements include the proposed Anti-Counterfeiting Trade Agreement.

Campaigning material produced by Health Gap (Global Access Project).

The Revised Drug Strategy:
Access to Essential Medicines, Intellectual Property,
and the World Health Organization

Ellen 't Hoen

The magnitude of the AIDS crisis has drawn attention to the fact that millions of people in the developing world do not have access to the medicines that are needed to treat disease or to alleviate suffering. The high cost of AIDS medicines has also focused attention on the relation between patent protection and high drug prices. The difficulties that developing countries experience in paying for new essential medicines have raised concerns about the effects of the 1995 World Trade Organization (WTO) Trade-Related Aspects of Intellectual Property Rights (TRIPS) Agreement, which sets global standards for the protection of intellectual property (IP). These standards derive from Western countries with a high level of industrial development, and though they are often referred to as "minimum" standards, they set the bar very high. The AIDS crisis gives us an alarming preview of the consequences of such intellectual property rules, which by no means are confined to AIDS medicines. All new health-care products may be affected by TRIPS and by the new patent rules that it imposes on almost every country in the world.

However the lack of access to medicines is not a recent problem for the developing world. For decades, countries have been dependent on Western companies for their supply of medicines, and they have at times suffered from it.[1] Western-style pharmaceutical patent-protection requirements are likely to increase developing countries' dependency further.

During the 1990s, in response to developing countries' need to increase the availability and improve the use of medicines, the World Health Organization (WHO) developed a medicine policy called the Revised Drug Strategy. The Revised Drug Strategy built strongly on the concept of "essential drugs" created in the late 1970s. (See the sidebar "The Concept of Essential Medicines.") After the adoption of the TRIPS Agreement, it became obvious that the Revised Drug Strategy needed to be adjusted to take into account the effects of intellectual property protection, and in particular of patenting, on the production and availability of medicines. In fact, the very concept of "essential drugs" brought the dilemma over intellectual property to the fore at the WHO. After all, the work of the program responsible for implementing the Revised Drug Strategy risked becoming obsolete, because its advice was based on a world where countries were at liberty to determine the kind of pharmaceutical patent regimes that they wished to have. The Revised Drug Strategy strongly recommended that countries adopt a national policy encouraging the use of generic medicines, a recommendation that ran counter to the new obligations that countries took on under the WTO TRIPS rules in 1995. Countries' struggle to access newer medicines such as antiretrovirals to treat people living with AIDS illustrated the need to adapt the Revised Drug Strategy to this new reality. The discussions on the Revised Drug Strategy came at a point in history where intellectual property became a controversial issue around which NGOs and health experts launched campaigns. The purpose of this article is to revisit this history to show how intellectual property became a political issue in the area of health, how NGOs became mobilized around the issue of intellectual property, and how the WHO became a major forum in which to discuss the impact of intellectual property legislation on people's lives.

THE ADOPTION OF THE REVISED DRUG STRATEGY

In November 1985, member states, the pharmaceutical industry, academia, consumer groups, and the WHO secretariat met in Nairobi under the leadership of then WHO director general Dr. Halfdan T. Mahler to discuss the WHO's strategy on medicines. At the 1984 World Health Assembly, the annual meeting of the WHO's member states, groups such as Health Action International, a network of consumer groups, public-interest NGOs, health-care providers, academics, media, and individuals in more than seventy countries advocating for "increased access to essential medicines and improved rational use of medicines,"[2] expressed concerns that the focus of the WHO's work on supply overshadowed the need to improve the use of medicines, to deal with unethical marketing practices, and to remove useless and dangerous products from the market. Health Action International felt that the WHO's involvement in medicine policy should go beyond listing what should be in every country's medicine chest and called for international rules that could help countries to intervene in how drug companies behave.

Director General Dr. Mahler announced at the meeting that he would develop "a strategy for strengthening WHO's activities in support of the action required to make drug use more rational throughout the world." He listed the key elements that such a policy related to access to medicines should contain: measures to

In the 1970s, in public-health circles, there was a strong move toward establishing primary health care. The selection and provision of essential medicines was then increasingly seen as a core function of governments in the context of primary care. As a result, in 1975, the World Health Assembly, the annual assembly of the WHO member states, adopted a resolution calling on the WHO to assist its members to select and procure essential drugs and assure that these drugs were of good quality and reasonable cost.[1] The first WHO Essential Drugs List, containing 207 items, was published in 1977. The purpose of the list was to identify those medicines "of utmost importance, basic, indispensable and necessary for the health and needs of the population."[2] Selection criteria included issues related to efficacy, quality, safety, and cost. Since then, fifteen editions of the list have been published. Today, the WHO Essential Drugs List—rebaptized the Essential Medicines List in 2002—contains around 325 items. The list is regularly updated to be able to respond to new needs, drug resistance, medical advances, scientific developments, and new evidence with regard to efficacy and safety. The Essential Medicines List is also useful to identify gaps in research and development of new essential medicines. For example, a few years ago, expert discussions over the Essential Medicines List highlighted the obvious need for pediatric formulations of AIDS medicines to treat children. Because pediatric AIDS is rare in rich countries, companies lacked the commercial incentive to develop easy-to-use medications for children.

Although the basic notion that some drugs are essential and others are not and the creation of an evidence-based list of essential drugs may not seem either controversial nor radical, the concept of essential medicines has been both. Dr. Ernst Lauridsen, the first director of the WHO Action Programme on Essential Drugs, described it as a "peaceful revolution in international public health."[3] However, the adoption of this new policy did not happen without controversy. The pharmaceutical industry saw the concept of essential medicines and WHO's work with governments on medicine policies as a frontal attack on its freedom to operate and feared that it would lead to government interference in the industry's marketing practices. Countries would indeed use the list to exclude drugs on the Essential Medicines List from patentability, as was the case in the Andean region; to remove medicines from the market; to establish price controls; and to develop their own national manufacturing capacity of essential medicines, thereby reducing dependence on Western companies.

NGOs have played an important role in protecting and promoting the concept of essential medicines. Health Action International, a global network of health and pharmaceutical groups and individuals, was a particularly key player in the advocacy for essential health policies at both the international and the national levels. This organization has its roots in the consumer movement in the Global South; its members saw the detrimental effects of the lack of essential medicines and the availability of dangerous and ineffective drugs on a daily basis.

In 1977, the International Federation of the Pharmaceutical Manufacturers Association, the main international lobbying organization for the multinational pharmaceutical industry, called the medical and economic arguments for the Essential Medicines List "fallacious" and claimed that adopting it would "result in substandard rather than improved medical care and might well reduce health standards already attained."[4] The industry was particularly concerned that the Essential Medicines List would become a global concept, applicable beyond the developing world, and, for example, be used for priority setting in the marketing approval of medicines or in the reimbursement decisions of health-insurance companies in industrialized countries. Such measures, they feared, would, for example, limit the industry's ability to market aggressively new medicines that have larger profit margins than older ones.

In 1982, a spokesman of the U.S. pharmaceutical manufacturers organization said that "the industry feels strongly that any efforts by the World Health Organization and national governments to implement this action program should not interfere with existing private sector operations," thereby signaling the industry's concern that industrialized countries would use the concept of essential medicines to introduce limited lists of prioritized medicines. The Italian drug industry put it more crudely in response to the Italian Senate's attempts to introduce an essential medicines list: "If they want to turn Italy into a Third World country, this is the way to go about it."[5] According to the drug industry view, the Essential Medicines List should be a tool only for the public sector in the poorest nations of the world.

This view has not changed much in the last twenty-five years. A 2002 International Federation of the Pharmaceutical Manufacturers Association issue paper on the concept of essential medicines expressed the belief that policies that extend restrictive drug policies to industrialized countries pose a serious threat to the delivery of effective health care and pose a threat to investment in drug research—a standard, multipurpose argument used by the pharmaceutical industry to object to any policy they see as detrimental to their interests.

1   World Health Assembly, Resolution WHA28.66.
2   World Health Organization, *The Selection of Essential Drugs: Report of a WHO Expert Committee*, WHO Technical Report Series 615 (Geneva: World Health Organization, 1977).
3   Andrew Chetley, *A Healthy Business?: World Health and the Pharmaceutical Industry* (London: Zed Books, 1990), p. 75.
4   Najmi Kanji et al., *Drugs Policy in Developing Countries* (London: Zed Books, 1992), p. 30.
5   *Scrip*, no. 1 (1987).

improve the way medicines are regulated, measures to improve the way they move in international commerce, and measures to improve the way they are advertised and used.[3]

Six months later, in May 1986, the World Health Assembly adopted the WHO Revised Drug Strategy, which prescribed a series of actions to ensure the availability of affordable essential medicines and strategies to improve the use of these medicines ("the rational use of drugs"). The Revised Drug Strategy would become the organization's policy on medicine for decades to come.

Participants in the Nairobi conference discussed high drug prices, addressed the fact that pharmaceutical research and development of medicines with real therapeutic value was lacking, compared with R&D on medicines that were only trivial advances, and the need to stimulate local production of essential medicines in developing countries to decrease the dependence on Western companies. For these countries, local production was also seen as an industrial-development objective as such. But notably, the subject of pharmaceutical patenting was nowhere on the radar screen of this gathering of the world's premier health experts and advocates.

Ironically, while the Revised Drug Strategy was being formulated, pharmaceutical companies were establishing an elite, high-powered lobbying group that worked to include intellectual property in the agenda of the General Agreements on Tariffs and Trade (GATT) framework, the predecessor of the WTO.[4] It would take another ten years before health advocates would make the link between access to medicines and the GATT talks. Nevertheless, unknowingly, by expanding the WHO's role in medicine policies, the Nairobi conference laid the foundation for later activism on intellectual property issues and access to medicines.

TRADE CONCERNS EMERGE

Ten years after the historic Nairobi conference, the International Conference on National Medicinal Drug Policies was held in Sydney, Australia, in 1995.[5] Delegates discussed most components that were considered essential to a sound medicine policy: how to select essential medicines, what measures to take to increase access to medicines, how to encourage the correct prescription by physicians and the proper use by consumers of medicines, how to establish government regulations that ensure that medicines are effective, safe, and of good quality, and how to regulate the drug industry and its drug-promotion practices.

It was at this conference that for the first time public-health advocates raised the concern that the globalization of new international trade rules and the harmonization of regulatory requirements would restrict countries' ability to implement drug policies that would ensure access to medicine for all. These concerns came in particular from speakers from Latin America and Asia, who drew attention to the long-term consequences of the introduction of twenty-year product patents— as required by the WTO—in countries such as India and Argentina, which were home to extensive generic-drug industries. Tellingly, some responded by suggesting that the effects of patents could be countered with policies for the substitution of generic drugs. Of course, a generic version of a product is precisely what a patent forbids. However, promoting generic medicines was a staple of the WHO's medicine policy as articulated in the Revised Drug Strategy, because by encouraging competition between producers, the substitution of generics can result in a reduction in the price close to marginal cost levels, and generic medicines are on the whole a fraction of the price of brand-name medicines. Such contradictory comments made it apparent that even drug-policy experts at the time had a very limited understanding of the ramifications of new international rules on intellectual property.

Nevertheless, over the course of this meeting, delegates did come to recognize their need to get a better sense of the consequences of new trade rules. And the meeting recommended that the WHO and local governments investigate and address the effect of agreements such as TRIPS on national medicine policies and take action to ensure that health policies are primary where trade-related policies are formulated.[6]

The debate quickly took hold when, at the next annual meeting of the WHO's member states in 1996, health ministers debated for the first time the effects of new WTO trade rules on access to medicines.[7] This debate was long overdue, considering that the WTO agreements were negotiated without input from health experts and had already gone into effect.

ESSENTIAL MEDICINES AND INTELLECTUAL PROPERTY AT THE WHO: FROM RED BOOK TO BLUE BOOK

The public-health advocates coordinated by Health Action International first raised concerns about the consequences of globalization and international trade agreements for drug access during the 1996 World Health Assembly. They sought to get the WHO to intervene in intellectual property issues because it became apparent that the GATT negotiators had drawn up the rules without any consideration for health issues. The assembly debated a resolution on the Revised Drug Strategy.[8] As a result of Health Action International's intervention, the resolution included a request for the WHO to study and report on the impact of the work of the WTO with respect to national drug policies and essential drugs and make recommendations for collaboration between the WTO and the WHO. This was important, because it gave the WHO a mandate to develop work on the effects of the new WTO rules, which was a new terrain for the organization.

This resolution gave the WHO the mandate to publish, in 1998, the first guide with recommendations to member states on how to implement TRIPS while limiting the negative effects of higher levels of patent protection on drug availability.[9] The response to this guide from the United States and a number of European countries was swift and fiercely negative. In particular, the United States, working very closely with drug-company lobby groups, pressured the WHO to withdraw the publication, calling the book "an outrageous and biased attempt to mold international opinion."[10]

Initially, the WHO withdrew the publication—which, because of its red cover, became known as "the red book"—and reissued it with some minor changes and an annex containing presentations by different parties reflecting different views on the issue of pharmaceutical patenting, but this time with a blue cover. Dr. Gro Harlem Brundtland, who had just been elected director general of the WHO, did not yield to the pressures and instead organized a meeting at which she invited a number of parties to express their views. She also dealt with the criticisms that the book held inaccurate information by inviting external reviews. The reviewers found very little wrong with the report. While the U.S. action caused a delay in publication, it did not succeed in suppressing the report. But sadly, in future WHO work on trade and intellectual property issues, this would be different.

The WHO's involvement in trade and intellectual property issues would remain highly controversial in the years that followed. The simple emphasis that the WHO placed on public-health needs over trade interests was perceived as a threat to the commercial sector of the industrialized world. In particular, a greater role for the WHO in issues related to TRIPS created considerable concern within the pharmaceutical industry, which lobbied hard against it.

A draft resolution discussed at the 1998 WHO Executive Board, the governing body of the WHO responsible for preparing the annual World Health Assembly, called on the WHO member countries to ensure that public health, rather than commercial interests, would have primacy in pharmaceutical and health policies. The resolution further referred to TRIPS and asked the WHO director general to analyze the effects of new trade agreements on health and to develop measures to counter these effects.

In 1998, in response to this draft resolution on the Revised Drug Strategy and in reference to "considerable concern among the pharmaceutical industry," the position of the European director general for trade of the European Commission's position was "No priority should be given to health over intellectual property considerations."[11]

The WHO Executive Board established an ad-hoc group chaired by France to prepare for the discussions on the Revised Drug Strategy at the World Health Assembly in 1999. The issue of trade agreements with regard to intellectual property and access to medicines had been put on the agenda of the WHO and was there to stay. The Executive Board ad-hoc group organized a five-day meeting, including a one-day hearing with interested parties. It concluded its work with a proposed resolution that was sent to the Fifty-Second World Health Assembly.[12]

That was also the year during which NGOs increased their involvement in the trade and health debates. In anticipation of the 1999 Seattle WTO ministerial conference, there was a flurry of activity and networking that strengthened the base of knowledge that NGOs had about intellectual property and their ability to mobilize quickly in relation to the issue, fuelled by the "health primacy" debates at the World Health Assembly and against the backdrop of a South African court case in which a group of thirty-nine pharmaceutical companies had taken South Africa to court over its Medicines and Related Substances Control Act, claiming that some of its provisions that could be used to supply patients with cheap medicines were not compliant with WTO standards.[13] A key coalition of groups consisted of Health Action International, the Consumer Project on Technology (CPTech, now Knowledge Ecology International, KEI), Act Up–Paris, the Health GAP coalition, Oxfam, and the Access to Medicines campaign of Médecins Sans Frontières (MSF). These groups worked in close collaboration with national treatment-action groups in various countries, notably in Thailand, Brazil, India, and South Africa.

The resolution adopted by the World Health Assembly in 1999 strengthened the WHO's role in intellectual property issues.[14] The text no longer called for the "primacy of health over trade," but noted the importance of "ensuring that public health interests are paramount in pharmaceutical and health policies." This is certainly a departure from the coalition's initial intention, but it did put the health advocates at the table of trade negotiations, as the subsequent developments at the WTO TRIPS Council and the Doha WTO ministerial conference would show.[15] The resolution also urged countries to look into the options they have under current trade rules to safeguard access to essential medicines, a clear reference to the flexibilities available under the TRIPS Agreement, such as compulsory licensing, which allows governments to overcome patents and produce, import, export, and market generic versions of a patented drug. Most importantly, the assembly requested that the WHO assess the health implications of trade agreements, which was understood to mean the WTO TRIPS Agreement, with a view to assisting countries to mitigate the negative effects of this agreement. This was in response to countries' calls on the WHO for technical assistance in implementing TRIPS flexibilities. In summary, the

UPDATING AND DISSEMINATING THE WHO'S MODEL LIST OF ESSENTIAL DRUGS

In 2001, the WHO, prompted by groups such as Médecins Sans Frontières' campaign for access to essential medicine and various academics,[1] embarked on a process to change the way new medicines were included in the WHO Model List of Essential Drugs. Over time, the requirement that an essential medicine be affordable had become a barrier to the inclusion of newer medicines. New medicines that are widely patent protected tend to be available at monopoly prices only because of the lack of generic competitors. In practice, this meant that the WHO was reluctant to label such products as "essential," because governments would not be able to afford them, and purchase of expensive medicines would be to the detriment of the treatment of other diseases. As a result, antiretroviral medicines needed to treat people living with AIDS were not on the list. In a world where over forty million people are infected with HIV and eight thousand die from AIDS every single day, maintaining the position that proven-effective antiretroviral medicines are "not essential" became absurd and risked making the WHO Essential Drugs List irrelevant.

By questioning the affordability criteria for including a medicine on the list, NGOs and health experts made two points: First, the primary criteria for defining an essential medicine should be the medical need for that product, and second, once a product is labeled "essential," it should be affordable and available to the individuals and communities that need access to it.

The 2002 definition of essential medicines changed. It stressed the need for essential medicines to be available at a price the individual and the community can afford.[2] The new definition implied that governments have an obligation to assure the availability and affordability of these products. And a high price was no longer a barrier for inclusion in the list. With this measure, the WHO anticipated that the Essential Drugs List, on which national essential drugs lists are based, could become a useful tool for selecting candidate drugs for compulsory licensing or other cost-containment measures.

---

1   Pierre Chirac and Richard Laing, "Updating the WHO Essential Drugs List," *Lancet* 357, no. 9262 (April 7, 2001): p. 1134.

2   The 2002 definition of essential medicines reads: "Essential medicines are those that satisfy the priority health care needs of the population. They are selected with due regard to public health relevance, evidence on efficacy and safety, and comparative cost-effectiveness. Essential medicines are intended to be available within the context of functioning health systems at all times in adequate amounts, in the appropriate dosage forms, with assured quality and adequate information, and at a price the individual and the community can afford. The implementation of the concept of essential medicines is intended to be flexible and adaptable to many different situations; exactly which medicines are regarded as essential remains a national responsibility." World Health Organization, "Essential Medicines," available on-line at http://www.who.int/topics/essential_medicines/en (last accessed February 26, 2010).

MILLIONS OF BABIES WON'T LIVE TO SEE THIS DAY.

Without treatment, half of all children with HIV/AIDS in developing countries will die before their second birthday. We desperately need diagnostic tests that work for babies, and pills that kids can swallow. HIV/AIDS is treatable, but millions of children are still waiting.

www.accessmed-msf.org

Médecins Sans Frontières Campaign for Access to Essential Medicines poster calling attention to the fact that patent-based companies neglect the needs of people in developing countries (Médecins Sans Frontières).

resolution amended the Revised Drug Strategy to enable the WHO to start work in an area previously the exclusive domain of trade negotiators and intellectual property lawyers.

Subsequent resolutions of the World Health Assemblies have further strengthened the WHO's mandate in the trade arena. In May 2001, the World Health Assembly adopted two resolutions in particular that had a bearing on the debate over TRIPS.[16] The resolutions addressed the need to strengthen policies to increase the availability of generic drugs and the need to evaluate the impact of TRIPS on access to drugs, local manufacturing capacity, and the development of new drugs. Each time, the adoption of these resolutions required massive mobilization by civil-society groups. The coalition of NGOs mentioned above had gained strength. Since the 1999 WTO Seattle conference, it was active on two fronts in Geneva: the WTO TRIPS Council and the WHO itself. The South African court case in which thirty-nine multinational drug companies sued South Africa over the access provisions in its Medicines Act helped to advertise globally the need to push back the commercial lobby in favor of a more health-oriented international trade agenda.

ESSENTIAL MEDICINES AND COMPULSORY LICENSING

According to the WHO Revised Drug Strategy, essential medicines should be available at a price that the individual and the community can afford. Before TRIPS, some developing countries assured cost containment by excluding medicines or "essential medicines" from patentability. For example, until 1991, only manufacturing processes for the preparation of medicines were patentable in the countries of the Andean Community—not the medicines themselves. Following the introduction of pharmaceutical product patents in 1991, the Andean Community adopted a declaration that provided that "inventions related to pharmaceutical products included in the WHO Model List of Essential Drugs"—the Essential Medicines List—should not be patentable.[17] This measure was taken to prevent abusive pricing of essential medicines that could result from the new patent rules.

Venezuela, with support from the Andean group and other developing countries, in particular South Africa, proposed at the Third Ministerial Conference of the WTO in Seattle in 1999 to amend TRIPS to create a new exception to patentability for medicines on the WHO Essential Medicines List.[18] A counterproposal led by the European Community was "to issue . . . compulsory licenses for drugs appearing on the list of essential drugs of the World Health Organization."[19] At that time, only about 11 of the 306 products on the WHO Model List of Essential Drugs were patented in certain countries.[20] The adoption of the EC proposal would have seriously limited the scope of compulsory licensing, because TRIPS does not limit such licensing to particular circumstances, as the EC proposal would have done. The Seattle WTO ministerial conference collapsed and never reached a conclusion. Nevertheless, since then, the effect of the globalization of patent rules on access to essential medicines has been on the agenda not only of the WHO, but of numerous trade and health forums, ultimately leading to the adoption of the declaration on TRIPS and public health at the Fourth Ministerial Conference of the WTO that took place in Doha, Qatar. The WTO Doha Declaration established the primacy of health over commercial interests after all.

CONCLUSION

Since 2001, as a result of the strengthened Revised Drug Strategy, the WHO's work program on pharmaceuticals and trade now includes the provision of policy guidance and information on intellectual property and health to countries for monitoring and analyzing the effects of TRIPS on access to medicines.[21] However, until today, the WHO leadership has been overly cautious in fulfilling this mandate. For example, the publication of guidance to countries about how to deal with pharmaceutical patents in case of access barriers remains problematic, and only one WHO staff member is working on intellectual property and medicines. The WHO director-general, Dr. Margaret Chan was initially critical of Thailand's decision in 2006 to issue compulsory licenses for three drugs on the national essential drugs list,[22] despite the fact that this is a decision Thailand can lawfully make under international and Thai law. She urged the Thai government to enter into negotiations with pharmaceutical companies, a line that was being pushed by the United States and not required by law.[23] She reversed her position after heavy criticism from developing countries, AIDS groups, and NGOs.[24]

Since then, Thailand has asked the WHO for technical assistance. The Thailand compulsory license case is illustrative of the failure of the WHO to this day to provide both technical and political support to the use of the TRIPS flexibilities. This is all the more alarming since numerous World Health Assembly resolutions have asked the WHO director general to step up work in this field, and this work has been formally part of the WHO medicine strategy since the 1999 revisions of the Revised Drug Strategy.

The story of the Revised Drug Strategy shows that even with all the right resolutions on the books, in the end, moving the health agenda forward also requires leadership and political courage.

NOTES

1   Andrew Chetley, *A Healthy Business?: World Health and the Pharmaceutical Industry* (London: Zed Books, 1990), pp. 94–106.

2   See the HAI Web site at http://www.haiweb.org/index.html (last accessed March 16, 2009).

3   WHO, *The Rational Use of Drugs: Report of the Conference of Experts*, Nairobi, November 26–29, 1985. (Geneva: World Health Organization, 1987).

4   See Peter Drahos's essay, "'IP World'—Made by TNC Inc.," in this volume.

5   "International Conference on National Medicinal Drug Policies—The Way Forward," *Australian Prescriber* 20, supplement 1 (1997).

6   The meeting recommended: "Global (for example, WHO), regional and country efforts should be made to analyze and address the consequences of international harmonization, macroeconomic changes, structural adjustment and international trade agreements (General Agreement on Tariffs and Trade/World Health Organization, Agreement on Trade Related Aspects of Intellectual Property Rights) on access, rational use of drugs, quality, safety and efficacy, local industrial development and other aspects of the national medicinal drug policy. Health issues should be considered as the policies are being formulated." *Ibid.*

7   In the words of Jonathan Quick, the director of WHO's Action Program for Essential Drugs,

"The conference provided one of the sparks that eventually led to a lively and frank debate on affordability, quality, the impact of world trade agreements and other issues related to the WHO's RDS at the 49th WHA in May 1996." *Ibid.*

8   World Health Organization, "Revised Drug Strategy Resolution," World Health Assembly Resolution WHA 49.14 (1996).

9   German Velasquez and Pascale Boulet, *Globalization and Access to Drugs: Perspectives on the WTO/TRIPS Agreement*, 2nd. ed. (Geneva: WHO, 1999).

10 Cable from the U.S. Mission in Geneva to the U.S. Department of State, May 27, 1998, available on-line at http://www.cptech.org/ip/health/who/confidential.rtf (last accessed March 17, 2009).

11 European Commission (DG1), "Note on the WHO's Revised Drug Strategy," doc. no. 1/D/3/BW D (98) (October 5, 1998), available on-line at www.cptech.org/ip/health/who/eurds98.html (last accessed March 17, 2009).

12 Executive Board, 103rd session, "Revised Drug Strategy. Report by the Chairman of the Ad Hoc Working Group." EB103/4, November 25, 1998, available on-line at http://www.cptech.org/ip/health/who/rds-report.html (last accessed March 17, 2009).

13 The case generated an intense mobilization in favor of patients and against the industry, both nationally and internationally. It appeared as a striking example of the companies' greed and extremist attitude. On April 19, 2001, as they seemed close to losing the lawsuit, the firms dropped it.

14 World Health Organization, Fifty-Second World Health Assembly, "Second Report of Committee A (Draft)," May 24, 1999, WHA A52/38, available on-line at http://ftp.who.int/gb/archive/pdf_files/WHA52/ew38.pdf (last accessed March 17, 2009).

15 See Sangeeta Shashikant's essay, "The Doha Declaration on TRIPS and Public Health: An Impetus for Access to Medicines," in this volume.

16 World Health Organization, "Scaling Up the Response to HIV/AIDS," World Health Assembly Resolution WHA 54.10 (2001); World Health Organization, "WHO Medicines Strategy," World Health Assembly Resolution WHA 54.11 (2001), both available on-line at http://www.cptech.org/ip/health/who/wha.html (last accessed March 18, 2009).

17 See Article 7(e) of Decision 344, Common Regime on Industrial Property, available on-line at http://www.sice.oas.org/trade/JUNAC/decisiones/DEC344e.asp (last accessed March 17, 2009).

18 World Trade Organization, "Preparations for the 1999 Ministerial Conference. Proposals Regarding the TRIPS Agreement...Communication from Venezuela, "August 6, 1999, available on-line at http://www.wtocenter.org.tw/SmartKMS/fileviewer?id=62714 (last accessed March 17, 2009).

19 Common Working Paper of the EC, Hungary, Japan, Korea, Switzerland, and Turkey to the Seattle Ministerial Declaration 3 (November 29, 1999), available on-line at http://europa.eu/rapid/pressReleasesAction.do?reference=DOC/99/15&format=PDF&aged=1&language=EN&guiLanguage=en (last accessed February 25, 2010).

20 Michael Scholtz, "Views and Perspectives on Compulsory Licensing," Conference on AIDS and Essential Medicines and Compulsory Licensing, Geneva, March 27–29, 1999, available on-line at http://www.haiweb.org/campaign/cl/scholtz.html (last accessed March 18, 2009).

21 World Health Organization, "Technical Cooperation Activities: Information from Other Intergovernmental Organizations," WHO doc. no. IP/C/W/305/Add.3 (September 25, 2001).

22 See Jiraporn Limpananont and Kannikar Kijtiwatchakul's essay, "TRIPS Flexibilities in Thailand: Between Law and Politics," in this volume.

23 Apiradee Treerutkuarkul, "WHO Raps Compulsory Licensing Plan: Govt. Urged to Seek Talks with Drug Firms," *Bangkok Post*, February 2, 2007, available on-line at http://www.aegis.com/news/bp/2007/bp070201.html (last accessed March 18, 2009).

24 Letter sent by Dr. Margaret Chan, director general of the World Health Organization, to Dr. Mongkol Na Songkhla, the Thailand minister of public health, February 7, 2007, available on-line at http://lists.essential.org/pipermail/ip-health/2007-February/010538.html (last accessed March 18, 2009); Piyaporn Wongruang, "Move to Break Drug Patents Lauded—Experts: WHO Should Back Thai Intentions," *Bangkok Post*, February 3, 2007, available on-line at http://www.aegis.org/news/bp/2007/BP070203.html (last accessed March 18, 2009); Paul Cawthorne, et al., "WHO Must Defend Patients' Interests, Not Industry," *Lancet* 369, no. 9566 (March 24, 2007): pp. 974–75, available on-line at http://www.thelancet.com/journals/lancet/issue/current?tab=past (last accessed March 19, 2009).

The Doha Declaration on TRIPS and Public Health: An Impetus for Access to Medicines

Sangeeta Shashikant

Today approximately two billion people worldwide—one-third of the world's population—do not have access to the essential medicines they need. In some of the lowest-income countries in Africa and Asia, this figure rises to more than half of the population.

These statistics reveal that despite the significant technological advances made by humankind in the medical field, getting medicines to those who need them remains a major challenge for the international community. Access to essential medicines, a fundamental element of the universal human right to health, depends on several factors, such as prices, rational medicine-selection processes, sustainable financing, and reliable health-care and supply systems.[1]

However, the price factor can be determinative all by itself, and price is literally a matter of life or death when a deadly disease is treatable. It also can determine whether the government will be able to provide treatment to its people or whether an individual will be able to obtain the treatment that he or she requires. The problem of high prices has been observed by the international community in the context of treatable infectious diseases such as HIV/AIDS and malaria. For example, in 2000, for a triple-combination antiretroviral treatment of stavudine (d4T) plus lamivudine (3TC) plus nevirapine (NVP), the price of the lowest-priced branded treatment was about $10,439 for a year's supply.[2] The high price tag meant patients living with HIV/AIDS would not be able to afford treatment and would be condemned to death. However, the availability of generic versions of branded medicines led to significant price reductions. In 2001, Cipla Ltd., a generic producer based in India, offered the same combination for $350. Over time, with more competition, this cost has been reduced to $99.[3] Reduced prices for antiretroviral treatment have been a crucial factor in the scaling up of HIV/AIDS treatment.

As can be seen from the example of HIV/AIDS, competition among multiple manufacturers is essentially the reason for reduced prices. However, the existence of competition has very much been threatened since the coming into force of the Trade-Related Aspects of Intellectual Property Rights (TRIPS) Agreement of the World Trade Organization (WTO) in 1995. TRIPS for the first time set out minimum standards and requirements for the protection of intellectual property rights—for example, trademarks, copyrights, and patents. It obliges all WTO members to adopt and to enforce high standards of intellectual property protection derived from the standards used in developed countries, except where provision for a transition period that delays the implementation of the agreement is made.[4]

Many development experts are of the view that TRIPS has very significantly tilted the balance in favor of the holders of intellectual property rights, most of whom are in developed countries, vis-à-vis consumers and local producers in developing countries and vis-à-vis development interests.[5] The minimum twenty-year patent protection required by TRIPS allows a pharmaceutical company monopoly over the production, marketing, and pricing of patent-protected medicines. This period can be further extended by the company through the use of various strategies, such as applying for patents on usage, dosage, or combinations of drugs —a practice commonly known as "evergreening,"[6] thus keeping the drug free from competition and enabling high pricing. TRIPS further mandates that patents have to be given for both products and processes in all fields of technology.[7] Whereas previously, many developing countries excluded crucial sectors such as medicines and chemicals from patentability, this is no longer an option. By virtue of TRIPS protection, no generic equivalent can come into the market until the twenty years of patent protection have expired unless TRIPS flexibilities—measures such as compulsory licensing or parallel importation of drugs, exceptions to patent rights, exclusions from patentability, and transition periods—are used,[8] thus denying patients cheaper alternatives.

While the situation was problematic prior to 2005, it is anticipated that it will worsen in the years to come. Médecins Sans Frontières (MSF) is already talking about the "return of the price crisis" that was seen in 2000, when life-saving anti-retrovirals were priced out of reach of those in need. For example, introducing more recent drugs in anti-AIDS combination therapy because of the emergence of resistance to older treatment would today increase the annual cost of treating an adult for one year in a developing country from $99 to $426. Since everyone in therapy today is expected to need these newer therapies at some point, the escalation in cost will have dire consequences for AIDS programs.

The main reason why cheaper generic alternatives were possible for older antiretroviral products is that there were no patents in some developing countries with vibrant generic pharmaceutical industries. India, for example, was free from product patents for medicines used to manufacture and supply generic medicines to the rest of the world. However, beginning in 2005, India, known as the pharmacy of the world, has had to comply with its TRIPS obligations and permit the patenting of pharmaceutical products. Therefore, the possibility of supplying affordable generic medicines in the future for new drugs seems rather bleak.[9]

Such concerns about TRIPS and its overall impact on access to affordable medicines sparked an international debate. Public-health crises afflicting many countries in the developing world, particularly in sub-Saharan Africa, and the strong-arming of developing countries by developed countries fuelled the debate, focusing intense public attention on the manner in which intellectual property protection affected people's lives and governments' ability to take measures to protect public health.

THE DOHA NEGOTIATIONS

The TRIPS Agreement is the result of a process of intense negotiations during the Uruguay Round of the General Agreement on Tariffs and Trade. It thus reflects an uneasy, delicate compromise between the developed countries, which sought high levels of intellectual property protection, and the developing countries, which sought to ensure that a degree of flexibility or policy autonomy was retained in interpreting and implementing TRIPS. Initially, the developing countries resisted inclusion of an agreement on intellectual property protection as part of the WTO agreements, but later accepted it in exchange for gains they hoped to obtain in other areas, such as agriculture.[10]

Thus, during implementation of the agreement, the differing socioeconomic and political interests of the WTO members resulted in differing interpretations of certain provisions in the TRIPS Agreement, leading to tensions between the developing countries, which wished to make use of flexibilities such as compulsory licensing, parallel importation,[11] and so on for purposes of improving access to affordable medicines, and the major developed countries, as well as their pharmaceutical industries, which did not wish to see the developing countries exercise their rights.

As a result of these tensions, in February 1998, the South Africa Pharmaceutical Manufacturers Association and thirty-nine other pharmaceutical manufacturers, mostly multinational, brought a lawsuit against the South African government for allegedly violating the TRIPS Agreement and the South African Constitution. The government had introduced an amendment to its 1997 Medicines and Related Substances Control Act to include provisions such as the substitution of generics for out-of-patent medicines, as well as transparent pricing and parallel importation.

Activists worldwide, led by the South African Treatment Action Campaign, an AIDS activist organization, rallied in support of the South African government. To protest against the lawsuit, these activists held rallies in different key cities, sent letters to the plaintiffs in the South African lawsuit and other influential officials, made joint statements, and held press conferences condemning the industry's attempt to derail implementation of the Medicines Act and demanding that the companies withdraw the lawsuit.

Various other groups also mounted direct-action campaigns against the companies. For example, activists from ACT UP New York, ACT UP Philadelphia, and the Health GAP (Global Access Project) Coalition occupied GlaxoSmithKline's investor-relations office in Manhattan, using chains to lock down the office.[12] GlaxoSmith-Kline was a lead plaintiff in the lawsuit. The suit soon became a public-relations nightmare for the companies, and they finally withdrew it in 2001.[13] Public pressure also forced developed-country governments such as the United States and those in Europe that were initially supportive of the industry's action to withdraw their support.

In one instance, strategically savvy AIDS activists disrupted Vice President Al Gore's presidential campaign with a series of protests over his support for the U.S. policy of pressuring countries such as South Africa not to use TRIPS flexibilities. Demonstrations organized by ACT UP and the national coalition AIDS Drugs for Africa saw activists waving banners dubbing the Gore 2000 campaign "Apartheid 2000" and declaring "Gore's Greed Kills."[14] "On one occasion," Karine Cunqueiro notes, "demonstrators displayed a life-size marionette of Gore, the strings of which were manipulated by effigies of drug-company executives."[15] These actions placed the issue in national media and on the national political scene, eventually leading to the U.S. government's withdrawal of its support for the lawsuit.[16]

In 2001, the United States initiated a complaint against Brazil in the WTO dispute-settlement system over Brazil's national law on compulsory licensing, which included a "local working" requirement. Under that provision, holders of patent rights in Brazil are required to manufacture the protected product in the country. If companies do not follow this requirement, after three years, Brazil can issue a compulsory license. The United States argued that the law violated the TRIPS Agreement by discriminating against U.S. patent owners and restricting patent holders' rights. Brazil responded that the law was consistent with the provisions and spirit of TRIPS, as well as with the Paris Convention for the Protection of Industrial Property.[17]

The complaint by the United States to the WTO was seen as a "warning shot" by the Bush Administration to the developing countries that had hopes of using the flexibilities provided by the TRIPS Agreement and South-South cooperation to develop local pharmaceutical production capabilities and to break their dependence on multinational pharmaceutical companies.[18] The U.S. actions brought fierce pressure from the international NGO community concerned about the negative effect of the complaint on Brazil's successful AIDS program and on South-South cooperation to ensure a sustainable supply of generic medicines.[19] MSF issued an international press release warning that the U.S. action at the WTO not only threatened Brazilian AIDS policy, but would "also intimidate countries which would like to take up Brazil's offer to help them produce AIDS medicines."[20] The Treatment Action Campaign also issued a statement denouncing the WTO complaint as an attempt "to destroy Brazil's generic pharmaceutical industry," charging that "it will not only hamper access to medicines for Brazil"s 500,000 people with HIV, but also many Third World countries which are hoping to import Brazil's cheap medicines and to accept Brazil's offer of knowledge transfer."[21] The US withdrew the complaint in June 2001.

These events are only two of the more prominent manifestations of the conflicts arising from differing interpretations of the TRIPS provisions and from political and economic pressure asserted by the United States and other developed countries against developing countries to change policies in favor of their pharmaceutical industries.[22] The conflicts and the vocal voice of the international NGO community highlighted the importance of reaching a common understanding about TRIPS and WTO members' right to take measures to promote public health. Pronouncements on the issue of trade and health by international organizations such as the WHO, the UN Sub-Commission for the Protection and Promotion of Human Rights, and the United Nations Development Program also added impetus to the movement for access to affordable medicines.[23]

On June 20, 2001, for the first time ever, the WTO Council for TRIPS held a special session on TRIPS and public health.[24] This historic meeting was a response to the Africa Group's call at the TRIPS Council to confront the problem of access to medicines due to high prices resulting from intellectual property protection and to discuss the interpretation and application of the relevant provisions of the TRIPS Agreement with a view to clarifying flexibilities by which members are entitled to gain access to medicines. Fifty developing countries put forward a joint paper presenting their common legal understanding of some of the TRIPS Agreement's key provisions (that is, its objectives, principles, nature, and scope), and of the agreement's requirements for the protection of undisclosed information and patent flexibilities such as compulsory licensing and parallel importation.[25] Such an initiative was a key move in a context in which, on a regular basis through the press and by other means, the developed countries and pharmaceutical companies were misrepresenting TRIPS flexibilities as much narrower than they were.

Zimbabwe, on behalf of the Africa Group, proposed that the Doha Ministerial Conference to be convened later in the same year issue a special declaration to affirm a common understanding that the TRIPS Agreement does not prevent members from taking measures to protect public health, adding that "this assurance and guarantee was needed to enable governments to adopt measures to protect public health, without fear of litigation, at national level or at the WTO, or bilateral pressures being applied on them."[26] NGO activities at national, regional, and international levels heightened the urgency of the need to heed the call of the developing countries. For example, on the eve of the special session, more than one hundred NGOs, led by MSF, Oxfam, and the Third World Network, called for a "pro-public health" interpretation of the TRIPS agreement and the use of TRIPS safeguards and exceptions. Some civil-society groups in the Global South went as far as to request that the TRIPS Agreement be taken out of the WTO.[27]

The reaction of the developed countries at the first special session to the stance taken by the developing countries was mixed, varying from acceptance to plain opposition. Norway was perhaps the most supportive of the developing countries' positions. On the links between patents, price, and access to medicines, it recognized that the price of medicines "does make a difference," especially in the case of poor people in developing countries who have to pay out of pocket for health care. It also agreed on the need for more legal clarity on the TRIPS provisions. The US took a hard-line position that strong patent regimes can produce benefits for developed and developing countries and refused to acknowledge the concerns of developing countries over TRIPS implementation and access to affordable medicines.[28] It also challenged proposals put forward by the developing countries. A U.S. representative is reported to have said that "as long as you cannot come up with concrete examples, we remain unconvinced of the problem."[29] The European Commission agreed on a number of points put forward by the developing countries. However its position was received with much skepticism by many developing countries and NGOs, because there was concern that the issue of TRIPS and public health would be used as part of a negotiating strategy to be traded against other issues during the Doha Ministerial Conference. Overall one clear message of the industrialized countries was that they would not agree to any diminution in the TRIPS standard of intellectual property protection.[30]

Despite the hard-line positions taken by some developed countries, determined developing countries with support from NGOs persisted jointly in advocating for a favorable outcome. At a meeting in September, the Africa Group, with nineteen other developing countries, presented a draft text for a ministerial declaration on TRIPS and public health. It proposed political principles that would ensure that TRIPS would not undermine the legitimate right of WTO members to formulate their own public-health policies and provided clarifications for provisions related to compulsory licensing, parallel importation, protection of undisclosed information, and production for export to a country with insufficient production capacity.[31]

The United States, Japan, Switzerland, Australia, and Canada circulated an alternate draft, stressing the importance of intellectual property protection for research and development and arguing that intellectual property contributes to public-health objectives globally. It further sought to limit the use of flexibilities to crisis and emergency situations.

The different proposed texts became the basis of engagement between a key group of some twenty delegations from developing and developed countries, but with little result. The parties repeatedly arrived at a deadlock. Major industrialized nations blocked language that would declare that "nothing in the TRIPS shall prevent Members from taking measures to protect public health" and instead insisted on formulations that would restrict flexibilities available to the developing countries.[32] The deadlock continued into the Doha Ministerial Conference, because the developing countries refused to be fobbed off with a declaration that had no value added and that in fact sought to restrict or to reduce the currently available flexibilities.[33]

During the Doha preparatory meetings, the United States and others (often sounded out for compromise by the WTO Secretariat), proposed language to the effect that the declaration would be "without prejudice to the rights" or would "preserve the rights" or should not be construed as "adding to or diminishing the rights" of the developed countries under the TRIPS Agreement for fear that the declaration could lead to changes in TRIPS. Key developing-country negotiators felt that if these terms were accepted, it would make "nonsense" of the declaration.[34] In the final agreed-upon text of the declaration, none of these elements are included. From the beginning, the aim of the developing countries was to obtain recognition that nothing in the TRIPS Agreement should be interpreted as preventing members from adopting measures necessary to protect public health.[35] They were frustrated by the opposition and pressure exerted on the developing countries by the pharmaceutical industry of the developed countries, backed by their governments.

NGOs such as MSF, Oxfam, and the Third World Network also kept pressure on the industrialized countries, charging them with echoing the views of the pharmaceutical companies and with frustrating developing-country efforts at the WTO to improve access to medicines in poor countries.[36]

The chair of the WTO General Council, Stuart Harbinson, presented a text with two options, which became the basis for Doha negotiations. The first option:

U.S. hypocrisy: willing to override patent for anthrax in 2001

> Nothing in the TRIPS Agreement shall prevent Members from taking measures to protect public health. Accordingly, while reiterating our commitment to the TRIPS Agreement, we affirm that the Agreement shall be interpreted and implemented in a manner supportive of WTO Members' right to protect public health and, in particular, to ensure access to medicines for all.
>
> In this connection, we reaffirm the right of WTO Members to use, to the full, the provisions in the TRIPS Agreement which provide flexibility for this purpose.

The second option:

> We affirm a Member's ability to use, to the full, the provisions in the TRIPS Agreement which provide flexibility to address public health crises such as HIV/AIDS and other pandemics, and to that end, that a Member is able to take measures necessary to address these public health crises, in particular to secure affordable access to medicines. Further, we agree that this Declaration does not add to or diminish the rights and obligations of Members provided in the TRIPS Agreement. With a view to facilitating the use of this flexibility by providing greater certainty, we agree on the following clarifications.

The first option was widely supported by the developing countries and public-interest civil-society groups. The United States, Switzerland, Japan, Australia, Canada, Korea, and some other developed countries supported the second option, because they viewed the first one as attempting to override the TRIPS rules.[37]

The second option, while seemingly meeting public-health concerns, reduces the rights of member countries to take actions on grounds of public health by narrowing those rights only to situations of "pandemics," which health specialists describe as diseases that are universal or affect populations across countries and continents. However, the position of the Western countries was hard to maintain for very long in light of the fact that prior to the Doha Ministerial Conference, Canada and United States had threatened to override Bayer AG's patent on the antibiotic ciprofloxacine (Cipro™) to deal with the shortage and high price of the product following letter-born anthrax attacks in 2001.[38] Brazil, India, and the Africa Group used the occasion to argue that they should be allowed the same discretion when it came to patented drugs for AIDS and other diseases. Health activists took the opportunity to highlight the hypocrisy and double standards of industrialized countries.[39]

After days of negotiation in Doha, members settled on a compromise text, which now forms paragraph 4 of the Doha Declaration:[40]

> We agree that the TRIPS Agreement does not and should not prevent Members from taking measures to protect public health. Accordingly, while reiterating our

commitment to the TRIPS Agreement, we affirm that the Agreement can and should be interpreted and implemented in a manner supportive of WTO Members' right to protect public health and, in particular, to promote access to medicines for all.

In this connection, we reaffirm the right of WTO members to use, to the full, the provisions in the TRIPS Agreement, which provide flexibility for this purpose.

The second part of the paragraph confirms one of the key points pushed by the developing countries: that in implementing the TRIPS Agreement at the national level, there is flexibility, and thus room to maneuver, to meet public health needs.

In the context of paragraph 4, the declaration then goes on to reaffirm countries' right to grant compulsory licenses, as well as "the freedom to determine the grounds upon which such licenses are granted," including "the right to determine what constitutes a national emergency or other circumstances of extreme urgency," both of which are grounds for issuing compulsory licenses. It adds, "public health crises, including those relating to HIV/AIDS, tuberculosis, malaria and other epidemics, can represent a national emergency or other circumstances of extreme urgency." It also reaffirms the right of members freely to establish their regimes for defining when the rights of a holder of intellectual property are exhausted.[41]

An issue that was not resolved in Doha, but that the Doha Declaration acknowledges, is the problem of WTO members with insufficient or no manufacturing capacities in the pharmaceutical sector facing difficulties in making effective use of compulsory licensing under the TRIPS Agreement.[42] The problem, which has since come to be known as the "paragraph 6 problem," is that most developing countries have inadequate or no manufacturing capacity, and those that do, once they implemented the TRIPS Agreement in 2005 (due to the expiration of the transition period) and thus allowed the patenting of pharmaceuticals, would not be able to meet the needs of other countries. This was because of a condition in the TRIPS Agreement that when a compulsory license is issued, the license shall be predominantly for domestic supply,[43] thus restricting the amount that may be exported.

Discussion on the solution to the paragraph 6 problem was the subject of heated debates in the WTO between 2001 and 2005.[44] Although the matter was eventually resolved in 2003 through a WTO decision of August 30, 2003,[45] which later, in 2005,[46] was accepted by WTO members as an amendment to the TRIPS Agreement, the solution agreed to by member states has been criticized severely by public-health groups for being burdensome to both exporting and importing countries.[47]

A major achievement, however was the agreement not to limit the declaration to a list of diseases. The final text recognizes "the gravity of the public health problems afflicting many developing and least-developed countries, especially those resulting from HIV/AIDS, tuberculosis, malaria and other epidemics."[48]

There were other victories for the developing countries in the declaration, as well. It extended the transition period for least-developed countries, so that they did not have to implement provisions on patent protection for pharmaceuticals and protection of undisclosed information until 2016, without prejudice to the right to seek other extensions.[49] Initially, the deadline for the transition period for least-developed countries was 2006. It also recognized concerns about the effects of intellectual property rights on prices, although on the insistence of developed countries, a statement about the importance of the intellectual property system prefaces the acknowledgement.

THE ROLE OF NGOS IN THE ACCESS DEBATE

It is undeniable that campaigning by the NGO community contributed significantly to greater awareness and heightened discussion about TRIPS and its effects on access to affordable medicines and to the Doha outcome on TRIPS and public health. In fact, one factor that led to the developing countries forming a coalition and making demands in the WTO was NGO activism and lobbying, as well as the media publicity surrounding the issue of access.

NGOs raised awareness internationally about high drug prices, about the reduced availability of quality generic alternatives, about inadequate research and development into tropical diseases, about bilateral pressures on the developing countries to adopt patent protection that would exceed the TRIPS requirements, and about the double standards practiced by the developed countries, as well as the bullying tactics of the pharmaceutical industry and several developed countries. They drew attention to TRIPS provisions that could be used to increase access, debunked myths put forward by the pharmaceutical industry and the developed countries, and shamed individuals, entities, and even countries that stood in the way of better access to affordable medicines for people living in developing countries.

In March 1999, in Geneva, NGOs (the Consumer Project on Technology, Health Action International, and MSF) organized the first international meeting specifically on the use of compulsory licensing to increase access to AIDS medicines.[50] Later, in November, the same group organized Increasing Access to Essential Drugs in a Globalized Economy, a conference in Amsterdam that brought together 350 participants from fifty countries on the eve of the Seattle WTO Ministerial Conference. Another significant event was the Oxfam workshop on TRIPS in Brussels in March 2001, which was attended by NGOs, experts, and diplomats.

These meetings are a part of many other important collaborative initiatives by national and international advocates from the Global North and South, such as

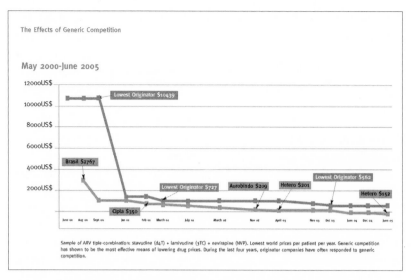

The Effects of Generic Competition

May 2000-June 2005

Sample of ARV tiple-combination: stavudine (d4T) + lamivudine (3TC) + nevirapine (NVP). Lowest world prices per patient per year. Generic competition has shown to be the most effective means of lowering drug prices. During the last four years, originator companies have often responded to generic competition.

This graph shows the impact of the introduction of generic medicines on drug prices (Médecins Sans Frontières, "Untangling the web of price reductions").

the Consumer Project on Technology, MSF, Oxfam, Act Up–Paris, Health GAP, the South African Treatment Action Campaign, the Third World Network, and others to raise awareness about concerns regarding TRIPS and to call for urgent action by WTO members. These meetings created awareness among NGOs and diplomats from Geneva, mobilized a variety of NGOs around the issue of intellectual property and access to medicines, and fostered a common NGO front.

Information disseminated by NGOs helped counter claims by the pharmaceutical industry and by industry-funded entities, as well as by developed-country governments; provided concrete arguments, as well as examples, concerning the threat posed by patents to access to affordable medicines, and raised awareness of the available flexibilities in the TRIPS Agreement, which proved invaluable, particularly in launching the access issue in the WTO and in the run-up to the Doha Declaration.

MSF issued a report on market and public-policy failures leading to research-and-development spending by the pharmaceutical industry on rich-country lifestyle preoccupations (obesity, impotence, etc.) or me-too drugs (medicines with very slight difference over existing compounds and involving no 'invention' or clinical advance) and a virtually empty pipeline of drugs for neglected diseases, thus countering claims that intellectual property protection encourages innovation and research into diseases and availability of medicines.[51]

Oxfam provided a briefing paper showing how the TRIPS rules would raise the costs of vital medicines, with potentially disastrous implications for poor countries.[52]

NGOs also argued that the majority of important HIV/AIDS drugs were actually developed by the public National Institutes of Health and funded by taxpayers' dollars,[53] and by way of concrete examples, they showed the threat of patents to access to affordable medicines,[54] rebutting claims that companies spend $500 to $800 million to develop a drug and that patents are not an important barrier to accessing antiretroviral drugs in African countries.[55]

To keep up the pressure on the developed countries that were taking a hard position during Doha negotiations, NGOs issued sign-on statements supporting developing-country positions, mobilized media to investigate the issue of patents and access and to write about it, and "named and shamed" countries, individuals, companies and any other entity that stood in the way of access to medicines. Through actions such as phone calls, e-mails, and faxes, NGOs also repeatedly placed pressure on the WTO Secretariat and its then director general, Mike Moore, to acknowledge the right of developing countries to make use of the TRIPS flexibilities. NGO actions grabbed media attention and created waves that Southern and Northern governments simply could not ignore. Developed countries learned that they could not get away with exerting trade pressures on developing countries or taking unfavorable positions without feeling repercussions, particularly back in their home countries.[56] The campaigns also had some effect in persuading developed countries such as the members of the European Commission to rethink the proindustry stance they had taken.[57]

The developing countries also relied substantially on the input and expertise of some civil-society groups in the formulation of papers and arguments for discussion and the drafting of texts during negotiations. The influence of documents prepared by civil-society groups led the United States to remark during the special session that members should "avoid documents circulated by other individuals and organisations that lack the WTO's expertise."[58] Without a doubt, close collaboration between civil-society groups and developing countries before and during the Doha Ministerial Conference was a pivotal factor in the success achieved there.

CONCLUSION: THE SIGNIFICANCE OF THE DOHA DECLARATION

The Doha Declaration represents a major political victory for the developing countries. Although it is only a political statement and does not modify the TRIPS Agreement in any way, it has important legal implications. It provides an understanding of the purpose of the TRIPS Agreement in relation to public-health issues that should guide any future rulings by WTO dispute-resolution panels dealing with such issues.[59] The declaration gave developing-country governments a degree of security in adopting national-level measures necessary to meet public-health objectives, and several developing countries, such as Malaysia, Thailand, Indonesia, Brazil, Zimbabwe, and Ghana, have since taken advantage of compulsory licensing to gain access to affordable generic medicines.[60] Many countries have also amended their laws to include the various TRIPS flexibilities.

To the NGO community, despite the disappointment that the outcome was not as strong or as legally binding as they had expected, the declaration was a big step forward in the battle for affordable medicines. Today, the declaration has become a common rallying platform for NGOs in persuading developing-country governments to take action to access affordable medicines and in holding developed-country governments accountable for what they agreed to in Doha, particularly that "the TRIPS Agreement does not and should not prevent Members from taking measures to protect public health" and that "the Agreement can and should be interpreted to protect public health and in particular, to promote access to medicines for all." The declaration is also invoked systematically by NGOs, policy makers, and others to counter developed countries' actions to pressure developing countries through trade, partnership, and investment agreements and their unilateral pressures to adopt intellectual property standards that go beyond the TRIPS Agreement.

The Doha Declaration was adopted for many reasons. However, a particularly notable reason is that a cohesive group of developing countries emerged to articulate their concerns about the effects of the TRIPS Agreement on access to medicines and to advocate a common position. These countries were well prepared, and with the support of NGOs, which maintained continuous pressure on the international community to do something concrete, mounted a strong case for urgent action in the WTO to address the TRIPS drugs issue and an interpretation of the TRIPS Agreement that enables national public-health measures.

The declaration evinces the possibility of winning a significant victory by advocates of access to knowledge, even in the face of strong opposition, when alliances are formed among developing countries and with concerned NGOs on major public-policy issues that need attention. The strategies and tactics used by NGOs and developing countries in the access debate, the collaboration between NGO groups from the Global North and the Global South, and the strategic collaboration between NGOs and developing countries provide very useful guidance and insight for those working on other issues in the access to knowledge movement.

NOTES

1. See World Health Organization, *The World Medicines Situation* (2004), ch. 7, available on-line at http://www.who.int/medicinedocs/collect/medicinedocs/pdf/s6160e/s6160e.pdf (last accessed March 19, 2009).

2. See Médecins Sans Frontières, *Untangling the Web of Price Reductions*, 10th ed. (July 2007) available on-line at http://www.msfaccess.org/fileadmin/user_upload/diseases/hiv-aids/Untangling_the_Web/UTW10_RSep_horizontal.pdf (last accessed March 19, 2009).

3. *Ibid.*, p. 5.

4. The TRIPS Agreement is available on-line at http://www.wto.org/english/tratop_e/trips_e/t_agm0_e.htm (last accessed March 22, 2009). Three transition periods are provided for in the Agreement: the period from 1995 to 2000, at the end of which developing countries were obliged to implement the TRIPS Agreement; the period from 2000 to 2005, which provided an additional period of five years in which to put in place product patent protection for pharmaceuticals or agrochemicals in those countries without such protection at the entry into force of the agreement; and the period from 1995 to 2006, after which least-developed countries would be required to implement their TRIPS obligations (Articles 65 and 66 of TRIPS Agreement). Presently only least-developed countries have transition periods after an additional extension was allowed in 2001. That is, a least-developed country need not apply TRIPS provisions until July 1, 2013 or until it ceases to be a least-developed country, if that period is shorter than the former; and least-developed countries do not have to implement and apply the TRIPS provisions on patents (Section 5) and on protection of undisclosed information (Section 7) until January 1, 2016. See World Trade Organization, Council for Trade-Related Aspects of Intellectual Property Rights, "Extension of the Transition Period under Article 66.1 for Least-Developed Country Members" (November 30, 2005), WTO doc. no. IP/C/40, available on-line at http://www.tripsagreement.net/documents/GATTdocs/Decision_of_the_Council_for_TRIPS_of_29_November_2005_E.doc (last accessed March 19, 2009) and World Trade Organization, "Extension of the Transition Period under Article 66.1 of the TRIPS Agreement for Least-Developed Country Members for Certain Obligations with Respect to Pharmaceutical Products," July 1, 2002, WTO doc. no. IP/C/25, available on-line at http://www.wto.org/english/tratop_e/TRIPs_e/art66_1_e.htm (last accessed March 19, 2009).

5. See Carlos M. Correa, *Intellectual Property Rights, the WTO and Developing Countries: The TRIPS Agreement and Policy Options* (London: Zed Books, 2000); Commission on Intellectual Property Rights, *Integrating Intellectual Property Rights and Development Policy* (September 2002), available on-line at http://www.iprcommission.org/graphic/documents/final_report.htm (last accessed March 19, 2009).

6. "Evergreening" is a term popularly used to describe patenting strategies that are intended to extend the patent term on the same compound. See World Health Organization Commission on Intellectual Property, Innovation and Health, *Public Health: Innovation and Intellectual Property Rights* (April 2006), p. 148, available on-line at http://www.who.int/intellectual-property/documents/thereport/ENPublicHealthReport.pdf (last accessed March 22, 2009).

7. See Article 27 of the TRIPS Agreement.

8. See Sisule Musungu and Cecilia Oh, *The Use of Flexibilities in TRIPS by Developing Countries: Can They Promote Access to Medicines?* (Geneva: South Centre and WHO, 2006) available on-line at http://www.southcentre.org/index.php?option=com_content&task=view&id=70&Ite

mid=67 (last accessed March 22, 2009). A compulsory license is a license issued by the government to allow the use of patented inventions without the consent of the patent holder. It is one of the flexibilities in the TRIPS Agreement. Parallel importation is the importation and resale in a country without the consent of the patent holder of a patented product that has been legitimately put on the market of the exporting country under a parallel patent. For example, if a patented antiretroviral cost $5.00 per tablet in Country X and the same patented medicine is sold in Country Y for $1.00 per tablet, then Country X could parallel import from Country Y the same medicine, since it is cheaper, and do so without the consent of the patent holder.

9   Médecins Sans Frontières, *Untangling the Web of Price Reductions*, 10th ed., p. 8. See Chan Park and Leena Menghaney's essay, "TRIPS Flexibilities: The Scope of Patentability and Oppositions to Patents in India," in this volume.

10   J. Michael Finger, *The Doha Agenda and Development: A View from the Uruguay Round* (Manila: Asian Development Bank, 2002), available on-line at http://www.adb.org/Documents/ERD/Working_Papers/WP021.pdf (last accessed March 22, 2009). "The Uruguay Round grand bargain was that developing countries would take on obligations in the new areas and in exchange developed countries would provide better access to their markets, particularly on agricultural products and on textiles and clothing. Comparing ... the net gains from changed patent obligations with the gains from Uruguay Round liberalization of tariffs on industrial goods by *all* WTO Members shows that TRIPS-patents are worth *thirteen times* more to the US than is the Uruguay Round tariff package on industrial goods" (pp. 11–12).

11   See Musungu and Oh, *The Use of Flexibilities in TRIPS by Developing Countries*.

12   ACT UP Press Release, *AIDS activists take over GlaxoSmithKline Investor Relations Office: Expose company-wide policy of blocking generic AIDS drug access* (February 2001) available at http://www.healthgap.org/press_releases/01/022101_AU_PR_GLAXO_NYC.html.

13   See Tido von Schoen Angerer, David Wilson, Nathan Ford, and Toby Kasper, "Access and Activism: The Provision of Antiretroviral Therapy in Developing Countries," *AIDS* 15, supplement 4 (2001): pp. S81–S90.

14   See Karine Cunqueiro, "Hostile AIDS Activists Target Gore over Patents," available on-line at http://www.twnside.org.sg/title/gore-cn.htm (last accessed March 22, 2009).

15   *Ibid.*

16   See Russell Mokhiber and Robert Weissman, "The Drug Lords Defeated," available on-line at http://www.twnside.org.sg/title/lords-cn.htm (last accessed March 22, 2009).

17   The Paris Convention for the Protection of Industrial Property is available on-line at http://www.wipo.int/treaties/en/ip/paris/trtdocs_wo020.html (last accessed March 23, 2009).

18   Gretchen Small, "Brazil Battles for Right of All Nations to Affordable Medicines," *Executive Intelligence Review*, March 23, 2001, available on-line at http://www.larouchepub.com/other/2001/2812BrazilAIDS.html (last accessed March 23, 2009). In 2000, at an international AIDS conference in Durban, South Africa, Brazil had offered to provide assistance to other developing countries committed to providing universal access to medicines by offering to help them build their own laboratories and to train people to run them. In December 2000, the health ministers of South Africa and Brazil had signed a letter of intent for cooperation.

19   See Ellen 't Hoen, "TRIPS: Pharmaceutical Patents and Access to Essential Medicines, Seattle, Doha and Beyond" (2003), p. 45, available on-line at http://www.accessmed-msf.org/fileadmin/user_upload/medinnov_accesspatents/chicagojournalthoen.pdf (last accessed March 23, 2009). Since the mid-1990s, Brazil has offered comprehensive AIDS care, including universal

access to antiretroviral treatment. An estimated 536,000 people are infected with HIV in Brazil, with 203,353 cases of AIDS reported to the Ministry of Health from 1980 through December 2000. In 2001, 105,000 people with HIV/AIDS received antiretroviral treatment. The Brazilian AIDS program reduced AIDS-related mortality by more than 50 percent between 1996 and 1999. In two years, Brazil saved $472 million in hospital costs and treatment costs for AIDS-related infections.

20 Quoted in Small, "Brazil Battles for Right of All Nations to Affordable Medicines."

21 Quoted in *ibid.*

22 Well-documented cases of countries that have been subjected to pressures are Thailand, Kenya, and Argentina. See Nathan Ford, "Patent, Access to Medicines and the Role of Non-governmental Organizations," *Journal of Generic Medicines* 1, no. 2 (January 2004): pp. 137–45; Frederick M. Abbott, "Doha Declaration on TRIPS and Public Health: Lighting a Dark Corner at the WTO," *Journal of International Economic Law* 5, no. 2 (2002): pp. 469–505.

23 The WHO Revised Drug Strategy Resolution, WHA 49.14, para. 2 (10) in 1996 led to the first guide by the WHO, German Velasquez and Pascale Boulet, *Globalization and Access to Drugs: Perspectives on the WTO/TRIPS Agreement*, 2nd ed. (Geneva: WHO, 1999), with recommendations to member states for implementing TRIPS while limiting the negative effects of higher levels of patent protection on drug availability. See also WHO Resolution WHA 54.10 (2001), "Scaling Up the Response to HIV/AIDS Resolution," available on-line at http://www.cptech.org/ip/health/who/wha.html, and WHO Resolution WHA 54.11 (2001), "Medicines Strategy Resolution," also available on-line at http://www.cptech.org/ip/health/who/wha.html (both last accessed March 23, 2009). See also Ellen 't Hoen's essay, "The Revised Drug Strategy: Access to Essential Medicines, Intellectual Property, and the World Health Organization," in this volume. On the UN Sub-Commission for the Protection and Promotion of Human Rights, see United Nations Economic and Social Council Commission on Human Rights Sub-Commission on the Promotion and Protection of Human Rights Resolution 2000/7, "Intellectual Property Rights and Human Rights," para. 2, UN doc. no. E/CN.4/SUB.2/RES/2000/7 (2000) (last accessed March 23, 2009). See also Someshwar Singh, "TRIPS Regime at Odds with Human Rights Law, Says UN Body," (August 2000), available on-line at http://www.twnside.org.sg/title/odds.htm (last accessed March 23, 2009). On the United Nations Development Programme, see *Human Development Report 1999* (Oxford: Oxford University Press, 1999), available on-line at http://hdr.undp.org/en/media/HDR_1999_EN.pdf (last accessed March 23, 2009).

24 The Council for TRIPS is a body within the WTO open to all WTO members and responsible for administering the TRIPS Agreement.

25 Drug regulatory authorities usually require pharmaceutical companies to submit data demonstrating the safety, quality, and efficacy (collectively known as "test data") of a pharmaceutical product as a condition for permitting the marketing of it. However, there are different opinions as to the obligation that the TRIPS Agreement places on countries with respect to the protection of test data. Article 39.3 of the TRIPS Agreement requires members to provide protection against "unfair commercial use" for undisclosed test or other data submitted for the purposes of obtaining marketing approval. Proponents of higher standards of protection argue that Article 39.3 requires the granting of exclusive rights over test data. This approach would prevent regulatory authorities from registering generic medicines on the basis of the test data provided by the originator companies and thus would hamper access to affordable

generic medicines. Developing countries in their submission to WTO stated that Article 39.3 of TRIPS does permit a national competent authority to rely on data in its possession to assess a second and further application relating to the same drug and that this would not imply any "unfair commercial use." See the submission by the African Group, Barbados, Bolivia, Brazil, Cuba, Dominican Republic, Ecuador, Honduras, India, Indonesia, Jamaica, Pakistan, Paraguay, Philippines, Peru, Sri Lanka, Thailand, and Venezuela, WTO doc. no. IP/C/W/296, available on-line at http://www.twnside.org.sg/title2/FTAs/Intellectual_Property/IP_and_Access_to_Medicines/TRIPSandPublicHealthWTOSubmission.doc (last accessed March 23, 2009).

26 Quoted in Cecilia Oh, "Developing Countries Call for Action on TRIPS at Doha WTO Ministerial Conference," available on-line at http://www.twnside.org.sg/title/twr131d.htm (last accessed March 24, 2009).

27 See Chakravarthi Raghavan, "NGOs Demand 'Pro-Public Health' Interpretation of TRIPS," available on-line at http://www.twnside.org.sg/title/pro.htm (last accessed March 24, 2009).

28 See Cecilia Oh, "U.S. Opposed to Moves to Address Public-Health Concerns about TRIPS," available on-line at http://www.twnside.org.sg/title/twr131f.htm (last accessed March 24, 2009).

29 Quoted in *ibid*.

30 Oh, "Developing Countries Call for Action on TRIPS at Doha WTO Ministerial Conference."

31 Submission by the African Group, Barbados, Bolivia, Brazil, Cuba, Dominican Republic, Ecuador, Honduras, India, Indonesia, Jamaica, Pakistan, Paraguay, Philippines, Peru, Sri Lanka, Thailand, and Venezuela, WTO doc. no. IP/C/W/296. See also subpara. 5 (d) of the Doha Declaration (see n.53 below).

32 Chakravarthi Raghavan, "Deadlock over Doha Public Health Declaration," *South North Development Monitor* (SUNS), no. 4985 (October 2001), available on-line at http://www.twnside.org.sg/title/deadlock.htm (last accessed March 24, 2009).

33 Chakravarthi Raghavan, "Deeply Divided WTO Faces Moment of Truth," *South North Development Monitor* (SUNS), no. 5008 (November 2001), available on-line at http://www.twnside.org.sg/title/truth.htm (last accessed March 24, 2009).

34 Chakravarthi Raghavan, "Talks on TRIPS and Public Health Break Down," *South North Development Monitor* (SUNS), no. 4996 (October 2001), available on-line at http://www.twnside.org.sg/title/breakdown.htm (last accessed March 24, 2009).

35 Carlos M. Correa, "Implications of the Doha Declaration on the TRIPS Agreement and Public Health," WHO, EDM series no. 12 (June 2002), available on-line at http://www.gefoodalert.org/library/admin/uploadedfiles/Implications_of_the_Doha_Declaration_on_the_TR.htm (last accessed March 24, 2009).

36 See Kanaga Raja, "NGOs Accuse Rich Nations over TRIPS and Public Health," *South North Development Monitor* (SUNS) no. 4971 (September 20, 2001), available on-line at http://www.twnside.org.sg/title/accuse.htm (last accessed March 24, 2009).

37 'T Hoen, "TRIPS: Pharmaceutical Patents and Access to Essential Medicines, Seattle, Doha and Beyond."

38 In 2001, over the course of several weeks, letters containing anthrax spores were mailed to several news media offices and U. S. senators.

39 Emma Clark, "America's Anthrax Patent Dilemma, BBC News Online, October 23, 2001, available at http://news.bbc.co.uk/2/hi/business/1613410.stm (last accessed March 24, 2009).

40 The Doha Declaration on the TRIPS Agreement and Public Health, November 14, 2001,

WTO doc. no. WT/MIN(01)/DEC/2, is available on-line at http://www.wto.org/english/ thewTO_e/minist_e/min01_e/mindecl_trips_e.pdf (last accessed March 24, 2009).

41 See subpara. 5 (c) and 5 (d) of the Doha Declaration. Subpara. 5 (d) pertains to the right of "parallel import." A country can provide "international exhaustion" which permits importation without the permission of the rightsholder after a first sale by that rightsolder anywhere in the world, "regional exhaustion" which permits importation after a first sale in the region, and "national exhaustion" which precludes parallel importation.

42 Para. 6 of the Doha Declaration.

43 Article 31 (f) of the TRIPS Agreement.

44 See 't Hoen, "TRIPS: Pharmaceutical Patents and Access to Essential Medicines, Seattle, Doha and Beyond"; Carlos M. Correa, "Implementation of the WTO General Council Decision on Paragraph 6 of the Doha Declaration on the TRIPS Agreement and Public Health" (2004), WHO doc. no. WHO/EDM/PAR/2004.4, available on-line at http://www.who.int/medicinedocs/collect/medicinedocs/pdf/s6159e/s6159e.pdf (last accessed March 24, 2009).

45 "Implementation of Paragraph 6 of the Doha Declaration on the Trips Agreement and Public Health, 30th August Decision of General Council," WTO doc. no. WT/L/540, available on-line at http://www.wto.org/english/tratop_E/TRIPS_e/implem_para6_e.htm (last accessed March 24, 2009).

46 On December 6, 2005, member states agreed that the August 30 decision would be translated into an amendment of the TRIPS Agreement. See WTO doc. no. WT/L/641 for the December 6, 2005 decision. See also Sangeeta Shashikant, "General Council Approves 'Permanent Solution' to TRIPS and Health," *South North Development Monitor* (SUNS), no. 5932 (December 8, 2005), available on-line at http://www.twnside.org.sg/title2/intellectual_property/wto_ trips/2005/SUNS.GCapprovespermanentsolutiontoTRIPSandhelath.doc.

47 See also Sangeeta Shashikant, "Rushing through a Permanent Solution for TRIPS and Health," *South North Development Monitor* (SUNS), no. 5931 (=December 7, 2005) available on-line at http://www.twnside.org.sg/IP_wto_2005.htm (last accessed March 24, 2009).

48 Para. 1 of the Doha Declaration.

49 Para. 7 of the Doha Declaration: "We reaffirm the commitment of developed-country members to provide incentives to their enterprises and institutions to promote and encourage technology transfer to least-developed country members pursuant to Article 66.2. We also agree that the least-developed country members will not be obliged, with respect to pharmaceutical products, to implement or apply Sections 5 and 7 of Part II of the TRIPS Agreement or to enforce rights provided for under these Sections until 1 January 2016, without prejudice to the right of least-developed country members to seek other extensions of the transition periods as provided for in Article 66.1 of the TRIPS Agreement. We instruct the Council for TRIPS to take the necessary action to give effect to this pursuant to Article 66.1 of the TRIPS Agreement."

50 'T Hoen, "TRIPS: Pharmaceutical Patents and Access to Essential Medicines, Seattle, Doha and Beyond."

51 Médecins Sans Frontières, "Fatal Imbalance: The Crisis in Research and Development for Drugs for Neglected Diseases," September 3, 2001, available on-line at http://www.msfaccess. org/fileadmin/user_upload/medinnov_accesspatents/fatal%20imbalance.pdf (last accessed March 25, 2009).

52 Oxfam, "Patent Injustice: How World Trade Rules Threaten the Health of Poor People,"

February 2001, available on-line at http://www.oxfam.org.uk/resources/policy/health/ downloads/patentinjustice.pdf (last accessed March 25, 2009).

53 Consumer Project on Technology, "Background Information on fourteen FDA Approved HIV/ AIDS Drugs," June 8, 2000, available on-line at http://www.cptech.org/ip/health/aids/ druginfo.html (last accessed March 25, 2009).

54 Consumer Project on Technology, Essential Action, Oxfam, Treatment Access Campaign, and Health GAP, "Comment on the Attaran/Gillespie-White and PhRMA Surveys of Patents on Antiretroviral Drugs in Africa," October 16, 2001, available on-line at http://www.cptech.org/ ip/health/africa/dopatentsmatterinafrica.html (last accessed March 25, 2009).

55 Amir Attaran and Lee Gillespie-White, "Do Patents for Antiretroviral Drugs Constrain Access to AIDS Treatment in Africa?" *Journal of the American Medical Association* 286, no. 15 (October 17, 2001): pp. 1886–92.

56 'T Hoen, "TRIPS: Pharmaceutical Patents and Access to Essential Medicines, Seattle, Doha and Beyond."

57 Abbott, "Doha Declaration on TRIPS and Public Health," pp. 469–505.

58 The submission by the African Group, Barbados, Bolivia, Brazil, Cuba, Dominican Republic, Ecuador, Honduras, India, Indonesia, Jamaica, Pakistan, Paraguay, Philippines, Peru, Sri Lanka, Thailand, and Venezuela, WTO doc. no. IP/C/W/296, states: "Finally, in civil society, a number of important non-governmental organizations, such as 'Médecins Sans Frontières', Oxfam and Consumers International also have emphasized their concern that the TRIPS Agreement may be applied in detriment to health policies." See Oh, "Developing Countries Call for Action on TRIPS at Doha WTO Ministerial Conference."

59 Correa, "Implications of the Doha Declaration on the TRIPS Agreement and Public Health."

60 Martin Khor, "Patents, Compulsory License and Access to Medicines: Some Recent Experiences," February 2007, available on-line at http://www.policyinnovations.org/ideas/policy_ library/data/patents_compulsory_license (last accessed March 25, 2009).

Student demonstration against software patents in Belgium, May 2005.

An Uncertain Victory:
The 2005 Rejection of Software Patents
by the European Parliament

Philippe Aigrain

On July 6, 2005, the European Parliament voted by a large majority to reject the Directive on the Patentability of Computer-Implemented Inventions proposed by the European Commission Internal Market directorate-general.[1] This event marked a milestone in the access to knowledge movement: For the first time, it obtained a major political decision after a mass mobilization of citizens and civil-society groups and a wide-ranging open debate reaching well beyond the action of specialized NGOs.

Software plays an essential role in many activities and fields of technology and science. Europe's legal recognition of software patents would have represented a very severe blow to the existence of a freely usable common body of knowledge.[2]

Software is information, expressed in a formal language, about how to process information. A computer program is a form of mathematical statement, and it is so regardless of whether it is used in a computerized pasta machine, for text processing, or to compute some exotic sort of number. Software has opened a new world of information processing that has deeply transformed human activities: thought, expression, communication, and creation. It has also changed the conditions of innovation in many fields of technology. Technology still deals with what Richard Stallman has called "the perversity of matter": the fact that material things break, heat up, wear out, are hard to manufacture consistently, and can be combined one with the other only at a very limited scale and with careful planning.[3] But these core technological challenges have been localized, broken down into their components. Their physical complexity has been confined. Some technical objects can be "reduced" to information-processing modules taking their input from receivers and sending their output to simple effectors. Sometimes, though, material, energetic, biological, or systemic complexity resists such reduction. These are the most important technical challenges in environmental or biological innovation, for instance. For example, a seed is more that just genetic material—it is also an environment in which the genes will be expressed and in which the future plant will start developing.

The case for patents as an incentive to innovation and the effects of granting them are radically different in the information domain and in the physical domain. Information-domain patents (software patents, genetic-information patents) lead to monopolies on the free reproduction of information and to arbitrary prices completely disconnected from production and even research costs. Actually, in the software domain, even much narrower monopolies, such as copyright, lead to extreme dominant positions when they are combined with network effects. In such cases, the effect of patents is to cement these monopolies. Because innovation in software is combinatorial (combining components) and incremental (refining functionality) and often results from transferring an idea from one domain to another, software patents block future innovation and its dissemination by creating patent thickets—accumulations of patents through which an innovator can no longer find a possible way to create an innovation without infringing on patents.[4] In contrast, for mechanical devices or chemical processes, patents can be worked around, and this often results in new ways of dealing with material challenges.

Described at this general level, information-domain patents are excellent for rent seekers, but useless, at best, and harmful most often, for innovation and access to knowledge. However, the promoters of software patents are not found only in pure-information industries such as proprietary software. Industry or research labs that are active in mixed domains, such as consumer electronics or mobile-phone devices, would like to have the best of both worlds: the plasticity and ease of innovating in the software domain and the patent protection that has been judged useful for material objects. They have summarized this view in a formula: "Why would it be impossible for us to patent a phone or hi-fi, now that there is plenty of software in it, when we were able to patent it before?" But what exactly do they want to patent? Is it the phone's physical components—for instance antennas, which remain necessary in software radio and whose patentability is not disputed—or a piece of software for the digital generation of a sine curve that is a pure mathematical method used in hundreds of fields other than telephony?[5] This distinction became the nexus of the software-patents debate, and one of the most surprising outcomes of the debate was to see a few members of the European Parliament becoming able to argue in detail with industry lobbyists on such complex issues.

In July 2005, after the vote to reject the proposed directive on patentability, there were shouts of victory from many sides. The almost unanimous vote was obtained by a mix of antisoftware-patent votes and prosoftware-patent votes. The former were pleased to reject the directive, since it did not appear possible to obtain majority for a text that would make a clear and updated statement that software and software-based information-processing methods are not patentable. The latter were resigned to rejecting the directive when it became clear that a prosoftware-patent text would never obtain majority. There is little doubt that at least the vote was a defeat for those who wanted to turn the practice of the European Patent Office of granting patents on software and software-based information-processing methods into law. However, the situation after this vote is one of great uncertainty, since the practice remains. This essay intends to help the reader understand what made possible this outcome and where things stand today in Europe.

BATTLES OVER SOFTWARE PATENTABILITY PRIOR TO 2005

To do so, we need to begin with a bit of perspective.[6] Ten years before the vote of July 6, 2005, the European Parliament had already rejected a directive extending the scope of patentability. On March 1, 1995, the European Parliament rejected by 240 votes to 188 (with 23 abstaining) a directive that permitted the patenting of gene sequences and of organisms that contain modified or otherwise patentable gene sequences. However, it took only three years for this victory to be reversed, with the adoption of Directive 98/44 by the European Parliament in 1998. During these three years, an innovative combination of lobbying techniques was put in place by industry players, a mix of agrifood biotech and pharma biotech companies that were interested in gene-sequence patentability.[7]

Part of the innovation in these efforts lay in the use of new forms of rhetoric. In the drafting of legal documents, the normative form is to define the scope of a permission or an interdiction by a sequence of alternate statements such as "Freedom of expression is a fundamental right, however, its exercise can be restricted by judicial authorities based on established reasons of national security or the protection of persons." When this form of legal discourse is used, the substance lies in the second provision. Directive 98/44 used this normative form by first stating that human gene sequences are not patentable inventions, because they are discoveries, but then claiming that they are patentable "when they are isolated from the human body or otherwise produced by means of a technical process." Because any gene sequence that is known is always isolated or otherwise produced by a technical process, this amounted to saying: "Human gene sequences are not patentable inventions, but are patentable inventions." Opponents denounced this rhetoric as analogous to Orwellian Newspeak, but were unable to prevent the directive from being adopted. However, civil-society groups quickly developed the ability to detect such rhetorical sleights-of-hand, and they were quick to detect its repeated use in the 2002 proposal for a directive on the patentability of computer-implemented inventions, in which the term "computer-implemented inventions," a neologism, was defined as referring to the underlying principles of software. This allowed those who drafted the proposal to say, in effect, "Software or algorithms remain unpatentable, but they can be patented under the name of computer-implemented inventions." Such Orwellian tactics were successful, however, and efforts to promote the patentability of software continued in Europe right up to the victory of patentability opponents in 2005.

Software patents were progressively recognized in the United States from the end of the 1980s on and became common in the 1990s.[8] The European Patent Office (EPO) therefore was subjected to increasing pressure from its customers to align the European practice with the U.S. standard of patenting software.[9] However, there existed a major obstacle to such an alignment: the provision in Article 52 of the European Patent Convention (EPC) that lists a number of things that cannot be patented because they are not inventions,[10] including computer programs, mathematical methods, and business methods, etc. In a series of cases (IBM 1997 and 1998, Philips 2000), the EPC therefore used its in-house Chamber of Appeal to create surrealistic case law that was soon incorporated in its examination guidelines. This case law used Article 52(3) of the EPC, which states that the exclusion from patentability applies only to the excluded entities "as such." It claimed that the excluded entities could be patented if they had "a technical effect" or if "technical considerations" were necessary to produce them.[11] According to this new case law, tens of thousands of software patents were granted by the EPO.[12]

However this home-made case law was fragile, since there is good evidence from managers of the EPO themselves that the EPC wording in the case of software was meant only to declare that a physical invention could still be patented, whether or not it contained software.[13] The EPO and its representatives within the European Commission consequently proceeded to make the law more explicitly favor patentability in accordance with practice by working along two parallel tracks.[14]

The first one was to hold a diplomatic conference for deleting the inconvenient exclusions from the EPC.[15] The initial proposal simply deleted all exclusions from patentability, including, for instance, exclusions for games or methods of teaching. After some debate developed, it was proposed to delete only the exclusion of computer programs. However, from the end of 1998 on, NGOs advocating for free or open-source software started alerting decision makers about the risks of software patentability for the freedom to innovate in software. This debate had reached a sufficient scale by 2000, when the diplomatic conference was held in Munich, to motivate national delegates to refuse to amend the convention until progress had to be made along the second track: producing European legislation on the patentability of software. The then fifteen countries of the European Union voted fourteen to one against deleting the exclusion... for the time being.

The proposal for a directive then was prepared by a number of steps that had been initiated from 1996 on. A green book on the future of patents in Europe was discussed,[16] mostly in specialized patent circles. In 1997, the European Commission published a communication on the follow-up to the green book that included an explicit mention of a directive to come. Until 1998, almost no software practitioners were involved in the debate. (The only one speaking at the London conference on March 23, 1998, took a clear stand against any form of software patentability.) However, from 1998 on, developers of free and open-source software, small and medium-sized shareware enterprises, and a number of academics started to alert the public and decision makers about the risks of accepting patents on software. These concerns were relayed within the European Commission by the Information Society general-directorate. A lively internal debate echoed the external debate that was developing in Europe. A provisional compromise was struck between the relevant commissioners: A new consultation of stakeholders and citizens would be launched on October 19, 2000. In parallel, some European Union members states such as the UK initiated a consultation of their own, while others, such as Germany, commissioned studies, and still others, such as France, created committees that were asked to recommend a policy.

The biased manner in which the Internal Market general-directorate handled the analysis of opinions submitted in answer to its consultation did a lot to weaken its case. The Foundation for a Free Information Infrastructure, an NGO dedicated to keeping innovation open in the software field, had asked stakeholders to transmit their opinion through them. This was an answer to the fact that the European Commission admitted nonpublic responses to its consultation. The Internal Market and Services directorate-general of the European Commission assigned a previously unknown consultant to produce an analysis of contributions. His report discarded 90 percent of the answers (all those—opposed to software patents—that were transmitted through the Foundation for a Free Information Infrastructure) as having been initiated by a specific party. Even then, half of the remaining answers were opposed to software patents. The report had to declare that those in favor were more significant in terms of sales and employment. Meanwhile, a large body of knowledge and evidence started to accumulate on the nature of software patents and their effects where they were already in place.

THE JUNE 25, 2002 DIRECTIVE

When the European Commission adopted a proposal for a Directive on the Patentability of Computer-Implemented Inventions on June 25, 2002, it was basically proposing to turn into law the existing practice of the EPO of granting patents on software and methods for processing information in the information domain. There was one difference, and a significant one, that testified to the effects of prior debates: The directive was not proposing to accept patent claims on software, "as this could be seen as allowing patents for computer programs 'as such.'"[17] The directive was presented as not following the U.S. practice of granting patents on business methods and claimed not to allow patents on algorithms. The former affirmation was quickly debunked when analysis of existing patents showed that it was enough for a business method to be implemented in software and to produce some improvement for it to be patentable. The latter claim was based on a radical misrepresentation of the relationship between algorithms and software, since algorithms are nothing other than the underlying principles of software, while the whole idea of patenting software is to grant monopolies on these principles. In fact, the use of "computer-implemented inventions" in the title was deceptive, because "computer-implemented inventions" were basically defined as software.[18]

The proposed directive then went through the complex European legislative process, consisting of two parallel readings in the European Council, which represents member States, and in the elected parliament. When both are in serious disagreement, the council has the stronger power, which means that the parliament could make its point only by rejecting the directive. It is generally reluctant to do such a thing, because a majority of its members committed to creating EU-level legislation. The European Council produced its first reading before the parliament did so. It was prepared by a "working party on intellectual property (patents)." In this group, more than half of the then fifteen member states were represented by patent offices, and representatives of the EPO sat on the commission bench. The council set out to amend the commission proposal by allowing software claims, thus aligning the directive with EPO practices. However, the council decided to wait for the parliament's reading before formally adopting its own position.

This position was adopted in a vote on September 24, 2003. It came as a thunderbolt. The parliament adopted amendments submitted by the Culture Committee (rapporteur, Michel Rocard, socialist), by the Industry, Trade, Research and Energy Committee (rapporteur, Elly Plooij van Gorsel, liberal) or by members of the European Parliament who often were drawing inspiration from proposals by civil-society groups. These amendments adopted a strict definition of what can be considered to be "technical," putting it in relation with physical devices and processes, and clarified that patents can be granted only when innovation lies in this physical, technical domain. Civil-society initiatives used the possibility for any European resident or group to petition the European Parliament on issues of its competence: Leading computer scientists signed a detailed analysis of the reasons to reject software patents,[19] while one hundred and fifty thousand citizens signed

Demonstration against software patents in Germany, June 2004.

a petition against software patents initiated by the Foundation for a Free Information Infrastructure. The amended text constituted a clear and detailed rejection
of all the mechanisms by which software patentability had been sneaked into the
practice of the EPO.

There was such a shock that patent lobbyists started expressing publicly the
view that patentability issues were truly too serious to be the object of democratic
decision making. Until then, prosoftware-patent lobbying had been restricted to
behind-the-doors contacts with the European Commission and members of the
European Parliament, while opponents argued on substance and conducted public
workshops. A significant change developed in the next two years, when advocates
for software patents developed an all-out lobbying campaign, including the estab-
lishmentment of a "Campaign for Creativity" that backfired when it appeared to be
a lobbying-consultant initiative funded by Microsoft and SAP, without any link to
real software practitioners.[20] Some opponents of software patents also adopted a
communication campaign, in particular, the NoSofthawarePatents.com campaign
conducted by Florian Müller with support from MySQL and Red Hat.[21] In the last
weeks before the July 2005 vote, communication efforts on both sides culminated

with distributions of free ice cream, demonstrations, and boat fights on the canals close to the European Parliament building near Strasbourg.

Before that climax, the reading of the proposed directive had proceeded with great pain in the council. A text was produced by the Irish presidency, under fierce criticism due to its interests as a tax haven for holders of intellectual property rights,[22] and a "political compromise" was recorded on May 18, 2004. It was a confusing text that basically reiterated the propatent, first-reading position, but installed it under smokescreens of complex language. Various opponents produced translations to normal language in the days that followed its adoption.[23] It took four meetings and several votes before a qualified majority was reached, on March, 7, 2005, to adopt this text formally. Whether there was a truly qualified majority is still open to doubt, because one country (the Netherlands) later changed its vote, and another (Poland) protested that its vote had not been properly recorded. The fragility of this decision eased the path toward rejection of this "compromise" position by the parliament. After the climax of lobbying mentioned above, it became clear that there was no majority in the parliament for adopting a text that would please the patent advocates and proprietary-software lobbies. So everyone rallied to reject the text, and each claimed that doing so was a victory for its views. The text was rejected by the unprecedented majority (for a rejection) of 648 in favor, 14 against, and 18 abstentions.

WHAT MADE THE "VICTORY" POSSIBLE?

How was such an unexpected result obtained? It resulted from the synergy between several movements, each of which had built a serious case in its domain. At the urging of Harmut Pilch, the Foundation for a Free Information Infrastructure accumulated a broad body of empirical knowledge on actual patenting practices in Europe that served as the basis for scholarly work in both Europe and the United States. It was not long before active opponents of patentability knew much more about what software patents looked like, who owned them, and how many of them there were than their defenders. This was useful for building four different cases: a case for innovation, a scientific case, a political case, and a case based on academic research into the actual effects of software patents, each of which mobilized different communities.

The case for innovation gave rise to a mass mobilization of software developers well beyond developers of free and open-source software. This group was by far the largest in terms of direct action. It included individuals who on their own initiative flew to Brussels to talk to members of the European Parliament. The members of parliament were not used to encountering twenty-year-old programmers wanting to give them pedagogic explanations of the impact of software patents on innovation, and they listened carefully. Hundreds of engineers of the large European companies that were supporting software patents signed the Foundation for a Free Information Infrastructure petition against the patentability of software.[24]

But there was also a scientific case, which mobilized fewer people, but which gave impressive intellectual credibility to the opposition.[25] The scholarly economics community was divided, but a leading group of economists signed a letter against software patents a few days before the vote. More importantly, the organizations of small and medium-sized European shareware enterprises made known their own opinion on the subject, making clear that they did not share the propatent view of the Union of Industrial and Employers' Confederations of Europe, the large-company employer organization. This had a major influence on bringing a small part of the conservative members of parliament, who traditionally speak for a lot of small and medium-sized shareware enterprises to a critical view of software patents. In reaction, some large companies created an ad-hoc organization of small and medium-sized shareware enterprises whose members were spin-off companies, directly or through university partnerships. In a similar move, Microsoft created an ad-hoc proprietary software-publisher organization when it became clear that the general software employer organizations and the professional societies were reluctant to support their view.

All this would have probably not been sufficient without a political case also being built. The European Parliament has a culture of cross-party work that rests significantly on the relationships between advisers, assistants, and sometimes members of parliament. The 2003 vote in which many parties split their votes (the conservatives, the socialists, the liberals) cannot be understood without reference to the lively discussions between young advisers and assistants in corridors, cafeterias, and Brussels pubs. These conversations took place in a context where public debate was also raging. The Green Party organized a number of seminars, some debates in which contradictory views were expressed, others more one-sided, but presenting the various facets of the antisoftware-patents movement.

In these seminars and more generally in the literature on software patents, scholarly work conducted in the United States had an important impact, building a case against patents based on academic research into the actual effects that the patents had produced. As I noted, the United States had introduced software patents at the end of the 1980s, and they were granted ever more massively, especially from 1994 on. The United States thus provided a real-life experiment, even if the true impact of changes in the scope of patents will in reality take much longer to be fully evident. A number of studies had a devastating impact. The Bessen-Maskin and Bessen-Hunt papers demonstrated an inverse correlation between an increase in software patenting and investment in research and development.[26] Work by Brian Kahin highlighted the huge costs of patent litigation and the increasing share of innovation budgets dedicated to patents and patent risks.[27] Evidence of a massive unbalance in the number of patents held by U.S.-based companies (and to a lesser extent Asian companies) compared with European companies also made obvious that from a specific European viewpoint, software patents were not more desirable than from a global viewpoint.

WHERE DO THINGS STAND?

The title of this essay, "An Uncertain Victory," calls for an explanation. After the vote of the European Parliament, we are in a regime of the status quo. The EPC still declares mathematical methods, computer programs, and so on to be not patentable as such. The EPO continues granting patents on software and software methods for processing information or doing business. Litigation and counterlitigation are limited, due to the obvious legal uncertainty: Companies are piling up software patents in Europe without using them, for the time being, while software developers keep ignoring them. Contrary to what happens in the United States, it is only in areas of standardization that the concrete effects of software patents are felt: Several standards have been blocked by patent jeopardy, for instance, JPEG 2000 and the internationalization of domain names in the Internet Engineering Task Force (IETF).

There are clear signs that patent-related institutions, the European Commission, and the propatent lobbies are busy working on other ways to give a firmer legal status or at least a stronger practical effect to software patents. The emphasis has first been on litigation and jurisdiction. The commission has been trying for ages to install a European Community Patent associated with a single European jurisdiction.[28] Critics fear that the creation of a specialized jurisdiction would have the same effect as when suchthe creation of the specialized Court of Appeal of the Federal Circuit through which software patentability was introduced in the United States in the 1980s and 1990s. This effort has been blocked so far by linguistic conflicts between member states, though the situation may change, since some opposing countries, such as France, have now seemingly decided to sign the London Protocol, an agreement that would allow institution of the European Community Patent to proceed. In parallel, the EPC is pushing for the European Patent Litigation Agreement, because this agreement would permit exporting the scope of decisions from one member state to another. Harmonization of patent examination in the Substantive Patent Law Treaty managed by the World Intellectual Property

Organization (WIPO) is another track by which the U.S. standard of software patentability could be exported to Europe. However, it seems to be blocked by the conscious opposition of emerging and developing countries in the organization.[29]

The trend toward modifying the substantive definition of rights indirectly (for instance, in scope) by acting on enforcement is not restricted to patents: One sees it also in the area of copyright, from the World Intellectual Property Organization Copyright Treaty to the U.S. Digital Millennium Copyright Act and the by-products of the European Copyright Directive or the proposed broadcasters' treaty.[30] It also uses instruments that apply to all intellectual property right titles, such as the intellectual property right enforcement directives and the recently initiated proposal for an international Anti-Counterfeiting Trade Agreement. These very abstract texts are much more difficult to debunk than texts extending the scope of intellectual property rights. It remains to be seen whether civil society, scholars, and public-interest-oriented policy makers will be able to make clear for all what is at stake in these more obscure corners. It may also be that a more frontal approach will be taken, for instance through a new diplomatic conference for the revision of the EPC. But the awareness built though the eight years that led to the July 2005 uncertain victory is still there.

During the period between the two votes in the European Parliament, the scale of the international access to knowledge movement changed. Prior to 2004, it was mostly an initiative of specialized international English-speaking NGOs, with some national counterparts in other countries. Today, it is a powerful coalition of better-coordinated NGOs and key emerging countries (Brazil, Argentina, India, and Chile), with growing support from other developing countries. It has obtained support from new segments of public opinion: scientists and policy circles well beyond those traditionally interested, including, for instance, those concerned with climate-change issues. The movement that led to the 2005 victory is one of the factors that helped access to knowledge to become credible in the public's mind and on the international scene.

NOTES

1   This title was in itself exemplary of the tactics put in place by the directorate-general when it proposed the directive. Because a strong opposition to patenting software existed, the drafters tried to hide the fact that the object of the directive was to recognize software patents. They did so by using the neologism "computer-implemented inventions," which was defined in the text as equivalent to software, but could be understood by some readers as

meaning physical inventions using software. See below for more on such tactics. The text of the proposal, 2002/0047/COD, is available on-line at http://eur-lex.europa.eu/LexUriServ/ LexUriServ.do?uri=CELEX:52002PC0092:EN:NOT (last accessed February 27, 2010).

2   On December 27, 2004, the Indian Parliament had adopted a last-minute amendment to the new Indian patent law, imposed by its obligations under the TRIPS agreement. This amendment rejected software patents that had been temporarily authorized in the case of embedded software by a governmental decree in 2002. The Indian rejection was clearer in its legal effect than the European Parliament vote, but it did not obtain the same publicity, because it was overshadowed by the acceptance of patents on chemical molecules.

3   See, for example, Richard Stallman, "Software Patents—Obstacles to Software Development," talk presented on March 23, 2002, at the University of Cambridge Computer Laboratory, available on-line at http://www.cl.cam.ac.uk/~mgk25/stallman-patents.html (last accessed March 25, 2009).

4   Recently, some analysts have put into question the risk of patent thickets blocking innovation in software, based on lack of evidence that innovation blockage has materialized in the United States. I claim that the case of standards provides evidence of adverse effects of patent thickets on the dissemination of innovation, if not on its initial stages, which generally proceed in total ignorance of patents. See Jim Bessen, "Software Patent Myopia," *Technology Innovation and Intellectual Property*, December 12, 2007, available on-line at http://www. researchoninnovation.org/WordPress/?p=90 (last accessed March 25, 2009).

5   This is a real example. See the WO2004082129 patent by Nokia: Methods, devices and a software product for generating a sinusoidal signal, available on-line at http://www.wipo.org/ pctdb/en/wo.jsp?wo=2004082129 (last accessed February 28, 2010). Do not imagine that the world "devices" in the title refers to anything physical. Claims include: "8. A software product for generating a sinusoidal signal of a desired frequency (f) at a sampling rate (fs), which software product comprises a program code for determining the nth sample of the first output sample sequence."

6   For a longer-term perspective on patentability issues, see the entry s.v. "Patentability" in the *Critical Dictionary of Globalization*, available on-line at http://mondialisations.org/php/ public/art.php?id=9274&lan=EN (last accessed March 25, 2009).

7   For an interesting account of the lobbying strategies, see Shail Thaker, "The Criticality of Non-Market strategies," KSM'03, available on-line at http://www.kellogg.northwestern. edu/biotech/faculty/articles/shail.pdf (last accessed March 16, 2010).

8   A massive increase came after the 1994 U.S. Patent and Trademark Office review. See "Working for Our Customers," 1994, available on-line at http://www.uspto.gov/web/offices/com/ annual/1994/pg1-5.pdf (last accessed March 26, 2009).

9   For a justification of the use of the word "customers," see my "11 Questions on Software Patentability Issues in the U.S. and in Europe," Software and Business Method Patents: Policy Development in the U.S. and Europe, Center for Information Policy, University of Maryland, December 10, 2001, available on-line at http://paigrain.debatpublic.net/docs/elevenquestions (last accessed March 26, 2009).

10   In patent law, inventions must be susceptible of industrial application, must be new, and must involve an inventive step. The statement that computer programs are not inventions in that sense refers to the term "industrial application" being understood (in European patent law) as industrially produced physical devices and physical processes in industry.

11 The actual details are more complex, since ever more adorned concepts were designed, such as "further technical effect," "technical considerations," etc. in order to open even wider the door to patentability.

12 Between twenty and thirty thousand, according to the database produced by the Foundation for a Free Information Infrastructure (FFII), available on-line at http://eupat.ffii.org/patents/stats/index.en.html (last accessed February 28, 2010).

13 For a remarkable account of the debates in the 1970s on software patentability, see Christian Beauprez, "In Defence of the Software Author: A Study of Copyright and Patent Law Interactions," August 2004, available on-line at http://circa.europa.eu/Public/irc/markt/markt_consultations/library?l=/copyright_neighbouring/legislation_copyright/beauprez_christian/_EN_1.0_&a=d (last accessed March 26, 2009). In the United States, the debate was initially not even about copyright, software, and sui generis protection, as many now believe, it was between copyright and "no IPR [intellectual property rights] at all." See also Gert Kolle, "Technik, Datenverarbeitung und Patentrecht—Bermerkungen zur Dispositionsprogramm—Entscheidung des Bundesgerichtshofs GRUR 1977–02," pp 58–74, available on-line at http://eupat.ffii.org/papers/grur-kolle77/index.de.html (last accessed March 26, 2009).

14 The assessment and evolution of patent law within the European Commission was mostly done by seconded experts from the EPO or from national patent offices. Even during the legislative process for the 2002 directive proposal, the representatives from the EPO representatives sat on the commission bench in the European Council working group and answered questions for the commission.

15 The EPO and the EPC are intergovernmental: some countries that are not members of the European Union are members of the EPO and parties to the EPC. A diplomatic conference had the great advantage of requiring neither a debate nor a vote in the European Parliament.

16 A green book is a document produced by the European Commission to solicit views of stakeholders on a topic or proposed legislation.

17 "Explanatory Memorandum: Objective of the Community Initiative" (regarding Article 5), available on-line at http://eur-lex.europa.eu/LexUriServ/LexUriServ.do?uri=CELEX:52002P C0092:EN:HTML (last accessed March 26, 2009).

18 See the discussion of Article 2 of the proposal in *ibid*.

19 A detailed comment on the vote on September 24, 2003, can be found in my September 30, 2003 speech in the Petition Committee of the European Parliament, available on-line at http://eupat.ffii.org/log/03/epet0929/aigrain/AigrainEpet030930.en.pdf (last accessed March 27, 2009), where I presented the petition by European computer scientists. On the date of this speech, I was no longer working with the European Commission, and I spoke as a simple member of the computer-science community.

20 See "Campaign for Creativity: EU Gene Patent Lobbyists Taking Up Software," available on-line at http://wiki.ffii.org/CampaignForCreativityEn (last accessed March 27, 2009).

21 See http://www.nosoftwarepatents.com (last accessed March 27, 2009).

22 Ireland has adopted a policy of low taxes on patent revenues, with a 9 percent tax in general and 0 percent in those geographical areas eligible for European Structural Funds, allocated by the European Union to provide support for the poorer regions of Europe and support for integrating European infrastructure. This has given rise to a massive delocalization of intellectual property assets to Ireland. For instance, in 1990, IP licensing between France and

Ireland was balanced. In 2005, there was a balance of eighteen thousand million euros in favor of Ireland. Ireland is now in competition with other IP tax havens, such as Estonia.

23 See, for instance, the analysis (in French) by François Pellegrini, available on-line at http://linuxfr.org/2004/07/27/16908.html, or mine (in English), available on-line at http://paigrain.debatpublic.net/docs/analysis-compromise.html (both last accessed March 27, 2009).

24 At the time, the engineers opposing patentability included people from systems integrators such Siemens and Thalès, from consumer electronics companies such as Philips, and from large telco suppliers such as Nokia.

25 "Petition to the European Parliament on the Proposal for a Directive on the Patentability of Computer-Implemented Inventions," in Philippe Aigrain and Jesus Gonzalez-Barahona (eds.), "Open Knowledge," special issue, *Upgrade, The European Journal for the Informatics Professional* 4, no. 3 (June 2003), available on-line at http://www.upgrade-cepis.org/issues/2003/3/up4-3Petition.pdf (last accessed March 28, 2009).

26 James Bessen and Eric Maskin, "Sequential Innovation, Patents and Imitation," *MIT Research Report* (2000); James Bessen and Robert M. Hunt, "An Empirical Look at Software Patents," Working Paper 3/17R, Federal Reserve Bank (2004), republished in *Journal of Economics and Management Strategy* 16, no. 1 (March 2004): pp. 157–89; these and many other papers are all accessible on-line at http://www.researchoninnovation.org (last accessed March 27, 2009).

27 Brian Kahin, "What's Wrong with the Development of Intellectual Property Policy?" Beitrag zur Konferenz der Heinrich Böll Stiftung "Die Zukunft der globalen Güter in der Wissensgesellschaft—Auf der Suche nach einer nachhaltigen Politik zum Schutz des geistigen Eigentums, November 8, 2002, Berlin, available on-line at http://www.wissensgesellschaft.org/themen/publicdomain/wrongipp.pdf (last accessed March 27, 2009). See also James E. Bessen and Michael J. Meurer, "The Private Costs of Patent Litigation," Boston University School of Law Working Paper no. 07-08, Second Annual Conference on Empirical Legal Studies, available on-line at http://papers.ssrn.com/sol3/papers.cfm?abstract_id=983736 (last accessed March 16, 2010).

28 In contrast with the European Patent, which rests on the intergovernmental EPC, the EPO would rest on European Union (community) law. The European Community Patent would theoretically be less expensive, and the single jurisdiction would ensure more consistent case law.

29 See Viviana Munoz and Sisule Musungu's essay, "A2K at WIPO: The Development Agenda and the Debate on the Proposed Broadcasting Treaty," in this volume.

30 The broadcasters' treaty is a text that would generalize specific rights for broadcasters over the signal they transmit (such rights exist for parties to the Rome Convention, which does not include the United States) and would create specific legal protection against circumvention of technical-protection measures similar to provisions in the World Intellectual Property Organization Copyright Treaty, the U.S. Digital Millennium Copyright Act, and the 2001/29 European Directive for other copyrighted works. Some would like to extend the scope of the treaty to Webcasting or at least simulcasting (the simultaneous transmission of video or sound on the Internet to many users). The proposed treaty is very vociferously debated, with opponents stressing the risks for democracy of digital locks on television that prevent fair use and criticism and the uselessness of creating a new propertylike right for broadcasters. The treaty is presently stalled in the WIPO.

A2K at WIPO: The Development Agenda and the Debate on the Proposed Broadcasting Treaty

Viviana Muñoz Tellez and Sisule F. Musungu

Some of the most important international discussions that affect access to knowledge (A2K) take place in a long-standing organization that is little known to the public—the World Intellectual Property Organization (WIPO). WIPO formally came into existence in 1970, and subsequently, in 1974, it became a specialized agency of the United Nations. Today, there are many other organizations involved in standard setting on intellectual property issues. Yet WIPO remains the main international intergovernmental organization responsible for the administration and negotiation of new intellectual property treaties and the provision of intellectual property–related technical assistance to developing countries. WIPO is therefore a major institutional player in the global governance and regulation of knowledge. Hence, the approach and discussions related to A2K in WIPO are of particular importance and interest to the A2K communities.

In recent years, WIPO has been undergoing a substantial transformation. This is due, in part, to the new, active participation by A2K communities and to demands by developing countries for a more inclusive and balanced approach to its norm setting and other processes. While intellectual property policy has traditionally been considered a complex and technical issue, mainly thought of as the competence of lawyers and transnational companies, the growing evidence of the impact of intellectual property on ordinary people has brought many new players to its discussions. The recent decision to launch a development agenda for the organization and the collapse of efforts to establish an exclusive rights-based treaty for the enhanced protection of broadcasting organizations are landmark events in the history of WIPO and the global debate on intellectual property policy.

This essay analyzes the A2K agenda and the role of the A2K movement within the process of establishing the WIPO Development Agenda and the discussions on the proposed WIPO Broadcasting Treaty. Looking at these two processes offers

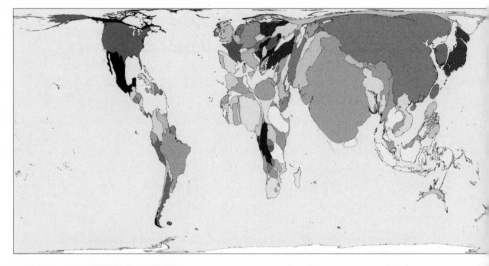

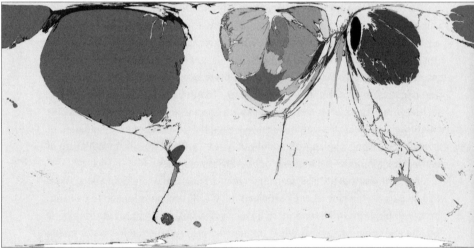

Map of global flows of royalties and licensing fees in 2002. In the top map, the size of each territory is determined by its population. In the bottom, the size of the territory shows the proportion of worldwide earnings (in purchasing power parity) from royalties and license fees. More than half of all such income was received by the United States (© Copyright 2006 SASI Group [University of Sheffield] and Mark Newman [University of Michigan] http:www.worldmapper.org/posters/worldmapper_map168_ver5.pdf).

important lessons for the A2K movement. In the main, it demonstrates workable strategies by which A2K advocates can introduce transformative ideas into the mainstream discourse on intellectual property and by which they can challenge rules and standards that negatively affect development and the public interest. The first part of the essay focuses on the WIPO Development Agenda and the second on the negotiations over the proposal for a WIPO Broadcasting Treaty. We conclude with reflections on the future of A2K at WIPO.

Understanding the events that took place in the context of the WIPO Development Agenda and the debate on the necessity and scope of enhanced protections for broadcasting organizations by granting exclusive rights requires a deeper look at the business of multilateral negotiation and its ways, a prospect that may at first seem off-putting, but that provides an ideal setting in which to analyze the evolution and growing impact of the A2K movement. This also helps us look more closely at the alliances and positions (connections, shifts, consolidations) of the players involved among the states, civil society, and industry.

THE WIPO DEVELOPMENT AGENDA: PATHWAY TO A2K

The agreement on a development agenda is one of the most significant developments in the recent history of WIPO. The WIPO Development Agenda is essentially an effort to reform the current structure of global intellectual property policy making. It is aimed at reshaping the organization to increase its ability to address concerns that had been historically relegated to obscurity or absent entirely from the WIPO policy discussions and activities, that is, development and public-interest concerns, as well as the concerns of new industries. The core objective was to ensure that WIPO activities and intellectual property discussions would balance business interests with broader consumer and public interests and would be in line with the broad mandate of the UN to support the development goals of its developing countries and least-developed countries.[1]

In the process of negotiating the elements of the WIPO Development Agenda, the most acknowledged and notable proposal on A2K was the attempt to negotiate, within WIPO, an A2K treaty. However, the A2K agenda in the development agenda by no means has been confined to the A2K treaty proposal.[2] As such, although the A2K treaty proposal did not become part of the final list of agreed-upon recommendations under the development agenda, various recommendations remain relevant to the A2K agenda. Many of these, however, are not under the rubric of access to knowledge. Rather, they are listed under the rubrics of norm setting, flexibilities, public policy, and the public domain. For example, under one proposal, the member states of WIPO commit themselves to initiate discussions on how, within WIPO's mandate, access to knowledge and technology for developing countries and least-developed countries can be fostered. To grasp the relevance of the recommendations that emerged from the WIPO Development Agenda as a pathway leading to possible future developments for the A2K movement, however, first we need to examine the history of the negotiations that led to the acceptance of the WIPO Development Agenda.

THE HISTORY OF THE WIPO DEVELOPMENT AGENDA

The establishment of a WIPO Development Agenda was formally approved by the 184 member states of WIPO in September 2007 after three years of discussion. The initial proposal was presented by Brazil and Argentina at the September–October 2004 session of the WIPO General Assembly.[3] The proposal was then cosponsored by twelve other countries known as the "Friends of Development" and strongly supported by all developing countries.[4] A wide range of public-interest groups and other civil-society stakeholders also backed the development agenda initiative and actively lobbied government representatives to support the proposal.

In fact, some of the ideas of the development agenda had been discussed in earlier discussions led by civil-society groups. The Future of WIPO meeting organized by the Trans Atlantic Consumer Dialogue (TACD) in September 2004 and the resulting Geneva Declaration on the Future of WIPO were key developments. The discussions at the conference and the list of signatories to the declaration helped demonstrate the widespread support for the underlying ideas of the development agenda among civil-society groups, academics, and other sectors that previously had not engaged closely in WIPO debates.[5]

The WIPO Development Agenda initiative was groundbreaking in several ways. For the first time in recent history, developing countries presented an encompassing, alternative agenda to guide international policy making at WIPO. The development agenda proposal asserted that the work of WIPO as a specialized agency of the UN needed to follow the UN-wide broad development objectives such as those elaborated in the Millennium Declaration adopted in 2000 and affirming the overall goals of the UN. It sought to reestablish the role and responsibility of WIPO as a member of the UN family, which until then was seen as a technical agency that should be concerned only with uncritically promoting global intellectual property protection. On the premise that WIPO had not systematically incorporated the development dimension into all of its activities, the proponents of the development agenda called for various internal structural and substantive reforms.

The proposal by the core group of countries known as the "Friends of Development" elucidated and brought together in the WIPO context various concerns and ideas that had matured as part of the growing global debate on intellectual

property policy. Attention focused on WIPO largely because of the startling findings of various research studies that pointed to significant problems in the current intellectual property system in areas such as agriculture and public health
and within WIPO's internal processes. Such studies included the Report of the UK
Commission on Intellectual Property Rights,[6] the papers presented at the Bellagio Dialogues on Intellectual Property and Development,[7] and a paper by Sisule F.
Musungu and Graham Dutfield, "Multilateral Agreement and a TRIPS-Plus World:
The World Intellectual Property Organisation (WIPO),"[8] which critically examined
the role and activities of WIPO in the era following the adoption of the World
Trade Organization (WTO) Agreement on Trade-Related Aspects of Intellectual
Property Rights (TRIPS), enabling many outsiders to understand the importance
of participating in WIPO processes.

Developing countries had been raising questions about the activities, norm
setting, and other decision-making processes of the organization. Some concerns
voiced by these countries and evidenced by the above studies and others on norm
setting related to the overly active approach of the WIPO Secretariat to advance
the Substantive Patent Law Treaty and a broadcasting treaty when no assessment
had been made as to the need for such initiatives or their impact. Other concerns
included the significant influence of private-interest groups in WIPO and therefore negotiation outcomes, evidenced by the direct advisory role of the Industry
Advisory Committee to the WIPO director general.[9]

THE NECESSITY FOR A DEVELOPMENT AGENDA FOR WIPO

The adoption and entry into force of the TRIPS Agreement substantially changed
the international intellectual property landscape. It established the rule that all

WTO members must adhere to minimum intellectual property standards. As a result, the last decade has seen a fierce global debate about the impact of TRIPS standards on development and on the public interest in developing countries. In the minds of many, the problems of intellectual property are therefore associated more with the WTO than with WIPO. However, while the "politics" of intellectual property have mainly taken place at the WTO, new intellectual property rules are not being debated and created at the WTO, but rather under the auspices of WIPO.[10] In this context, activities at WIPO continue greatly to influence the shape of the international intellectual property system.[11] It is essentially the recognition of the importance of WIPO, in this sense, that spurred the actions related to the introduction of the original proposal on the WIPO Development Agenda.

The ideas and proposals suggested for the WIPO Development Agenda largely stem from the international debate on the current functioning and evolution of the intellectual property system in both developed and developing countries and the impact of that debate on different stakeholders. It is for this reason that the development agenda gathered significant momentum and the necessary political and technical support.

Two key questions are at the center of the current global intellectual property debate. The first concerns the costs and benefits of intellectual property protection in light of changing patterns of innovation and creative activity. The second concerns the impact of intellectual property rights on development and public-interest concerns such as access to medicines, access to knowledge, sustainable agriculture, nutrition, and the protection of biodiversity.[12] The far-reaching impact of the intellectual property system brought to the debate voices of a wide range of nontraditional stakeholders, including farmers, students, scientists, consumers, people suffering from life-threatening diseases, software developers, and innovative and creative businesses making use of alternative models of innovation.

The previously successful campaigns on intellectual property and access to medicines and discussions of intellectual property and biodiversity in different national contexts and multilateral organizations, such as the Convention on Biological Diversity and the WTO, had helped deepen the understanding among states and other stakeholders of the impact of intellectual property on public-interest and development concerns. One of the most notable achievements coming out of the earlier debate was the 2001 WTO Doha Declaration on TRIPS and Public Heath,[13] which effectively reaffirmed the primacy of public-health objectives over intellectual property protection.

Nothing similar had occurred in the recent history at WIPO. Historically the norm-setting and other activities of WIPO had focused on strengthening the protection for intellectual property rights and on advancing the global harmonization of standards of protection. Moreover, although developed countries are small in number, compared with developing countries, the former have remained the most active and influential participants at WIPO, alongside industry, lawyer associations, and other rights-holder interest groups. In terms of the role and responsibility of WIPO to contribute to the broader development goals of the UN system, the organization has focused on promoting the use and protection of intellectual property as "a tool for development" and the implementation of intellectual property–related obligations through the provision of technical assistance.

The approach of the WIPO in support of exclusive intellectual property rights can be explained by four factors. First, developed countries are the more powerful parties in WIPO, representing strong intellectual property–based industries, and as such are responding to those industries' demand for a strong, harmonized global system that facilitates and reduces the cost of intellectual property protection and enforcement around the world. Rights-holder groups from developed countries have been able to form strong alliances within their home countries and between developed countries and have built strong and long-lasting relationships with the WIPO Secretariat.

Second, until recently, developing countries did not make changing the rules of the game a priority. This was especially true for the least-developed countries, which are not required to implement the commitments under the TRIPS Agreement until the end of the transition period available to them.[14] In the post-TRIPS era and until a couple of years before the introduction of the development agenda proposal, the main concern and focus of the demands of developing countries at WIPO was to access technical and legal assistance to implement the obligations acquired under the 1995 TRIPS Agreement and subsequent WIPO treaties.[15] That remains a priority for a number of developing countries. Some have perceived that strengthening intellectual property protection remains in their best interests.

However, a growing number of developing countries are increasingly wary of the assumed positive correlation between intellectual property and development and concerned that the rigidity of the intellectual property system may affect their ability to address public-policy issues such as access to medicines and access to knowledge. The changing preferences of developing countries, particularly with respect to the work of WIPO, have increased the divergence among the preferences of developed and developing countries.

Third, the national and international debate on the costs and benefits of intellectual property, especially for developing countries, is a recent one. Only a few years ago, the notion that strengthened intellectual property–rights protection promotes development remained largely uncontested. The preferences and interests of WIPO, developing countries, and some developed-country member states have evolved along with the debate.

Finally, industry, lawyer associations, and other rights-holder interest groups historically have enjoyed a strong presence and deep influence at WIPO, reflecting the interests of the countries in which they originate. It is only in recent years that the participation of development-oriented and public-interest nongovernmental organizations at WIPO has increased significantly.[16]

THE FORMAL PROCESS FOR ESTABLISHING A DEVELOPMENT AGENDA FOR WIPO

The process for establishing a development agenda for WIPO was formally initiated when the WIPO General Assembly unanimously agreed at its 2004 annual session to consider the proposal by the "Friends of Development" and other proposals that other member states might submit on the subject.[17] Further, the General Assembly agreed on the future process to examine the specific suggestions contained in the original development agenda proposal and any additional proposals.[18] Two years later, a multiplicity of proposals for a WIPO Development Agenda were on the table.[19]

The 2006 General Assembly then faced the daunting task of determining how to move forward on the basis of a list of the 111 proposals submitted by member states. The proposals were grouped under six rubrics: technical assistance and capacity building; norm setting, flexibilities, and public policy and the public domain; technology transfers, information and communication technology, and access to knowledge; assessments, evaluations, and impact studies; institutional matters, including mandates and governance; and other issues.[20] The multiplicity of proposals made consensus building difficult, even among developing countries. Coalition building between the proponents and other interested stakeholders, such as civil-society organizations, also became a challenge. Given that the members could not come to any agreement, the General Assembly renewed the mandate of the Provisional Committee on the development agenda for the committee to accelerate their deliberations, report back, and make recommendations to the 2007 General Assembly.

After three years of intense debate and negotiations, member states finally agreed, in September 2007, on the need and method to establish permanently the development agenda for WIPO. Forty-five recommendations or proposals were necessary to mainstream development into the different WIPO program and activities.[21] The next phase will be to implement the proposals effectively, including defining the expected outcomes and deliverables and providing financing for the respective activities. To do so, the Committee on Development and Intellectual Property, which began meeting in early 2008, was established. Progress, however, has been slow, and no concrete implementation plan had been agreed upon at the time this text was written, in August 2008.

The committee has three main tasks: to develop a work program for implementation of the adopted recommendations; to monitor, assess, discuss, and report on the implementation of all recommendations adopted and for that purpose coordinate with relevant WIPO bodies; and to discuss intellectual property issues and development-related issues as agreed upon by the committee, as well as those decided by the General Assembly.

CONCEPTUAL FRAMING AND PROCESS STRATEGIES

How could the WIPO Development Agenda succeed, given the prior dominance of wealthy countries and business interests in its standard-setting and rule-making processes? One key ingredient of the success of the development agenda was the solid conceptual framework on which the original proposal was built. The proposal was framed by the "Friends of Development" in a way designed to reduce divergence among the preferences and interests of the powerful states and rightsholder groups, on the one hand, and developing countries and public-interest groups, on the other.

More specifically, the original development agenda proposal was built on two main concepts that challenged the general view hitherto widely held at WIPO—that neither intellectual property protection nor harmonization of intellectual property laws leading to higher protection in all countries, irrespective of their levels of development, can be seen as ends in themselves and that WIPO, as a specialized UN agency that is mandated to promote technological innovation and the transfer of technology, must explicitly support the UN's broader development goals. The power of these concepts is signaled by the fact that they were not contested by any member state. The kind of conceptual framework on which the original proposal was built was vital to achieve a positive response from whole membership of WIPO to the discussion of the proposal for the establishment of a development agenda.

Another ingredient of the proposal by the "Friends of Development" was the clarity they offered regarding the core problem and the measures they proposed to address it. For this reason, these countries could no longer be dismissed as simply complaining, but offering no direction.

The original development agenda proposal identified five main areas to be prioritized in reforming WIPO into a development-friendly organization: WIPO-mandate and norm-setting activities; transfers of technology; the implications of intellectual property enforcement; technical cooperation and assistance; and the concerns of all stakeholders, in particular, those of groups representing civil society. Within each of the areas, the perceived problems were presented together with the measures considered necessary to redress them. In a subsequent submission, the "Friends of Development" elaborated further elements and measures. The identification and prioritization of issues helped to refine the position of the proponents and the expected outcomes. It also served to rally increased support from other developing countries and from groups outside of WIPO.

To succeed in establishing a development agenda also required sustained and coordinated leadership by the those making the demands for it. While the development agenda sought to institute reforms primarily aimed at benefiting developing countries and other stakeholders from civil society that previously had not been admitted into the process of setting the WIPO agenda, a core group of member states, including Argentina, Brazil, Egypt, and South Africa, in fact led the initiative, expending the most political capital. Over time, leadership also emerged among the developed countries, the United Kingdom and Netherlands being notable here.

Despite the framing of the proposals as "development proposals," however, there was no guarantee that developing countries would agree on the elements of the development agenda—or even that they would all support it. Many countries did not formally sign onto the various documents setting forth the demand for the new agenda or actively participate in the formal deliberations. And while most developing countries did agree on the basic elements of the new agenda, they disagreed on the specifics. How important was it to change how WIPO did "technical assistance," compared with the goal of reforming its norm-setting processes? How quickly should the changes be implemented? Differences on issues such as these led countries to submit competing proposals, diluting the strength of development agenda proponents and diffusing their demands at various points.

THE ROLE OF NONGOVERNMENTAL ACTORS

One of the most important factors that contributed to advancing the development agenda process was the active engagement of a broad range of nongovernmental stakeholders. While the proposal to incorporate the "development dimension" in WIPO was led by a group of developing-country governments, the initiative received important support and input from a broader constituency in both developing and developed countries. The proposal for the WIPO Development Agenda was taken as an opportunity to consolidate and give coherence to the multiple initiatives and campaigns to reform the global governance of knowledge and technology. One of the important inputs to the development agenda process was the September 2004 "Geneva Declaration on the Future of WIPO."[22]

The declaration, as already noted, was drafted after a meeting in Geneva organized by the TACD that brought together various stakeholders from civil society, including nongovernmental organizations, public-health activists, consumer groups, academics, scientists, Nobel Prize laureates, and businesses. The declaration argued, among other things, that the WIPO Development Agenda created the first real opportunity to debate the future of WIPO.

The original development agenda proposal was also identified by various civil-society groups as part of a broader agenda to reform global institutions and regimes as they affect innovation, access to knowledge, and creative activity. Accordingly, the declaration broadened the conceptual basis of the WIPO Development Agenda beyond the emphasis on development as an issue of primary concern for developing countries. Development concerns were brought together with a wide range of public-interest and other concerns shared by constituencies in both the Global North and the Global South. The A2K proposals are an example. The declaration also affirmed that the WIPO Development Agenda was an agenda not only for developing countries, but for everyone.

The engagement with and support for the WIPO Development Agenda process by civil-society groups in the North helped in dealing with some of the difficult developed countries, such as the United States. Such member states, though powerful, could not ignore their own citizens and the local interests represented at WIPO through civil-society groups. These groups also brought important technical expertise to the debate. Important collaboration established among developing countries, particularly the "Friends of Development," and civil-society stakeholders ensured that the concerns of civil-society groups found their way into the specific proposals of the WIPO Development Agenda, such as the initiative for a treaty on A2K and commitments to increase efforts to bring civil-society groups into the WIPO discussions and to more open consultations and events in which civil-society groups could present their views to member states.

THE EFFECT OF THE DEVELOPMENT AGENDA ON A2K GOALS
While it is premature to evaluate the actual impact and/or success of the WIPO Development Agenda, given that the process for its implementation formally began only in 2008, the initiative has already brought about significant changes in the dynamics in the organization. In terms of substance, the development agenda process has served to attenuate the historical bias in WIPO policy making toward developed-country and rights-holder interests, as well as the dogmatic discourse on the benefits of strong intellectual property protection and harmonization. It has also allowed and stimulated a more open, participatory, and fact-based debate on the relationship between intellectual property and development and the public interest, as well as on the concerns voiced by the A2K communities. Ongoing norm-setting processes and discussions at WIPO, such as the proposed treaty on the protection of broadcasting organizations, have also been influenced by the debate on

establishing a development agenda. In terms of process, developing countries have taken a more active role in the discussions and in articulating their demands. Likewise, a broader range of civil-society actors are effectively engaging in WIPO.

In the longer term, the WIPO Development Agenda has the potential to do four things: establish a set of general principles on knowledge governance and intellectual property, provide a substantive program of work for WIPO, ensure good governance and the democratization of WIPO, and establish a basis for evidence-based standard setting and rule making in the organization.[23] The continued participation of constituencies concerned with A2K will be critical in realizing this potential.

THE PROPOSED WIPO TREATY ON THE PROTECTION OF BROADCASTING ORGANIZATIONS

In the post-TRIPS period, as already noted, WIPO has continued to advance new intellectual property norms and standards. One such area of work has taken place under the so-called "digital agenda." The digital agenda in WIPO has focused on adapting copyright and related proprietary forms of protection to the digital environment.[24] In the past ten years, the international protection of copyright and related rights has been expanded significantly to include new rights, extended terms of protection, and new subject matter, such as computer programs and databases. Under the auspices of WIPO, the scope of copyright protection has also been extended to create paracopyright regimes allowing copyright and related rights holders to make use of digital technology to gain greater control over the access, use, and distribution of content in electronic and digital form.

The most notable example to date is the WIPO Copyright Treaty and the WIPO Performances and Phonograms Treaty, which came into force in 2002. The WIPO Copyright Treaty and the WIPO Performances and Phonograms Treaty—commonly referred to as the "Internet Treaties"—extend protection for copyright and related rights in the digital environment and create new legal obligations to support the protection of on-line works via technological means. The new legal framework created by the Internet Treaties to control access to works in electronic and digital form effectively gives rights holders greater control over content.[25]

One of the main concerns, from an A2K perspective, is that government-backed technological-protection measures may render inapplicable the limitations and exceptions to access and use of works protected by copyright and related rights, such as for noncommercial research and educational purposes, and that, according to the WIPO Copyright Treaty and the WIPO Performances and Phonograms Treaty, may be devised as appropriate for the digital environment. This is because technological protection measures effectively block access to works, irrespective of the reason why access is sought, given that the technologies cannot distinguish whether the circumventing purpose is lawful or not. The problem becomes more acute when national legislation implementing the respective WIPO Copyright Treaty and the WIPO Performances and Phonograms Treaty obligations go as far as prohibiting not only the act of circumventing a technological-protection measure, but also the manufacture of and trade in devices that may be used to circumvent technological-protection measures, as in the case of the United States 1998 Digital Millennium Copyright Act. The experience of countries implementing paracopyright legislation shows that even when limitations and exceptions are defined in national law, technological-protection measures can prevent their exercise.

Notwithstanding these concerns, WIPO continued to be engaged in efforts to create additional rights for new players in the name of adapting existing rights to the digital environment. It is in this context that we examine the demand for and the debate relating to the protection of broadcasting organizations.

THE DEMAND FOR FURTHER PROTECTION FOR BROADCASTING ORGANIZATIONS

After the adoption of the WIPO Copyright Treaty and the WIPO Performances and Phonograms Treaty, audiovisual performers and broadcasting organizations demanded negotiations on new international treaties to extend and update the protection they enjoyed in the same manner that the two treaties had done for authors of creative works and the music recording industry. In the history of the initiative for the protection of broadcasting organizations, we can see a microcosm of the challenges to and the potentially revolutionary nature of the development agenda.

Starting in 1998, discussions commenced at WIPO on a proposed treaty on the protection of broadcasting organizations. Evidence of the direct influence of broadcasting organizations on the discussions is that the first treaty proposal was made by a coalition of broadcasting unions, which, according to the WIPO structure, fall under the category of nongovernmental organizations—NGOs. (NGOs are the only category under which any type of nongovernmental actor, whether industrial, commercial and/or nonprofit, noncommercial, can participate at WIPO as an observer.) Although according to WIPO rules, it is only member states that can officially submit proposals. In 1999, the WIPO Secretariat placed on the agenda of the second session of the Standing Committee on Copyrights and Related Rights the treaty proposals submitted by groups of broadcasting organizations, most from Europe and Japan.[26] At the time, the submissions by member states on the issue were not drafted in treaty form.

As of 2001, the question of improving the protection of the rights of broadcasting organizations by way of a proposed new international treaty became the main item on the agenda of the standing committee. The longstanding chair of the committee, Mr. Jukka Liedes, a representative of the government of Finland, pressed member states to submit proposals for the treaty's language. The subsequent basis for the discussions was a compilation of these proposals from member states by the WIPO Secretariat. The second compilation included proposals by Argentina, Cameroon, the European Communities and its Member States, Honduras, Japan, Kenya, Mexico, Paraguay, Tanzania, Ukraine, and, at a later stage, the United States.[27] By November 2003, the Standing Committee on Copyrights and Related Rights had agreed to continue discussions in April 2004, based on a consolidated draft text with explanatory comments that were prepared and distributed by the chair of the committee. Ultimately, it was up to member states to consider whether to convene a diplomatic conference, the last step in the treaty-making process at WIPO. It is at this stage that the initiative failed.

THE ROLE OF THE A2K MOVEMENT

The initiative failed because the consolidated text presented by the chair of the Standing Committee on Copyrights and Related Rights in no way reflected a consensus among member states. In a meeting in June 2004, it became apparent that there was no agreement on the objective of the treaty, or on the nature and scope of the protections it would offer for broadcasters, or even on evidence for the need for the specific rights proposed. However, its main proponents, including the European Union and Japan, rights-holder groups (particularly broadcasting organizations), and the very active WIPO Secretariat and chair of the committee continued to push strongly for the conclusion of the treaty.

Some contentious issues in the proposed treaty included proposals for extension of the scope of coverage of the rights currently granted to broadcasting organizations under the Rome Convention;[28] additional rights, such as the exclusive right to authorize or prohibit the retransmission of broadcasts following what is known as the "fixation" of such broadcasts — their reduction to material form; the addition of new beneficiaries of protection, not only traditional broadcasting organizations, but also cablecasting organizations and possibly simulcasting and Webcasting organizations; more restrictive limitations and exceptions to the rights conferred in the treaty, compared with the WIPO Copyright Treaty and the WIPO Performances and Phonograms Treaty; and more stringent obligations on technological-protection measures and digital rights management (DRM) than those contained in those treaties.[29]

In previous meetings of the Standing Committee on Copyrights and Related Rights, many member states, particularly developing countries, had not actively participated in the discussions on the protection of broadcasting organizations.

Moreover, a review of the reports of the committee and the discussions on the proposed Broadcasting Treaty in the WIPO General Assembly reveal that the participation and technical input of nongovernmental observers in the discussions was largely dominated by rights-holder groups, particularly groups representing broadcasting organizations.[30]

From June 2004 onward, however, there was a significant change. In particular, there was significant increase in the participation of consumer, public-interest, and development-oriented nongovernmental observers and representatives of the technology industry.[31] The participation of new players with technical expertise and the addition of the voices of different stakeholders brought about an important change in the dynamics and substance of the deliberations. This was a very important change, since no serious, in-depth debate on the implications of the proposed treaty provisions had ever taken place in the committee. Among other things, these groups questioned the broad scope of the treaty, the nature of the proposed rights, and their duration, and they highlighted possible unintended consequences, especially for the business models of the technology industry.

The new players created coalitions among public-interest and consumer groups in both the North and South and had a significant impact on the positions of key member states concerned with the potential negative impact of the proposed treaty on access to information and knowledge and on technological innovation. One of the main achievements of the new players was to bring representatives of the information technology, electronics, and telecommunications industries together with groups representing performer groups, library associations, development activists, public-interest and consumers on the basis of these concerns to build together a dynamic and broad-based coalition to oppose the proposed treaty in its current form.[32] AT&T, Dell, Intel, Verizon, and Sony were among the corporations that joined the coalition.

The involvement of the industry players was significant because it blunted the rhetoric of the broadcasting lobby and their allies. It dawned on many, even the staunchest supporters of the treaty within governments, that if some of the most successful companies in the digital era had a problem with the treaty, one needed to stop and think again. It was no longer easy to dismiss criticism as simply the work of a few NGOs opposed to intellectual property rights.

The parallel discussions on the WIPO Development Agenda also had an important impact on the subsequent debate on whether the Broadcasting Treaty was necessary. More member states demanded further deliberations on the proposed provisions, impact assessments, and studies before moving forward. These were all ideas that had been raised in the context of the discussion on the development agenda.

Ultimately, as a result of all the above factors, the 2006 WIPO General

Assembly rejected the recommendation of the Standing Committee on Copyrights and Related Rights to call for a diplomatic conference and instead asked for the convening of two special sessions of the committee to deliberate further on the essential elements of the proposed treaty—namely, its objectives, specific scope, and objects of protection. The General Assembly also clarified that the scope of the treaty would be limited to the protection of broadcasting signals from piracy, as opposed to granting broadcasting organizations additional rights. The two special sessions of the committee did not lead to any agreement on the basic elements of the proposed treaty, and consequently, the 2007 WIPO General Assembly decided that a diplomatic conference would not be convened in the near future.

The result is that the issue is no longer at the top of the agenda of the Standing Committee on Copyrights and Related Rights. Though the proponents have not given up their quest, it is likely to take some time before there can be another attempt to craft a treaty with such a broad scope, in light of there being little evidence either of a need for it or of its potential impact.

THE FUTURE OF A2K IN WIPO

The A2K movement has made an important contribution to the more systematic introduction of public-interest concerns into the deliberations of WIPO, particularly with respect to access to and sharing of on-line works for educational and research purposes. The inclusion of important A2K-related proposals in the WIPO Development Agenda and the halting of the discussions on the proposed WIPO Broadcasting Treaty clearly demonstrate the success achieved. However, looking forward to the future of the A2K agenda in WIPO, many challenges still lie ahead.

While it is unlikely that new momentum will emerge for the proposed WIPO Broadcasting Treaty, there is a need to continue to advance the A2K agenda in the Standing Committee on Copyrights and Related Rights, as well as in other WIPO bodies, on issues such as limitations and exceptions to intellectual property rights. The implementation of the A2K proposals of the WIPO Development Agenda will also require significant work to identify clear and specific actions that will enable the proposals to be realized. To maximize the potential impacts of the A2K initiatives in WIPO, the A2K movement will also need to work toward bringing greater coherence to related initiatives being pursued in other UN agencies and other international organizations.

Both the development agenda and the experience with the proposed treaty on broadcasting organizations demonstrate the flaws in WIPO's approach to intellectual property protection and provide a new opportunity for exploring alternative models of innovation and collaboration.

The future of A2K at WIPO is therefore bright, provided that the A2K movement and those concerned with A2K issues double their efforts to put on the table proposals that can benefit the needs of creative and competitive industries and businesses and the public interest, including development and consumer interests generally. Only the battles of this decade have been won. The war of the century—for the control of knowledge—may just be beginning.

NOTES

1   World Intellectual Property Organization, "Proposal to Establish a Development Agenda for the WIPO: Elaboration of Issues Raised in Document Wo/Ga/31/11," (April 6, 2006), WIPO doc. no. IIM1/4, para. 2, available on-line at http://www.wipo.int/edocs/mdocs/mdocs/en/iim_1/iim_1_4.pdf (last accessed April 4, 2009).

2   The proposal to establish or to adopt a treaty on access to knowledge and technology was originally proposed by the Group of the "Friends of Development." See Provisional Committee on Proposals Related to a WIPO Development Agenda, "Working Document for the Provisional Committee on the Proposal Related to a WIPO Development Agenda (PCDA)" (February 20, 2007), WIPO doc. no. PCDA/3/2, Annex B, proposal 34, dated February 20, 2008, available on-line at http://www.wipo.int/edocs/mdocs/mdocs/en/pcda_3/pcda_3_2.pdf (last accessed April 4, 2009).

3   World Intellectual Property Organization, "Proposal for the Establishment of a Development Agenda for WIPO," (April 6, 2005), WIPO doc. no. WO/GA/31/11, available on-line at http://www.wipo.int/edocs/mdocs/mdocs/en/iim_1/iim_1_4.pdf (last accessed April 5, 2009). The proposal was then developed further in "Proposal to Establish a Development Agenda for the WIPO: Elaboration of Issues Raised in Document Wo/Ga/31/11," (April 6, 2006), WIPO document IIM1/4 and World Intellectual Property Organization, "Proposal for the Establishment of a Development Agenda for WIPO: A Framework for Achieving Concrete and Practical Results in the Near and Longer Terms" (February 12, 2006), WIPO doc. no PCDA/1/5 (last accessed April 5, 2009).

4   The "Friends of Development" group included Argentina, Bolivia, Brazil, Cuba, the Dominican Republic, Ecuador, Egypt, Iran, Kenya, Peru, Sierra Leone, South Africa, Tanzania, and Venezuela. The majority of WIPO developing-country members, though refraining from official cosponsorship, expressed support for the proposal.

5   See the "Geneva Declaration on the Future of WIPO," available on-line at http://www.cptech.org/ip/wipo/futureofwipodeclaration.pdf (last accessed April 6, 2009). See also James Boyle, "A Manifesto on WIPO and the Future of Intellectual Property," *Duke Law and Technology Review* 9 (2004), available on-line at http://www.law.duke.edu/journals/dltr/articles/2004dltr0009.html (last accessed April 6, 2009).

6   Commission on Intellectual Property Rights, *Integrating Intellectual Property Rights and Development Policy* (2002), available on-line at http://www.iprcommission.org/papers/

pdfs/final_report/CIPRfullfinal.pdf (last accessed April 5, 2009).

7   Information on the Bellagio Dialogues is available at http://www.iprsonline.org (last accessed April 6, 2009).

8   Sisule F. Musungu and Graham Dutfield, "Multilateral Agreement and a TRIPS-Plus World: The World Intellectual Property Organisation (WIPO)," *TRIPS Issues Paper* 3, Quaker United Nations Office, Geneva, and Quaker International Affairs Programme, Ottawa (2003), available on-line at http://www.quno.org/geneva/pdf/economic/Issues/Multilateral-Agreements-in-TRIPS-plus-English.pdf (last accessed April 6, 2009).

9   For detailed explanations of these concerns, see Musungu and Dutfield, "Multilateral Agreement and a TRIPS-Plus World," and Boyle, "A Manifesto on WIPO and the Future of Intellectual Property."

10 As illustrated by the adoption, in 1996, of the WIPO Copyright Treaty and the Performances and Phonograms Treaty.

11 For detailed discussions see Carlos M. Correa and Sisule Musungu, "The WIPO Patent Agenda: Risks for Developing Countries," *T.R.A.D.E Working Papers* 12, South Centre, Geneva, available on-line at http://www.southcentre.org/index.php?option=com_content&task=view&id=76&Itemid=67 (last accessed April 7, 2009).

12 The 2002 report of the UK Commission on Intellectual Property Rights, *Integrating Intellectual Property Rights and Development Policy*, is one of the core works examining the relationship between intellectual property rights and development. The report outlined the real impacts of intellectual property, both positive and negative, on various areas, including health and agriculture. A key conclusion that surfaced from the report is that the impact of intellectual property rights and their use as an instrument of public policy will vary significantly among countries, depending on the level of scientific and technological capability and socioeconomic circumstances. Other important work on development and intellectual property included the various studies discussed and resulting from the Bellagio Dialogues on Development and Intellectual Property, as well as those published by the Quaker United Nations Office in Geneva and the Quaker International Affairs Programme, including Sisule F. Musungu, "Rethinking Innovation, Development and IP in the UN: WIPO and Beyond," *Issues Papers* 5, Quaker International Affairs Programme, Ottawa, 2005, available on-line at http://www.quno.org/geneva/pdf/economic/Issues/TRIPS53.pdf.

13 The Doha Declaration on the TRIPS Agreement and Public Health, November 14, 2001, WTO doc. no. WT/MIN(01)/DEC/2, is available on-line at http://www.wto.org/english/theWTO_e/minist_e/min01_e/mindecl_trips_e.pdf (last accessed March 24, 2009). See also Sangeeta Shashikant's essay, "The Doha Declaration on TRIPS and Public Health: An Impetus for Access to Medicines," in this volume.

14 The current transition period for least-developed countries runs until July 1, 2013 or until the date on which the country graduates from least-developed-country status. There is a separate timeline with respect to patent protection for pharmaceuticals, which runs until 2016.

15 WIPO, under a special agreement with the WTO signed in 1995, is the main multilateral provider of intellectual property technical assistance to WIPO and WTO developing-country and least-developed-country members.

16 See Ahmed Abdel Latif's essay, "The Emergence of the A2K Movement: Reminiscences and Reflections of a Developing-Country Delegate," in this volume.

17 *Report of the WIPO General Assembly: Thirty-First (15th Extraordinary) Session, September*

27–October 5, 2004 (October 5, 2004), WIPO doc. no. WO/GA/31/15, para. 128, available on-line at http://www.wipo.int/edocs/mdocs/govbody/en/wo_ga_31/wo_ga_31_15.pdf (last accessed April 8, 2009).

18   It also agreed that WIPO would organize together with the United Nations Conference on Trade and Development, the United Nations Industrial Development Organization, the World Health Organization and the World Trade Organization an international seminar open to all stakeholders to discuss intellectual property and public policy and intellectual property and development. This took place on May 2 and 3, 2005. While the event was the first of its kind at WIPO, it is unclear whether it influenced the Development Agenda process thereafter. The program is available at http://www.wipo.int/edocs/mdocs/mdocs/en/isipd_05/isipd_05_inf_1_prov.pdf (last accessed April 8, 2009).

19   In 2005, the proposals were discussed in the newly created Inter-sessional Intergovernmental Meeting on a Development Agenda for WIPO. In 2006, the meeting ceased to exist, and the discussion moved to the newly created Provisional Committee on Proposals Related to a WIPO Development Agenda until September 2007. The proposals are contained in the submissions by the following members: Argentina and Brazil, cosponsored by the "Friends of Development" (WIPO doc. nos. WO/GA/31/11, WO/GA/31/14, both available on-line at http://www.wipo.int/meetings/en/details.jsp?meeting_id=6309, and IIM/1/4, available at http://www.wipo.int/edocs/mdocs/mdocs/en/iim_1/iim_1_4.doc); the African Group (WIPO doc. no, IIM/3/2, available at http://www.wipo.int/edocs/mdocs/mdocs/en/iim_3/iim_3_2.doc); Bahrain, cosponsored by Jordan, Kuwait, Lebanon, Libyan Arab Jamahiriya, Oman, Qatar, Saudi Arabia, the Syrian Arab Republic, the United Arab Emirates, and Yemen (WIPO doc. no. IIM/2/2, available at http://www.wipo.int/edocs/mdocs/mdocs/en/iim_2/iim_2_2.doc); Colombia (WIPO doc, no. PCDA/1/3, available at http://www.wipo.int/edocs/mdocs/mdocs/en/pcda_1/pcda_1_3.doc); the United States (WIPO doc. no. IIM/1/2, available at http://www.wipo.int/edocs/mdocs/mdocs/en/iim_1/iim_1_2.doc); Mexico (WIPO doc. no. IIM/1/3, available at http://www.wipo.int/edocs/mdocs/mdocs/en/iim_1/iim_1_3.doc); and the United Kingdom (WIPO doc. no. IIM/1/5, available at http://www.wipo.int/edocs/mdocs/mdocs/en/iim_1/iim_1_5.doc) (all last accessed April 8, 2009).

20   The list of 111 proposals can be found in World Intellectual Property Organization, *Report of the WIPO General Assembly: Thirty-Third (16th Extraordinary) Session, September 25–October 3, 2006*, Annex A, WIPO doc. no. WO/GA/33/10, available on-line at http://www.wipo.int/edocs/mdocs/govbody/en/wo_ga_33/wo_ga_33_10.doc (last accessed April 9, 2009).

21   The final list of forty-five agreed recommendations for the WIPO Development Agenda, including nineteen recommendations for immediate implementation, are contained in World Intellectual Property Organization, *Report of the Assemblies of the Member States of WIPO: Forty-Third Series of Meetings, September 24–October 3, 2007*, Annexes A and B, WIPO doc. no. A/43/16, available on-line at http://www.wipo.int/edocs/mdocs/govbody/en/a_43/a_43_16-main1.pdf (last accessed April 9, 2009).

22   See n.5 above.

23   For detailed discussions see Sisule F. Musungu, "The Development Agenda: The Implications for IP Governance and the Future of WIPO," *Bridges* 11, no. 7 (2008).

24   The digital agenda was the initiative of WIPO Director General Kamil Idris. One of the core components of the digital agenda was to promote the entry into force of new WIPO treaties, the WIPO Copyright Treaty and the WIPO Performances and Phonograms (WPPT), before

December 2001. A second core component of the digital agenda was to "promote adjustment of the international legislative framework to facilitate e-commerce through i) the extension of the principles of the WPPT to audiovisual performances, ii) the adaptation of broadcasters' rights to the digital era, and iii) progress towards a possible international instrument on the protection of databases." For a discussion of the WIPO Digital Agenda, see Musungu and Dutfield, "Multilateral Agreement and a TRIPS-Plus World."

25   For further analysis of the WIPO Copyright Treaty and the WIPO Performances and Phonograms Treaty, including their practical implications, see, for example, Ruth L. Okediji, "The International Copyright System: Limitations, Exceptions and Public Interest Considerations for Developing Countries," *Issue Paper* 15, International Centre for Trade and Sustainable Development, Geneva, available on-line at http://www.iprsonline.org/unctadictsd/docs/ruth%202405.pdf (last accessed April 9, 2009). In 2010, the WIPO Copyright Treaty was in force in eighty-eight WIPO member states, while the WIPO WPPT was in force in eighty-six WIPO member states.

26   World Intellectual Property Organization, "Agenda Item 4: Protection of the Rights of Broadcasting Organizations: Submissions Received from Non-governmental Organizations by March 31, 1999," April 7, 1999, WIPO doc. no. SCCR/2/6, available on-line at http://www.wipo.int/edocs/mdocs/copyright/en/sccr_2/sccr_2_6.pdf (last accessed April 9, 2009).

27   World Intellectual Property Organization, "Protection of the Rights of Broadcasting Organizations: Comparison of Proposals of WIPO Member States and the European Community and Its Member States Received by April 15, 2003," WIPO doc. no. SCCR/9/5, available on-line at http://www.wipo.int/edocs/mdocs/copyright/en/sccr_9/sccr_9_5.doc, last accessed April 9, 2009).

28   Members of WIPO agreed to the Rome Convention for the Protection of Performers, Producers of Phonograms and Broadcasting Organizations on October 26, 1961. It extended copyright protection from the author of a work to the creators and owners of the physical implementations of that intellectual property, including the producers of audiocassettes.

29   For a detailed history of the proposed WIPO broadcasting treaty and an analysis of the draft text, see Viviana Muñoz Tellez and Chege Waitara, "A Development Analysis of the Proposed WIPO Treaty on the Protection of Broadcasting and Cablecasting Organizations," *Research Papers* 9 (January 2007), Geneva, South Centre available on-line at http://papers.ssrn.com/sol3/papers.cfm?abstract_id=1039301 (last accessed June 3, 2010).

30   A notable exception was the role of the United Nations Educational, Scientific and Cultural Organization (UNESCO), as an intergovernmental observer.

31   These included various groups, part of the Civil Society Coalition: the Electronic Frontier Foundation; Knowledge Ecology International, formerly the Consumer Project on Technology; European Digital Rights; IP Justice; and the Union for the Public Domain, among others.

32   See for example, "Joint Statement Opposing Broadcast Treaty by Broad-Based Coalition," available on-line at http://drn.okfn.org/node/135 (last accessed April 9, 2009).

PART THREE

THE CONCEPTUAL TERRAIN OF A2K

Demonstration during the sixth round of negotiations for a free-trade agreement between the United States and Thailand, Chiang Mai, January 11, 2006 (Gaëlle Krikorian).

"IP World"—Made by TNC Inc.

Peter Drahos

We live in a world where the rules of intellectual property (IP) and the intellectual property generated using those rules are globally pervasive phenomena. For example, in the nineteenth century, two important multilateral agreements on intellectual property were negotiated by some states: the Berne Convention for the Protection of Literary and Artistic Works (1886) and the Paris Convention for the Protection of Industrial Property (1883). Today, the World Intellectual Property Organization (WIPO) administers some twenty-three treaties on intellectual property.

The global quantity of intellectual property being generated under the rules of intellectual property cannot really be accurately quantified, but it is vast. By way of illustration, in 2004, there were about 5.5 million patents in force around the world.[1] There were at least another 5 to 6 million unexamined patent applications. In 2006, the U.S. Patent and Trademark Office reported that there were 1,332,155 active certificates of trademark registration in the United States. There are many other forms of intellectual property that one would have to add to a global stock-taking of intellectual property, including the number of works protected by copyright, the number of plant variety registrations, the number of registered designs, the number of protected circuit layouts, and so on.

One important issue is whether globalizing the rules of intellectual property and encouraging the production of more and more intellectual property under those rules will lead to a continuous increase in social welfare. In a moment, we will see that as a matter of theory, more intellectual property does not necessarily mean more social gains. This leads to the question that is the focus of this paper: If there are dangers and risks in continuing to expand both the rules of intellectual property and the production of intellectual property, why is this expansion occurring? The answer can entail explanations of the structural kind or of the agent-centered kind or something in between. This paper focuses on agents in the form of companies and individuals, but especially on transnational corporations—TNCs. A focus on agents raises the possibility for social action. If the intellectual property world that we have today reflects certain choices and actions taken by one group of actors, can other actors with different views about the desirability of intellectual property change the direction of its growth? Our answer is a qualified yes.

TOO MUCH INTELLECTUAL PROPERTY?

Economic theory suggests that a society that had no intellectual property protection at all would almost certainly not be allocating resources to invention and creation at an optimal level.[2] But equally, a society that went to extremes of protection would almost certainly incur costs that would exceed the benefits. Intellectual property rights permit owners to exclude people from the use of socially valuable information. At some point, allowing intellectual property owners to exploit this power of exclusion becomes too costly in terms of social welfare. The rules of arithmetic, for instance, can be used and reused endlessly. The costs of excluding people from the use of these rules would be very high in economic terms and in terms of basic human freedoms. The diagram below illustrates the proposition that one can have too much intellectual property protection. It also suggests that there is an optimal level of intellectual property protection.

Like most abstractions, Diagram 1 does not capture the real-world dynamic complexity of the way in which intellectual property rights and the growth of knowledge actually interact. For example, it implies nothing about the mix of intellectual property systems that a society should employ. A patent system, for example, might not be part of an optimal mix. In the nineteenth century, both Holland and Switzerland were able to industrialize without a patent system. A patent

DIAGRAM 1 The strength of intellectual property rights (IPR) standards and social welfare.

system might be part of an optimal mix, but whether it is or not depends in part on the scope of patentability. For example, as other chapters in this book illustrate, the efficiency of extending patents to software, business methods, and pharmaceutical products is highly debatable.

When we come to think about optimal levels of intellectual property protection in the context of a world of interdependent nation-states, it is clear that there is not one level of protection that is universally optimal for all states. It is clear that imitative production and learning are important to developing countries. TNCs operating in developing countries typically do so with higher levels of knowledge assets than domestic firms, for example. There is scope for domestic firms to benefit from this positive externality.[3] But, whether domestic firms make productivity gains is profoundly affected by the property rules that govern imitative production. Imitative production and learning require an appropriately designed set of intellectual property rights (for example, rules that permit some degree of reverse engineering). We know, for example, that Japan for a large part of the twentieth century designed and used a patent system that placed the emphasis on the diffusion of knowledge, rather than on the right to appropriate knowledge.[4]

Imitative production typically requires less capital, a factor that is important in developing countries. If, with Ronald H. Coase, we think of property rights as a factor of production, it follows that those property rights should be designed in ways that match the comparative advantage that a country has in other factors of production.[5] This suggests that there will be real long-run costs for developing countries if they continue to participate in a global regime of intellectual property rights that continues to ratchet up standards of protection. Much the same conclusion follows from the theory of comparative capitalism.[6] This theory suggests that countries must choose their system for regulating intellectual property with an eye to how it will fit other crucial legal and industry policy institutions, from competition policy to labor-market policy. Property and these other institutions form an organic whole. Whether or not particular property rights contribute to the well-being of the whole is a matter of careful diagnosis. Crucially, just like a physician, countries must have the freedom to design the right treatment once the diagnosis has revealed the source of the problem. As Jeffrey Sachs says, development economics must strive to be more like clinical medicine in its approach to problems.[7]

The idea that there are different optimal points of intellectual property protection for different countries is captured in Diagram 2 below. Even if there are benefits for New Guinea in having a patent system (and this is an open question), an optimally designed patent system for New Guinea is likely to be very different from that of an optimal system for that of the United States. In Diagram 2, Country B's optimal point of intellectual property protection is well and truly passed by the standards of protection required in order for Country A's optimal point to be reached. If Country B is required to harmonize with Country A's standards of protection it is likely to be made even worse off.

Like Diagram 1, Diagram 2 abstracts from a much more complex empirical reality. At a given point in time in a country's development history, the wrong set of institutional choices when it comes to intellectual property rights may drive it into negative territory when it comes to the welfare impacts of intellectual property rights. For example, a country such as New Guinea, which has a weak manufacturing base and a minerals-based economy, has virtually nothing to gain from adopting a patent system. Yet in order to meet its World Trade Organization (WTO) obligations, it has adopted a patent law based on a WIPO model law. It also has a growing HIV/AIDS crisis. Depending on what happens in the next decade, New Guinea may find that as a result of its membership in the Patent Cooperation Treaty, it ends up being designated for pharmaceutical patent applications. Such patents may well complicate the New Guinea government's capacity to access the cheapest medicines. There are other kinds of complex interdependencies at work. New patent laws in countries such as India and China, which have been a source of low-cost pharmaceuticals, when combined with the patent law in New Guinea, may also complicate access. The curve for New Guinea for patents might take on the shape in Diagram 3 below.[8]

This brief analysis of the economics of intellectual property in the context of economic development suggests that it would be prudent for states to retain design sovereignty over intellectual property rights. Moreover, given the differences in development among nations, one might expect to find a real diversity of standards of intellectual property protection. When we look at the intellectual property world, however, instead of finding diversity, we find an increasing

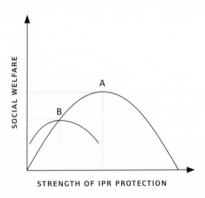

DIAGRAM 2 Different optimal points for different countries.

convergence on standards of intellectual property protection. For example, all the members of the WTO have to comply with the standards of protection that are set down in the Agreement on the Trade-Related Aspects of Intellectual Property Rights (TRIPS). Since TRIPS came into operation, states have signed hundreds of bilateral agreements, many of which include provisions that deal with intellectual property and that set standards of protection that are higher than required under the minimum standards of TRIPS.

Accompanying this global spread of intellectual property standards has been a vast growth in the bureaucracies that administer intellectual property rights. Patents, trademarks, and designs are registration systems and therefore require a bureaucracy that examines applications for the rights, decides on eligibility, and maintains a register of the rights. Patent offices are costly operations. The U.S. Patent and Trademark Office has a staff of some sixty-five hundred, the Japanese Patent Office some twenty-five hundred, and the European Patent Office approximately five thousand.

The costs of creating intellectual property rights do not end with administration. Property rights that cannot be enforced are worth little. Enforcement requires the participation of civil courts and specialist tribunals. Increasingly, criminal-law-enforcement agencies have begun to play a much greater role in enforcement as states have moved down the path of criminalizing the infringement of intellectual property.

Administering and enforcing intellectual property is particularly costly for developing countries. Should they direct their scarce scientific resources into patent examination? In order to save on the costs of patent administration, they may be tempted to rely on the work of offices such as the European Patent Office or the U.S. Patent and Trademark Office, but will the work of these offices meet the

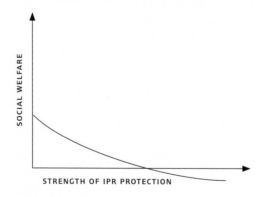

DIAGRAM 3 Losses and no gains.

French police tracking IP infringement (Sirpa-gendarmerie).

needs of developing countries? Similarly, there is a real issue as to whether developing countries should devote scarce criminal-justice resources to enforcing what in the end are private monopoly privileges and what historically have been the subject of civil proceedings.

THE ORIGINS OF THE IP WORLD: FROM THE MEDIEVAL PERIOD TO THE 1980S

Historically speaking, two types of actors have been key in the processes that have led to the globalization of intellectual property rules and the production of intellectual property: states and companies. The sovereigns of newly emerging states in medieval Europe well understood the importance of capturing resources for the benefit of the states, resources that included knowledge. There was widespread warfare between the powers of Europe, of which the Hundred Years' War between England and France was but one example. Natural disasters such as the Black Death and crop failures were other sources of instability. Sovereigns found themselves having to compete for skilled artisans who could bring commercially and militarily important goods and techniques to their territories. To some extent, the comparative advantage of nations and city states was locked up by them in the guilds that formed around all important technologies, such as mining, the making of sailcloth, machines for milling or weaving, and so on. Local guilds could not, however, provide all the innovation that the emerging states of the time demanded. Using the privilege system to entice foreign skilled workers to defect from their guilds and relocate to another territory was a natural step for sovereigns to take. It was a way of building comparative advantage and robbing others of theirs. For this reason, many monopoly privileges of the Middle Ages went to foreigners.

Out of the medieval system of privileges that existed across Europe there evolved statutory forms of intellectual property, copyright and patents being the earliest examples. At the same time, the laws relating to the formation of corporate personality were also evolving and being used for the purposes of business and trade. The links between intellectual property and the economic interests of companies existed early on in the history of intellectual property (for example, the Stationers Company and printing privileges in sixteenth-century England). The large-scale use by companies of systems of intellectual property evolved more slowly. States did not really invest in the creation of the administrative infrastructure needed to run systems of intellectual property rights till the nineteenth century. So, for example, patent offices were modernized and patent fees were reduced. Companies in Europe and the United States began to see that patents could confer business opportunities not just in their domestic markets, but in markets abroad. The United States and the lead industrial states in Europe saw in intellectual property, especially patents, a means by which to increase control over resources that mattered in the final instance to state power.

By the 1880s, the United Kingdom, Germany, France, and Belgium led Europe in terms of industrialization, and Europe led the world. For example, it accounted for 63 percent of the world's steel production.[9] Across the Atlantic, the United States was making giant strides in industrialization. By 1913, its manufacturing output matched that of the United Kingdom, France, and Germany combined, accounting for some one-third of world production.[10] Nation-states, which had become the principal unit of political and economic organization, saw that their economic supremacy depended on their capacity to compete in the heavy industries of coal, iron, and steel, as well as in the new industrial technologies based on chemicals and electricity. One of the important features of this period of industrial growth in Europe and the United States was the increase in monopolistic business combinations in the form of cartels, trusts, or syndicates.[11] Put simply, firms colluded and cooperated in the marketplace. There were, of course, some attempts by governments to deal with this, the passage of the Sherman Antitrust Act in 1890 in the United States being the single most notable example. In Europe, some industries rapidly organized themselves into national cartels. In Germany, for example, the firms in the coal-tar dye industry formed special agreements to regulate production and the exchange of patented knowledge. By 1913, 88 percent of world chemical dye production was controlled by German industry, with the companies being part of one of two agreements that regulated the industry and that were merged into one in 1916.[12]

The institutions of intellectual property were regularly the subject of attack by skeptics. In the nineteenth century, the patent system narrowly survived an attack by a coalition of free traders, economists, and some politicians. The arguments against the patent system were much the same as today. These included the view that prizes and other payments are a superior way to create incentives for inventors, that the monopoly costs of the patent system outweigh its incentive effects, that there are doubts about its incentive effects in any case, that patents inhibit trade across borders, and that patents are not natural rights.[13] Fritz Machlup and Edith Penrose suggest that one of the main ways in which defenders of the patent system prevailed was by the use of sophisticated techniques of propaganda. It is a point that remains relevant today. Critics of intellectual property rights often find themselves embroiled in propaganda wars in which criticism of the design of intellectual property is framed as an attack on private property rights and the rights of investors. Corporate intellectual property owners use this rhetorical framing technique to shroud the fact that they are pushing states into expanding and enforcing private monopoly rights.

In any case, because the lead industrial states at the beginning of the twentieth century realized that their military and economic power depended on the key industries of coal, iron, steel and chemical production, they concluded that if the lead industrialists in these sectors supported the patent system, so would they. Not for the last time, states bought into the belief that strong intellectual property rights would make for a strong state.

For the first half of the twentieth century, states concentrated on developing the two nineteenth-century pillars of the international framework for intellectual property—the Berne Convention and the Paris Convention. In general, the companies that participated in domestic and international processes of intellectual property lawmaking tended to view those processes from the perspective of national businesses protecting national or regional interests. American publishers, for example, were not a strong force for encouraging the United States to join the Berne Convention (the U.S. did not join till 1988). The publishing cartels that were formed in the first half of the twentieth century between U.S. and UK publishers (known as the British Publishers Traditional Markets Agreement) were more defensive in nature, dividing up the world into territories where one would agree not to trespass on the business interests of the other.

In some industries, the chemical and pharmaceutical industries being an example, some companies did have an aggressive international focus. The German chemical industry employed thousands of chemists, and their output was measured by thousands of patents. Companies such as Bayer and Badische Anilin Fabrik held hundreds of patents in America. German industry held in total approximately forty-five hundred U.S. patents, creating a "colossal obstacle to the development of the American dyestuff industry."[14] But there was also sufficient flexibility in the international system for states to be able to defend their national interests along with companies that saw themselves as nationally, rather than globally based. A good example of the scope of this flexibility was the change that the United Kingdom made to its patent law in 1919 preventing the patentability of chemical compounds. Chemical processes remained patentable. Fearing the might of IG Farben, British industry pursued a strategy of free riding by concentrating on inventing better processes that duplicated German dyestuffs.

For the most part, companies and industries took an interest in those areas of intellectual property that directly affected their particular business model and did not operate in lobbying terms across all of intellectual property. The publishers were active in copyright, pharmaceutical companies were active in patents, and a variety of industries were active in trademarks. A cross-cutting, unified approach to intellectual property by companies did not take place until the 1980s, when they united on a common agenda for an international intellectual property framework in the context of the General Agreement on Tariffs and Trade Uruguay Round of Multilateral Trade Negotiations.

TNCS AND TRIPS

The antecedents of this unified approach to intellectual property lie in the rise of transnational corporations after the Second World War. TNCs are characterized by the fact that their investment strategy takes the form of foreign direct investment in production, sales, and distribution. The vehicles of this foreign direct investment are foreign affiliates that allow the TNC to manage a centrally coordinated foreign direct investment strategy across a large number of countries. TNCs are companies that have a genuinely global investment philosophy. At base, TNCs evaluate the regulatory systems of nation-states in terms of the impact of those systems on their ability to make, control, and manage their investments in those states. It was this investment philosophy that ended up unifying different TNCs during the course of the Uruguay Round on the crucial issues of trade in services and intellectual property.

The U.S. pharmaceutical industry, and Pfizer, in particular, turned out to be leaders in the Uruguay Round because pharmaceutical companies were among the first companies to change into genuine TNCs. Pfizer, facing strong domestic competition in the production of penicillin after the end of World War II, moved to a program of expansion into developing country markets. Pfizer's move into overseas markets was the idea of John "Jack" Powers, Jr., assistant to the president, then president himself of the company, who in effect globalized Pfizer as a firm. Out of his initiative was born Pfizer International. Manufacturing plants and distribution networks were established "in countries ranging from Argentina to

Australia and Belgium to Brazil."[15] By 1957, Pfizer International had achieved more than its target of $60 million overseas sales.

Pfizer's investment in developing countries sensitized it to the threat to international markets that generic manufacturers in countries such as India posed for the pharmaceutical research-and-development industry. It also saw that developing countries were increasingly using their superior numbers in WIPO to put forward initiatives that favored their own position as net importers of foreign technology. During the early 1980s, a small group of Washington-based policy entrepreneurs had conceived the idea of linking the intellectual property regime to the trade regime. Pfizer executives, including the CEO Edmund Pratt, were among the leading proponents of this idea. Essentially, their policy idea was to get an agreement on intellectual property into the General Agreement on Tariffs and Trade (GATT). Among other things, such an agreement would be enforceable under GATT dispute-resolution procedures. Existing intellectual property treaties such as the Berne Convention lacked meaningful enforcement mechanisms. Moreover, the wide membership of GATT meant that the enforcement mechanism would be potentially available for use against more states.

Pfizer executives used their established business networks to disseminate the idea of a trade-based approach to intellectual property. Pratt began delivering speeches at business forums such as the National Foreign Trade Council and the Business Round Table, outlining the links between trade, intellectual property, and investment. As a CEO of a major U.S. company, he could work the trade-association scene at the highest levels. Other Pfizer senior executives also began to push the intellectual property issue within national and international trade associations. Gerald Laubach, president of Pfizer Inc., was on the board of the Pharmaceutical Manufacturers Association and on the Council on Competitiveness set up by President Ronald Reagan; Lou Clemente, Pfizer's general counsel, headed up the Intellectual Property Committee of the U.S. Council for International Business; Bob Neimeth, Pfizer International's president, was the chair of the U.S. side of the Business and Industry Advisory Committee to the Organisation for Economic Co-operation and Development. The message about intellectual property went out along the business networks to chambers of commerce, business councils, business committees, trade associations, and peak business bodies. Thus, Pfizer executives who occupied key positions in strategic business organizations were able to recruit the support of more and more organizations for a trade-based approach to intellectual property. With every such enrollment, the business power behind the case for such an approach became harder and harder for governments to resist.

Pfizer also managed to gain representation on a key committee, the Advisory Committee on Trade Negotiations, created in 1974 by the U.S. Congress under U.S.

trade law as an organization of numerous private-sector advisory committees with the Advisory Committee on Trade Negotiations at its apex. The purpose of this committee was to ensure a concordance between official U.S. trade objectives and the U.S. commercial sector. Pratt, with the assistance of other senior executives within Pfizer, began to put himself forward within business circles as someone who could develop U.S. business thinking about trade and economic policy. In 1979, Pratt became a member of the committee and in 1981 its chairman. During the 1980s, representatives from the most senior levels of big business within the United States were appointed by the president to serve on the committee. Out of this business crucible came the crucial strategic thinking on the trade-based approach to intellectual property.

With Pratt at the helm and the CEOs of IBM and Du Pont Corporation serving on the committee, it began to develop a sweeping trade and investment agenda. John Opel, the then chairman of IBM, headed this task force. During Pratt's six years of chairmanship, the Advisory Committee on Trade Negotiations worked closely with William E. Brock III, the U.S. trade representative from 1981 to 1985, and with Clayton K. Yeutter, the U.S. trade representative from 1985 to 1989, helping to shape the services, investment, and intellectual property trade agenda of the United States.

The committee's basic message to the U.S. government was that it should pull every lever at its disposal in order to obtain the correct result for the United States on intellectual property issues. U.S. executive directors of the International Monetary Fund and World Bank could ask about intellectual property when casting their votes on loans and access to bank facilities; U.S. aid and development agencies could use their funds to help spread the intellectual property gospel. Over time, the message was heard and acted upon. Provisions protecting intellectual property as an investment activity were automatically included in the Bilateral Investment Treaty program that the United States was engaged in with developing countries in the 1980s. Means of influence of a personal and powerful kind also began to operate. George Shultz, the secretary of state, discussed the intellectual property issue with Prime Minister Lee Kuan Yew of Singapore, according to Jacques Gorlin in his 1985 analysis of the trade-based approach to intellectual property. President Reagan in his message to Congress of February 6, 1986 entitled "America's Agenda for the Future" proposed that a key item in that agenda should be greater protection for U.S. intellectual property abroad. This was consistent with the recommendation of the Advisory Committee on Trade Negotiations that the development of a U.S. strategy for intellectual property be endorsed by the president and the cabinet. The ground was being prepared for intellectual property to become the stuff of big-picture political dealing, and not just technical trade negotiation. The ground was being prepared for the TRIPS Agreement.

The detailed story of how TRIPS came to be part of the Final Act of the Uruguay Round has been told elsewhere.[16] Key to the achievement of TRIPS was the formation of the Intellectual Property Committee. The Intellectual Property Committee was an ad hoc coalition of thirteen major U.S. corporations; Bristol-Myers, DuPont, FMC Corporation, General Electric, General Motors, Hewlett-Packard, IBM, Johnson and Johnson, Merck, Monsanto, Pfizer, Rockwell International, and Warner Communications. It described itself as "dedicated to the negotiation of a comprehensive agreement on intellectual property in the current GATT round of multilateral trade negotiations."[17]

Europe was the key target for the committee. Once Europe was on board, Japan was likely to follow, or at least would not raise significant opposition. The support of European and Japanese corporations was crucial. What followed was a consensus-building exercise carried out at the highest levels of senior corporate management. CEOs of U.S. companies belonging to the Intellectual Property Committee would contact their counterparts in Europe and Japan and urge them to put pressure on their governments to support the inclusion of intellectual property in the Uruguay Round. Ultimately, the linkages that were created between U.S., European, and Japanese companies led to the joint release in 1988 of a suggested draft text of an agreement on intellectual property.

TRIPS was a stunning negotiating victory that was made possible because a small group of individuals in the 1980s saw the possibilities of networked governance, especially when those networks could capture and deploy a "big stick" in the form of U.S. trade threats. Within these intersecting TNC networks, there were pools of technical expertise upon which to draw for the purposes of producing a draft agreement, while other networks steered the draft through a multilateral trade negotiation involving more than one hundred states that lasted from 1986 to 1993. Important to this achievement were a small number of business actors who created ever-widening circles of influence that enrolled more actors in networks that had TRIPS as their mission.

POST-TRIPS

The post-TRIPS era has seen a shift to bilateral trade agreements as the principal means for spreading intellectual property norms by means of the trade regime. These agreements contain standards that are either the same as or higher than those to be found in TRIPS. In the United States, TNCs continue to monitor these agreements through a U.S. trade representative advisory committee called IFAC-3 (the Industry Functional Advisory Committee–3). IFAC-3 is made up of twenty members drawn from Industry Sector Advisory Committees and another twenty

Cartoon of a negotiation discussion (*Diario La Republica*, Lima, September 27, 2006, www.grain.org/photos).

members drawn from private-sector areas who provide the committee with a large pool of expertise in intellectual property. The private-sector members are: the International Intellectual Property Alliance; The Gorlin Group; Pfizer, Inc.; the Law Offices of Hope H. Camp, representing Eli Lilly and Company; the Pharmaceutical Research and Manufacturers of America; Cowan, Leibowitz and Latman, P.C.; the Anheuser-Busch Companies, Inc.; Merck and Company, Inc.; the National Foreign Trade Council, Inc.; Powell, Goldstein, Frazer and Murphy, LLP, representing the Biotechnology Industry Organization; Time Warner Inc.; the International Anti-Counterfeiting Coalition; the Recording Industry Association of America; the Intellectual Property Owners Association; and Levi Strauss and Company.

IFAC-3 works across all U.S. trade initiatives on intellectual property, whether bilateral, regional, or multilateral. It is thus able to coordinate at a technical level the work it does across these different forums, thereby ensuring that U.S. trade-negotiating initiatives push intellectual property standards in the direction that U.S. industry would like. IFAC-3's technical expertise, as well as the expertise available to it from its members' corporate legal divisions, means that, for example, it can evaluate a country's intellectual property standards in detail when that country seeks accession to the WTO, and it can provide detailed assessments of the standards that U.S. trade representative negotiators must bring home in a negotiation.

The other striking feature of the post-TRIPS era has been the increasing involvement of civil-society NGOs in intellectual property policy. Today, there are thousands of NGOs working on issues such as access to medicines, access to knowledge, biopiracy, indigenous intellectual property rights, licensing, Internet governance, copyright-user rights, software freedom, and so on. The presence of NGOs working on a range of intellectual property issues provides scope for an alliance between developing states and NGOs. United minority factions can, under certain conditions, secure global regulatory change, the Declaration on the TRIPS Agreement and Public Health of 2001 being an example. Western NGOs are at their most effective when they can capture Western media interest and publicity. Often this requires a crisis of some kind. It has taken literally millions of deaths in Africa in order for the Western media to become interested in the links between patents, price, and AIDS drugs, despite the fact that cartelism in the pharmaceutical industry has been a problem for the health-care system of developing countries for decades.[18]

THE FUTURE

The possibility of securing change that benefits citizens in the context of intellectual property rights should not be overestimated. For the most part, intellectual property policy ends up mired in complex debates over rules and systems that only a few insiders really understand. Ignorant or corrupt politicians will nine times out of ten listen to the TNC representative who promises that bad things will happen to investment if policy X, which favors stronger intellectual property rights, is not followed. Of course, disagreements over the rules of intellectual property do break out among TNCs. A good example is the recent conflict over the rules that regulate the use of continuations in the U.S. patent system. Continuations are applications for inventions that have already been claimed in earlier applications. They are a way of keeping the application process going. Continuations are used most heavily in the biotechnology and chemical fields. The lack of restrictions on their use means that examiners have to devote time to reworking applications already examined, time that could be used to deal with new applications.[19]

As part of its attempt to reduce its volume of applications, the U.S. Patent and Trademark Office issued rules placing limits on the use of continuations.[20] This rule change was supported by a number of large companies, including Intel. Intel has a patent strategy based on filing for many patents and obtaining them as quickly as possible. Rules that allow an applicant to play for delay hold no advantage for it and in fact hurt its patent strategy, because continuations divert scarce examination resources to applications with earlier priority dates. (Applications forming part of a continuation chain get the benefit of earlier priority dates.) Intel thus supported the patent office rule change.[21] Members of the biotechnology industry, on the other hand, are in the unfortunate position of not ever being sure what they have invented. The complexity of the biochemical world means that they are often left guessing, but they nevertheless file patents early and then use the continuation process to refine their original application.[22] The biotech industry came out against the patent office's rule changes. GlaxoSmithKline was part of a group of plaintiffs that were successful in temporarily stopping the patent office from implementing the rules on November 1, 2007.

The above example shows how a reform desired by one TNC player may be seen as damaging by another. Compromises and incremental rule changes are the usual results of this kind of conflict. But the divisive politics that are generated by specific rules of intellectual property should not be confused with the global politics that surrounds the institution of intellectual property. Here, TNCs remain united, because despite their different business models and strategic uses of intellectual property, they understand that the globalization of intellectual property is consistent with their long-term investment strategies. The source of their unity does not lie in any of the abstract philosophies of intellectual property, such as natural property rights, utilitarian-based cost-benefit approaches, or personality theories, because all of these would set limits on the corporate ownership of intellectual property. Rather, TNCs are unified by the belief they will all do better in a world where states and citizens have embraced an ideology that favors hyperstrong intellectual property rights because that ideology enables those TNCs to invest in turning knowledge from a public good into a private good and to set the terms of access to it.

The claim being made here about the unity of TNCs should not be read as a claim about specific rules of intellectual property. TNCs will often be opponents in the context of a given set of international business rules because of the way those rules limit or increase their individual strategic opportunities. Other essays in this book have drawn attention to the divisions among TNCs in the context of software patents and the Broadcasting Treaty. The unity of TNCs does not operate at the level of rules, but rather at the level of deep ideology, because it is deep ideology that defines the evolutionary space in which some institutions flourish and others do not. Even if, for example, an Intel and a GlaxoSmithKline disagree about the reform of continuations in the context of the U.S. patent system, they remain unified on the need for a strong patent system to be spread to as many countries as possible and for that system to be made cheap and easy to use so that they can pursue their respective global patenting strategies. The inevitable disagreements among TNCs over the specifics of intellectual property rules should not blind us to their deeper-level ideological unity over the constraints to be placed on the evolutionary space of intellectual property institutions.

A good example of the unified TNC ideology that surrounds the institution of intellectual property is the proposal by Japan, the European Community, the United States, and Switzerland for a new plurilateral Anti-Counterfeiting Trade Agreement. The idea behind the proposal is to forge new standards for the enforcement of intellectual property rights to combat global counterfeiting and piracy. All TNCs have been calling for some time for more to be done on the enforcement of intellectual property, calls that have been broadcast through state mouthpieces such as G8 gatherings. The same states and TNCs that pushed for TRIPS are now pushing a global enforcement agenda that will deeply affect the rights and privacy of citizens in developed and developing countries alike. The enforcement push is real, and it will profoundly affect policing resources in developing countries over the coming decades. The e-mail message below that circulated recently in South Africa might also be seen as a future leitmotif in which states have been persuaded to rearrange their criminal enforcement priorities:

> Please take note as this is scheduled for the weekend. . . .
>
> Please note that SAPS [the South African Police Service] are having roadblocks where they will check all CD Recordables in your car. If they find any, you get arrested and taken down to the police station so that all of them can be checked for pirated copies.
>
> The minimum fine is R3,000 and you have to pay it immediately or they will detain you until you come up with the money. So if you have any pirated CDs please discard them and if you have empty recordable CDs or CDs that have information other than music and movies then keep them out of your car.
>
> Don't say you were not warned.[23]

The philosophy that unites TNCs in the institutional politics of intellectual property is a form of absolutism that elevates the rights of investors above all else. A world in which investor absolutism drives the making of intellectual property law is a world in which the welfare of all citizens will be diminished in vital areas such as education, health, and privacy.

Developing countries and civil society can fight these TNC agendas, but in order to do so effectively, they have to form much closer political bonds than they have to date. While a few developing countries can resist developed-country intellectual property agendas, they are doing so in a nuanced way, picking and choosing their issues and interests. India, for example, in the Uruguay Round of trade negotiations, concentrated its resistance on the patents part of TRIPS and not on copyright, even though the latter has just as many implications for access to knowledge. Vietnam, when it signed a bilateral trade treaty with the United States in July 2000, accepted a chapter on intellectual property, but was successful in keeping out a provision that would have limited its capacity for the parallel importation and resale of goods without the consent of the patent holder. Small to medium-sized parties involved in a negotiation with the United States or the European Union on intellectual property issues tend to adopt a harm-minimization strategy, agreeing to intellectual property standards that they believe will not hurt them too much.

There is, however, a collective cost for developing countries in adopting this harm-minimization approach. As they become integrated into the global architecture of intellectual property by means of free-trade agreements, they create for themselves an institutional box beyond which it becomes more difficult to experiment with real alternatives to the existing system. Having entered a web of international obligations (TRIPS, the Patent Cooperation Treaty, the International Union for the Protection of New Varieties of Plants, and so on), a web dotted with enforcement spiders (for example, the WTO's dispute-resolution mechanism), they become more cautious. When confronted with a radical model of access to knowledge, the first reaction of developing-country officials will be to ask, "Is this consistent with our international obligations?"—the answer to which will keep many lawyers joyfully occupied for a long time.

The need for a cooperative multilateral leadership on intellectual property by developing countries has become increasingly urgent. The monopoly control of the production of oseltamivir (Tamiflu) by Roche and the lack of global coordination by countries in dealing with the problem of inadequate stockpiles of oseltamivir, especially in high-risk developing countries, shows that the patent system has become a factor in the management of pandemic risk. The World Health Organization (WHO) had recommended that countries stockpile oseltamivir. Yet because of the patent price, many countries could not afford to build a stockpile. As a study showed, this produced, in risk-management terms, the absurd situation of poor countries (e.g. Vietnam, Cambodia) that were also high risk in terms of the flu pandemic breaking out having the smallest stockpiles, while the lowest-risk countries (the United States and the European Union countries), which were also the richest, having the largest stockpiles.[24] Moreover, the possibility of generic production was severely hampered by the fact that Roche did not disclose the patent position of oseltamivir, leaving public-health officials uncertain about what they could do in terms of importing or manufacturing it. Similarly, the diffusion of climate-change technologies will be crucially affected by intellectual property rights over those technologies.

Clearly, developing countries should be collectively thinking about ways in which to manage intellectual property in the context of global risks such as pandemics and climate change. Their current philosophy of mild cooperation in multilateral forums while defecting to short-term gains in bilateral contexts is inconsistent in the management of risk. More generally, if developing countries really wish to change the evolutionary space of intellectual property institutions so that real alternatives can flourish, they will have to design much better and stronger coalitions than they have to date. It is not good enough, for example, for developing countries to unite in the WTO on a disclosure obligation with respect to the patenting of genetic resources and then for some of those developing countries to agree to free-trade agreements that do not support that WTO position. Obviously this kind of coalition breakdown simply creates incentives for the United States to continue to operate outside of multilateral forums.

For civil-society groups working on intellectual property issues, the good news is that there are more groups engaged across a broader spectrum of issues than ever before. Among other things, this increases the possibilities of coalition building and, as the negotiations on the Doha Declaration on TRIPS and Public Health showed, a coalition of state and civil-society actors can be forged and wielded successfully. But just as civil-society actors can build coalitions and networks, so can the TNCs, with very different power outcomes. For civil-society actors, the prescription is to continue to invest in the creation of alternative models of knowledge creation and to float these in various national and international policy forums. Many of these will be ignored or will fail to gain wide support. But some will take hold, especially in times of crisis, such as a pandemic or the environmental crises that are predicted to accompany climate change, when state actors are desperately looking for solutions. For civil-society actors, floating new models for the growth and diffusion of knowledge, creating coalitions around those models, and acting in times of crisis are the basic elements of a strategy to change the evolutionary space of intellectual property.

NOTES

1   See the *Trilateral Statistical Report*, 2005 edition, p. 5 available on-line at http://www.trilateral.net/statistics/tsr/2005.html (last accessed March 1, 2010).

2   William M. Landes and Richard A. Posner, *The Economic Structure of Intellectual Property Law* (Cambridge, MA: Belknap Press of Harvard University Press, 2003).

3   Holger Görg and Eric Albert Strobl, "Multinational Companies and Productivity Spillovers: A Meta Analysis," *Economic Journal* 111, no. 475 (2001): pp. 723–39, available on-line at http://econpapers.repec.org/article/ecjeconjl/default111.htm (last accessed April 10, 2009).

4   Janusz A. Ordover, "A Patent System for Both Diffusion and Exclusion," *Journal of Economic Perspectives* 5, no. 1 (1991): p. 43.

5   Ronald H. Coase, "The Problem of Social Cost," *Journal of Law and Economics* 3, no. 1 (1960): pp. 1–23.

6   Peter A. Hall and David Soskice, "An Introduction to Varieties of Capitalism," in Peter A. Hall and David Soskice (eds.), *Varieties of Capitalism* (New York: Oxford University Press, 2001), pp. 1–68.

7   Jeffrey D. Sachs, *The End of Poverty* (New York: Penguin Press, 2005), p. 75.

8   My thanks go to Amy Kapczynski for suggesting a diagram of this kind.

9   J. M. Roberts, *Europe: 1880–1945* (London: Longman, 1967).

10 *Ibid*.

11 This feature of capitalism was noted by observers at the time. See Rudolf Hilferding, *Finance Capital: A Study of the Latest Phase of Capitalist Development*, trans. Morris Watnick and Sam Gordon (Vienna, 1910; London: Routledge and Kegan Paul, 1981).

12 Gary Herrigel, *Industrial Constructions: The Sources of German Industrial Power* (Cambridge: Cambridge University Press, 2000), p. 66.

13 For a full description see Fritz Machlup and Edith Penrose, "The Patent Controversy in the Nineteenth Century," *Journal of Economic History* 10, no. 1 (1950): pp. 1–29.

14 Abraham S. Greenberg, "The Lesson of the German-Owned U.S. Chemical Patents," *Journal of the Patent Office Society* 9, no. 1 (1926–27): pp. 19–35, 20.

15 Originally posted on the Pfizer Web site, now posted at http://www.scripophily.net/pfizerinc.html (last accessed April 10, 2009).

16 Peter Drahos, with John Braithwaite, *Information Feudalism: Who Owns the Knowledge Economy?* (London: Earthscan, 2002); Susan K. Sell, *Private Power, Public Law: the Globalization of Intellectual Property Rights* (Cambridge: Cambridge University Press, 2003).

17 Intellectual Property Committee, *Accomplishments and Current Activities of the Intellectual Property Committee* (June 14, 1988).

18 See G. Gereffi, *The Pharmaceutical Industry and Dependency in the Third World* (Princeton, NJ: Princeton University Press, 1983).

19 See the U.S. Patent and Trademark Office's comments on the effect of continuations on its workload at 71 Fed. Reg. 48–49 (January 3, 2006), available on-line at http://www.uspto.gov/web/offices/pac/dapp/opla/comments/fpp_continuation/vanvoorhies.pdf (last accessed March 1, 2010).

20 U.S. Patent and Trademark Office, "Changes to Practice for Continued Examination Filings, Patent Applications Containing Patentably Indistinct Claims, and Examination of Claims in Patent Applications (Claims and Continuations Final Rule)," 72 Fed. Reg. 46716 (August 21, 2007), available on-line at http://www.uspto.gov/web/offices/pac/dapp/opla/presentation/ccfrslides.html (last accessed April 11, 2009).

21 See Intel's comments in support at http://www.uspto.gov/web/offices/pac/dapp/opla/comments/fpp_continuation/continuation_comments.html (last accessed April 11, 2009).

22 The comments of the Biotechnology Industry Organization in favor of continuations can be found at http://www.uspto.gov/web/offices/pac/dapp/opla/comments/fpp_continuation/continuation_comments.html (last accessed April 11, 2009).

23 Forwarded to the author by a criminologist based in South Africa on November 9, 2007.

24 Buddhima Lokuge, Peter Drahos, and Warwick Neville, "Pandemics, Antiviral Stockpiles and Biosecurity in Australia: What about the Generic Option?" *Medical Journal of Australia* 184, no. 1 (2006): pp. 16–20, available on-line at http://www.mja.com.au/public/issues/184_01_020106/lok10852_fm.pdf (last accessed April 11, 2009).

Extract from Keith Aoki, James Boyle, and Jennifer Jenkins, *Bound by Law? (Tales from the Public Domain)* (Durham, NC: Center for the Study of the Public Domain, 2006) (available at http://www.thepublicdomain. org/wp-content/uploads/2009/04/bound-by-law-duke-edition.pdf).

The Idea of Access to Knowledge
and the Information Commons:
Long-Term Trends and Basic Elements

Yochai Benkler

In the global networked information economy, the constituent elements of human welfare and development depend on information and knowledge. Well-designed health systems and practices, research on disease and health, and access to medical innovation and its products go directly to the ability of people to live a long and healthy life. So, too, agronomic and biological research and learning, which have contributed significantly to food productivity in some regions, have the potential to reduce the prevalence of chronic malnutrition and hunger. Information and communications technology, books, educational materials, and learning practices promise improved literacy and educational attainment so that people around the world can live more engaged and fulfilling lives. Better access to distributed media hold out the promise of a more participatory public sphere, greater accountability of governments, and at least attenuation of the hold that authoritarian governments have over what their citizens know and think.

In the past few years, a diverse coalition of movements, political and economic actors, NGOs, scientists, and other academics have begun to coalesce around the idea, or the catch phrase, "access to knowledge" — A2K. The coalition is diverse. It includes activists concerned with access to AIDS treatments alongside European free-software developers concerned with software patents and digital rights management (DRM). It includes large, developing nations such as Brazil and Argentina alongside large multinational corporations such as Cisco Systems and IBM. It includes scientists concerned with open-journal publications and NGOs concerned with information and communications technology. The basic claim of this unlikely coalition is that information policy, on a global scale, is of central importance to a wide range of human values. Economic concerns with innovation and growth, on the one hand, and the core political values of human development, justice, and freedom, on the other, are being affected by a set of policies historically thought of in technical terms, but now increasingly seen and engaged for what they really are: policies that are of central importance to political economy and the moral quality of contemporary society.

WHY NOW?

The emergence of the access to knowledge movement is usefully understood in light of four long-term intellectual and material-historical trends. The first of these is the arc of the self-understanding of postcolonial societies, especially regarding strategies of economic development. In the 1950s and 1960s, the period of decolonization led to the creation of a large number of countries, some through violent liberation fights, others through the more or less voluntary acquiescence by colonial powers in the loss of empire. Throughout this period and increasingly in the 1960s and 1970s, the question of how these newly emerging societies were to manage their economies and societies was framed by the terms of self-determination and independence that were so central to the struggle for liberation more generally. At a practical level, this led to the widespread adoption of autarky, or self-sufficiency—not only in the newly independent countries, but in older, but nonetheless poor countries, most prominently in Latin America. The adoption of economic self-sufficiency as a core mechanism of independence led to the pursuit of import substitution (replacing imports with domestically produced goods) and rapid industrialization as core goals, often implemented through national ownership and high tariffs to protect locally owned industries. Interlaced with the pursuit of economic self-sufficiency was the ideological battle between capitalism and Communism. The latter's basic commitment to forced redistribution was, unsurprisingly, congenial to at least some segments of the impoverished former colonies. In opposition, one also saw the rise of nationalism as an alternative totalizing ideology to Communism, as internal elites and popular movements battled in many of the still-unstable new countries. The consistent use of local battles as proxies for the Cold War did not help.

By the middle to late 1980s, however, import substitution and self-sufficiency had come to be seen as failed economic strategies. Increasingly, and with greater speed after the collapse of the Soviet Union, integration into a liberal, global trading system came to be seen as the sole or at least the dominant strategy available to countries, poor as well as rich, to promote growth and development. Dissenting voices continued to be heard, but very few governments followed them throughout the late 1980s and through the 1990s.

The second, much longer-term trend underlying the A2K movement is the shift from industrial to information economies. The history of the Western shift from agrarian to industrial, from industrial to service-based, and from service-based to information-based economies has been the subject of countless books and articles. The emphases differ, but the basic arc is not contested. This longer-term history has been telescoped in curtailed form in many of the newly independent states of the latter half of the twentieth century. In these countries, rapid industrialization was characterized by the same dislocation and misery that accompanied European industrialization in the nineteenth century. This industrialization came to fill the place of some of the production in the industrialized economies as these economies moved toward information manipulation as their core activities. But industrial production under these terms was dependent upon the information-rich inputs—innovation, financial capital, and marketing—that dominate the wealthy economies. The more recently emerging economies, as well as their poorer followers, are themselves trying to move into the later stages of capitalism at a more rapid pace. In a world trade system typified by industrialized peripheries and an information-rich core, the rules regulating the production and exchange of information, knowledge, and culture have therefore become a major battleground over competitiveness between the already-developed and the newly developing countries, or between the postindustrial and the newly industrialized economies.

Access to information and knowledge as basic inputs into processes of learning and innovation thus have become a central pathway to growth and development and ultimately to competition with the wealthiest economies. The newly emerging economies need access to the existing stocks of knowledge—such as the existing outputs of science and technological innovation—to speed up their ability to achieve something like parity in the global knowledge economy, while poorer developing economies and poorer regions in unevenly developing countries require that access as part of obtaining the preconditions to human development.

This long term-trend toward industrialization and beyond it toward a global knowledge economy is therefore at the very root of what has become the A2K movement. It both necessitates the emergence of access to knowledge as a core element of concern with human development and gives it a focal point in those global institutions that have been the locus of the assertion and institutionalization of control over knowledge flows: most prominently, the exclusive rights regimes usually collected under the umbrella term "intellectual property." These have become a central part of the basic legal underpinnings of wealth and poverty, productivity and development, in the contemporary global economy and have therefore unsurprisingly become the locus of contestation around which the A2K movement is coalescing.

The first two long-term trends combined to underwrite the integration of the international intellectual property system into the global trade system. They have also been the reason that the ways information, knowledge, and culture are produced and managed have become so central to questions of global justice and development. They undergird the development of the idea of "intellectual property" and the rise of the Trade-Related Aspects of Intellectual Property Rights (TRIPS) Agreement. In brief, over the course of the late nineteenth century and throughout most of the twentieth, copyright and patents were treated, in the Berne and Paris treaties, in particular, as distinct regimes, and their international enforcement was largely a system of reciprocity and mutual recognition of national policies. These agreements were largely peripheral to the international trade system and had practically no teeth. In the 1970s, some of the countries that were focused on import substitution and the development of local industries passed laws, such as India's 1970 Patent Act, that withdrew patent protection in core industries, such as pharmaceuticals, allowing their indigenous industries to displace imports. As Peter Drahos and John Braithwaite have documented, in the 1980s and until the mid-1990s, the core industries—orchestrated primarily by the pharmaceutical industry, but joined by other players such as the Hollywood studios and proprietary software companies—shifted the global regulation of information and innovation away from the global copyright and patent regimes centered on the World Intellectual Property Organization (WIPO), and toward the trade regime.[1]

This push was concluded with the entry of TRIPS into force in 1995 as part of the Uruguay Round of General Agreement on Tariffs and Trade (GATT) that created the WTO. In the decade that has followed, this push has been complemented by the internal drive within WIPO to save itself from obsolescence by offering services to those actors who successfully pushed the TRIPS Agreement. Beginning in the mid-1990s, WIPO became a forum for pushing new, extensive intellectual property rights and for offering technical assistance that would speed up adoption of highly protective property rights regimes throughout the world. During this period too, the TRIPS agreement became a baseline of minimal protection, rather than a standard, while the coalition that pushed for its initial adoption further pushed both through WIPO and through U.S. bilateral free-trade agreements for the adoption of the even more restrictive "TRIPS-plus" protections. This steady trajectory is a product of the combination of the increasing importance of information to economic growth as well as to capturing value from economic production and of the wide perception that integration into the global trade regime is the only option for nations wishing to avoid stagnation and underdevelopment.

The most important institutional and intellectual moves in this period were the creation of the idea of "intellectual property" itself and its inclusion in the trade regime, the weaving of diverse types of mechanisms to increase the degree of exclusivity everywhere, and the abstraction of the protected category. The first of these equated innovation with strong intellectual property rights and wealth with the export of goods that were intellectual property. Providing industries that depend on exclusivity with a stronger enforcement mechanism globally thus became a core trade goal. The creation of the idea of intellectual property and its inclusion in the trade regime also shifted the institutional base of the relevant national negotiators from more locally protective and development-sensitive government ministries such as those of culture or education to ministries more oriented toward global trade and industry. The negotiating dynamics of the terms of trade therefore were easier to tilt in favor of the intellectual property exporters, in exchange for concessions, real or imagined, on agriculture, textiles, and so on.

The second move consisted of weaving unilateral, bilateral, and multilateral mechanisms together to form a net that could be used to ratchet up the level of protection everywhere. The industries pushing for stronger exclusive rights regimes were able, over the course of this period, to identify various weak spots, in terms of political economy, where it was possible to achieve higher protection. It might be the U.S. trade representative or the European Commission; it might be WIPO or the WTO; or it might be a bilateral trade agreement with a country that had much to gain in areas other than the information economy by agreeing to a particularly broad set of protections. In each case, victory in one arena was available as a baseline for renegotiating the terms in other arenas and for generalizing the practice globally. This playable international system assures that there is no clear bottleneck to ratcheting up protection while at the same time placing international harmonization requirements as a backstop against the "loss" of protections already agreed to in some other forum.

The third major move was an intellectual one of generalization or abstraction: The various different industries such as Hollywood, pharmaceuticals, and semiconductors came to be seen not as discrete industries with special issues, but as instances of "the intellectual property industries." This, in turn, pushed governments to move from seeing intellectual property policy as involving a series of discrete policy issues that represented industry-specific tradeoffs to viewing these problems as a broad project of setting industrial policy in a global information economy, one in which they were information exporters. And finally, this allowed the United States, the European Union, and Japan to move from seeing each other purely as competitors to seeing themselves as having a common interest as information exporters, forming a formidable interest bloc in the institutions of the world trade system.

However, the rise of the information economy has also played a role in fostering the counterforce that has today resulted in the emergence of the access to knowledge movement. It combined with two further long-term trends to form a response to the rise of the global trade and intellectual property system.

The third long-term trend that has helped give rise to the A2K movement is the shift from mass-mediated culture and monopoly telecommunications systems to the networked information society. Mass media initially emerged with the rise of electrically driven presses and automated typesetting and newspaper folding, complemented by the rise of professionalized journalism and telegraph-based news services around the middle of the nineteenth century. As rail travel and steam-based trade increased the size of the relevant (that is, interdependent) societies and economies over the course of the nineteenth century, high-cost communications facilities led to the organization of communications and the public sphere around large aggregations of capital. First in the telegraph, then in telephones, long-distance communications were either monopolized by market players, as in the United States, or nationalized, as in most other countries. In either case, only large organizations with the capacity to amass capital were able to build systems. As radio and later television joined the press, the capital costs of producing and disseminating information, knowledge, and culture to the relevant communities continued to be high. These formed the basis for the relatively concentrated media environment typical of most countries in the world, whether the concentration was market-based or state owned.

The period beginning in the late 1980s saw rapid changes in the communications and media environment. Initially, we saw the introduction of competition from new, but still large-scale players introducing a more competitive market into telecommunications, both wired and wireless, and into the cultural industries and media.

More dramatically, beginning in the mid-1990s, we saw the rise of Internet-based communications and the emergence of a networked information society and economy based on the radical decentralization of information production. This trend created new opportunities and new social forces that did not exist before or at least that would not have been aligned before. Firms that dedicated themselves to providing communications and computation found themselves aligned with software programmers who wanted to participate in the free software movement; citizen journalists saw themselves aligned with Wikipedia editors; NGOs found themselves more effective than they were before and aligned with scientists, who found that through networked communication they could sequence and annotate the human genome faster than their commercial, proprietary competitors. The rise of decentralized peer production and of nonmarket production in general and the increased efficacy and practices of those who participated in the networked information economy provided some of the intellectual framing, as well as the surprising alliances, that seem to characterize the A2K movement.

The fourth and final long-term trend that has propelled the A2K movement is the shift in the global ideological framing of questions of justice and human freedom. As the failures and excesses of both Communism and autarkic statism as viable and attractive alternatives to capitalist democracies came to be too painful to ignore, so, too, did the limitations of simple realism in the international relations sense, in which nation-states interact solely within a framework of *machtpolitik*. Instead, we have seen the gradual rise of human rights, human dignity, and participatory politics as the more or less universal ideals toward which most societies in the world aspire—if not in practice, in principle.

There is, obviously, no simple, linear progression toward the adoption of human rights as a framing ideal. To see this, one need look no further than the rise of fundamentalism as a rejection of the modern, as one sees in contemporary U.S. politics or in the Muslim world, or as a challenge to the liberal demand of treating others with equal dignity, as arguably was demonstrated by the late 1990s rise of Hindu nationalism. But the majority of countries and the majority of discourse focused on engagement in the global system, rather than disengagement, has had to accept some form of a human rights framework. In particular, in the last decade and a half, we have seen the rise of the idea, associated primarily with Amartya Sen, of development as freedom, which integrates in important and interesting ways both the civil-political and social-economic rights of the international human rights system.

These four trends—the rise of a globalized, liberal trading system, the rise of the information economy, the subsequent genesis of a networked information society in which information, knowledge, and culture have become central to human welfare and economic growth and in which the production of all three increasingly has become Internet based, radically decentralized, the domain of nonmarket or small, independent actors, and finally, the rise of human rights in general as an ideal and the idea of development as freedom—these undergird the rise of the A2K movement.

In the more immediate history since the mid-1990s, the convergence of all these trends has manifested itself in the convergence of several initially independent and disjointed efforts. The first was the access-to-medicines movement, which during the 1990s received a major boost in visibility through its relation to the battles over research-and-development priorities triggered by the HIV/AIDS pandemic. To some extent, this was due to the sheer immensity of the devastation wrought by the pandemic. But it was also likely due to the fact that HIV/AIDS is a disease that strikes not only at the poor of Africa, but also at the very heart of the cultural elites of the United States and Europe. As combination therapies that could halt the progress of the HIV virus were developed in the late 1990s, the stark disparity between outcomes for the wealthy and the poor became harder to ignore.

During the same period, but independently, the explosive growth of Internet usage spawned two movements that were initially only very loosely linked. These were the movements for Internet freedom, anchored in concerns over encryption, privacy, and antipornography regulation, on the one hand, which dominated the concerns of the computer geeks, and the information commons movement, populated initially by librarians, academics, and different groups of geeks, which responded to efforts by Hollywood and the recording industry to rein in the Internet and to stem what these corporate entities saw as a threat to their entire production model, but also as an opportunity to turn the Internet into a global pay-for-play jukebox, by rapidly expanding copyright and a variety of copyright-reinforcing mechanisms. In both cases, there were relatively few companies systematically involved in resisting the expansion of exclusivity or censorship, although opportunistically, the telecommunications carriers cooperated with these civil-society efforts in order to avoid regulatory burdens aimed at forcing them to enforce the various content restrictions sought. By the late 1990s, the free and open-source software development communities began to grow from the engaged technical communities they had been before into politically mobilized groups. The open-source community focused on expanding the acceptability of this approach among businesses and forged the affinity alliances with business that are becoming important in the present coalition, while the free-software movement focused on the political mobilization of participants and on affinity alliances with the global left. Together, these have become genuine grassroots movements around questions of DRM and software patents, in particular, with hundreds of thousands of participants around the world, and have played significant roles in policy making in the European Union concerning software patents, in the United States concerning trusted systems, and in Brazil, at least, concerning development initiatives.

As the 1990s came to a close, a completely different set of actors began to organize around the threats of enclosure, or the expansion of copyright and other exclusive rights to ever-broader domains and uses of information and knowledge. Scientists, on two fronts, began to see intellectual property as a hindrance, rather than a help. On one very publicly visible front, the Human Genome Project captured the imagination, because the prospect of patenting human genes led to extensive public debate. But while the public at large was concerned with metaphysical questions about owning human beings, in some sense, scientists were mostly worried that they would be shut out of the ability to do research. A major international effort incorporating academic scientists, government and nonprofit funders, and even some pharmaceutical companies that were worried about upstream patents became engaged in an unprecedented effort to sequence as many genes as possible, as quickly as possible, and to publish them freely as

DefectiveByDesign.org is a broad-based, antidigital-rights-management (DRM) campaign that targets media, manufacturers, and distributors who facilitate DRM. The campaign seeks to discourage companies from bringing DRM-enabled products to market, and to identify "defective" products for consumers (Andrew Becherer, license: http://creativecommons.org/licenses/by/3.0/).

quickly as possible to preempt their appropriation by Celera Genomics and other private, proprietary efforts to sequence all or parts of the human genome. The result was a mobilized segment of the scientific community.

Over the same period, many of the same academics saw the rising costs of academic journals, primarily scientific journals, and realized that although they were writing the papers and providing peer review of them (typically for free), they were required to pay high access fees to read those same materials because of the highly concentrated nature of the journal-publishing industry. Scientists began to adopt a wide range of open publishing efforts, beginning with ArXiv.org in physics, e-Biomed in science, and later, the Public Library of Science, the Budapest Open Access Initiative, and self-archiving. Parallel and independent of these were efforts by librarians and archivists to deal with questions of digital archiving, obtaining and structuring materials that could be archived and presenting them on the Internet.

On the infrastructure side, two distinct movements were present. The first was the more traditional, development-focused work on information and communications technologies for development. Here, traditional aid agencies and development economists, often in response to the global digital divide, were concerned with computers, kiosks, and network connections. At the same time, beginning in the late 1990s and picking up in the first half of the 2000s, a movement around open-spectrum policy developed to question the whole approach of spectrum management as property. Wireless communications long had been regulated as a form of public property, and as privatization took hold across many domains, the wireless spectrum, too, was subject to a sustained critique by economists, so that by the mid-1990s its conversion to a private property regime was widely accepted. By that point, however, the model of regulating wireless communications via an understanding of "spectrum" as a "resource" that needs to be managed by either the state or a company had become obsolete. Instead, it had become feasible to permit the deployment of wireless equipment that would enable users to own their own network connections and to circumvent the bottleneck that traditional providers of last-mile Internet connectivity held, and continue to hold, over Internet access. Originating mostly in the United States and receiving a major global push with the adoption of Wi-Fi, more municipalities, companies, and, increasingly, nations and aid agencies are working on solutions to provide decentralized, ubiquitous Internet broadband access over wireless networks using off-the-shelf equipment that uses a spectrum commons, which no one owns, rather than following the expensive traditional path of licensing or its very close twin, spectrum auctions.

Since 2004, these diverse groups of actors and movements have begun to find common cause, to see common themes, and to coalesce around a set of ideas, organizations, and conferences to form what appears increasingly like a global social movement. They interact with the growing normalization of cooperative, nonmarket social practices such as Wikipedia, with the increasing political and practical consciousness that finds the Creative Commons initiative as its focal point, with the fact that many more commercial entities are beginning to find ways to interact productively and profitably with commons-based production, and with the newly invigorated efforts, of developing nations, headed most prominently by Brazil, to shift the agenda of international exclusive rights regimes away from ever-increasing harmonized protection toward a more context-dependent and development-oriented policy. A major catalyst in the mutual recognition of these diverse groups and actors has been a series of conferences organized by the Trans Atlantic Consumer Dialogue (TACD), where these various actors have come to meet, talk, and understand their mutual agenda.[2]

A2K, COOPERATION, AND THE INFORMATION COMMONS

The ideas of the information commons and the use of networked cooperation have been central to discussions within and about the emergence of the A2K movement. In the remainder of this essay, I will explore why this might be. My claim is that these ideas subvert the traditional left-right divide, form the foundation for some of the most interesting and unusual alliances, and provide the platform on which political and economic interests meet around a common institutional and organizational agenda.

Recall that the networked information economy is built on an inversion of the capital structure of the production of information, knowledge, and culture. For the first time since the Industrial Revolution, at least, the most important inputs of core economic activities are broadly distributed throughout the populations of the most advanced economies and in significant segments of emerging economies. These inputs include computation, communications, and storage capacity and human intuition, creativity, and wisdom, which are personal, nonfungible, and uniquely held by individuals. General Motors did not have to worry about competition from amateurs getting together on a weekend, because the cost of an assembly line was too high for their efforts to matter in the market. The same is not true of Microsoft or Britannica. The widespread distribution of material and human resources has meant that behaviors that have always been central to human sociality—from real friendship to simple decency toward a stranger in a chance encounter—have moved from being socially important, but economically peripheral, to being centrally effective in the economy as a modality of production.

The already-existing fact that creativity and wisdom are distinctly individual and human, together with the new and radical decentralization of physical capital, has located the practical capacity to act effectively in the hands of individuals and of loosely coupled cooperative groups in society. In acts ranging from individual authorship of Web sites or blogs, to the small-group authorship of blogs, to massive collaborative efforts such as Wikipedia or the Linux kernel, production based on social motivations and signals, rather than on price signals or hierarchical commands, engaged in as both individual and peer production, has become a significant force in contemporary economies, societies, and cultures.

The importance of the information *commons*, in particular, is anchored in the nature of the existing universe of information, knowledge, and culture. In order for a person to act effectively, both authority (under whatever system of prescription is applicable to the action) and the practical capacity to act have to be located in the same place. Effective, large-scale patterns of human action will emerge only through the actions of those actors who have both the practical capacity and the authority to act. And it is here that the information commons enters. Both individuals, who are now made more capable and potentially effective by the decentralization of material capital, and the newly feasible networks of cooperation that are so central to this new, effective agency need a universe of existing information resources on which they have the authority to act. Exclusive rights, such as copyrights or patents, are designed to remove the general authority to act on a given information or cultural resource and instead locate the authority to permit learning and to use a given information "bit" in the hands of a given agent. Permissions from that agent then form the basis for a particular kind of market in permissions to use the information resource. But information is a public good, in the economic sense. It is what economists call "nonrivalrous"—it can be consumed by one consumer without preventing simultaneous consumption by others. Its marginal cost is zero. Any market that imposes a positive price on information therefore leads to underutilization of the information. And in a setting where information is itself used as a productive input, not only as what is consumed, this underutilization is not merely what is known as a "deadweight loss" in terms of efficiency (those who would benefit from it at a price less than it would cost to deliver to them do not get it), but actually inhibits innovation and new creativity.

All this is well known, but the critical point here is that enclosure of information through exclusive rights regimes locates the authority to act with and upon covered information and culture with the rights holder, rather than with whoever has the practical capacity and insight to do something useful and interesting with the information—even if that person is entirely willing to pay the actual social cost of using the information, that is, their own time and attention in using it. Exclusive rights regimes pose a particular and heightened threat to innovation and creativity as noncommercial and nonproprietary production increases in general, and as cooperative peer production increases in particular. Peer production thrives on combining a wide range of contributions from diversely motivated individuals scattered across the globe. If each participant were required to pay a license fee, even one that was "reasonable" by the lights of a commercial producer, even one low enough to be reasonable to a highly committed amateur, still there would be a large number of smaller-scale contributors whose contributions would be critical to sustaining the cooperative project as a whole, but who would be priced out of the market. The denial of a general authority to learn from and to be creative with the existing universe of information, treated as a commons, therefore inhibits creativity and innovation when carried out in peer-production endeavors and limits the human agency, the freedom, that such creativity both expresses and enables.

The productivity of the commons and counterproductive effects of property-mimicking regulations such as exclusive rights regimes, the increasing recognition of the value and importance of nonmarket action generally and of cooperation, in particular, by commercial actors such as IBM, which has developed a substantial "Linux services" business in collaboration with the free and open-source software development community, by civil-society organizations, and by loose alliances of individuals practicing these forms of social production has set the stage for a new and interesting set of intellectual shifts.

Because of its capital cost structure, the industrial economy promoted a binary view of effective action focused on the two mechanisms available for raising sufficient capital to be effective: the market and the state. Because effective action by a significant number of people required sustained commitments of time and focus, often in conjunction with large-scale capital investment, groups that were formed to undertake such action were seen as stable, and the binary opposition of solidarity and individualism likewise was seen as stable. The state of organization theory was such that hierarchy in the early twentieth century was seen as the epitome of effectiveness, whether it came in the form of Taylorism, Fordism, Weberian bureaucracy, New Deal progressivism, or Communist Party discipline.

Commons-based information production and peer production destabilize these binaries. These new forms of production are based neither in the state nor in the market. The most prominent among them are either structurally participatory and self-governing, like Wikipedia, or at least drastically more dialogic and persuasion-based than earlier organizational models, even when they are not formally participatory, like the Linux kernel development community. In this context, much looser associations can retain efficacy, rendering the individualism/solidarity choice less stark and stable. These new forms of production enable and thrive on flat organizational structures with large amounts of authority for individuals to self-assign tasks, sense the environment for opportunities for action, act, communicate with others, and repeat. This is precisely what makes these approaches valuable—their advantage as large-scale systems of learning through initiative, trial, error, communication, and adaptation. They support—indeed, they require—a more cooperative view of human action, without also requiring a strong commitment to a view that privileges solidarity over individualism.

The destabilization of these industrially derived intellectual binaries makes networked cooperation using commons-based strategies for resource management an attractive modality of production within the framework of the unusually broad range of views that characterized the political-theoretical map of the nineteenth and twentieth centuries. Almost the entire range of liberal traditions, from laissez-faire to progressive liberalism or social democracy, can find information-commons-based cooperation attractive. The left, too, can find in these practices one way out of the dead end that state socialism proved to be. Libertarianism, of both right-wing, market-oriented, and left-wing, anarchistic varieties, likewise finds attractive narratives to tell about cooperation in the networked commons. Adherents to this broad range of views can then, as a practical matter, ally with market actors who eschew political views altogether and who are focused on survival, innovation, and growth in an increasingly competitive global economy where learning and adaptation are an imperative. Needless to say, some of this congruence is temporary and ad hoc. Some, however, represents a real change in conditions and intellectual alignments.

Take, for example, a question such as the European consideration of software patents, opposed widely because of the possibilities of strategic holdup—the concern that patent owners would temporarily hide their rights, but then exploit the interdependencies of software and standards to demand excessively high payments once their software became integrated into a standard—and for its effects on free software development. At a basic strategic level, this opposition aligns companies that, rather than selling software as "goods," sell software services and computerized enterprise solutions (these businesses account for over three-quarters of the software business; IBM is the leading example) with free and open-source software developers and activists concerned with constraining the scope of expansion of patents or exclusive rights regimes in general. They can all converge around the basic critique of intellectual property or exclusive rights regimes in terms of efficiency and innovation policy. In this case, because the software market is so heavily pervaded by nonexclusion-based business models and because patents have been applied so poorly in the United States, the minimal, functional case forms a foundation for a broad, tactical alliance, and when enacted at a higher theoretical framing as being about "intellectual property" versus "commons-based strategies," as it often is, this tactical alliance can be part of a broader strategic alliance between firms in the information technologies sector and the A2K movement. This tactical and strategic alliance is the least interesting theoretically, but is of enormous importance politically.

Moving one level deeper, free and open-source software (the political and apolitical names for the phenomenon, respectively) and commons-based peer production can be framed as attractive to libertarians, liberals, the postsocialist left, and anarchists, though in each case for different reasons and viewed through different lenses.

Laissez -faire liberals and libertarians can see in open-source software development an instance of people acting according to their own preferences, unforced, to produce together. They need some more or less fancy story about motivation and why people would do this. They need some clear specification of how people ultimately make money. These tasks have been taken on by economists studying this problem. But the basic framing is congenial to market-centric liberalism and property-based libertarianism: People are using their propertylike rights—either copyrights or simply their right to be free in their bodies to work on whatever project they choose—to adopt business models and strategies, often implemented through licenses, with firms that engage in this activity as a strategic option, producing in ways that they deem useful. When the state comes and tries to extend patent rights that cover the object of action, particularly given the background understanding of information as a public good in the strict economic sense, patent law comes to be framed for laissez-faire liberals and libertarians as a regulatory intervention.

A central claim of the information-commons movement has been precisely to emphasize the regulatory nature of exclusive rights regimes, resisting and undermining the move to unthinking application of the "intellectual property" label. In other words, the state has a model of how software development goes (or encyclopedia writing, or video entertainment, in the case of copyright and paracopyright), and it is intervening in what seems to be a perfectly functional innovation system, imposing new rules that are upsetting a whole set of freely chosen business practices already in place. Needless to say, this is not the only way to view what is happening, but it is a sufficiently plausible characterization that many libertarians and laissez-faire liberals in fact understand what is happening in these terms. The rhetorical foundation of the "open-source software" movement was precisely to frame the practice in these terms of free choice, innovation, and business benefits.

The left sees in the information commons very different things. Here they see proof that when people own the means of production, they can cooperate outside of the market (both in the sense that the outputs need not be sold in the market and in the sense that labor is not commodified), without reliance on property as the organizing principle to achieve productive goals. It is proof that there is no one right path of capitalism. Here, even more importantly, is a vector through which the existing distribution of power can be resisted: power not only in the political sense, but in the economic sense and the cultural sense, as well. This is where the commitment to *free* software offers an important rhetorical marker of a basic underlying observation. The central distinctive commitment of the left has been resistance to the dehumanizing application of power by economic production systems and with it by culture and society, religion and the family being the two main loci of illegitimate power and coercion. The necessity of sustaining economic production and its former dependence on large-scale capital aggregation led the left to give the state an enormously powerful and ultimately corrosive role in achieving freedom from this power. But in peer production, we are seeing an avenue of resistance to the hierarchical exercise of economic power that does not flow through the state. More Kropotkin than Lenin, this source of power in the hands of people networked together is, I think, the single most attractive feature of the information commons to the left. There are, of course, some on the left who will continue to see the distribution of material goods as central and who will be skeptical of the importance of information as the locus of egalitarian production.

The debate here, from the perspective of the A2K movement, will be over the relative centrality of the distribution of dynamic, productive capabilities for learning, growth, and effective production in the domains of information and knowledge as engines of justice over time in the distribution of social and economic power and material goods.

For liberals, free software and open-source software, the commons, and peer production offer ways of deepening individual freedom, improving democratic participation and the accountability of both government and corporate power, providing new avenues for human community and sociality without imposing the constraints of conservative social forms, and offering a basis for a more participatory public culture—all this without the need to resort to rejection of the market qua market and without subjecting the individual to collective or solidaristic claims, at least not to claims that are not freely chosen, negotiated, taken up, and capable of being renounced when the individual desires to do so.

Needless to say, if cooperation in the information commons were in fact all these things to all people in all these ideological camps, we would have indeed come to a certain kind of end of history. There are, of course, market liberals and libertarians who see peer production and the commons as the left does and either disbelieve it or resist it on principle, or both at different times. There are those on the left who emphasize the disparities of power between those few million who are newly empowered, perhaps, and the billions for whom things have not changed at all. And there is a strong, central strand in liberalism that sees the role of an effective, constitutionally limited, deliberatively legitimate *state* to be central both to liberal thought and to individual freedom.

All these views—both those that embrace cooperation in the commons and those that express skepticism about it—are correct. All are incomplete. That is why this moment calls for a theoretical engagement with the possibility of free, nonhierarchical, flow-based, and networked, rather than stable-structured and institutionalized social forms. The A2K movement is at the heart of dealing with the main limitation of commons-based and peer production from both the left and liberal perspectives—its application to justice, both local and global. The freedom to act, alone and in loose cooperation with others, in effective forms free of hierarchical power depends on the distribution of basic capabilities and authority to engage in open and collaborative modalities of production. The A2K movement can and must play a political and social role in assuring the global distribution of access to the basic conditions—both material and institutional—that enable the decentralization of practically effective human agency and sociality. Whether and to what degree the contemporary partial intellectual alliance can be sustained between the left and left liberals, on the one hand, and market liberals and some libertarians, on the other, depends upon how power in the market and power in the state can be reinterpreted, reconceived, and restructured as a matter of practical programs.

Freedom, justice, and efficacy are the core interfaces for the realigned map of political theory that we can draw and that we need to draw if we are to change the power alliances that have for so long disabled a more egalitarian global distribution of capabilities and opportunities. The core interface of the A2K movement with libertarianism is in the area of individual freedom. The centrality of practical freedom to explore, experiment, and adapt, and hence to learn and innovate, converts freedom into efficacy and becomes the core interface with market liberalism. Freedom and efficacy, then, will be the interface with both liberalisms, market and social. Justice and freedom, in the sense of the dissipation of structured, stable hierarchical power, will be the interface between liberalism and the left. And all three—freedom, justice, and efficacy—will be the interfaces with the social, pragmatic, liberalism that has occupied the center in the United States, Europe, and gradually, since the end of World War II, much of the rest of the world.

At a programmatic level, the core foci of the A2K movement lend themselves well to characterization through these conceptual interfaces. The policy goals of the movement can be, and indeed are, couched in terms of justice, freedom, and efficacy. The mainstream understanding of the economics of information and innovation lends itself to complementary, rather than competing rhetorics of access. The public-goods nature of information in the technical economic sense supports limiting the scope and reach of patents and copyrights. The character of information as both the input and the output of its own productive process provides the foundation for an argument about diverse rules for diverse economies, industries, and activities, and all these can be made in terms of efficacy, or in this case, growth and innovation, to push back on the rhetoric of harmonization that has been so central to the strategy of ratcheting up exclusive rights around the world. At the same time, the centrality of individual freedom and social cooperation to the efficacy sought—learning and innovation—aligns the programmatic concerns for a more expansive commons and more limited exclusive rights with the interfaces to liberals, social democrats, and the left. One reason, perhaps, that the movement was able to coalesce as it did around patents and copyrights was that this programmatic focus was so easily transferred across theoretical divides.

But access to the information commons in the abstract is insufficient. As I discussed above, freedom, justice, and innovation all require *effective* agency, not merely formal permission to act. And effective agency in the domains of information and knowledge production requires access to material means, as well as to a knowledge commons—not the same access as was required in the industrial age, but access to a minimal set of material capabilities and educational faculties nonetheless. Here, the alliances must come under some pressure, particularly in the interface with libertarians. Still, the centrality of innovation and information production, as well as the widespread recognition among market liberals that integration into the global economy requires investments in infrastructure, suggest that at least some of the congruence can be kept as we move to infrastructure.

On spectrum policy, in particular, the drive to a spectrum commons is wholly couched in the United States in terms of efficiency and growth, and, if successful, will create a large market pull for the creation of devices capable of creating an infrastructure merely by their local deployment by users. This is one path that is both radical in its implications—enabling the development of a free and open infrastructure owned by its users—and capable of being couched along any of the conceptual interfaces. The drive to deployment of broadband capabilities, neutral or open-access telecommunications networks, and open personal computers and mobile platforms all have that similar feature: They are debated in the United States and Europe in terms of innovation and growth—that is, of efficacy—but have obvious and direct effects for, and framings in, freedom and justice.

In the debates over patents in medicines, framing a congruent agenda and understanding is a bit harder. The language of justice is most easily available and has been dominant. But a particular historical contingency has made some alliances at least feasible, if not easy, on practical and theoretical bases. The political dynamic that has driven the patent system to excess, particularly in the United States, has put pressure on companies that are not in the pharmaceutical industry from "patent trolls"—persons or companies who exact payments for the use of patents they hold without intending to use them productively themselves— and high transaction costs associated with operating in too restrictive a knowledge environment, polluted by too many unnecessary patents. These industries, in particular the high-technology and consumer electronics industries, have begun to push back on the pharmaceutical industries on reform of the basic patent law. The points of contention are very different from those of the access-to-medicines movement, but the overall direction is congruent, and the timing and common understanding of the need for a very powerful push provide a moment of opportunity for creating alliances around the issue of patent reform that could sweep in the concerns of very different parts of the movement. As everywhere, however, the risk of this kind of opportunistic alliance formation is that the partnership dissolves as some, especially those who are powerful and interest driven, obtain what they need and leave. This is a risk that needs to be addressed by continuous engagement and framing of the ideas around the long-term, stable congruence of interests well beyond the opportunity of the moment. Indeed, it is in order to stabilize the alliances that make the A2K movement and its agenda feasible that the exercise of self-definition and theorization is important.

Looking at the long-term trends that I described in the beginning of this essay, the task of conceptual integration is neither incoherent nor impossible. The rise of the networked information economy has created the material conditions for the confluence of freedom, justice, and efficacy understood as effective learning and innovation. The decline of statism and the more or less global consensus on at least the inevitability of some form of market-based economy has eliminated what was a core unbridgeable gap between liberalism in its right and left forms and the left. We have seen this in the "Third Way" literature for over two decades. The emergence of networked cultural and information networks has provided the mechanism for dialogue about what is to be done and for collective action to organize to do it. And the development of the idea of freedom to extend to human rights and development has created a framework for bridging justice-seeking and freedom-seeking discourses. But to say that the task is neither incoherent nor impossible is not to say that it is easy. It is, nonetheless, necessary if the alliance represented by the A2K movement or by the information commons, free culture, and similar aligned movements is to become the basis of a new political alignment, rather than a temporary marriage of convenience.

NOTES

1   Peter Drahos, with John Braithwaite, *Information Feudalism: Who Owns the Knowledge Economy?* (London: Earthscan, 2002). See Peter Drahos's essay, "'IP World'—Made by TNC Inc.," in this volume.

2   See the essays by Manon A. Ress, "Open-Access Publishing: From Principles to Practice," and Ahmed Abdel Latif, "The Emergence of the A2K Movement: Reminiscences and Reflections of a Developing-Country Delegate," in this volume.

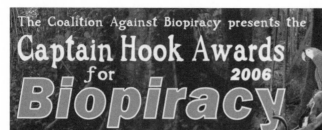

The Coalition Against Biopiracy presents the

Captain Hook Awards
for 2006
Biopiracy

BIOPIRACY refers to the monopolization (usually through intellectual property) of genetic resources and traditional knowledge or culture taken from peoples or farming communities that developed and nurtured those resources.

Worst Threat to Food Sovereignty:
Syngenta

For its Terminator-like patent designed to prevent potatoes from sprouting, despite the company's pledge not to commercialize technologies involving sterile seed. US patent 6,700,039 describes a genetic modification method that prevents sprouting unless an external chemical inducer is applied.

And for Syngenta's multi-genome patent applications on thousands of gene sequences vital for rice breeding and extending to dozens of other plant species.

Greediest Biopirate:
J. Craig Venter

For undertaking, with flagrant disregard for national sovereignty over biodiversity, a US-funded global biopiracy expedition on his yacht, Sorcerer II, to collect and sequence microbial diversity from the world's oceans and soils. The genetic material will play a role in his most ambitious project to date: building an entirely new artificial organism.

Biggest Threat to Genetic Privacy:
Google Inc.

For teaming up with J. Craig Venter to create a searchable online database of all the genes on the planet so that individuals and pharmaceutical companies alike can 'google' our genes – one day bringing the tools of biopiracy online.

Extreme Makeover Award:
Delta & Pine Land Co.

For vowing, since 1998, to commercialize Terminator technology. Initially, D&PL promoted genetic seed sterilization for use in the South to prevent farmers from re-using seed. After massive protest, the company changed its tune and said Terminator was primarily intended for Northern farmers. Now the company is greenwashing Terminator by promoting it as a biosafety tool to contain gene flow – for farmers everywhere!

Most Shameful Act of Biopiracy:
US Government

For imposing plant intellectual property laws on war-torn Iraq in June 2004. When US occupying forces "transferred sovereignty" to Iraq, they imposed Order no. 84, which makes it illegal for Iraqi farmers to re-use seeds harvested from new varieties registered under the law. Iraq's new patent law opens the door to the multinational seed trade, and threatens food sovereignty.

Worst Déjà Vu:
Human Genographic Project

For resurrecting the old (much discredited) Human Genome Diversity Project with new corporate money. IBM and the National Geographic Society are spending $40 million dollars and establishing ten research centers around the globe to collect and analyze more than 100,000 DNA samples from Indigenous Peoples, claiming this will help them understand their ancestry.

Access of Evil Award:
Canada, Australia, New Zealand

For repeated attempts to undermine the *de facto* moratorium on Terminator technology at the Convention on Biological Diversity (CBD). And for their betrayal of Indigenous Peoples at the CBD's Working Group on 8(j) in Spain.

Biggest Tiny Claim On Nature:
Nanosys Inc.

For securing a US patent on 'metal-oxide nanorods' covering more than a third of the chemical elements of the periodic table.

Worst Betrayal:
Genencor et al.

For patenting, cloning and selling "extremophile" microorganisms that were collected from lakes in Kenya without the permission of Kenyan authorities or the collaborating Kenyan researcher. The microorganisms produce industrially-important enzymes (used to fade blue jeans) that reap millions for industry but nothing for Kenya.

Most Hypocritical: Joint Winners:
University of California-Davis

For patenting a blight-resistant gene extracted from a rice variety developed by the Bela peoples of Mali, and for failing to deliver on the Genetic Resources Recognition Fund to benefit Mali's farmers. The Philippines-based public plant breeding institute – the International Rice Research Institute – handed over the blight resistant rice sample to UC-Davis researchers in 1990. But when IRRI requested access to the blight resistant gene derived from the sample, UC-Davis demanded a $10,000 fee.

The Biotechnology Industry Organization (BIO)

For writing Bonn-inspired bioprospecting guidelines for use by BIO member companies and then inviting the companies to ignore them.

Captain Hook Awards, 2006, granted by the Coalition Against Biopiracy, an informal group of organizations that first came together at the 1995 Conference of the Parties to the Convention on Biological Diversity meeting in Jakarta, Indonesia (http://www.captainhookawards.org/winners/2006_pirates).

Access to Knowledge:
The Case of Indigenous and Traditional Knowledge

Carlos M. Correa

The A2K movement generally aims at an information society where knowledge is openly accessible to the benefit of all. Despite a broad convergence on the ultimate objectives to be achieved, the positions of civil-society groups, governments, scholars, and other individuals that participate in such a movement are probably not unanimous.

Divergences are most likely to arise with regard to the role, if any, of intellectual property rights. On the assumption that the more widely that a piece of knowledge is distributed, the better it is for the society, both for the utilization of the knowledge and for its further refinement and development,[1] some A2K supporters advocate for alternatives to the system of intellectual property rights. Others, however, find some space for the use of such rights. For instance, although the Adelphi Charter on Creativity, Innovation, and Intellectual Property postulates that humans' creative imagination "requires access to the ideas, learning and culture of others, past and present" and that "human rights call on us to ensure that everyone can create, access, use and share information and knowledge, enabling individuals, communities and societies to achieve their full potential," the charter also indicates that "creativity and investment should be recognised and rewarded. The purpose of intellectual property law (such as copyright and patents) should be, now as it was in the past, to ensure both the sharing of knowledge and the rewarding of innovation."[2] The charter therefore does admit the idea that intellectual property rights may be granted under certain circumstances.

Similarly, the free- and open-source software movement promotes "free" access to software as a means of furthering its diffusion and improvement, but "free" in this context does not necessarily mean that a particular piece of software is in the public domain, since the system relies on copyright licenses to require that modified versions also be freely available. Likewise, the Creative Commons scheme utilizes licenses that forbid many of the same acts that copyright law does.[3]

Another area in which the supporters of the A2K movement may disagree relates to the protection of traditional knowledge. As examined in an abundant literature,[4] there are a number of initiatives either to extend existing intellectual property rights to traditional knowledge or to create new, "sui generis" rights conferring exclusive rights over such knowledge. In fact, some countries, such as Panama, have already passed legislation that recognizes some form of exclusive rights to registered or unregistered traditional knowledge.[5] Such legislation reveals a considerable diversity in the approaches followed, the objectives pursued, the scope of protected knowledge, and the rights conferred, among other differences.[6]

There is no agreed-upon definition of traditional knowledge.[7] The World Intellectual Property Organization (WIPO) uses the term to refer to tradition-based literary, artistic, or scientific works, performances, inventions, scientific discoveries, designs, marks, names and symbols, undisclosed information, and all other tradition-based innovations and creations resulting from intellectual activity in the industrial, scientific, literary, or artistic fields.[8] Notably, traditional knowledge includes environmental or ecological knowledge and plant-based therapies ("traditional medicine"). An operational concept of traditional knowledge may be based on the source of the knowledge (traditional and indigenous communities)[9] and on its cultural specificity, rather than on the specific content of its components.

The United Nations Declaration on the Rights of Indigenous Peoples, adopted on September 7, 2007, has confirmed indigenous peoples' rights over their knowledge:

> Article 11
> 1. Indigenous peoples have the right to practise and revitalize their cultural traditions and customs. This includes the right to maintain, protect and develop the past, present and future manifestations of their cultures, such as archaeological and historical sites, artefacts, designs, ceremonies, technologies and visual and performing arts and literature.
> 2. States shall provide redress through effective mechanisms, which may include restitution, developed in conjunction with indigenous peoples, with respect to their cultural, intellectual, religious and spiritual property taken without their free, prior and informed consent or in violation of their laws, traditions and customs.[10]

While different reasons justify the search for the protection of traditional knowledge, equity considerations have largely dominated the debates on the matter, especially in the light of the numerous reported cases of misappropriation (or "biopiracy") without any recognition or compensation to holders of traditional knowledge.[11]

Legal regimes aimed at preventing the misappropriation of traditional knowledge seem fully compatible with the general philosophy of the A2K movement to the extent that their intended objective is to avoid the creation of exclusive rights over knowledge. More controversial may be those regimes granting some forms of exclusive rights over such knowledge. If granted, such rights may be exercised so as to impede the use of the protected knowledge by third parties without the rights holders' authorization. While this may be regarded as antithetical to the A2K open-access goal, the A2K movement may be understood not just as a libertarian agenda, but as a quest for the realization, in the area of knowledge generation and sharing, of the fundamental principles of justice and economic development. Granting rights to holders of traditional knowledge may, in some circumstances, be required purely for equity reasons or to improve their living conditions.

This paper examines the extent to which intellectual property rights protection of traditional knowledge is compatible with the paradigm envisioned by the A2K movement. It is not so clear exactly what the underpinnings of A2K are, and there is some anxiety about where traditional knowledge fits. Hence, it is important to discuss how the claims for traditional knowledge protection by indigenous/traditional communities may be reconciled with the main philosophical approach of the A2K movement. The paper considers, first, whether, in the context on national laws, traditional knowledge may be deemed a part of the public domain. Second, it analyzes the principles emerging from international treaties and other instruments, particularly as they may limit the options available with regard to the legal treatment of traditional knowledge. Finally, the paper elaborates on the implications of the analysis for the conceptions underpinning the A2K vision.

TRADITIONAL KNOWLEDGE AS PART OF THE PUBLIC DOMAIN

A2K advocates expanding the public domain. Although this concept seems simple at first sight, defining what information is actually part of the public domain is a complex task. In particular, the situation of traditional knowledge is unclear. Can traditional knowledge that is not protected under intellectual property rights be considered to belong to the public domain?

There are at least three concepts of "the public domain" employed in the context of intellectual property law that lead to different scenarios regarding the way the protection of traditional knowledge can be approached.[12] First, in accordance with a commonly used concept, information in the public domain is information whose intellectual protection rights have expired, information for which protection would be appropriate, but has been lost due to a failure to comply with certain formal requirements of intellectual property law, and information outside the scope of legislation on intellectual property because it is not eligible for protection according to the law.

In this conception, the public domain thus encompasses the entire *pool* of works and knowledge, including factual and scientific information, that is not subjected to intellectual property rights,[13] as well as any information that was not or could have not been subjected in the past to intellectual property rights due to a lack of eligibility for protection.

Under this definition, with a few exceptions, traditional knowledge would be considered part of the public domain. Hence, no authorization would be needed to use it, nor should any compensation be paid for doing so. Indigenous and traditional communities would have no right to prevent the use of the knowledge they hold.

It is worth noting that, as an exception, in some countries (including Algeria, Argentina, Benin, Bolivia, Burkina Fasso, Cameroon, Congo, Ivory Coast, Hungary, Italy, Mexico, Rwanda, Senegal, and Uruguay) the reproduction of works of art that have fallen into the public domain are subject to a payment to the state. This is called a "paying public domain" (*domain public payant*).[14]

In the second concept of the public domain, the notion is regarded more restrictively. Strictly construed, the concept of the public domain does not refer to works that are inherently unprotectable, but only to subject matter that could have had intellectual property protection that for some reason was not obtained and that was previously protected and has later fallen into the public domain. The latter is the sense of the term that is invoked in the Agreement on the Trade-Related Aspects of Intellectual Property Rights (TRIPS).[15] This narrower concept of the public domain excludes all material that was never eligible for protection. This would leave out, for example, purely fact-based information, unoriginal works, and nonpatentable techniques. Likewise, traditional knowledge that is not susceptible to protection under the conventional forms of intellectual property rights would not be part of the public domain defined in this way. Holders of traditional knowledge therefore might exercise the rights that national legislation would recognize, if any, over the knowledge they possess and require authorization or payment for its use.

In a third conception, information in the public domain may be broadly understood as information that can be freely used without effectuating payment to third parties or obtaining authorization from them.[16] The public domain in this sense is "a collection of things available for all people to access and consume freely."[17] Although works protected by copyright may be freely accessible under "fair use" or other exceptions, this may not be deemed to put such works into the public domain, because only limited uses are allowed under the conditions determined by the applicable law.[18] The extent to which traditional knowledge may be considered freely accessible and usable would determine whether it may be considered to be part of the public domain or not under this definition.

Western intellectual property systems have regarded traditional knowledge as information freely available for use by anybody. As a result, traditional knowledge has often been published or exploited without any recognition, moral or economic, to those who originated or preserved the relevant knowledge. Further, diverse components of traditional knowledge have been appropriated under intellectual property rights regimes by researchers and commercial enterprises without the prior consent of and any compensation to the knowledge creators or holders. Well-known examples include U.S. patent 5,304,718 on a quinoa variety granted to researchers of the Colorado State University and U.S. patent 5,401,504 relating to the wound-healing properties of turmeric, as well as a diversity of patents on products based on plant materials and local or indigenous communities' knowledge such as the neem tree, kava, barbasco, maca, and endod.[19]

Traditional knowledge has been considered de facto as freely usable and appropriable. A survey of scientific journals indicated that articles in twenty-five journals in English, French, and Chinese made explicit references to traditional medicinal uses of the substances described.[20] University-based authors from both developed and developing countries accounted for an overwhelming 81 percent of such publications. Among developing countries, the leading producers of ethno-medical publications were India (twenty publications), Brazil (nineteen), Mexico (ten), Argentina (ten), South Africa (nine), Turkey (nine), and Nigeria (six).

These differing interpretations of the scope of the public domain in relation to traditional knowledge have been one of the hurdles confronting the still-ongoing debates about protection of and compensation for traditional knowledge.[21] What is in the public domain is determined, in the last instance, by national laws, in accordance with the principle of territoriality applicable in the area of intellectual property law.[22] In fact, "information is not in the public domain because of its nature as a public good or even its governmental origin but as a result of a network of formal and informal social agreements, explicit or implicit but entrenched in the common law and in the culture of a society."[23] Therefore, the limits of the public-domain spectrum can be greater or smaller, depending on the types and degree of appropriation determined by the law of a particular state. Such limits depend, in the last instance, on debates and decisions at the national and international levels.

Unlike the public domain under administrative law, which is subject to limits established by the state, such as the authorization to use assets privately under governmental control as part of public-services concessions, the public domain under intellectual property law is, in principle, absolute and mandatory,[24] meaning that it cannot be the object of private appropriation unless a new law expands the limits of what may be appropriated, as the European Directive 96/9/CE did, for example, for the protection of nonoriginal databases. It is also possible for information in the public domain to have its protection restored, as stipulated, for example, by Article 70.2 of the TRIPS Agreement. In applying this article in 1994, the U.S. Uruguay Round Agreement Act restored authors' rights for foreign works, such as movies and music, that had not been protected earlier in the United States.

The legal status of traditional knowledge as a part of the public domain thus depends, in principle, upon determinations made under national laws. However, national legal solutions vary considerably today. Some countries have adopted, as noted above, sui generis legislation that clearly removes traditional knowledge from the public domain by conferring exclusive or remuneration rights of various kinds.[25] One example is Panama's Law No. 20 of June 26, 2002, which established a special regime of intellectual property on the collective rights of indigenous peoples for the protection and defense of their cultural identity and traditional knowledge. The subject matter protected under this law encompasses customs, traditions, beliefs, spirituality, cosmovision (the worldviews of the Mesoamerican peoples), folkloric expressions, artistic manifestations, traditional knowledge, and any other type of traditional expression of indigenous communities that are part of their cultural assets—their cultural heritage. In order to be protectable, the subject matter must be collective in nature, capable of commercial use, based upon tradition (although it need not be "old") and fit within the classification system established by the law. "Collective intellectual property rights" and "traditional knowledge" under this law are embodied in creations such as inventions, models, designs, and drawings, in innovations contained in images, figures, graphic symbols, petroglyphs and other material, and in cultural elements of history, music, the arts, and traditional artistic expressions. The collective rights granted under this regime permit rights holders to prevent the use, commercialization, industrial reproduction, or acquisition of exclusive intellectual property rights over the subject matter and allow for the certification of cultural expressions as works of indigenous traditional art or handicraft and as handmade by natives.

In the case of Peru, in addition to recognizing the indigenous people's ownership of intellectual property and associated rights, the Law Establishing a Regime of Protection of the Collective Knowledge of Indigenous Peoples Related to Biological Resources of 2002 provides that if collective knowledge has passed into the public domain in the last twenty years, a percentage of the value, before taxes, of the gross sales resulting from the marketing of products developed from this knowledge is to be set aside for the Fund for the Development of Indigenous Peoples. The fund will also receive a minimum 10 percent of the gross sales, before taxes, resulting from the marketing of products developed from collective knowledge.[26] This regime has attracted little interest from indigenous peoples so far, despite the efforts of the government to promote its use.

In other countries, unpublished traditional knowledge is deemed to belong in the public domain and may be appropriated without the consent or compensation of its holders. This is notably the case in the United States, where disclosure of the claimed invention in a nonwritten form is not an obstacle to patenting. According to Article 102 of the Patent Law (35 United States Code):

A person shall be entitled to a patent unless—

(a) the invention was known or used by others in this country, or patented or described in a printed publication in this or a foreign country, before the invention thereof by the applicant for patent, or

(b) the invention was patented or described in a printed publication in this or a foreign country or in public use or on sale in this country, more than one year prior to the date of the application for patent in the United States.[27]

This concept of relative novelty implies that traditional knowledge, even if publicly used, but not documented in a foreign country, is patentable in the United States. As a result, several patents relating to or consisting of genetic materials or traditional knowledge acquired in developing countries have been granted to researchers or firms by the U.S. Patent and Trademark Office. A classic example is the patent— regarded as outrageous by some indigenous communities in Amazonia—covering a variety of the ayahuasca vine (*Banisteriopsis caapi*). In 1986, after research in Ecuadorian Amazonia, a U.S. scientist was granted U.S. plant patent 5,751. Ayahuasca is the vernacular name for the plant among the Amazon Quichua people, in whose language ayahuasca means "vine of the spirits." It is used for many medicinal and ritual purposes. Although the validity of the patent was legally challenged, it was confirmed by the U.S. Patent and Trademark Office in 2001.[28]

In sum, there is no unique response to the legal status of traditional knowledge as part of the public domain. Its legal treatment is determined by national law— subject to the territoriality principle—and by applicable international law.

THE LEGAL STATUS OF TRADITIONAL KNOWLEDGE UNDER INTERNATIONAL LAW

The legal status of traditional knowledge under international law is also considerably uncertain. Article 8(j) of the Convention on Biological Diversity requires the contracting parties to "respect, preserve and maintain knowledge, innovations and practices of indigenous and local communities embodying traditional lifestyles relevant for the conservation and sustainable use of biological diversity and promote their wider application with the approval and involvement of the holders of such knowledge, innovations and practices and encourage the equitable sharing of the benefits arising from the utilization of such knowledge, innovations and practices." The Convention on Biological Diversity qualifies this recognition by indicating that a state must do this only "as far as possible and as appropriate" and "subject to its national legislation."[29] Although the adoption of this provision gave an unprecedented impulse to international discussions on the protection of traditional knowledge because it signaled the interest of the international community in protecting that knowledge, it does not bind states to protect traditional knowledge in their own territories. In accordance with the Convention on Biological Diversity, access to genetic resources is subject to the consent of the state (Article 15.5). Some states have passed access regulations (for example, the Philippines and the Andean Community)[30] that require prior informed consent of the relevant traditional communities or indigenous peoples, but this is beyond what the the Convention on Biological Diversity requires.

Likewise, the Food and Agriculture Organization International Treaty on Plant Genetic Resources for Food and Agriculture recognized, but deferred to national governments the implementation of "Farmers' Rights." In Article 9.2 of the treaty:

> The Contracting Parties agree that the responsibility for realizing Farmers' Rights, as they relate to plant genetic resources for food and agriculture, rests with national governments. In accordance with their needs and priorities, each Contracting Party should, as appropriate, and subject to its national legislation, take measures to protect and promote Farmers' Rights, including:
> (a) protection of traditional knowledge relevant to plant genetic resources for food and agriculture.

Like the provision in the Convention on Biological Diversity, this article does not set a uniform standard, because protection of traditional knowledge is to be provided by a contracting party "as appropriate, and subject to its national legislation."[31]

The international conventions on human rights also contain elements relevant to the analysis of the legal status of traditional knowledge. The Universal Declaration of Human Rights established in Article 27(2) that "everyone has the right to the protection of the moral and material interests resulting from any scientific, literary or artistic production of which he is the author."[32] Although this proclamation refers to the "author," there is no reason to presume that it does not apply to cases of collective authorship, as in the case of indigenous or traditional communities' cultural expressions.

The International Covenant on Economic, Social and Cultural Rights contains a similar clause in Article 15(c). It affirms that everyone has the right "to benefit from the protection of the moral and material interests resulting from any scientific, literary or artistic production of which he is the author."[33] Significantly, the Committee on Economic, Social and Cultural Rights, the body charged with interpreting the International Covenant on Economic, Social and Cultural Rights, in its "General Comment 17" on article 15(c), specifically states that "the moral and material interests resulting from one's scientific, literary and artistic productions safeguards the personal link between authors and their creations and between peoples, communities, or other groups and their collective cultural heritage." It also affirms that states "should adopt measures to ensure the effective protection of the interests of indigenous peoples relating to their productions, which are often expressions of their cultural heritage and traditional knowledge."[34]

It is to be noted that the Committee on Economic, Social and Cultural Rights does not refer to "intellectual property rights" but more generally to "interests." In "General Comment 17," it draws a distinction between human rights, which "are fundamental as they are inherent to the human person as such," and intellectual property rights, which "are first and foremost means by which States seek to provide incentives for inventiveness and creativity, encourage the dissemination of creative and innovative productions, as well as the development of cultural identities, and preserve the integrity of scientific, literary and artistic productions for the benefit of society as a whole."[35] The comment adds:

> In contrast to human rights, intellectual property rights are generally of a temporary nature, and can be revoked, licensed or assigned to someone else. While under most intellectual property systems, intellectual property rights, often with the exception of moral rights, may be allocated, limited in time and scope, traded, amended and even forfeited, human rights are timeless expressions of fundamental entitlements of the human person. Whereas the human right to benefit from the protection of the moral and material interests resulting from one's scientific, literary and artistic productions safeguards the personal link between authors and their creations and between peoples, communities, or other groups and their collective cultural heritage, as well as their basic material interests which are necessary to enable authors to enjoy an adequate standard of living, intellectual property regimes primarily protect business and corporate interests and investments. Moreover, the scope of protection of the moral and material interests of the author provided for by article 15, paragraph 1(c), does not necessarily coincide with what is referred to as intellectual property rights under national legislation or international agreements.[36]

As this comment suggests, although traditional knowledge does not need to be protected under intellectual property rights, the moral and material interests of those who create and maintain traditional knowledge need to be respected as human rights. An implication of this statement is that the misappropriation of traditional knowledge, for instance through patents by those lacking the right to apply for them (or by those who have failed to compensate and acknowledge the contributions of others), violates a fundamental right. Moreover, traditional knowledge may not be considered freely available and usable by any party. Hence, it cannot be regarded as integrated with the public domain in the sense of information free to be used and consumed.

Finally, the United Nations Declaration on the Rights of Indigenous Peoples, adopted by the UN General Assembly in September 2007, recognizes that "respect for indigenous knowledge, cultures and traditional practices contributes to sustainable and equitable development and proper management of the environment." Specifically, Article 31 of the declaration states the following:

1. Indigenous peoples have the right to maintain, control, protect and develop their cultural heritage, traditional knowledge and traditional cultural expressions, as well as the manifestations of their sciences, technologies and cultures, including human and genetic resources, seeds, medicines, knowledge of the properties of fauna and flora, oral traditions, literatures, designs, sports and traditional games and visual and performing arts. They also have the right to maintain, control, protect and develop their intellectual property over such cultural heritage, traditional knowledge, and traditional cultural expressions.

2. In conjunction with indigenous peoples, States shall take effective measures to recognize and protect the exercise of these rights.[37]

In more straightforward wording than that used in the Convention on Biological Diversity and the International Covenant on Economic, Social and Cultural Rights, the declaration affirms that indigenous peoples have the "right to maintain, control, protect and develop" their knowledge and "also the right to ... their intellectual property." Consistently with "Comment 17," the declaration does not subsume all rights over traditional knowledge into the categories of intellectual property. This means that these rights exist independently from their formal recognition as intellectual property. The declaration thus also seems to support the view that traditional knowledge is not a *res nullius* that everyone may use and eventually appropriate to his or her own benefit.

It is beyond the scope of this paper to discuss the legal value or status of the referred-to provisions in the Convention on Biological Diversity, the International Covenant on Economic, Social and Cultural Rights, and the UN declarations. Whatever that legal value is, however, they state unequivocally that traditional knowledge may not be considered to belong to the public domain or to be open for free and/or uncompensated use. The discussed precedents of international law suggest some limits to nations' freedom with regard to traditional knowledge. As a minimum, they cannot treat traditional knowledge as freely available and appropriable information, nor can they legitimize the misappropriation of traditional knowledge on the basis of legal fictions, such as assuming that information not published within a given territory is "novel" and hence susceptible of being patented by the person who disclosed it in a patent application.

IMPLICATIONS FOR A2K

The preceding analysis indicates that traditional knowledge cannot, in accordance with accepted principles of international law, be deemed part of the public domain if "the public domain" is defined as the pool of information that is freely usable. It may, however, be considered part of the public domain if the concept is more narrowly interpreted as including information not covered by intellectual property rights, but not necessarily freely usable for this reason.

Is this conclusion consistent with the initiatives aiming at promoting access to and wide diffusion of knowledge? This seems to be the case if it is accepted that such initiatives consent to some form of "balanced" intellectual property rights, protection of traditional knowledge does not necessarily entail the granting of exclusive intellectual property rights, and protection of traditional knowledge is justified, among other reasons, by considerations of equity and human development.

In effect, the majority of the actors in the A2K movement do not seek the abolition of all forms of intellectual property rights, but the proper balance between public and private interests. The Adelphi Charter, for instance, points out that "the public interest requires a balance between the public domain and private rights. It also requires a balance between the free competition that is essential for economic vitality and the monopoly rights granted by intellectual property laws."[38] James Boyle, in his "Manifesto on WIPO and the Future of Intellectual Property," also calls for balance:

> As intellectual property protection has expanded exponentially in breadth, scope and term over the last 30 years, the fundamental principle of balance between the public domain and the realm of property seems to have been lost. The potential costs of this loss of balance are just as worrisome as the costs of piracy that so dominate discussion in international policy making. Where the traditional idea of intellectual property wound a thin layer of rights around a carefully preserved public domain, the contemporary attitude seems to be that the public domain should be eliminated wherever possible. Copyrights and patents, for example, were traditionally only supposed to confer property rights in expression and invention respectively. The layer of ideas above, and of facts below, remained in the public domain

for all to draw on, to innovate anew. Ideas and facts could never be owned. Yet contemporary intellectual property law is rapidly abandoning this central principle. Now we have database rights over facts, gene sequence, business method and software patents, digital fences that enclose the public domain together with the realm of private property . . . the list continues.[39]

Protection of traditional knowledge,[40] on the other hand, may be conceived of as a means to prevent different modalities of the misappropriation of traditional knowledge (sometimes called "defensive" protection), rather than as a tool for the granting of positive rights (often called "offensive protection"). Moreover, positive rights may not confer exclusivity. They may be based on a remuneration right or a liability rule,[41] in line with the paradigm of the Convention on Biological Diversity, which does not require or suggest the establishment of exclusive rights.[42] Indeed, many indigenous communities reject the very idea of exclusive property over their knowledge or of obtaining payment for its use. In addition, if intellectual property rights were established for traditional knowledge, their holders might encounter enormous difficulties in enforcing them, given the need to comply with certain formalities (except in the case of copyright) and, above all, the high cost of enforcing rights in courts. Enforcement procedures are generally long and may be prohibitively costly for holders of traditional knowledge, particularly if litigation is to take place in foreign countries.[43] In some cases, such as in the case of the U.S. patent on an ayahuasca variety, NGOs assumed the defense of the interests of traditional knowledge holders, and this may be the only feasible approach in many situations.

Finally, one of the main reasons for seeking protection for traditional knowledge (often implicit in various analyses and proposals on the subject) is the lack of equity in current relations between indigenous/traditional communities and the rest of society. A main objective of such protection would be to obtain moral recognition or some economic compensation for the commercial use of traditional knowledge, or both. In addition, the protection of traditional knowledge may be a component of policies aimed at preserving the cultures of those communities while ensuring possession of their lands and participation in decisions that affect the use of resources under their control. If properly designed and implemented, traditional knowledge protection thus may be instrumental to human development and the realization of human rights.

CONCLUSION

A number of ongoing initiatives aim at broadening A2K. Given the importance of traditional knowledge for developing countries and the imperative to ensure an equity-based utilization of that knowledge, it seems necessary to clarify its legal status and the conditions under which it may be eventually appropriated or shared.

In accordance with Western intellectual property rights laws and principles, knowledge created and held by indigenous or traditional communities may be deemed to belong to the public domain if understood as the pool of knowledge that is not subject to existing modalities of intellectual property rights. This would mean that traditional knowledge could be freely used without prior consent from or compensation to their holders.

National law determines what does and does not belong in the public domain. In some cases, national solutions permit the appropriation of traditional knowledge by individuals or companies that have obtained access to traditional knowledge, even without the consent of its holders. A number of provisions in international instruments, however, recognize rights in favor of such communities. Although such rights do not necessarily pertain to one of the categories of intellectual property rights, they would clearly exclude traditional knowledge from the realm of freely usable knowledge.

The need to protect traditional knowledge may be justified, among other reasons, on the grounds of equity and development. Protection for intellectual property rights does not seem incompatible with the philosophy that underpins the A2K movement, particularly if such protection is conceived in defensive terms, with the intention of preventing misappropriation, rather than asserting positive rights.

NOTES

1   On the relationship between innovation and diffusion see, for example, Organisation for Economic Co-operation and Development, *Technology and the Economy: The Key Relationships* (Paris: Organisation for Economic Co-operation and Development,1992).

2   See http://sitoc.biz/adelphicharter/adelphi_charter_document.asp.htm (last accessed April 20, 2009). This charter certainly cannot be considered "the" charter of the A2K movement, because it is only the consensus view of a number of individuals, but it may be deemed as representative of the thinking behind the movement.

3   See, for instance, the description of Creative Commons licenses at http://creativecommons. org/about/licenses/ (last accessed April 21, 2009).

4   See, for example, Sophia Twarog and Promila Kapoor (eds.), *Protecting and Promoting Traditional Knowledge: Systems, National Experiences and International Dimensions* (2004), available on-line at http://www.unctad.org/en/docs/ditcted10_en.pdf. See also the selection of legislative texts in World Intellectual Property Organization, *Legislative Texts on the Protection of Traditional Knowledge and Traditional Cultural Expressions (Expressions of Folklore)*

and Legislative Texts Relevant to Genetic Resources, available on-line at http://www.wipo. int/tk/en/laws (both last accessed April 21, 2009).

5   See Carlos M. Correa, "Protecting Traditional Knowledge: Lessons from National Experiences," in Twarog and Kapoor (eds.), *Protecting and Promoting Traditional Knowledge*.

6   *Ibid*.

7   See Chidi Oguamanan, *International Law and Indigenous Knowledge: Intellectual Property, Plant Biodiversity and Traditional Medicine* (Toronto: University of Toronto Press, 2006), p. 15.

8   World Intellectual Property Organization, Intellectual Property Needs and Expectations of Traditional Knowledge Holders (Geneva: WIPO, 2001), p. 25, available on-line at http://www. wipo.int/tk/en/tk/ffm/report/index.html (last accessed April 21, 2009).

9   The distinction between "indigenous" and "traditional" knowledge has been deemed superficial, because both share common characteristics. See Oguamanan, *International Law and Indigenous Knowledge*, p. 20.

10 United Nations Declaration on the Rights of Indigenous Peoples, available on-line at http:// www.un.org/esa/socdev/unpfii/documents/DRIPS_en.pdf (last accessed April 21, 2009).

11 For a discussion on the subject, see Carlos M. Correa, *Protection and Promotion of Traditional Medicine: Implications for Public Health in Developing Countries* (Geneva: South Centre, 2002), available on-line at http://www.who.int/medicinedocs/collect/medicinedocs/pdf/s4917e/ s4917e.pdf (last accessed April 21, 2009).

12 For a discussion of different definitions of the public domain, see Pamela Samuelson, "Enriching Discourse on Public Domains," *Duke Law Journal* 55 (2006), available on-line at http://people.ischool.berkeley.edu/~pam/papers/enriching%20discourse%20on%20 public%20domains.pdf (last accessed April 21, 2009). The notion of the "public domain" is also used in administrative law to refer to things under the control of or subject to the property of the state.

13 On facts as part of the public domain, see Corynne McSherry, *Who Owns Academic Work?: Battling Over the Control of Intellectual Property* (Cambridge, MA: Harvard University Press, 2001), who argues that "data, especially scientific data, are classic public domain material" (p. 191). On scientific information as part of the public domain, see J. H. Reichman and Paul F. Uhlir, "Database Protection at the Crossroad: Recent Developments and Their Impact on Science and Technology" (1999), available on-line at http://eprints.law.duke.edu/1518/1/14_ Berkeley_Tech._L.J._793_(1999).pdf (last accessed April 21, 2009).

14 See Delia Lipszyk, *Derecho de autor y derechos conexos* (Paris: United Nations Educational, Scientific, and Cultural Organization, 1993).

15 The agreement uses the term "public domain" only once, in relation to protectable material that is no longer protected (Article 70.3). The TRIPS Agreement is available on-line at http:// www.wto.org/english/tratop_e/trips_e/t_agm0_e.htm (last accessed March 22, 2009).

16 William van Caenegem, "The Public Domain: Scientia Nullius," *European Intellectual Property Review* 24, no. 6 (2002): p. 324.

17 Inge Kaul, Pedro Conceição, Katell Le Goulven, and Ronald U. Mendoza, "Why Do Public Goods Matter Today?" in Inge Kaul, Pedro Conceição, Katell Le Goulven, and Ronald U. Mendoza R. (eds.), *Providing Global Public Goods: Managing Globalization* (New York: Oxford University Press, 2003), p. 8.

18 On the "fair use" exception, see Carlos M. Correa, "Fair Use in the Digital Era," *International Review of Industrial Property and Copyright Law* 33, no. 5 (2002).

19 The neem tree (*Azadirachta indica*) is a tree in the mahogany family native to South Asia that is a major component in ayurvedic medicine and that is said to treat several different diseases. Kava (*Piper methysticum*) is used traditionally in the western Pacific as a mild tranquilizer. Barbasco (*Dioscorea mexicana*) is an inedible wild Mexican yam from which progesterone can be synthesized. It is also the common name of a South American evergreen tree (*Lonchocarpus urucu*) that is a natural source of rotenone, a fish poison. Maca (*Lepidium meyenii*) is a herbaceous plant of the Andes with several claimed health benefits. Endod (*Phytolacca dodecandra*), a plant that occurs throughout sub-Saharan Africa and parts of South America and Asia, is a potent molluscicide and is used to kill snails.

20 Russel Barsh, "Who Steals Indigenous Knowledge?" *Proceedings of the 95th Annual Meeting of the American Society of International Law* (2001): pp. 153–61. The largest number of articles were found in the *Journal of Ethnopharmacology* (128 articles, or 51 percent of the total), *Pharmaceutical Biology* (50 articles, or 19 percent), *Economic Botany* (14 articles, or 6 percent), and *Phytomedicine* (12 articles, or 5 percent).

21 Note, for example, the failure to deliver concrete outputs on the subject of traditional knowledges by the WIPO Intergovernmental Committee on Intellectual Property and Genetic Resources, Traditional Knowledge and Folklore, which held its eleventh session on July 3 to 12, 2007, in Geneva.

22 In conformity with this principle, the existence and validity of intellectual property rights are to be judged in accordance with the law of the jurisdiction where protection is obtained. One corollary of this principle is that decisions, for instance on the validity of a patent in one country, do not affect an equivalent patent in other countries.

23 Clemente Forero-Pineda, "Scientific Research, Information Flows, and the Impact of Database Protection on Developing Countries," in Julie M. Esanu and Paul F. Uhilr (eds.), *Open Access and the Public Domain in Digital Data and Information for Science: Proceedings of an International Symposium* (Washington, D.C.: The National Academies Press, 2004), p. 40, available on-line at http://books.nap.edu/catalog.php?record_id=11030#toc (last accessed April 21, 2009).

24 Stéphanie Choisy, *Le domain public en droit d'auteur* (Paris: Litec, 2002), p. 53.

25 See Correa, "Protecting Traditional Knowledge."

26 See Begoña Venero Aguirre, "The Peruvian Law on Protection of the Collective Knowledge of Indigenous Peoples Related to Biological Resources," in Christophe Bellmann, Graham Dutfield, and Ricardo Meléndez-Ortiz (eds.), *Trading in Knowledge: Development Perspectives on TRIPS, Trade, and Sustainability* (London: EARTHSCAN, 2003).

27 U.S. Patent and Trademark Office, "35 U.S.C. 102: Conditions for Patentability; Novelty and Loss of Right to Patent —Patent Laws," available on-line at http://www.uspto.gov/web/offices/pac/mpep/documents/appxl_35_U_S_C_102.htm#usc35s102 (last accessed April 21, 2009).

28 See Correa, *Protection and Promotion of Traditional Medicine*.

29 The Convention on Biological Diversity, "Article 8. In-situ Conservation," available on-line at http://www.cbd.int/convention/articles.shtml?a=cbd-08 (last accessed April 21, 2009).

30 See Carlos M. Correa, "The Access Regime and the Implementation of the FAO International Treaty on Plant Genetic Resources for Food and Agriculture in the Andean Group Countries," *Journal of World Intellectual Property* 6, no. 6 (November 2003).

31 International Treaty on Plant Genetic Resources for Food and Agriculture, Article 9.2, available on-line at http://www.sipo.gov.cn/sipo/ztxx/yczyhctzsbh/zlk/gjgy/200508/

P020060403601427961427.pdf (last accessed April 21, 2009).

32   The Universal Declaration of Human Rights is available on-line at http://www.un.org/Overview/rights.html (last accessed April 22, 2009).

33   The International Covenant on Economic, Social and Cultural Rights is available on-line at http://www.unhchr.ch/html/menu3/b/a_cescr.htm (last accessed April 22, 2009).

34   See Committee on Economic, Social and Cultural Rights, "General Comment No. 17 (2005)," November 21, 2005, E/C.12/GC/17, January 12, 2006, para. 32, available on-line at http://portal.unesco.org/culture/en/files/30545/11432142221comment_en.pdf/comment_en.pdf.

35   *Ibid.*, para. 1.

36   *Ibid.*, para. 2.

37   United Nations Declaration on the Rights of Indigenous Peoples, Article 31.

38   The Adelphi Charter on Creativity, Innovation, and Intellectual Property, principle 3.

39   James Boyle, "A Manifesto on WIPO and the Future of Intellectual Property," *Duke Law and Technology Review* no. 9 (2004): p. 2, available on-line at http://www.law.duke.edu/journals/dltr/articles/PDF/2004DLTR0009.pdf (last accessed April 22, 2009).

40   See the analysis of the various meanings of "protection" in Correa, *Protection and Promotion of Traditional Medicine*.

41   See on this concept Jerome H. Reichman, "Of Green Tulips and Legal Kudzu: Repackaging Rights in Subpatentable Innovation," *Vanderbilt Law Review* 53, no. 6 (November 2000): pp. 1743–98.

42   The UN declaration uses the word "control," but this concept is not equivalent to exclusivity. See, for instance, Article 39.2 of the TRIPS Agreement, where this distinction becomes apparent.

43   For instance, in the United States, according to the 2006 *Economic Report of the President*, median litigation costs average $4 million each for the plaintiff and defendant when more than $25 million is at stake in a patent suit. Executive Office of the President of the United States, *Economic Report of the President Transmitted to the Congress February 2006 Together with the Annual Report of the Council of Economic Advisers*, available on-line at http://www.gpoaccess.gov/eop/2006/2006_erp.pdf (last accessed April 22, 2009).

Undermining Abundance:
Counterproductive Uses of Technology and Law in Nature, Agriculture, and the Information Sector

Roberto Verzola

After World War II, the chemical industries of the West shifted their attention back to civilian applications, including the large-scale production of synthetic urea, organochlorines, and other fertilizers and pesticides. These agrochemicals were marketed supposedly to provide additional nutrition for farmers' crops and to kill crop pests. However, farmers and governments did not realize that these products also killed, incapacitated, weakened, or otherwise made life difficult for very important but little-known creatures: soil organisms that turn organic matter into natural plant food and friendly organisms such as predators and parasites that keep pest populations in check. These creatures constituted a vast, largely invisible, and unrecognized commons into which all farmers unknowingly tapped every time they planted seeds and grew crops. In their defense, the chemical industry might claim that they did not know, either, which would be an admission of recklessness, if not negligence. But this excuse was untenable by the 1960s, when the chemical industry viciously attacked Rachel Carson and her book, *Silent Spring*,[1] which called attention to the harmful effects of DDT and other agrochemicals on nontarget organisms, including human beings.[2]

In effect, the chemical industry was selling farmers and governments a deadly technological Trojan horse, an antiabundance poisoned pill. Agrochemicals appeared to offer more abundant harvests. In truth, their deployment would gradually weaken and take the life out of the farmers' biological support systems, including natural sources of plant food and the enemies of pests. As more agrochemicals were used, the diverse soil populations dwindled, the soil became less fertile, and farmers' crops starved. To keep the plants from starving, more synthetic fertilizers had to be added, which caused the living soil populations to dwindle even further. As the predator and parasite populations likewise dwindled, pest populations went up. So farmers had to spray more pesticides, which then killed even more predators and parasites. More recent studies based on the theory of trophobiosis—the theory that the relations between plant and parasite are essentially nutritional in nature—suggest that synthetic fertilizers actually make plants more attractive to pests.[3] Farmers who took the poisoned pill were caught in the trap and fell into agrochemical addiction, draining life out of the soil and around the crops.

UNDERMINING ABUNDANCE, CREATING SCARCITY

The poisoned pill of agrochemical fertilizers and pesticides is just one example of the ways in which technology and, as we will see, the law are increasingly used to undermine processes of abundance intrinsic to nature and agriculture and even processes intrinsic to the information sector as well. A variety of techniques based in both technology and law, separately or in various combinations, are being intentionally used counterproductively by businesses and governments to undermine abundance and create artificial scarcity. In the examples that follow, technological approaches such as copy protection, copy restriction, copy identification, and user restriction often are combined with legal restrictions such as the enforcement of patents, copyrights, and plant-variety protection and with bans on simple copying, seed sales, and seed exchanges. They are also combined with laws to protect technological copy-protection schemes from being bypassed or to mandate its use. These technologies are actively promoted by governments through incentives such as low-cost credit, subsidies, and other forms of support, while similar incentives are withdrawn from competing technologies. If the poisoned-pill strategy or these other strategies fail to work, the abundant resource and related know-how often are simply ignored or suppressed as much as possible.

AGRICULTURE AND THE LAW: SEED DEPENDENCE

In the 1960s, the International Rice Research Institute introduced IR-8, the first of a series of new "high-yielding varieties" of rice whose high yields partly came from their better responsiveness to chemical treatment.[4] Farmers were wary, and few were willing to let go of their traditional varieties. Drawn by aggressive government subsidies and lending programs, however, more and more farmers switched. As they did, they also stopped planting their heirloom varieties, which were soon lost, because the old seeds they had saved dried up and died. As the heirloom varieties disappeared and dependence on high-yielding varieties grew, farmers also lost their selection and breeding skills.

Agrochemicals and the new chemically responsive varieties would eventually be promoted as the "Green Revolution."[5] Even today, this technological poisoned pill continues to keep millions of farmers addicted to agrochemicals, mired in poverty and debt.

Another facet in the technological substitutions of this period in nature and agriculture was the gradual replacement of work animals by farm machinery. In the Philippines, for instance, carabaos—a domesticated subspecies of water buffalo—were the farmers' main source of mechanical power. Carabaos also grazed the less fertile areas around the farm, their dung enriching the soil. The animal usually recovered by itself from injury or sickness. Even more—perhaps the most amazing thing of all—the female carabao can give birth to another carabao every two years or so. Yet through the same poisoned-pill strategy, farm machinery suppliers and the government eventually managed to get many farmers to switch to a mechanical power source that cost a fortune, was fueled by expensive imported gasoline instead of free grass, gave out noxious pollutants instead of milk and natural fertilizer, required a skilled technician and costly spare parts if it stopped working, and of course never gave birth to its own replacement.

Also in the 1960s, another development would worsen this slippery slide toward seed dependence. U.S. seed companies introduced their commercial version of the F1 corn hybrid developed decades earlier in the public sector.[6] ("F1" means the first filial generation after crossing two different parental lines.) Unlike heirloom varieties, F1 hybrids did not breed true. When their seeds are replanted, the offsprings' characteristics segregate, and the desirable traits are expressed weakly or irregularly in subsequent generations. So regardless of the benefits that the current crop might offer, saving seeds becomes pointless.

Corn farmers now had to buy hybrid seeds from the seed suppliers every planting season. Obviously they still had the option to go back to traditional varieties, but government technicians promoted the hybrid varieties aggressively and extended highly subsidized credit to the farmers who used them. So the use of F1 hybrids among corn farmers grew.

As more farmers abandoned their traditional corn, these varieties became scarce and gradually disappeared. Commercial hybrid corn varieties eventually dominated the seed-corn market, the way the high-yielding varieties did among rice farmers—but with a difference. Seed buying had been an occasional purchase in the past, when seeds produced their own kind, but hybrids led to repeat sales, season after season, turning seeds into highly profitable commodities.

As the seed business became more profitable, giant agrochemical firms began buying up the seed companies that had established themselves in the market. A similar corporate trend toward F1 hybrids emerged in the vegetable sector and, later, in the rice sector, a trend that continues today.[7]

F1 hybrids mark the beginning of corporate efforts to gain full control over seeds, especially in major staple crops and vegetables. They also represent the first technology in agriculture explicitly meant to end the farmers' age-old practice of saving part of their harvest to use as seed in the next planting season. This counterproductive technology strikes at the very heart of sustainability and the seed commons.

Commercial seed breeders took care that nonhybrid varieties would remain under their control, too. Their demand for exclusive rights over varieties that they developed eventually gave rise to the 1961 Convention for the Protection of New Varieties of Plants. This convention defined plant breeders' rights, mandated plant-variety protection, and established an international union, the UPOV, to work for plant breeders' interests. As countries acceded to UPOV agreements, they moved to adopt counterproductive national seed laws that limited the freedom of farmers to exchange seeds or to sell them. Subsequent UPOV agreements (in 1972, 1978, and 1991) became more and more restrictive of farmers' rights.[8] The result was a two-pronged offensive against seed saving and exchange: promotion of the technology of hybrids and new laws and international agreements restricting farmers' options over seeds.

In the early 1980s, seed companies learned to modify plant genomes directly through genetic engineering.[9] Then they patented the modified genes, using the patent system—originally meant for industrial inventions and designs—to claim exclusive rights over seeds and plants with the patented genes.[10] This new weapon in the growing corporate arsenal of counterproductive practices was even more restrictive than plant-variety protection: The novelty of the technology itself now justified excluding everyone by law from using patented seeds unless they paid some kind of royalty or technology fee.

The first commercially successful applications were soya and canola plants that incorporated herbicidal resistance and corn plants that incorporated pesticidal toxins. For the first time, seed companies held the power to sue farmers who saved the seeds of these crops and planted them in a subsequent season, simply on the strength of the patents they held over the genes incorporated in these seeds.

Genetically engineered corn was also a poisoned pill, engineered to produce a modified version of a pesticidal toxin from the soil bacterium *Bacillus thuringiensis*. Organic farmers had used *Bacillus thuringiensis* for decades to control corn pests, prudently spraying the cultured bacteria only if pest damage reached significant levels. When the *Bacillus thuringiensis* gene was inserted into the corn plant, the resulting *Bacillus thuringiensis* corn now expressed the toxin throughout the plant's life, making it more likely for *Bacillus thuringiensis* resistance to develop rapidly among the target pests and sabotaging a resource that organic farmers—the nemesis of the agrochemical/genetic-engineering industry—had used for decades.

Counterproductive technologies now in the pipeline are taking to higher levels

Campaigning material produced by Progressio (www.polyp.org.uk/Progressio; www.progressio.org.uk).

the bizarre goal of attacking natural abundance to create artificial scarcity in agriculture. This opens a market for substitute products and leads to a supply system completely under corporate control through various technological and legal mechanisms.[11]

The precursor of these technologies is the "Terminator Technology," which genetically modifies plants to make their seeds sterile, ending the 350-million-year-old process of reproduction through seeds. Truly, it is the "death of birth."[12] U.S. patents were granted, though commercial applications seem a long way off. The real question was: Would farmers use them? The idea was so outrageous that its promoters backtracked for a while, trying to find a spin that would make their idea more publicly palatable.

They soon found one. Engineered seeds lead to a seemingly intractable problem: genetic contamination. Engineered soya and canola, which had survived despite herbicide applications, were showing up in places where they were neither expected nor wanted—on farms that had used no engineered seeds, especially organic farms where strict safety standards prohibit such seeds. So on the strength of their patent claims, Monsanto sued. The farmers insisted that they had used no engineered varieties. Yet some plants on their farms tested positive for

Monsanto's patented genes. Many farmers, intimidated by Monsanto's legal and financial muscle, paid the fines and suffered the consequences, such as losing their organic certification. However, in one celebrated case that dragged on for years, Canadian farmer Percy Schmeiser stood his ground and fought the legal battle to the end. The Canadian Supreme Court issued an ambiguous decision which each side interpreted as its victory.[13] Promoters of the Terminator Technology now say that their technology can prevent genetic contamination from engineered crops by further modifying these crops to produce sterile seeds.

New ideas in the pipeline fine tune the concept further to allow finer-grained control of sterility. Known as genetic use-restriction technologies, these will enable the seed companies to control seed sterility in the field through external triggers such as chemicals—presumably patented, too. By spraying this chemical on a genetic use-restriction-modified plant, the plant can be induced to turn its sterility (or fertility) on or off. Scarcity and abundance thus can be marketed under full corporate control. A similar technology can also be used for turning genetically engineered traits themselves on or off.

The use of hybrids and genetic engineering have been justified in the interest of "feeding the world." Yet a U.S. Department of Agriculture study in 2006 found that 10 percent of U.S. adults and 17 percent of children occasionally went hungry for lack of food.[14] If they cannot even feed all Americans sufficiently, how can they feed the world?

A VIRTUAL CORNUCOPIA OF SOFTWARE

The same approach of attacking abundance in order to cause artificial scarcity and create a market for substitute products in a supply system under corporate control occurred in the domain of computer software via the intellectual property laws. In the 1980s and early 1990s, in many countries, for a very affordable fee, one could copy from computer shops almost any Apple or IBM PC software that was also available in the United States. Students, new graduates, and enthusiasts bought cheap IBM clones and practiced basic computer operations, word processing, presentations, spreadsheet uses, database management, and programming. There was no Internet then—in the Philippines, a 64 Kbps connection ushered in the Internet in 1994—but it did not matter. A de facto software commons was maintained in computer shops and on electronic bulletin-board systems that made software quickly and efficiently available to students and computer enthusiasts. Many computer professionals today, who now form the backbone of their country's computer industry or who enjoy well-paying jobs abroad as overseas workers, regularly dipped into this cornucopia and acquired their computing skills thanks to the software abundance of that period.

Back in the United States, software developers tried various copy-protection schemes, from nonstandard disk formats to hardware dongles.[15] But the best minds of the U.S. software industry were no match for the resourcefulness of hackers and altruists who wanted to keep the abundance coming. Some U.S. companies even specialized in software that duplicated copy-protected software. Other software developers abandoned copy protection to gain a competitive advantage, and consumers responded favorably. Eventually, the U.S. software industry gave in and, except for some niche markets, abandoned technical copy-protection schemes altogether.

Invoking copyright laws did not help much. Though software was legally protected by copyright laws and international agreements, many countries did not take these seriously, preferring to let their citizens enjoy the abundance that then prevailed. People likewise knew that governments enforce laws selectively anyway, whether they are laws on the minimum wage, corruption, pollution, taxes, elections, or copyrights. In the eighteenth and nineteenth centuries, the United States itself was a center of piracy of British books and publications. Subsequent experiences in Japan, Taiwan, Hong Kong, and other countries and territories likewise showed that copying is a necessary stage in national development. Furthermore, the countries that complained most loudly about the piracy of their intellectual property were themselves most guilty of pirating intellectuals such as doctors, nurses, and engineers from the Third World. The latter was deemed a more malignant case of piracy because it took away the original and left no copy behind. Finally, how can a government clamp down on its citizens when commercial software is freely copied between government computers?[16]

Things began to change after the 1994 formation of the World Trade Organization (WTO). This global system adopted effective mechanisms to enforce its highly protectionist provisions on intellectual property. An international legal infrastructure was gradually built that, combined with strong diplomatic pressures and economic threats, started to turn the tide for copyright holders.

In the Philippines, a turning point occurred in 1998, when Microsoft chairman Bill Gates visited President Fidel Ramos.[17] Gates offered to recognize as legal all copies of Microsoft products installed in government computers. In return, Ramos promised to enforce copyright laws, now that the government's copies were "legal." The United States still needed to direct a whole series of economic, political, and diplomatic pressures on the administrations that followed Ramos's, but the days of software abundance in the Philippines appeared to be numbered.

As copyright enforcement began in earnest, CDs, video CDs, and DVDs were introduced in the 1990s and early 2000s. For a while, the industry managed to prevent copying and to restrict the use of DVDs by geographical region. However, this was eventually thwarted by a combination of dedicated hacking, the technical savvy of rising industrial giant China, and plain consumer freedom of choice.

The Philippine case is probably typical: When illegal CD, video CD, and DVD discs began to circulate, rumors spread that these discs could damage the disc player. The original players made in the United States, Europe, or Japan were so expensive that owners would not risk damage from discs of unknown quality. So those who bought the original players stuck to expensive original discs and suffered under ridiculous geographic restrictions: DVDs sent home by U.S.-based or Middle East–based relatives were unreadable, and players they sent or brought home could not play locally available DVDs.

Enter China. Cheap DVD players that could play discs from any geographic region and priced at one-fifth or less of their competitors flooded the Asian market, including the Philippines. Another rumor—perhaps apocryphal—began to circulate: that original DVDs might damage *these* players. Between China-made machines that played cheap, unauthorized discs and branded players that played only high-priced discs that were also geographically restricted, it was no contest. With the further entry of low-cost CD/DVD burners, duplicating these read-only discs became easy.

So Asia remains a flourishing market of China-made DVD players and unauthorized CDs and DVDs, creating a new abundance of cultural fare for Asians. Many of the DVDs are adult material or otherwise of doubtful cultural value. But most regular movies are available, too, as are, increasingly, movie classics and truly educational collections of documentaries from the Discovery Channel, National Geographic, and similar cable channels—software, too. In some countries, the materials are made more accessible to ordinary people by translations into the local language.

To suppress the new abundance, special government police and private detectives from the United States now regularly conduct surprise raids not only against the disc vendors and distributors, but also against businesses, schools, computer shops, and Internet cafés that use unauthorized software. These highly disruptive raids have driven CD/DVD and software copying underground, where it flourishes unabated, thanks to cheap China-made disc burners.[18]

In the United States, another round of efforts against unauthorized copying was launched under the banner of digital rights management (DRM), consolidating counterproductive technological and legal measures for finer-grained control of copying and access to materials in digital media and on the Internet. DRM includes content encryption, digital signatures, digital fingerprinting, digital watermarks, digital serial numbers built into central processing units and computer motherboards, and miscellaneous authentication systems. They involve such concepts as

Communist remixes for the people (CRFTP).

conditional access systems, remote revocation of use rights, and other means to ensure that scarcity and abundance remain under tight corporate control. They may be aptly called digital use-restriction technologies, after their genetic counterparts for controlling seed reproduction, the genetic use-restriction technologies.

The U.S. remains ahead in the development of digital use-restriction technologies and genetic use-restriction technologies, having the most corporate interests to protect, especially in the information sector. The U.S. Digital Millennium Copyright Act now mandates and protects digital use-restriction technologies themselves, making it illegal to construct devices that bypass or disable these technologies. Citizens' groups in the United States such as the Electronic Frontier Foundation and Public Knowledge are concerned about the impact of DRM and the Digital Millennium Copyright Act on privacy, political freedoms, and human rights.[19]

The increasing availability of high-quality free and open-source software, however, has pulled the rug from under the argument that creativity can be encouraged only by granting creators statutory monopolies through intellectual property rights. In the information sector, as well as in the agriculture sector, the see-saw between abundance and scarcity, between markets and commons, continues through skirmishes on the technology front, in the legal arena, and, of course, in the market.

CREATING ARTIFICIAL SCARCITY ELSEWHERE

Counterproductive efforts to control abundance and produce scarcity have occurred in other fields, as well. Drug laws make medically effective herbal preparations inaccessible to many. Ironically, herbs easily grown in backyards and community gardens, whose preparations would be illegal if prescribed by traditional healers, are often the basis for very expensive patented drugs manufactured by pharmaceutical firms.[20] It is not a coincidence that many of these firms are owned by the same agrochemical companies that control the seed industry.

Through misleading advertising and collusion with hospitals and medical professionals, formula milk companies have managed to undermine mothers' confidence in their own breast milk. This had led to a decline in breast-feeding in a number of Asian countries.[21] As mothers try substitutes, their production of milk slows down and eventually stops, creating a vast new market for formula milk.

A traditional Filipino song about plants found around the hut, "Bahay Kubo,"[22] taught to every child in grade school, enumerates eighteen food plants that include legumes, greens, root crops, seeds, nuts, and spices. The song omits many more. Filipinos have become so fixated on Western foods and diets that they overlook the great variety of indigenous food sources, many of which simply grow untended like weeds in their backyards. The monoculture mindset treats these food sources as indeed weeds that must be suppressed. Razed by farm mechanization and the use of herbicides, most of them have now disappeared from people's backyards, from their diets, and from their consciousnesses, creating real food scarcity and malnutrition.

Organic products are scarce and expensive because a system biased toward chemicals imposes on organic producers the burden of proof: detailed record keeping, testing, inspection, certification and labeling. What if, in accordance with the "polluter pays" principle, producers of chemically treated crops and foods, not organic producers, were required by law to keep detailed records of chemical treatments, get their products regularly inspected and tested by accredited laboratories for minimum residue levels, undergo third-party certification, and follow mandatory labeling requirements to identify to which chemicals and by what amounts their food products have been exposed? If this were so, the price tags of both organic and chemically treated foods would change dramatically in favor of organics.

A low-power radio station that can serve a large community or a small town now costs only about as much as a laptop. Yet such stations continue to be a rarity, because most governments make it nearly impossible to meet all the legal requirements to operate one. As communications expert and president of the World Association of Community Radio Broadcasters Steve Buckley writes, "it is the policy, legal and regulatory framework that remains the single most persistent obstacle" to such stations.[23]

Internet service providers continue to charge exorbitant rates for static Internet Protocol (IP) numbers, arguing that they are running out of these numbers. Yet by simply upgrading to IP Version 6, every person on Earth can be assigned hundreds of IP numbers each, with a lot more to spare.

The sun cannot be hidden, suppressed, made illegal, or otherwise made scarce. Instead, this universal source of abundance has been largely ignored—intentionally, it has been argued—as energy industries focus on energy sources easier to privatize and to control, such as fossil and nuclear fuels.[24]

These examples suggest that the phenomenon of abundance in the natural world and in human societies should not be taken for granted. We need to study it, learn its dynamics, and tap it for human good.

ABUNDANCE IN THE AGRICULTURE AND INFORMATION SECTORS

Creating abundance is a matter of reproducing a good over and over again until more than enough is available for everyone's need or even for everyone's capacity to consume. In nature, the tendency toward bountiful abundance is obvious, especially where seasonal variations highlight the contrast between abundance and scarcity. Prehistoric artifacts of fertility goddesses as well as harvest festivals and rituals still practiced today show the extent to which abundance has been recognized and sought.

Abundance is inherent in the reproductive processes of life. Natural abundance is simply life reasserting itself through the endless cycle of reproduction of its own kind by every life form. This is the wellspring of abundance in nature and in agriculture. The process is self-limiting, too. As every available ecological niche is filled up, species gradually form a food web and settle into a dynamic balance, with closed material cycles ensuring that the balance is maintained. This enables the processes of abundance to continue indefinitely.

Abundance in the domain of information is different. Sharing information does not diminish or deplete it, but rather multiplies and enriches it. Shared information begets more information. The wellspring of information abundance is the inherent human desire to communicate, to seek information and knowledge, and to share them, an urge that gets more fully expressed as the cost of sharing goes down.[25] The cost of reproducing electronic signals is now approaching zero. With digital technology, books, artworks, music, and video can now be stored in the same format as software and databases, as a long string of binary values. From these ones and zeroes, with the right equipment and algorithm, an exact copy of the digital original or a faithful copy of the analog original can be reconstructed. Once stored digitally and made available in easily searchable form on a global network, an unlimited number of users may now get any number of exact copies of the work. Who cannot recognize the abundance of human knowledge, experience, and creative work made possible by the Internet? As more and more people discover its possibilities for sharing freely, the whole range of human skills, thought, and feeling is now being made available through this medium.

From an information perspective, abundance in nature and in agriculture, which is driven by the inherent program within genetic information to reproduce itself, is constrained by material limits, because it must eventually express itself in terms of biomass. Information abundance, on the other hand, is of the nonmaterial variety. Thus, information goods offer the promise of practically unlimited abundance, constrained mainly by the limits of human creativity, the storage capacity of media, and the availability of electricity to power servers on the Internet twenty-four hours a day.

WHO WANTS ABUNDANCE RESTRICTED—AND WHY?

Abundance helps to meet human needs and wants and should therefore be welcomed. Who, then, could be interested in restricting it? As we have seen, attacks against abundance have been mostly initiated by business firms or by governments. Where governments have undertaken these measures, however, they have done so at the instance of business firms, which in the final analysis have reaped the benefits of the government measures.

Looking more closely at the logic of business firms, it is obvious that the immediate effect of restricting abundance is to reduce supply and increase demand. This in turn raises prices or keeps their levels high. If the costs of production change little or not at all and prices go up, profits go up. This is the logic behind corporate efforts to develop technologies and influence state policies that give them closer control over the abundance and scarcity of goods: to create the best conditions for maximizing profits.

Indeed, restricting abundance may maximize profits, but may not necessarily be the best way to encourage creativity. Free and open-source software and farmer-bred plant varieties show that creativity can continue to flourish even without the attraction of monopoly earnings.

Shouldn't this selfish end give way to higher societal goals? The economist's answer is that society's higher goals are indeed served when everyone pursues his or her own self-interest in free competition with others. In fact, economists argue, the competitive pursuit of individual gain accomplishes overall social goals *better*, even if this was no part of the individual's intention, than when the individual consciously tries to advance society's higher goals. The idea that individual pursuit of self-interest not only leads to but is actually the *best* path toward overall social good became the moral basis for capitalist society. This was programmed into business firms as an urge to maximize gain, and they do so by controlling abundance and scarcity in their favor. This is the driving force behind restrictions on abundance.

Because individual human beings are a complex bundle of urges, emotions, and motivations who often act irrationally (that is, regardless of self-interest) from an economist's perspective, corporations are the ideal economic agents, pursuing nothing but maximum gain for themselves based on the economic theory of laissez-faire capitalism.[26] They are therefore driven to undermine abundance and create artificial scarcity as an unintended, but logical consequence of their internal programming, creating a modern class of rentiers who accumulate wealth by charging fees for access to the resources they control.[27]

Viewed more broadly, economics has always assumed a condition of scarcity and defined its goal as the efficient allocation of scarce resources relative to unlimited human wants. Nowhere does abundance figure in the definition or goals of economics. Practically all economic textbooks are premised on scarcity. Check any index: "Scarcity" will be found in the early pages—in the first chapter, probably— and "abundance" will be missing. In the classic introductory textbook *Economics*, Paul Samuelson and William Nordhaus write on page 2: "At the core, [economics] is devoted to understanding how society allocates its scarce resources. Along the way to studying the implications of scarcity, economics tries to figure out the 1001 puzzles of everyday life."[28] Some books might refer to "overproduction," suggesting an anomaly to be avoided or corrected. Misunderstanding abundance as overproduction logically leads to counterproductive measures restricting abundance, a misapplication of concepts developed under assumptions of scarcity.

Yet once we open our minds, we should see abundance all around us. Solar energy has been with us from the beginning. So have clean air and water, plants and animals, soil life, forests, and the astounding variety of life on Earth, now threatened. Since the Internet emerged, we have also seen an extraordinary abundance of information and knowledge and no lack of people willing to share them freely. Just look at the World Wide Web, Yahoo!, Google, Wikipedia, YouTube, and all the lesser-known, but incredibly useful efforts to make information and knowledge freely available on the Internet. New technologies promise even more abundance: in bandwidth through fiber optics, in air time through spread-spectrum technology, and in storage through new media.

YES! Magazine graphic, 2007 (Worldwatch Institute, Institute for Policy Studies, PDF with legends available at http://www.yesmagazine.org/article.asp?ID=1832#commonspdf).

Clearly, abundance is as much a feature of the real world as scarcity is. To understand this blind spot of economics and harness it fully for the human good, we need to construct theories of abundance to complement the theories of scarcity that dominate economics today. In fact, economists who talk of "relative scarcity" only need a minor leap of logic to recognize "relative abundance." After all, a glass that is half empty is also half full.

CONSTRUCTING A THEORY OF ABUNDANCE

CONSIDER THE VARIATIONS IN ABUNDANCE

It can be precarious (with collapse imminent), temporary (lasting less than a lifetime), short term (lasting a few lifetimes), medium term (lasting many lifetimes), or long term (lasting longer than human existence). It can be relative (enjoyed by a limited number), local (confined to a specific area), or universal (accessible to all). The abundance of solar energy and other energy forms associated with it, such as hydro, wind, and wave energy, is obviously long term. Solar energy is universal, while hydro, wind, and wave energy are more local. Coal's abundance is medium term, if the estimates are correct that the world's reserves may last for several hundred years more (that is, for many human generations). Oil, which is perhaps good for another generation or two at current extraction rates, is short term. In addition, fossil-fuel abundance is relative, because it is not accessible to all, but only to large firms with enough financial, technical, and human resources. While universal abundance can have free and open access, other forms may need some kind of management. Those who depend on local resources may need to restrict or even exclude outsiders. Extraction rates may need to be regulated. Moratoriums may even have to be imposed on threatened resources. The ultimate goal of any management regime should be to ensure against any failure of abundance by pursuing the following specific goals.

MAKE THE RESOURCE ACCESSIBLE TO A GREATER NUMBER OF PEOPLE—IDEALLY, TO ALL

This is merely a restatement of the goal of social justice. Potable water, for instance, is so important to human survival that this goal should be paramount for this resource, whether it is abundant or not. For water—and for land, as well—Gandhi's observation rings true: "There is enough in the world for everyone's need, but not for everyone's greed." These resources can become abundant for all or scarce for many, depending on how they are managed. In a country such as the Philippines, land seems scarce to the millions who do not own a home lot, because the ownership structure allows a few to own thousands of hectares of land. Agrarian reform is, in effect, an effort to keep land abundant for every rural household that is willing to farm land. Some have also argued that family-size farms can be as productive and efficient, if not more so, than huge, corporate-held tracts.[29]

MAKE SURE THE RESOURCE WILL LAST FOR GENERATIONS, PREFERABLY INDEFINITELY

This means turning limited, temporary or short-term abundance into long-term abundance. This is also a restatement of the goal of sustainability. Rain forests, for instance, have been providing countless generations of indigenous tribes everything they have needed for survival. At current rates of depletion, however, our generation has turned rain forests into a short-term or temporary resource that will be gone in a few generations, if not within our generation. Economists should be familiar with the difference between income and capital, between natural-resource stocks and flows. In the rain forest case, ensuring long-term abundance means limiting the consumption of forest products to the natural income we get out of the forest and refraining from eating into the capital stock. Strategies for managing nonrenewable resources or information resources would of course be different.

BUILD A CASCADE OF ABUNDANCE

Abundance in one sector (or of one good) can help create abundance in another sector (or of another good). The food chain is a good example of abundance at one level (solar energy) supporting abundance at the next level (plants), which supports abundance at a higher level (herbivores), and so on. By building linkages among farm components, permaculture teaches how one type of abundance can be made to support another through conscious design.[30] A similar cascade occurs on the Internet, which supports the Web, which in turn supports search engines and new applications such as wikis and blogs, one abundance building on another. The sun is a flexible energy source that can provide, through collectors and concentrators, a wide range of temperatures to match various end uses. By tapping it more, industry can harness potentially huge amounts of energy for various productive activities, opening up possibilities for creating abundance in many other sectors. Photovoltaic cells made from silica, also an abundant resource, can transform sunlight into cheap electricity for industrial, commercial, and home use. This can make viable the electrolytic extraction of hydrogen and oxygen from water, another abundant resource. These can be stored and later used in fuel cells, holding the promise of a pollution-free, hydrogen-based economy.

Most computer equipment, which is silicon-based, such as photovoltaic cells, has either been halving in price or doubling in capacity every few years or so. Liquid-crystal-display projectors now sell for a fifth of their price ten years ago. If photovoltaic cell prices follow suit, perhaps due again to China's entry into the global marketplace, we can look forward to a cascade of solar-based abundance in the future.

Eventually, we should be able to recognize the conditions that lead to abundance and then learn how to create more abundance. We already have a rough idea how abundance happens in nature, in agriculture, and in the information sector. We simply need to nurture the forces that generate such abundance. One challenge is how to emulate ecological processes such as the cyclic loops of nature to create a similar material abundance in the industrial sector without disrupting natural cycles.

DEVELOP AN ETHIC THAT NURTURES ABUNDANCE

To manage abundance well, its community of beneficiaries must adopt a behavioral rule set and corresponding enforcement mechanisms. It is desirable eventually to turn this rule set into a mind set, similar to Aldo Leopold's land ethic and Sandra Postel's water ethic,[31] that is, into an ethic that makes the other goals of social justice, sustainability, cascading abundance, and dynamic balance second nature to all.

ATTAIN DYNAMIC BALANCE

In a finite world, material abundance cannot grow indefinitely. Nature shows us how abundance can instead be sustained indefinitely through a dynamic balance (a harmony) of abundant elements connected in closed material cycles. Citing permaculture again as example, a similar balance can be attained in a farm by modeling it after long-lived, self-regenerating ecological systems to design what are, in effect, forests or ponds of food and cash crops. After we learn to design similar closed loops in industry, we can bring this sector back into harmony with the rest of the living world.

At least four major sources of imbalance threaten our world today. The first is the current reliance on a nonrenewable energy base. Although the size of the world's fossil fuel stock may be debatable, its rate of exploitation will sooner or later surely fail to keep up with rising demand, causing major economic disruptions.

The second source of imbalance is the linear production processes of the industrial sector. The industrial sector uses raw materials from nature and agriculture and turns them into finished products. Whether these goods are durable, reusable, or disposable, they are eventually thrown away as waste. Unlike the closed cycles of nature, this is a linear process that consumes biomass, dead matter, and energy at the input end and that produces synthetic, often nonbiodegradable and even toxic goods and wastes at the output end. This one-way transformation constantly disrupts the dynamic balance and closed loops of the natural world. Eventually, the finished goods reach the end of their useful life—quickly, if they are disposable or one-time-use goods—and become wastes, too. If these wastes enter the body of any living organism, including humans, they can seriously disrupt its health. In effect, fueled by an ideology of accumulation, industry is transforming the natural world into a synthetic and ultimately unlivable place. The solution, as Barry Commoner proposed,[32] is to turn linear industrial processes into closed material loops and recycle all industrial wastes as well as goods that have reached the end of their useful lives back into the production process.

The third source of imbalance is the unchecked growth of the human population. For most biological forms on Earth, at least one more life form exists—feeder, predator, or parasite—that limits the former's population and keeps it in balance with the rest of the living world. This food chain creates an energy pyramid that is wide at its base, where plants directly tap solar energy, and that becomes narrower toward the top, as it tapers from herbivores and to predators. There is one exception: The human population at the apex of this biological food chain has grown disproportionately larger than the rest of the pyramid, appropriating to itself much of the Earth's livable habitat as well as its production of energy and biomass. With no natural enemies to limit our population effectively, we have to discover other means to do so. (Perhaps the global drop in sperm counts is nature's own response?) Because the growth of the human population involves the fundamental biological urge to reproduce one's kind, the issues are complex, and the debates rage on. But solutions we must find.

The fourth source of imbalance that threatens our world today is the unlimited corporate drive for profit. The business firm is programmed to maximize its return on investment—no more, no less. This simplistic programming as a profit seeker driven purely by self-interest has made it better adapted than the individual human being to the world of markets, competition, and capital accumulation that economists have defined the world to be. Being better adapted, corporations have become the dominant economic player in our world. Because under our legal systems corporations are legal persons distinct from their board of directors and shareholders, corporations have now acquired a life of their own. They can feed themselves, regenerate, reproduce, make plans in pursuit of their internal urges, and hire people to execute these plans. Using their superior economic power, they have also acquired political power and taken over media and education. They have become so well entrenched and their accumulated economic, political, and cultural powers have become so extensive that if they were counted as a distinct species, they would now be considered the dominant species on this planet, having managed to domesticate the great *Homo sapiens* itself. As corporations relentlessly pursue their internal programming, seeking profits without limit, they are causing huge global imbalances that threaten the survival not only of human societies, but of many other species, as well. Displacing these runaway automatons from their dominant status and reprogramming them with more benign goals (Isaac Asimov's laws of robotics,[33] for instance) has become the greatest challenge of our era.

RELIABILITY AND THE PRECAUTIONARY PRINCIPLE

Corporations maximize their gain (profits) through efficiency and scale. Another concept, however, could be more important than efficiency. This is the concept of reliability, the quality of "being available when needed," of "lasting for a long time." This common concept may further clarify how the two goals of social justice and sustainability can be met.

When abundance fails and becomes unavailable to some sectors of society or to subsequent generations, this failure is a loss of reliability. Reliability is measured in terms of mean time before failure (or mean time between failures). Improving reliability means reducing the risk of failure. A more familiar formulation is the "precautionary principle."

To prevent abundance from turning into scarcity, maximizing gain (efficiency) should give way to minimizing risks (reliability) from threats to the sources of abundance. This suggests a risk-averse strategy, which precisely is a strategy common among ancient tribes and traditional societies. Perhaps, they have learned over centuries that their goal was to preserve the natural abundance that sustained them and to minimize any risk that may cause such abundance to end.[34] Under conditions of abundance, the ideal economic agent is not the gain maximizer competing out of self-interest and incidentally making markets efficient, but the risk minimizer cooperating with others intentionally to make their common resources more reliable.

Often, a resource that a community considers optimally used because the risk of failure has been minimized will appear underutilized to a corporation because gain is not being maximized. This is probably the cause of resource conflicts in many areas, especially where corporations intrude into community resources.

To get optimum yield, gain maximizers keep increasing production toward the "carrying capacity" of the resource. However, imperfect knowledge, uncertainties, and lags inherent in natural systems can lead to oscillatory behavior and overshoots. Exceeding carrying capacity, even temporarily, can trigger a major mindset shift that can lead to a race that ends up in a breakdown of the commons.

Guided by the precautionary principle, risk minimizers focus not on carrying capacity, but on the impact of extraction on the resource. Individuals evaluate the negative impact as risk to their perpetual source of abundance—risk being the probability of failure times the present value of their income stream that would be lost—and weigh this against their own need. This self-regulating mechanism, where individuals limit their gain as they minimize the risk of losing a perpetual source of abundance, can keep the system in equilibrium. Even pure self-interest should drive them to cooperate with others to make sure the rate of extraction stays well below carrying capacity, which represents a nearly 100 percent risk of failure. Should dire need push one to extract beyond acceptable risk, he or she will have to contend with the wrath of others, whose perpetual income streams are also being put at risk. Or perhaps everybody else will cooperatively chip in to help meet a member's dire need, given their common interest to protect the resource that gives each of them a perpetual income stream.

ABUNDANCE CREATES COMMONS

If we were to review history, and perhaps prehistory, as well, we would see that abundance has often led to the creation of commons. In communities that respond to abundance by treating it as a common pool resource, community members tend to act cooperatively to manage the commons so that the goals of social justice and sustainability are met and the risk of failure in abundance is minimized.

Commons management involves not only economic rules, but also cultural and political factors such as conscious community decisions, appeals to the common good, and the values of sharing, cooperation, altruism, and community spirit. It often relies not only on prices, but also on restrictions, prohibitions, and taboos. Ancient tribes and other traditional societies have evolved complex social norms of behavior and hierarchies of communal use and access rights that have served them well in managing abundance and the commons for many generations. Similar norms have likewise evolved among successful modern commons, such as free and open-source software and Wikipedia.

Their institutions and methods for governing the commons have proved even more useful for threatened resources, as well as for resources that have actually become scarce, by helping meet the goals of social justice and sustainability. In a number of instances, fishing grounds and forest reserves have been nursed back to abundance, thanks to the proper management of these commons.

Thus, a rich heritage of theory and practice in managing abundance and coping with scarcity exists and may be found in the literature of the commons. This heritage was overlooked for several decades by many after Garrett Hardin observed in 1968 that a "tragedy of the commons" ensues when rational gain maximizers exploit the commons in pure pursuit of self-interest.[35] This has led governments to take over these commons as state property or to turn them over to corporate interests through privatization, often creating worse tragedies. What can be worse than the tragedy that befell Russia when the common wealth of its people—literally the product of their sweat, tears, and blood—became the private property overnight of party bureaucrats turned capitalists? Subsequent studies have since shown that Hardin's "tragedy" is by no means universal and that successful practices in managing the commons continue to serve many communities today.[36]

Hardin's analysis of herders and a common pasture was too simplistic. Hardin argued that a rational herder would gain for himself one unit of the commons per additional head of cattle herded, and split with other herders the unit of damage to the pasture. He concluded that the positive net gain would drive every herder to keep adding heads of cattle to the pasture until the commons collapsed. Hardin's risk-blind herder does not take into account the risk to his own perpetual income stream created by each additional head of cattle he puts to pasture. A risk-wise herder, weighing the gain from each additional head against the increasing risk of losing his perpetual income stream, will stop adding heads before the probability of losing that income stream reaches 100 percent, which occurs as carrying capacity is exceeded. Every herder will get a clear signal as the risk increases, because he will be getting less gain per unit effort as the pasture deteriorates. Here is a self-regulating system that requires no unrealistic assumptions, such as perfect knowledge or perfect competition.

A foolhardy herder who needs the plus-one-unit gain badly enough may still insist on risking not only his own, but also everyone else's perpetual income stream. Since each one could, one day, face a similar situation of urgent need, they may eventually realize that it would be better for each herder to contribute a small amount to raise the increment. This suggests, as a long-term solution, a system of insurance or social security, a type of commons that reduces individual risk by pooling resources.

CONCLUSION

The following table shows how a focus on abundance creates a mind set that is orthogonal to one that focuses on scarcity:

ABUNDANCE	SCARCITY
Commons	Markets
Community	Corporation
Common good	Self-interest
Cooperation	Competition
Culture	Commerce
Balance	Growth
Stewardship	Exploitation
Minimizing risk	Maximizing gain
Reliability	Efficiency

The three major sectors of the economy—the agriculture, industrial, and information sectors—present us with a complex mix of markets and commons of scarce and abundant goods. We need to tap into the vast pool of historical as well as current insight, knowledge, and experience to develop a modern theory of political economy that can cope with both abundance and scarcity.

NOTES

1   Rachel Carson, *Silent Spring* (Boston: Houghton Mifflin, 1962).

2   Kira Gould and Lance Hosey, *Women in Green: Voices of Sustainable Design* (Bainbridge Island, WA: Ecotone Publishing, 2007), p. 20.

3   Francis Chaboussou, *Healthy Crops: A New Agricultural Revolution* (Charlbury, UK: Jon Carpenter Publishing, 2004).

4   The International Rice Research Institute is a Philippines-based, Rockefeller-funded research center on rice breeding. See http://www.irri.org (last accessed April 22, 2009). For a comprehensive critique of the institute's approach, see Nicanor Perlas and Renee Velvee, *Oryza Nirvana?* (Quezon City: Southeast Asian Regional Institute for Community Education, 1997).

5   For a critique of the Green Revolution, see Andrew Pearse, *Seeds of Plenty, Seeds of Want* (Oxford: Oxford University Press, 1980). See also Vandana Shiva, *The Violence of the Green Revolution: Third World Agriculture, Ecology and Politics* (London: Zed Books, 1991).

6   Mark Mikel and John Dudley, "Evolution of North American Dent Corn from Public to Proprietary Germplasm," *Crop Science* 46, no. 3 (May–June 2006): pp. 1193–1205, available on-line at http://crop.scijournals.org/cgi/content/full/46/3/1193 (last accessed April 22, 2009).

7   See for instance, Wayne Wenzel, "Syngenta Buys Garst," *Farm Industry News*, May 12, 2004, available on-line at http://farmindustrynews.com/news/Syngenta-buys-Garst; Matthew Dillon, "Monsanto Buys Seminis," *Organic Broadcaster* (March–April 2005), available on-line at http://www.newfarm.org/features/2005/0205/seminisbuy/index.shtml; and Carrey Gillam, "Monsanto to Buy Vegetable Seed Company," April 1, 2008, available on-line at http://www.planetark.com/dailynewsstory.cfm/newsid/47728/story.htm (all last accessed April 22, 2009).

8   See The Crucible Group, *People, Plants and Patents: The Impact of Intellectual Property on Trade, Plant Biodiversity, and Rural Society* (Ottawa: International Development Research Centre, 1994).

9   "Top Awards for Scientist Who Developed the First Transgenic Plant," *Pesticide Outlook*, June 2002, available on-line at http://www.rsc.org/delivery/_ArticleLinking/DisplayArticleForFree.cfm?doi=b205176c&JournalCode=PO (last accessed April 22, 2009).

10   See for instance, U.S. patents 4900676, 4970168, 5276268, 5312912, 5382429, 5503999, 5498533, 5633434, 5648249, 6043409, 6130368, 6204436, 6495745, 6646184, 6791009, and 7314973, available on-line via the U.S. Patent and Trademark Office Web site at http://www.uspto.gov/# (last accessed April 22, 2009).

11   A thorough discussion can be found in Vandana Shiva, *Protect or Plunder: Understanding Intellectual Property Rights* (London: Zed Books, 2001).

12   This term comes from Paul Hawken, *The Ecology of Commerce (A Declaration of Sustainability)* (New York: HarperCollins, 1993).

13   E. Anne Clark, "So, Who Really Won the Schmeiser Decision?" *Crop Choice* 13 (June 2004), available on-line at http://www.plant.uoguelph.ca/research/homepages/eclark/pdf/sc.pdf (last accessed May 19, 2009).

14   Food Research and Action Center, "Hunger and Food Insecurity in the United States," available on-line at http://www.frac.org/html/hunger_in_the_us/hunger_index.html (last accessed April 23, 2009).

15   A hardware dongle is a small device plugged into a computer that enables an application program to run.

16   See a fuller discussion of these arguments in Roberto Verzola, *Towards a Political Economy*

of Information: Studies on the Information Economy (Quezon City: Constantino Foundation, 2004), available on-line via a link at http://rverzola.wordpress.com/2008/01/26/towards-a-political-economy-of-information-full-text.

17 See Philippine Greens, "Philippine Greens Protest the Visit of #1 U.S. Cyberlord Bill Gates," March 20, 1998, available on-line at http://www.hartford-hwp.com/archives/29/048.html (last accessed April 23, 2009).

18 For a fuller discussion, see Alan Story, Colin Darch, and Debora Halbert (eds.), *The Copy/South Dossier: Issues in the Economics, Politics and Ideology of Copyright in the Global South* (Kent, UK: University of Kent, 2006), available on-line at http://www.gutenberg.org/files/22746/22746-p/22746-p.pdf (last accessed April 24, 2009). See also Pradip Ninan Thomas and Jan Servaes (eds.), *Intellectual Property Rights and Communications in Asia: Conflicting Traditions* (New Delhi: Sage Publications, 2006).

19 Adam Thierer and Wayne Crews (eds.), *Copy Fights: The Future of Intellectual Property in the Information Age* (Washington D.C.: Cato Institute, 2002).

20 See, for instance, Tiahan Xue, "Exploring Chinese Herbal Medicine Can Foster Discovery of Better Drugs," *The Scientist* 10, no. 4 (February 19, 1996): p. 9.

21 Simon Montlake, "Milk Formula Goes on Trial in Asia," *Christian Science Monitor*, June 22, 2007, available on-line at http://www.csmonitor.com/2007/0622/p05s01-woap.html (last accessed April 24, 2009).

22 For an English version, see http://rverzola.wordpress.com/2007/12/27/bahay-kubo-english-translation (last accessed April 24, 2009).

23 Steve Buckley, "Community Radio and Empowerment," May 1, 2006, available on-line at http://portal.unesco.org/ci/en/files/22022/11472542151Steve_Buckley.doc/Steve%2BBuckley.doc (last accessed April 24, 2009).

24 See Ray Reece, *The Sun Betrayed* (Boston: Southend Press, 1979). See also Daniel Berman and John O'Connor, *Who Owns the Sun?* (White River Junction, VT: Chelsea Green Publishing, 1997).

25 For a full discussion, see James Boyle, *Shamans, Software and Spleens: Law and the Construction of the Information Society* (Cambridge, MA: Harvard University Press, 1996).

26 See David Korten, *When Corporations Rule the World* (West Hartford, CT: Kumarian Press, 1996).

27 How this happens in the information economy is discussed in Peter Drahos, with John Braithwaite, *Information Feudalism: Who Owns the Knowledge Economy?* (New York: New Press, 2002). See also Verzola, "Cyberlords: Rentier Class of the Information Sector," in *Towards a Political Economy*, pp. 145–61.

28 Paul Samuelson and William Nordhaus, *Economics*, 14th ed. (New York: McGraw Hill, 1992), p. 2.

29 Peter M. Rosset, "The Multiple Functions and Benefits of Small Farm Agriculture in the Context of Global Trade Negotiations," Food First/The Institute for Food and Development Policy, September 1999, available on-line at http://www.foodfirst.org/files/pb4.pdf (last accessed April 25, 2009). For a comprehensive discussion of the sustainable agriculture approach for small farms, see P. G. Fernandez, A. L. Aquino, L. E. P. de Guzman and M. F. O. Mercado (eds.), *Local Seed Systems for Genetic Conservation and Sustainable Agriculture Handbook* (Los Banos, Laguna: University of the Philippines College of Agriculture, 2002).

30 Permaculture (permanent agriculture) is a system of designing farms that minimizes non-renewable energy requirements, ensures the continuous cycling of biomass, fills as many

ecological niches as possible with food and cash crops, and provides for a wilderness area within the farm. See Bill Mollison, *Permaculture: A Designer's Manual* (Hyderabad: The Deccan Development Society, 1990).

31 "A thing is right when it tends to preserve the integrity, stability, and beauty of the biotic community. It is wrong when it tends otherwise." Aldo Leopold, *A Sand County Almanac* (New York: Ballantine Books, 1966). A water ethic would "make the protection of water ecosystems a central goal in all that we do." Sandra Postel, *Last Oasis: Facing Water Scarcity* (New York: W. W. Norton, 1997).

32 Barry Commoner, *The Closing Circle: Nature, Man, and Technology* (New York: Bantam Books, 1971). See also Barry Commoner, *Making Peace with the Planet* (New York: Pantheon Books, 1990).

33 Isaac Asimov's Three Laws of Robotics are: a robot may not injure a human being or, through inaction, allow a human being to come to harm; a robot must obey orders given to it by human beings, except where such orders would conflict with the First Law; a robot must protect its own existence as long as such protection does not conflict with the First or Second Laws. Programming firms with these robotic laws would have made corporations easier to control.

34 For a fuller discussion, see Verzola, *Towards a Political Economy of Information*, pp. 170–90.

35 Garrett Hardin, "The Tragedy of the Commons," *Science* 162, no. 3859 (December 13, 1968).

36 Elinor Ostrom, Thomas Dietz, Nives Dolšak, Paul Stern, Susan Stonich, and Elke Weber (eds.), *The Drama of the Commons* (Washington, D.C.: National Academy Press, 2002). See also Elinor Ostrom, *Governing the Commons: The Evolution of Institutions for Collective Action* (Cambridge: Cambridge University Press, 1990).

The Man Who Mistook His Wife for a Book

Lawrence Liang

A number of critical interventions in debates on the impact of intellectual property on knowledge and culture are framed by a critique of the expansionist tendency of the global intellectual property regime. There are calls for an institutional overhaul of the intellectual property regime to make it more equitable and to enable greater access to knowledge and culture. What seems to be missing in these debates in the registers of development discourse or in the mandates of liberal reformist agendas are epistemological challenges to intellectual property. The success of intellectual property as a concept perhaps lies more in its ability to have established itself in a universally intelligible narrative concerning what constitutes selfhood and owner-ship than in specific instances of its enforcement within countries. Ideas of prop-erty are centrally tied to larger ideas of personhood and hence are far too impor-tant to be left to policy makers.

The globalization of intellectual property is hence as much about the creation of a mind-set that sees all forms of creative activity as property as it is about uni-versal norms. It also articulates a particular idea of authorship tied closely to the idea of an individual's relation to the world of knowledge and culture. This paper argues that unless we understand the epistemological challenges posed by the idea of intellectual property, we are left with limited corrective measures to a system that threatens to destroy the diversity that marks our relation to the world of ideas and consequently our relation to others and to ourselves.

I will be examining the link between ideas of personhood and self that under-lies Western liberal property regimes. Philosophers such as John Locke played a key role in creating a link between the "self" and "ownership." This paper examines alternative ideas of the self within Western and non-Western metaphysics and argues that a relational conception of the self helps us rethink our assumptions about property and personhood, especially in relation to the world of knowledge

Extract from Keith Aoki, James Boyle, and Jennifer Jenkins, *Bound by Law? (Tales from the Public Domain)* (Durham, NC: Center for the Study of the Public Domain, 2006) (available at http://www.law.duke.edu/ cspd/comics/; License: http://creativecommons.org/licenses/by-nc-sa/2.5/).

and culture. The political implications of such a shift are many and can be best seen in the normative aspirations of the access to knowledge movement, which attempts to destabilize the language of exclusive rights and property and to focus on the ideas of responsibility and obligation as part of the ecology of knowledge.

My account of the puzzle of property and personhood in relation to intellectual property begins with what seems to be a standard copyright dispute. In 1999, three members of the 1980s band Spandau Ballet sued Gary Kemp, the fourth member of the band, for not sharing the royalties to the band's songs, which they claimed they had jointly authored.[1] Kemp claimed that he was the sole author of the songs and that he was not obliged to pay them any share of the royalties. The aggrieved members of the band argued that while Kemp presented the "bare bones" of a tune to the other band members, the band went through a process of jamming, whereby "someone started to play and the rest joined in and improvised and improved the original idea"—in other words, that the creation of the songs was a collaborative process and that the songs should be considered a work of joint authorship.[2]

The court analyzed the manner in which music was created and, while acknowledging that there was a collaborative process that went into the composing of the music, they held that it was Kemp who should be considered the sole creator of the songs. They argued that Kemp "developed, and fixed in his musical consciousness, the melody, the chords, the rhythm or groove, and the general structure of [each] song from beginning to end" before playing it to the band and inviting the band as a whole to rehearse its performance as an ensemble with a view to recording it. The judge accepted that the other band members' vocal and instrumental performances were skilful individual interpretations of the musical works that Kemp had composed. However, he held that an interpretation of a musical work was not the kind of contribution that the law of copyright could accept as sufficient to constitute the interpreter an author of that work: "the contributions need to be to the creation of musical works, not to the performance or interpretation of them."[3]

The Spandau Ballet case illustrates an interesting and in many ways typical problem that copyright law faces in its adjudication of claims of authorship and creativity. My interest in the case emerges from the ways in which the case attempts to deal with the questions of collaboration, property, and personhood. In the Spandau Ballet case, there seem to be three distinct kinds of claim made about the relationships involved in these questions: Kemp's claim (affirmed by the court) that the songs were written solely by him and hence are his *own* songs (a claim based on the songs' relationship to the self), the claim that as a result, Kemp *owns* the songs exclusively (a claim based on his relationship to the work), and the claim that as an *owner*, Kemp is entitled to exclude others from a share in the royalties arising from the songs (a claim based on his relationship to others).

At the heart of the problem, and of our understanding of the philosophical divide that exists in debates on intellectual property, lies the issue of property and personhood. The language of property narrates the individual as a proprietor of one's own person, and it gives rise to a theory of personal identity in which the self and what it owns are often treated as being the same, or at least as existing within the same orbit of meaning, whereby the one can be used interchangeably with the other. This discourse flows directly from classical liberal political theory, in which every individual is considered to be the proprietor of his or her own person.

LOCKE AND THE EQUATION OF SELF AND OWNERSHIP

The philosopher who is most often identified with this theory of property and the self is of course John Locke. Many of our ideas of selfhood emerged in the seventeenth and eighteenth centuries, and in many ways, the question of personal identity was the prime question that motivated Locke's inquiries. His theories set the stage for the philosophical and juridical establishment of what C. B. Macpherson calls the theory of "possessive individualism."[4] The question of personal identity troubled many philosophers before Locke, but it was with the publication of Locke's *Two Treatises of Government* and *An Essay Concerning Human Understanding* that the most coherent argument linking theories of identity to property emerged. It is interesting to note that Locke initially did not have a chapter on consciousness and identity in the Essay, and it was at the suggestion of William Molyneux that he included a section on the *principium individuationis* to the second edition.

Consider, for instance, the following statement in Locke's journals: "Identity of persons lies not in having the same numerical body made up of the same particles, nor if the mind consists of corporeal spirits in their being the same. But in the memory and knowledge of one's past self and actions continued on under the consciousness of being the same person whereby every man own's himself."[5] For Locke, consciousness is a question of mental operations that appropriate the self to itself, where to appropriate means to identify with or to make a property of. The use of the word "own" is both as an adjective (as in "my own thought") and as a verb (to confess). The relationship between the self and the own is therefore dependent on a circularity whereby ideas of identity and identification on the one side and appropriation on the other continuously exchange their function and become virtually equivalent. The relationship between the self and the own is dependent on a self-fulfilling prophesy in which "what I can consider as me, myself, is my self and 'my' self is some 'thing' that I own, or that I must own (confess) is mine, was done or thought by me, has become my own because I appropriated it to me by doing it or thinking it consciously."[6]

This circularity also informs much of Western metaphysics, in which consciousness sets the criteria of personal identity and of a political theory in which the possessive individual is generalized or universalized because any individual ought to be considered as proprietor of his or her own person or as a self-owning personality to the extent that he or she is such a proprietor. In a fascinating rereading of Locke, Étienne Balibar, citing Jacques Derrida, claims that the reason for this equivalence is the "metaphysics of (a)propriation," in which linguistic expression is provided by the circularity of meanings between "my self" and "my own," or the fact that you can explain self only by referring to what is your own and your own only by referring to yourself. This is at the heart of European psychological, moral, juridical, and political individualism and on the surface of it, it does seem that my self and my own are one and the same thing.

The circular relationship between the self and the own appears at first glance to pose a problem of translatability. For instance, if you attempt to translate the terms "self" and "own" into French, while "self" can more or less accurately be translated as *moi* or *soi*, the closest French word for "own" is *le propre* or *propre*, with its very close relation to property. The pair "self"/"own" and *moi/propre*, however, cannot be considered as accurate equivalents. One could blame the inherently flawed project of translation, with the problem seen as being on a par with other conceptual/linguistic problems that have plagued philosophers involving the precise meaning of a word in different languages. However, Balibar sees it as a far more serious problem. He wonders if it is the easy semantic coincidence available in English that enables the easy linkage of the self and the own that allows for Locke's theory of identity and property.

To test Balibar's hypothesis, I attempted to look for an equivalent in Hindi of the ideas of "my self" and "my own." The closest translation that I could find emerges from the phrase "mere apne" which is the equivalent of "my own." The word *apna* refers to the idea of owning, but not merely in terms of possession. The phrase "mere apne" could refer to something as being mine, but at the same time, this claim is not limited only to an assertion of delineation and exclusion, but refers instead to a certain idea of relationship of proximity between the self and an other. The word *apnaapan*, for instance, translates as "closeness," so that "mere apne" is a reference to the idea of a relational proximity.

This is interestingly mirrored by Balibar's reading of "By the Fireside," a poem by Robert Browning:[7]

> My own, confirm me! If I tread
> This path back, is it not in pride

To think how little I dreamed it led

To an age so blest that, by its side,

Youth seems the waste instead?

Balibar initially reads the poem as being addressed to oneself, or as self-interpellation, and as an appeal to memory, but realizes his mistake when he encounters the next lines and then a subsequent stanza:

My own, see where the years conduct!

At first, 'twas something our two souls

Should mix as mists do....

My perfect wife, my Leonor,

Oh heart, my own, oh eyes, mine too,

Whom else could I dare look backward for,

With whom beside should I dare pursue

The path grey heads abhor?

So what might be only a linguistic dilemma—is this an address to a self or an address to an other?—leads us back to the foundational question of the nature of self and subjectivity that is invoked when we speak of something being "our own." The idea that "my own" could possibly refer not merely to a sovereign claim of the individual self but also to relationality involving others conflicts with the world of property norms in which a to say "my own" is an act that makes a claim of absolute possession, that declares the ability to exclude others, and that asserts the legal ability to alienate what you own.[8]

RELATIONAL PROXIMITIES AND THE REWORKING OF THE SELF

Thinking of our relation to the world of knowledge and culture via the trope of proximity enables us to rethink our relations to our work, to our selves, and to each other, not as distinct sets of legal relations bound together by the idea of rights, but as a continuum that blurs the boundaries between rights, obligations, and relationalities. Consider, for instance, the following statements, each of which refers to certain claims that sound deceptively similar, but that in fact exist in very different ethical and legal registers.

This is my pen.

This is my friend.

This is my poem.

The first statement refers to the classical conception of the claims of possessive individualism, in which the self and the owner exist as interchangeable concepts. This is "my" pen, and hence I own it. The second statement takes us into the domain of relational proximities, where an assertion of someone being your friend does not lead to an assumption, either of ownership or of exclusion, but into the domain of the closeness/*apnaapan* that you share with your friend. Thus, the statement "This is my friend" could well be mapped in terms of its presence in Hindi as "mere apne," with a sense of "owning" that leads to an understanding of how close you are to someone.

The third statement is perhaps the most deceptive, because to assert "This is my poem" within the social imaginary of intellectual property is to make a claim that sounds very much like "This is my pen," whereas in fact, it might be more accurate to think of its claim as the same as "This is my friend." And it is in this liminal space where poems look like pens that friendships get lost and property takes over.

What is it about the logic of property and the language of rights in the domain of intangibles that creates this act of misrecognition? The ontological character of information, knowledge, and cultural practices provide them with an unbounded-ness, and very much like the world of social relations, they are not exhausted by acts of circulation. When was the last time we heard of the problem of someone having too many friends? The imposition of strictly defined norms of property rights, with its imagination of legitimate rights holders and trespassers, enforces a transition that converts the possibilities of friendship into acts of hostile takings.[9]

The role of intellectual property and the language of rights creates a normativ-ized and legalized domain in which our experience of social relations, with their attendant complexities, is unavailable to us except as juridically defined sets of relations. It would be useful at this stage for us to turn to our title character, who is derived from one of Oliver Sacks's case studies, that of "Dr. P," the man who mis-took his wife for a hat. Sacks informs us that Dr. P suffers from a peculiar neurolog-ical disorder that affects his ability to retain visual recognition while retaining this ability to discern abstract figures, leading to a series of misrecognitions in which he is unable to distinguish his foot from his shoe and his wife from a hat. Sacks writes:

> By and large, he recognized nobody: neither his family, nor his colleagues, nor his pupil, nor himself. He recognised a portrait of Einstein because he picked up the characteristic hair and moustache; and the same thing happened with one or two other people. 'Ach, Paul!' he said, when shown a portrait of his brother. 'That square jaw, those big teeth, I would know Paul anywhere!' But was it Paul he recognised, or one or two of his features, on the basis of which he could make a reasonable guess as to the subject's identity? In the absence of obvious 'markers', he was utterly lost. But it was not merely the cognition, the gnosis, at fault; there was something radically wrong with the whole way he proceeded. For he approached these faces—

even of those near and dear—as if they were abstract puzzles or tests: He did not relate to them, he did not behold. No face was familiar to him, seen as a 'thou', being just identified as a set of features, an 'it'. Thus there was formal, but no trace of personal, gnosis.[10]

Dr. P provides us with a fascinating case study of how a neurological condition may completely alter our abilities to see and to relate to the phenomenological world. We can perhaps think of intellectual property rights as a similar affliction, founded on very particularized ideas of property and personhood, but narrated as universal truths, that prevents us from seeing our acts of reading, writing, creating, sharing, and borrowing in terms of the relational world that they occupy. Instead, we see them abstracted of their social relations.

The equivalent of Dr. P in the world of ideas is Daniel Defoe, the great chronicler of piracy's golden era, who writes that "A Book is the Author's Property, 'tis the Child of his Inventions, the Brat of his Brain; if he sells his Property, it then becomes the Right of the Purchaser; if not, 'tis as much his own, as his Wife and Children are his own." Defoe was of course writing at a time when wife and children could indeed be owned as property. But we now know better and understand that you cannot own your wife or your child, but you can feel that they are your own.

So here we have before us the case study of Daniel Defoe, the man who mistook his wife for a book. His condition (unlike Dr. P's) is not an isolated malady, and an increasingly large number of people are showing symptoms similar to Defoe's, encouraged and enabled by the large institutional sponsors of the malady such as the World Intellectual Property Organization (WIPO), a malady whose contagion is ensured by instruments such as the TRIPS Agreement.

DIFFERENT TRADITIONS OF THE SELF

J. G. A. Pocock says that if property is both an extension of personality and a prerequisite of it, then we should be aware of the possibility that different modes of property may be seen as generally encouraging different modes of personality.[11] One way in which we can rethink the idea of our relationship to what we create is not through terms of ownership, but through how close we are to it—through proximity. Proximity to people and things creates a relationship of care and responsibility, and when thought of in terms of things that we create, it allows us to create a different ethical register through which we can examine the relationship between property and personhood. For Locke and many other thinkers within the Western metaphysical tradition, the idea of a distinct self serves as the basis for a range of concerns, from self-identity, to moral agency, to property. This account of the self within the tradition of possessive individualism has been challenged both within the Western tradition and by non-Western accounts of the self.

Proximity may indeed be the basis on which alternative accounts of the self may be forcefully articulated, because it is accompanied by a whole host of ethical principles such as generosity and obligation that may help us order a different mode of dealing with what we "own." Proximity or closeness is marked by a relationship of care, and if we are to revisit the three modes of relationality that I invoked in the Spandau Ballet case, we see that proximity reworks the way we see the three relations: the relation to the self (to be an author is not just to own a work, but to own up to the work), the relation to the work (taking care of what you own, or the duty of care that emerges from proximity), and the relation to others (a relation predicated on an ethical bond).

The opposite of an ethic of care and proximity is the violence and brutality that motivated Daniel Defoe to mistake his wife for a book. One consequence of the idea of a relational self is that it does not make sense to speak of an essential core that is the basis of a sense of unified self and self-identity. Let us consider two challenges to the idea of the unified self that informs Western metaphysics. The first challenge emerges from the contrast between the idea of the self in Western philosophy and its absence in the non-Western tradition, while the second challenge emerges from the Western philosophical tradition itself.

S. N. Balagangadhara, a philosopher whose work focuses on Indian traditions, argues that the basic idea of the self in Western cultures consists of a sense of "an inner core which is separable and different from everything else. In such a culture, when one speaks of 'finding oneself' one means that one should look inside oneself, get in touch with an inner self that is there inside oneself, and peel everything away that surrounds this core. To such a self, even its own actions can appear strange."[12]

Furthermore, Western culture allows each of us a self—a self waiting to be discovered within each one of us, something that can grow and actualize itself, that either realizes its true potential or fails to do so. Such a versatile self has various properties. One of them is its reflexivity: The self is aware of itself as a self, or it has self-consciousness. Consequently, human beings who are endowed with such selves are all self-conscious beings. As we know, most philosophers are agreed that self-consciousness typifies the uniqueness of human beings, and that this self-consciousness distinguishes humans from the rest of nature.

Balagangadhara contrasts the idea of the self in non-Western cultures with that of the West by using an interesting example. He says that if you were to look at the different ways in which a culture talks about persons, you would often find that in Western cultures, in answer to a question such as "What kind of a person is he?" you would find straightforward answers such as "He is a friendly person."

However, the same question can elicit a different response in a non-Western culture, such as "He comes home every week to enquire after my health." According to Balagangadhara, while this initially seems like a wrong answer or an indirect answer, answers of this kind are very typical in a country such as India. He says that by reading these answers as in fact direct answers, we can see the answer asserting an identity relation between actions and persons. That is, Indian culture does not draw a distinction between an agent who performs an action and the action that the agent performs. An agent is constituted by the actions that a he or she performs, or an agent is the actions performed and nothing more.

Thus for Balagangadhara, the self of a person is nothing other than the actions that the person performs. But these actions do not exist in isolation and are dependent in turn on how another person construes them: Person Y constructs person X's self, just as person X constructs person Y's self. Person Y is crucial for the construction of X's self, because in the absence of Y, the actions that X performs are meaningless. That is, Y is required so that X's actions may be seen as some specific type of action. If we were to restrict ourselves to X in order to talk about his or her self so as to contrast this notion with that of the West, we could say that the Western self consists of a bundle of meaningless actions. Because of this, the self of X depends upon continuously being recognized as such by Y. According to Balagangadhara, there is nothing unusual about this, and it gestures toward the fact that we are all relational selves, and you are only a son, a daughter, a father, a friend, and so on to the extent you are so recognized. And you can be thus recognized only when you perform those actions that are appropriate to the station of a son, a daughter, a father, a friend, and so on.[13]

Let us now turn to another attempt at characterizing this idea of relationality or proximity, this time within more contemporary Western philosophy. Emmanuel Levinas is one of the key thinkers working with the idea of proximity or "being with" within the Western philosophical tradition, and his work has inspired a range of ethical philosophers as well as legal scholars to think through questions of the obligations that we may have to others. For Levinas, proximity implies a "closeness to others who can be approached but never reached. We are never exactly the same as another person, and in the trauma of that distance lies summoned our soul" and likewise our sense of responsibility.[14] It is clear that unlike the non-Western idea of relationality that arises from a close sense of relationships, Levinas has a more expanded idea of relating. For him, "The relationship of proximity cannot be reduced to any modality of distance or geometrical contiguity, nor to the simple 'representation' of a neighbor; it is already an assignation, an extremely urgent assignation—an obligation, anachronously prior to any commitment."[15] For Levinas, the ability to be in a proximate relationship is what "intimates" an other, and this intimation forms the essence of who we are and why we have a responsibility to others. Levinas also sees proximity as a complete experience that in many ways exceeds our theories and ideas of the world.

The similarities between the idea of the relational self in non-Western thought and proximity in Levinas's work is striking. Levinas is impressively nonchalant about other pressing concerns in Western metaphysics, including intentionality or the existence of an essential self that defines our being. He is instead more concerned with the domain of experience and how we act responsibly. The ability to act responsibly is in turn dependent on the ability to respond adequately, and the instantiation of a response dilutes any unified sense of self, since neither the self nor intentionality makes any sense outside of its relation to another. I'm already obligated and called before any decision on my part. Therefore, there is no point in asking whether or not my act(ion) of responsibility is free or voluntary. If responsibility is prior to freedom, neither chosen nor not chosen, it is out of the question to ask under what circumstances I am responsible.

REVISITING LOCKE

While it is tempting to contrast the idea of the relational self with the Lockean idea of the autonomous individual, the task turns out to be difficult. If we consider, for instance, Locke's theory of relations, there are certain productive contradictions that emerge. Objects, according to Locke, are related to each other by the mind. He argues that there may be certain properties, for instance that of being white, that may not be a relational fact. But when we think of a relational idea, Locke suggests, that is, when we think of someone as a husband or as whiter than someone else, the mind is actually going beyond the particular to some other person or persons distinct from the self, and a relation is the result of this activity of the mind, which has simultaneously considered and compared two distinct things.

For Locke, some relational terms such as "father and son," "lesser and bigger," and "cause and effect" are self-evident and can exist only together and explain each other. These correlative pairs "reciprocally intimate" each other, but it is not exactly clear whether there is a common reciprocal relational tie connecting these correlative pairs, a common relation that each member of the pair has toward the other that makes this correlation possible. This limitation seems to emerge from the fact that Locke sees the ability of naming and identifying relations as emerging solely from the mind. In other words, the ontological status of relations seems unequivocally mind-dependent for Locke. The question that then logically arises is, how does the mind create a set of relationships to itself?

It is perhaps useful at this stage to return to the site of our original problem: the equation of self and owner that emerges in Locke's theory of property and personhood. We began by locating the conceptual problem within a specific linguistic dilemma: the manner in which English produces a reciprocal duality of the self, a self who is owning and a self who is owned. Balibar says that while it could be argued that this linguistic dilemma could be dismissed as a performative contradiction, it would be more useful to look at the productive nature of this contradiction and the manner in which it resolves the contradiction in Locke by introducing another element—that of uneasiness.

Balibar argues that for Locke, the process of identification or self-interpellation (I address myself) or the performative contradiction is already taken into account. For instance, in his segment on consciousness, Locke argues that there is no consciousness that is not associated with desire and at the same time troubled and pushed by it toward ever new contents or ideas, so that the notion of consciousness as a fixed or stable identity is a contradiction in terms. Consciousness is by its nature restless. It must escape itself toward new contents, and its identity is associated with a perpetual flow, escape, or train of ideas. The category that names this intrinsic association of consciousness is "uneasiness." Balibar argues that in light of this, we may return to Locke's identification of the self and owning and what it means to understand them as being exactly the same thing. He says that what is owned by me inasmuch as I own it (speaking, thoughts, actions) is the uneasiness of this relation and the fact that the identity or sameness of the self and its own does indeed exist, but only as an uneasy one.[16]

He argues, by returning to his reading of Browning's poem, that the critical element causing the uneasy appropriation of identity is the element of sexual difference. It is the other with whom I make one and the same precisely because we can never become wholly identified, indistinguishable, with whom I experience the uneasy relation of identity and difference, not only because it is conflictual. but because the identification of what is shared or what is the same and of what is separated or different can never be established in a clear-cut and stable manner: We "should mix as mists do." The name of this uneasy experience conventionally is "love." But we know that love is anything but a simple thing, perhaps because in love, there is precisely so much consciousness associated with so much desire.

The implications of this line of thought are immense for a rethinking of the idea of knowledge creation in terms of proximity. If my own work can exist only in a certain relationship with others, then can I ever claim what is my own in a manner that seeks to exclude any claims that others may have?

RETURNING TO THE INTELLECTUAL PROPERTY DEBATE

Let us now return to the domain that motivated this brief enquiry. The global acceleration of intellectual property norms in recent times is critically linked to a new articulation of our relationship to our selves, our work, and to others. The response that is required cannot be limited within the terms of political realism, whether of the left or the right. And the real potential of new modes of knowledge production and sharing (free software, open access) stem not from their status as solutions to the problem of the knowledge or information deficit alone, but from their rearticulation of alternative relationalities that do not generate untroubled and easy cohesive accounts of the self and its own. They instead offer us an opportunity to think about the ways in which acts of sharing create new forms of intimacies and of relating to each other. The global access to knowledge movement currently is as much premised on the language of rights and equity as it is dependent on acts of generosity and giving. In an era when the language of theft, mistrust, and panic marks our relation with the world, it is all the more important to recall the ethical basis of our relationships with each other and with the world.

Our global contemporary era is marked by all kinds of turbulences that momentarily dislodge our stable notions of the nation, identity, property, stability, friend and enemy, self and other. The experience of turbulence in an airplane induces the stranger next to you momentarily to become your most intimate human contact, while the uncertainty of the moment causes you to reach out to an unfamiliar, but reassuring hand. Only later does the uneasy recognition occur of a compact of unfamiliarity having been breached. Our abilities to communicate and to share ideas in ways that were hitherto unthought of provides us with an opportunity to rework our accounts of ourselves and the possible horizons of the relations we inhabit. Michel Foucault asks:

> what would be the value of the passion for knowledge if it resulted only in a certain amount of knowledgeableness and not, in one way or another and to the extent possible, in the knower's straying afield of himself? There are times in life when the question of knowing if one can think differently than one thinks, and perceive differently than one sees, is absolutely necessary if one is to go on looking and reflecting at all.[17]

Seen in this light, what is narrated as transgressions in the world of ideas reappear as explorations and reinventions of the self, a curiosity about the other and perhaps a way of looking for your wife in what appears to be a hat or a book. "Hospitality," "gratitude," "friendship," "caring," "owning"—these almost sound like archaic words from a distant time in a period when juridical relations replace social relations and contracts of adhesion are more powerful than word of mouth. If gratitude is the moral memory of mankind, as Georg Simmel claimed,[18] then it is perhaps time to refresh our memory dulled by the ubiquity of property and contracts. And frankly, it does not matter where these mnemonic tools emerge from, either temporally or spatially. If our sense of self has been narrowed by its linguistic affinity with the need to own, let's start exploring other semantic worlds where we can multiply and expand the idea of the self. Here are some with which to begin.

The etymology of "data" comes from the Latin *datum*, which means a thing that is given, the neuter of which is "to give." Similarly, the word *daata* in Hindi/Sanskrit is taken to mean "the giver," which suggests that we must always be generous with information and make gifts of our code, images, and ideas. To be stingy with data is to violate an instance of the secret and sacred compacts of homophonic words from different cultural/spatial orbits (*daata* in Hindi and "data" in English) as they meet in the liminal zone between languages, in the thicket of the sound of quotidian slips of the tongue.

Another entry point is provided by the common root words that bind the words "owe" and "own," so if intellectual property is about thinking of ways of owning the future, perhaps we need to start thinking not only about how we own, but also how we owe the future. The point is not to take these systems of administering the world of knowledge and ideas as a given, but to think about the ways in which they enable us to work out our relation with ourselves. As Patricia Williams reminds us, "The task...is not to discard rights, but to see through or past them so that they reflect a larger definition of privacy and property: so that privacy is turned from exclusion based on self-regard into regard for another's fragile, mysterious autonomy; and so that property retains its ancient connotation of being a reflection of that part of the universal self. The task is to expand private property rights into a conception of civil rights, into the right to expect civility from others."[19]

I end this piece with a small parable that many of us will have read while we were children. The story is from Antoine de Saint Exupéry's tale *The Little Prince*. The Little Prince visits a number of planets and encounters a range of different characters. On the fourth planet, he meets a businessman who owns millions of stars, and the reason why he owns them is because he was the first one to think of owning the stars. The Little Prince is perplexed, because he can't seem to find a reason for owning the stars beyond the fact that they can be put in a bank to enable the businessman to buy more stars. The Little Prince tells the businessman that "I own a flower myself, which I water every day. I own three volcanoes, which I rake out every week. I even rake out the extinct one. You never know. So it's of some use to my volcanoes, and it's useful to my flower, that I own them. But you're not useful to the stars."[20]

NOTES

1 *Hadley v. Kemp,* Entertainment and Media Law Reports (UK) 589 (1999).

2 For an excellent account of this case and the problems that copyright law has with musical works, see Anne Barron, "Copyright Concepts and Musical Practices: Harmony or Dissonance," *Social and Legal Studies* 15, no. 1 (2006). For an interesting account of similar issues among musicians in India see Rajesh Mehar, "Understanding Notions of Creative Ownership among Contemporary Musicians in India," available on-line at http://community.livejournal.com/whosemusic (last accessed May 1, 2009).

3 *Hadley v. Kemp.*

4 C. B. MacPherson, *The Political Theory of Possessive Individualism: Hobbes to Locke* (London: Oxford University Press, 1962).

5 John Locke, journal entry, for June 5, 1683, quoted in John Marshall, *John Locke: Resistance, Religion and Responsibility* (Cambridge: Cambridge University Press, 1994), p. 153.

6 See Étienne Balibar, "My Self and My Own: One and the Same?" in Bill Maurer and Gabriele Schwab (eds.), *Accelerating Possession: Global Futures of Property and Personhood* (New York: Columbia University Press, 2006).

7 Robert Browning, "By the Fireside," in *The Complete Poetic and Dramatic Works of Robert Browning* (Boston: Houghton Mifflin, 1895), pp. 185–87.

8 And yet at the same time it seems that there indeed does exist a large set of claims within diverse cultural traditions (including, as Browning shows, in English) in which a claim about something or someone may lie more in the domain of your relation with the person or object than as a claim of possession. In *Nehiya* (Cree), to refer to something as "mine" does not necessarily imply ownership, but refers instead to a relational proximity to objects (animate and inanimate) and to beings, along with the accompanying responsibilities and obligations that emerge from such a relational proximity.

9 For an interesting contrast, see Stanley Cavell, *A Pitch of Philosophy: Autobiographical Exercises* (Cambridge, MA: Harvard University Press, 1994), where he argues that "those who are too sure ideas cannot be stolen like to say that ideas are not private property. But my feet are not my property, yet they are mine, and you are not to step on them. The punch line I have set up is not my property, but you are not to preempt it. My turn is not my property, but you are not to take it. Justice is not solely a measure of property rights" (p. 37).

10 Oliver Sacks, *The Man Who Mistook His Wife for a Hat and Other Clinical Tales* (New York: Perennial Library, 1987), p. 13.

11 J. G. A. Pockock, "Tangata Whenua and Enlightenment Anthropology," *New Zealand Journal of History* 26, no. 1 (April 1992): pp. 28–33.

12 S. N. Balagangadhara, "Comparative Anthropology and Moral Domains: An Essay on Selfless Morality and the Moral Self," *Cultural Dynamics* 1, no. 1 (1988): pp. 98–128.

13 *Ibid.*

14 Desmond Manderson, *Proximity, Levinas, and the Soul of Law* (Montreal: McGill-Queen's Press, 2007), p. 14.

15 Emmanuel Levinas, *Otherwise Than Being; Or, Beyond Essence*, trans. Aphonso Lingis (The Hague: Martinus Nijhoff, 1981), pp. 100–101.

16 Balibar, "My Self and My Own."

17 Michel Foucault, *The History of Sexuality, Volume 2: The Use of Pleasure*, trans. Robert Hurley (New York: Vintage, 1990), p. 8.

18 Georg Simmel, "Faithfulness and Gratitude," in *The Sociology of Georg Simmel*, ed. and trans. Kurt H. Wolff (New York: Free Press, 1950), p. 388.

19 Patricia Williams, *The Alchemy of Race and Rights* (Cambridge, MA: Harvard University Press, 1991), pp. 164–65.

20 Antoine de Saint-Exupéry, *The Little Prince*, trans. Richard Howard (New York: Houghton Mifflin Harcourt, 2000), pp. 39–40.

Free-Trade Agreements and Neoliberalism: How to Derail the Political Rationales that Impose Strong Intellectual Property Protection

Gaëlle Krikorian

Free-trade agreements are at the forefront of efforts to increase intellectual property protection. Their chapters on intellectual property regularly enforce intellectual property owners' most restrictive limitations and have become one of the
most efficient ways in which developed countries are able to increase intellectual property protection in the developing world. They also exemplify the kind
of relations that predominate between the rich countries and the Global South:
the imposition of an apparatus of legal administrative rules that mostly favors the
rich countries. The intervention of trade negotiators allows intellectual property
exporters to maintain control over the sources of economic accumulation—information and knowledge—and to prevent their appropriation by others. During the
era of colonization and under both the mercantile and industrial economies, developing countries were considered mainly to be sources of raw materials.[1] In the
knowledge economy of the information society, it is now the populations of developing countries that are exploited—relegated to the status of a workforce and,
when they can afford it, a source of simple consumers, but commonly excluded
from access to knowledge.

   During the colonial era, a number of developing countries under colonization
had to adopt legislation on intellectual property similar to those of the colonizing country.[2] After decolonization, however, many developing countries reviewed
and modified their laws in order to introduce more flexibilities and implement
standards more favorable to their development and to the fostering of their local
industries. The result was a nascent resistance in developing countries to the
imposition of stronger and what they considered to be unbalanced intellectual
property regimes.

   Indian policy makers, for example, drafted and passed a new patent law in 1970
allowing patents on processes used to fabricate pharmaceutical products, but not

on the products themselves, which made it possible for local industries to manufacture and market any existing drug as long as they could establish their own production process. Many other countries, such as Argentina or Mexico, also modified their legislation to limit the scope of patentability after decolonization. A study conducted by the World Intellectual Property Organization (WIPO) in 1988 showed that forty-nine countries among the ninety-eight states that were then signatories of the Paris Convention for the Protection of Industrial Property excluded pharmaceutical products from protection.[3]

In the 1960s, developing countries had organized to revise the Berne Convention for the Protection of Literary and Artistic Work, originally signed in 1886, so that it would take into account their educational needs and favor their economic and industrial-development goals and technology transfers from developed countries.[4] Countries such as India, Brazil, and Korea developed an offensive position from the 1960s to the 1980s, leading developing countries to mobilize against the increasing protection of intellectual property, both in international arenas and in their own national policies and laws.

During an attempt to revise the Paris Convention in the 1980s,[5] when the United States tried to obtain higher standards of protection, developing countries acting as a bloc mobilized to lower them. They formalized the G-77, which was established in 1964 by seventy-seven developing countries at the end of the first session of the United Nations Conference on Trade and Development in Geneva, and some of them maintained a firm position through the Uruguay Round negotiations of the General Agreement on Tariffs and Trade (GATT), notably against the inclusion of an agreement on intellectual property, and later in favor of retaining flexible compulsory licensing provisions in the agreement.[6] However, in the end, there was no revision of the Berne Convention, and in 1994, the Trade-Related Aspects of Intellectual Property Rights Agreement (the TRIPS Agreement) was finally adopted by 128 countries, including many developing countries, ultimately strengthening intellectual property rights for all the members of the newly created World Trade Organization (WTO).

The promulgation of that agreement is one indication that the movement undertaken by the developing countries in opposition to increasing intellectual property protection was facing a powerful counterforce promoting stronger protection of intellectual property, which soon became dominant. A neoliberal "revolution," as David Harvey characterizes it,[7] reshaped the priorities of world leaders and imposed neoliberalism as the way forward for the world. It is in this context that, despite the free-market rhetoric that accompanied them and contrary to their very name, free-trade agreements started to be employed as a way to increase restrictions on intellectual property.

This essay analyzes the political logic in which the increase of intellectual property protections through the negotiation of free-trade agreements is embedded. Neoliberalism can be seen as many different things, among them, as a way to think about power and to regulate behaviors, as an ideology offering people a fiction that appears coherent, and as a set of public policies adopted and implemented under the effective collective action of the ruling class. Without trying to reconcile the existing interpretative frameworks of neoliberalism, what follows will consider some of the characteristics of this political regime, assess its strength and weaknesses, and consider the strategies and tactics that a mobilization such as the A2K movement can take to overthrow its dominant logic and to counter its actions in the global field of intellectual property protection.

THE LOGIC OF THE U.S. FREE-TRADE AGREEMENTS

Many of the world's countries, whether or not they are members of the WTO, have levels of intellectual property protection in their national laws that are much more elevated than what is required by the TRIPS Agreement. One may wonder why, since few countries routinely export intellectual property goods and thus few benefit directly from these legislations.

Intellectual property owners, whether private companies or the governments that support them, have developed strategies to increase intellectual property standards globally. As a result, after the adoption of the TRIPS Agreement in 1994, intellectual property protection expanded considerably around the world while intellectual property owners' rights were extended in different directions.

The United States has played and still is playing a leading role in this process, and the free-trade agreements that it is promoting are key elements of a step-by-step strategy. Peter Drahos and John Braithwaite have described the "forum-shifting" tactic in which the intellectual property movement shifts from one forum of negotiations to another where it is more likely to succeed, from bilateral talks to the WTO, from the WTO to WIPO, and so on.[8] Hence, it is possible to follow the history of the increase in the standards of intellectual property protection by monitoring U.S. intellectual property demands in free-trade negotiations over time in different international and bilateral or regional forums.[9]

Within this system, free-trade agreements represent a parallel venue to the multilateral arena that has been repeatedly used by the United States trade representative, for they offer several advantages over the multilateral negotiations, including the fact that they deal with powers that are unequal economically and politically in ways that are advantageous to the United States. Before the adoption of the TRIPS Agreement, the United States exploited bilateral negotiations to facilitate acquiescence to TRIPS—once countries had accepted provisions similar to those proposed within TRIPS, they were less likely to oppose multilateral U.S. demands and might even support them during international discussions. After TRIPS was signed, the United States used bilateral agreements to develop higher standards.[10]

The tactic of using various forums for negotiations rested on the development of legal instruments providing the U.S. negotiators with tools to facilitate the action of the U.S. trade representative and to influence its trade partners. This tactic also hinged on threatening countries with commercial retaliation or using economic benefits as an incentive to impose stronger intellectual property protections, a carrot-and-stick approach that started in the early 1980s. The Caribbean Basin Economic Recovery Act signed by President Ronald Reagan in 1983 is one of the earliest examples of this. It stated that a Caribbean country "would be given duty-free privileges for their goods in the US market if they met certain criteria" and would not if they "had taken steps in relation to intellectual property that amounted to the nationalization or expropriation of that property."[11] The economic interests at stake were not significant, but movie copyright owners had succeeded in convincing Reagan that trade and intellectual property could and should, in some cases, be linked.

In 1984, similar language was reused in an amendment of the Trade Act of 1974.[12] The Trade and Tariff Act linked trade and intellectual property and authorized the withdrawal of trade benefits from a country or the imposition of duties to goods exported to the United States if a U.S. trade representative deemed the country insufficiently protective of U.S. intellectual property assets. In 1988, a new amendment led to what is known as the "Special 301" provision, which requires the U.S. trade representative to identify "foreign countries that deny adequate and effective protection of intellectual property rights or fair and equitable market access for U.S. persons that rely on intellectual property protection."[13] Depending on the level of dissatisfaction of the U.S. trade representative and based on U.S. stakeholders reports, the countries are placed on a "Watch List," on a "Priority Watch List," which entails greater scrutiny, or on a "Priority Foreign Countries" list, reserved for the worst cases, where countries can be subjected to a Section 301 investigation and face the threat of trade sanctions, including through the Generalized System of Preferences and the "Section 306 Monitoring" list.[14] Over time, the objectives of the U.S. negotiations concerning intellectual property became more explicitly described in the law. The Trade Act of 2002 (H.R. 3009), in outlining the U.S. trade representative's goals, states that "the principal negotiating objectives of the United States regarding trade-related intellectual property are . . . to further promote adequate and effective protection of intellectual property rights, including through ... ensuring that the provisions of any multilateral or bilateral trade agreement governing intellectual property rights that is entered into by the United States reflect a standard of protection similar to that found in United States law."[15]

We can distinguish several periods of bilateral trade negotiations over the last few decades of U.S. history. The first one began in the 1980s. As developing countries levied escalating criticism against international intellectual property conventions, the United States faced mounting resistance in multilateral forums such as WIPO and the United Nations Conference on Trade and Development. These constraints, as well as the emergence of a new strategy developed by industry, led the U.S. government to press for the integration of intellectual property rights into the General Agreement on Tariffs and Trade (GATT) and the Uruguay Round negotiations that ended with the creation of the WTO.

Starting in the mid-1980s, a number of bilateral agreements on intellectual property were signed,[16] compelling partner countries to accept intellectual property rights standards similar to those outlined in the TRIPS Agreement. In the meantime, bilateral investment treaties requiring "adequate and effective" intellectual property rights were signed. In November 1983, President Reagan and Israeli Prime Minister Shamir launched the negotiations of a U.S.-Israeli free-trade agreement, including provisions on intellectual property. This agreement concluded in February 1985 was a first for the United States, in terms of both scale and content.[17] A few years later, in 1990, the negotiations of the North American Free Trade Agreement between the United States, Mexico, and Canada started. Concluded in August 1992, it became a landmark for future U.S. negotiations and a baseline for intellectual property demands. This period, which saw the emergence of a new U.S. policy articulating a relationship between intellectual property and trade, ended with the successful ratification of the TRIPS Agreement in 1994.

Developing countries expected that once the United States had achieved its objectives at the WTO, the pressure would fade, but the post-TRIPS era did not bring a slowdown of U.S. bilateral activity or a diminution of its demands. Talks for a Free Trade Area of the Americas involving thirty-four countries began in December 1994. Bilateral negotiations with Vietnam, Laos, Singapore, Jordan, and Chile started toward the end of the 1990s and at the dawn of the new millennium. In 2002, the Trade Act restored the fast-track authority of the president to negotiate international trade agreements,[18] which had expired in 1994, sparking intense negotiations with Central American countries (the Central American Free Trade Agreement, including the Dominican Republic), with Morocco and Australia, with the countries of the Southern African Customs Union,[19] with Bahrain, Oman, the United Arab Emirates, and with Bolivia, Ecuador, Panama, Thailand, South Korea, Malaysia, and Kuwait. Negotiations have also been considered or planned with a number of countries, including Qatar, Indonesia, and New Zealand. As Jeffrey Schott of the Institute for International Economics puts it, this "spurt of negotiating activity—in parallel with the Doha Round of multilateral trade negotiations in the World Trade Organization (WTO)—is unprecedented in postwar US trade policy."[20]

In recent years, some of these negotiations have stopped because the obstacles to reaching an agreement were too numerous to be overcome within a limited time or because the political context was not favorable. Negotiations for the Free Trade Area of the Americas came to a standstill in March 2004. Talks with Thailand were suspended two years later, in March 2006, followed shortly thereafter by a definitive stalling of Southern African Customs Union negotiations in April and the suspension of deliberations with Ecuador in May. Negotiations with Qatar have been on hold since 2006. However, most of the agreements that the United States was negotiating have been signed. Several countries are still in the midst of negotiations, including Malaysia and the United Arab Emirates, but since fast-track authority expired in July 2007, the intensity of trade negotiations has slowed down significantly.

As I began by noting, the free-trade agreements negotiated by the United States during this period require levels of protection that are more stringent than the standards required by the TRIPS Agreement of the WTO.[21] They contain several types of provision that increase protection and/or strengthen monopolies and that, for instance, limit or preclude the introduction of competitive generic products or delay the entrance of creative work into the public domain. They play out in two different ways: by increasing the measures of protection or by reducing the possibility of using flexibilities—measures that can be used to facilitate access to knowledge and knowledge goods under TRIPS—to limit exclusive rights.[22]

Among the intensification of protections, the provisions of these free-trade agreements include the expansion of patentability criteria and the limitation of exceptions to patentability (for example, the patentability of new uses of known medicines and methods for the treatment of plants, animal, seeds, genes, and so on). These measures are responsible for the increase in the number of unessential patents, an increase that adds new barriers to the production of and access to generic medicines, food, medical technologies, and so on. At the same time, several agreements include a ban on what is known as "pregrant opposition," which would allow third parties to oppose the granting of a patent while it is still under review by the patent office and which is an efficient tool to curtail the proliferation of petty patents. Other measures extend the term of patent protection beyond the twenty years imposed by the TRIPS Agreement, using the pretext of delays during the patent-granting procedure and/or the marketing-authorization procedure. Although patent rights are private rights, the provisions of U.S. free-trade agreements link patent status to drug-marketing approval and compel regulatory agencies to play the role of defenders of patent owners' interests. They request that the regulatory agency check the patent status of products and that patent owners be informed if or when generic competitors request marketing approval. Free-trade agreements also create exclusive rights over marketing-approval data that prevent the introduction of generic versions of pharmaceutical products into the market, even in the absence of a patent. In order to secure marketing approval, companies have to provide regulatory agencies with clinical data to prove the safety and efficacy of their products. Data-exclusivity provisions prohibit relying on these data to approve generic versions of the original product. Not only does this measure prohibit generic competition even in the absence of patent protection, but it establishes a monopoly that, due to the absence of clear mechanisms to do so, is even more difficult to challenge than a patent.

Measures included in free-trade agreements also limit the flexibilities available to developing countries, such as compulsory licensing, whereby the use of a patent can be allowed without the consent of its holder. They do so by restricting the grounds on which the compulsory license can be issued and through data-exclusivity provisions that indirectly prevent countries from taking full advantage of compulsory licensing by blocking marketing approval for generics produced or imported under compulsory licensing. Likewise, they prohibit parallel imports—the importation and resale of a patented product in a country without the consent of the patent holder, which allows the purchase of cheaper products from a foreign country—by imposing a national or regional regime of rights exhaustion.

Copyright laws are targeted by free-trade-agreement provisions, as well, resulting in the obstruction of fair use and the limitation of access to all sorts of materials (in hard-copy or digital form), including the prevention of the online distribution of software, music, or publications, all of which amounts to an obstruction to education, research, technology advancement, and publishing, but which also endangers the global architecture of the Internet and its freedom as a public space. Free-trade agreements request copyright protection similar to or stronger than what prevails in U.S. copyright laws, such as the 1998 Digital Millennium Copyright Act. Many of them extend the term of copyright protection to the life of the author plus seventy years or to seventy years from the publication or creation of the work. Because in many countries, the copyright protection is fifty years, the extension further delays the entrance of works into the public domain. Just as they prohibit parallel imports of patented products such as medicines, free-trade agreements prohibit parallel imports of copyrighted works that have been lawfully sold in foreign markets. They also impose technical-protection measures on copyrighted works and create obligations to prevent the circumvention of such measures, and enhance enforcement obligations that go beyond the TRIPS requirements, including, in some cases such as Korea, very specific unilateral obligations to prosecute Internet piracy, impose liabilities on Internet service providers whose networks are used to distribute copyright-infringing material, and shut down offending Web sites. This is in line with the adoption of criminal sanctions or increased criminal penalties for a range of activities that the United States also promotes in international negotiations, such as those that took place over the Anti-Counterfeiting Trade Agreement. Free-trade agreements even enlarge the concept of infringement, for example to acts of reproduction or the use of copyrighted material that would not be considered as such under the fair use limitations of the U.S. copyright law. This includes changing the definitions of "reproduction" and "material form" to cover transitory reproductions, shifting the burden of proof onto the defending party to show that the activity is noninfringing, and a number of similar measures.

In addition to all this, U.S. free-trade agreements require adherence to international treaties such as the WIPO Copyright Treaty and the WIPO Performances and Phonograms Treaty, which require that countries adopt additional measures to impose, increase, and protect the rights of copyright holders.

With the development of new technologies and the emergence of the knowledge economy, the race for economic dominance became a technology race between the rich countries that have control over the production of scientific and technical knowledge and the less developed countries trying to acquire this knowledge and to develop their own potentials. As Ha-Joon Chang notes, "knowledge had always flowed from where there is more to where there is less,"[23] and thus the task for the United States is to stem the flow of scientific and technical knowledge. This is precisely what is at stake with the restrictions on intellectual property in the U.S. free-trade agreements.

Since the Democratic Party took control of the U.S. House of Representatives and the Senate in 2006, Democrats have raised concerns about the country's policy on free-trade agreements.[24] One consequence has been a new trade arrangement between Congress and the White House, adopted in May 2007 in the context of the approaching Congressional passage of free-trade agreements signed with Peru, Panama, and Colombia. Several aspects of the TRIPS-plus provisions in these agreements were revised: Patent extensions and mandatory linkages were eliminated, while data exclusivity was limited to five years. These changes affected only the three agreements pending at the time, which, regardless of these improvements, nevertheless still remain more protective than the WTO standard. They did, however, mark the first time since the United States embarked on its post-TRIPS pursuit of free-trade agreements that the U.S. trade representative officially and publicly reversed the policy of pursuing increased levels of intellectual property protection. Nevertheless, the fact that an agreement was ultimately reached between the Congress and the White House also demonstrates that the current U.S. trade policy, with minor changes such as these, is still a bipartisan one.

U.S. FREE-TRADE AGREEMENT POLICY: THE POLITICAL CONTEXT

Peter Drahos and Susan K. Sell have both shown how a handful of corporate leaders got together to convince the U.S. administration and successive U.S. governments, as well as the international business community and the governments of other developed countries, of the necessity of formally linking trade with intellectual property and of including intellectual property in the GATT negotiations.[25] As a result of this lobbying effort, the line of demarcation between U.S. corporations and the U.S. government became blurred. This is not to say that the U.S. administration strictly follows the same agenda as the business community, blindly obeys it, or is completely corrupt,[26] but U.S. corporate representatives gradually have found ways to come closer to policy makers, establishing regular contacts and collaboration. Corporations thus have become not only useful collaborators, but indispensable in the process of policy making, creating interdependency between them and the policy makers.[27] In this collaborative dynamic, policy makers slowly have incorporated business logic and goals into the process, and these have become part of those of the government's own logic and goals once they were included in its administrative and legal language.

The history of international intellectual property negotiations involving the United States provides many examples of how this dynamic was established and how it unfolded. Drahos and Braithwaite recount the example of how, as the prospect of the negotiations on intellectual property during the Uruguay Round loomed, the U.S. trade representative asked U.S. businesses to provide negotiating objectives and concrete demands, which resulted in strong similarities between official U.S. documents and the companies' own "blueprint" for trade negotiators, the *Basic Framework of GATT Provisions on Intellectual Property: Statement of Views of the European, Japanese and United States Business Communities*, when U.S. officials were not simply borrowing parts of this document.[28] The same kind of phenomenon can be observed more recently between the annual U.S. trade representative's 301 report and the submissions that the industry provides every year.

Of course, U.S. business goals in many ways reflect the goals and interests of multinational corporations, regardless of where the headquarters or production facilities of these corporations are, a phenomenon reinforced as economic globalization intensifies. This is not to say that a company's nationality, or more precisely, the nationalities of its executive decision makers, no longer matters. As Chang notes, most top executives are nationals from the country in which the firm originates.[29] Capital stays rooted geographically, and its logic does not override the logic of national interests. However, following Jeff Faux, we can consider the logic of company executives to be similar to a class logic,[30] which illustrates itself in its ability to define common objectives, to then promote them within governments and among political leaders, inspire reforms, and in some ways structure society beyond national borders. The revolving-door practice, in which employees shift back and forth between the public and the private sector, which is so common in the United States, is part of how that system homogenizes the way leaders see issues and define policy goals. Money changes hands in some cases, but most of the time, things happen more through the exchange of favors and the sharing of common views and interests.

But the phenomenon observed in the field of intellectual property is not only the result of the collective action of a fraction of the world population. It is also in line with a broader change in politico-economic governance that has unfolded since the end of the 1970s: the emergence of neoliberal rationality, which has favored both a new attitude by business toward governments and the redefinition of these two entities, their roles, and the relations between them. With neoliberalism, political affairs, like all sectors of society, have become subjected to economic rationality, while in turn, the new political rationality has been institutionalized and spread.[31] It has embraced the state, which has redefined itself according to market logic as subject to profitability criteria and thus seeks to favor entrepreneurial interests in its public-policy decisions and uses the law as an instrument to serve this purpose.[32]

Neoliberalism provided the conditions to impose a new global intellectual property system in the 1990s: a new definition of the state and its prerogatives, a new relationship between the public and private spheres, and the expansion of free-trade agreements as vehicles for promoting strong intellectual property protection provisions. In turn, stricter intellectual property protections have been one of the ways in which the dominant class spreading neoliberalism has reinforced its power by securing its control of the assets that fuel capitalist accumulation in the knowledge economy.

While their theoretical foundations are based on the classical liberal economics of Adam Smith and its followers, neoliberals stray from their ancestors in that they do not limit the influence of market logic to the economic sphere, but at the same time do not take the market as a natural fact.[33] Neoliberals disparage state intervention when the state acts as a social and economic safety net, but encourage the state to support and protect economic activity and to guarantee the conditions of competition and free trade. The free market, unhampered by government interference in its effects, but assisted by a range of state interventions, including deregulation and privatization, is supposed to be able to reach maximal economic profitability. Neoliberals, unlike classical liberals, are not bothered by monopolies as long as they emerge within the market, for they believe they will be automatically regulated by the economic dynamic, but they reject, in theory at least, institutionalized forms of monopolies, whether public or private.[34] Their deep enthusiasm for strong intellectual property rights—the justification being that insufficient protection of intellectual property rights harms free trade by introducing trade distortions—introduces contradictions into this orthodoxy, since it results in extensive institutionalized monopolies and heavy regulation hedging a market economy.[35] That the TRIPS Agreement, with its substantial protections for intellectual property, is part of the WTO can be seen as one of the paradoxes that illustrates existing tensions between the theory of neoliberalism and the reality of its practices. But such contradictions have not prevented the business and economic elites in the United States and the United States government from pressing for greater restrictions on intellectual property in the past decades.

Intellectual property protection can be seen as a natural extension of the proprietary system promoted by classical liberalism as adapted to the growing knowledge economy. It is a key factor in controlling the conditions of production and commodification of what are currently the most valuable goods within the capitalist order. Neoliberals obviously have an interest in making the control of information technologies a major source of accumulation—the motivation to do so was increased by the belief in the risk of the decreased competitiveness of the United States and the fear of an economic decline that spread among U.S. policy makers and economists at the end of the 1970s and the beginning of the 1980s.[36] Intellectual property was seen as important to securing the position of U.S. industries, and a strong intellectual property regime was seen as central to securing the profitability of their technologies and innovations.

In the United States, the lax domestic patent environment gave way in the 1980s to a resurgence of patent rights, while antitrust laws and policies, on the other hand, were weakened.[37] This change in the U.S. patent regime attests to a more general change in the thinking about intellectual property and its role in the economic life of the United States. At stake was the ability to prevent foreign economies from autonomously absorbing "U.S." technologies and their benefits. To use Drahos and Braithwaite's portrayal in *Information Feudalism*, "old protectionism was about keeping your rival's goods out of your domestic market. New protectionism in the knowledge economy was about securing a monopoly privilege in an intangible asset and keeping your rival out of the world markets."[38]

As we began by noting, until the late 1970s, developing countries were able to rely on domestic market-protection measures and subsidies to undergird industries or specific economic sectors and to relax intellectual property rules in order to achieve better economic growth and development—in other words, they were doing what developed countries had done in their time as developing countries, too. This rapidly changed with the new direction encouraged by the United States and other wealthy countries. Developing countries were now bound by the imperative to adopt the rule of free markets and free trade while securing stronger rights for intellectual property owners.

Under the influence of the "Chicago School" of free-market economics, Ronald Reagan, who first took power in 1981, and his administration played a fundamental role in the changes in U.S. intellectual property policy. An efficient network of individuals in American institutions was gradually put in place through which significant legal changes were introduced domestically. In 1981, Reagan appointed William F. Baxter as head of the Antitrust Division of the Department of Justice, and Baxter introduced several legal changes aimed at increasing intellectual property protection while deregulating competition.[39] In 1982, Reagan signed the bill creating the Court of Appeals for the Federal Circuit, which became an important tool for patent owners to promote their interests,[40] to mention only two examples.

On the international level, with the adoption of the Trade and Tariff Act of 1984, Reagan became the first U.S. president to have a direct say on the behavior of foreign countries toward intellectual property and their access to Generalized System of Preferences benefits.[41] He rapidly made use of this new prerogative: In 1988, he announced that Taiwan, South Korea, Singapore, and Hong Kong were going to be removed from that program the following year.[42] This use of actual sanctions had the important effect of giving credibility to latter threats, even though, in the future, inscription of countries on the 301 list rarely led to real punitive measures. In 1986, Reagan set up the Council on Competitiveness, which gathered together corporate CEOs, university presidents, and labor leaders and which became one of the trade institutions within which executives from pharmaceutical companies introduced and pushed the issue of intellectual property.[43]

On February 6, 1986, in a message to Congress, the president declared: "Trade is the life blood of the global economy. Growing world markets mean greater prosperity for America and a stronger, safer, and more secure world for the family of free nations. We will continue to work to promote a free, fair and expanding world trading system.... In addition, we will propose legislation to strengthen and broaden protection of intellectual property."[44] Intellectual property was officially now on the U.S. political agenda.

Globally, Ronald Reagan was a key player in the neoliberal revolution, striving with great success, along with Margaret Thatcher, to make what had been minority positions become the mainstream intellectual, political, and ideological framework.[45] Once in place, this new framework made it difficult for their successors, regardless of their own beliefs or political orientations, to escape the new political rationality or to change the course of policies or to extract themselves from the web of relations and obligations to which they were linked. In this regard, the resumption by the Clinton Administration of the free-trade agenda that Reagan had launched illustrates how, during the 1990s, neoliberalism was so largely embraced that it became the new orthodoxy.

Bill Clinton declared during his presidential campaign in 1992 that he would not support the North American Free Trade Agreement (NAFTA) unless social protection was included in it, but that did not prevent him from becoming the master of ceremonies for its signing in November 1993, once he was elected. The idea for NAFTA was first proposed by Ronald Reagan in 1979. The negotiations that started in 1990 were conducted by the first Bush administration. It was then signed and sent to Congress for approval by Bill Clinton. To emphasize what was framed as a bipartisan victory, Clinton appeared at the signing ceremony together with former presidents George H. W. Bush, Jimmy Carter, and Gerald Ford. The message was clear: NAFTA was above ideology. As Vice President Al Gore stated during the ceremony:

> There are some issues that transcend ideology. That is, the view is so uniform that it unites people in both parties. This means our country can pursue a bipartisan policy with continuity over the decades. That's how we won the Cold War. That's how we have promoted peace and reconciliation in the Middle East. And that's how the United States of America has promoted freer trade and bigger markets for our products and those of other nations throughout the world. NAFTA is such an issue.[46]

Free-trade agreements are a very efficient tool in the process of promoting neoliberal policies. They are effective not only because they are promoted by the most powerful political and economic country of their time, but also—an illustration of the expression of a hegemonic system—they become perceived in many political and intellectual circles as an option that is nonideological. The consequences of signing a free-trade agreement thus seem to be apprehended outside and independently of the issue of the political system from which it is generated and that it reinforces. This is shown by the fact that even government leaders with leftist political orientations are involved in free-trade-agreement negotiations—or want to be.

A principal tenet of this pervasive hegemonic system is the assumption that securing private property is the most efficient means to ensure productivity and

The Market, UK antipiracy advertisement (available on-line at http://www.youtube.com/watch?v=yXiHlY6iHqk; www.officialdvd.co.uk) (Red Planet Accidental 2007).

make citizens responsible—the language of individual responsibility being one of the key terms of neoliberalism. Individuals are responsible for what they own: This line of argument is used to discredit publicly owned companies and public services as being inefficient and underproductive. This framing, used in the context of the Cold War to decry the Communist economic system, was later "naturally" incorporated into neoliberal discourse and disseminated as the first waves of massive privatization took place in Britain under Thatcher in the early 1980s. The belief—practically a doctrine—that private equals good, while public leads to failure, became common rhetoric of neoliberal propaganda. One can easily imagine how, as a further evolution of this equation, the conventional wisdom could soon become that free (as in free of charge) equals bad—or dangerous. Lawrence Liang describes very convincingly the sort of insecurities that grow regarding knowledge goods while intellectual property protections increase.[47] The fear or apprehension to which he refers, the fear of trespassing or unlawful appropriation, could easily expand and apply to all free goods. Are they really free? Am I doing something wrong by downloading them? If they are free, are they safe?

The propaganda advocating increases in protections for intellectual property rights commonly plays on this sort of insecurity. Public-relations offensives of brand-name pharmaceutical companies, for instance, recurrently state or imply that generic medicines are of low quality. Initiatives such as the Anti-Counterfeiting Trade Agreement, an international agreement currently under expedited negotiation,[48] use semantic shifts to undermine the image and use of certain goods and to extend the realm of illegality to them. An obvious example is the use of the term "counterfeit" to designate generic products. For its part, the entertainment industry multiplies commercials and campaigns accusing pirated DVDs of being bad copies and of supporting terrorism, describing those who copy goods as criminals and those who buy them and use them as thieves.[49] These campaigns use threats of imprisonment, social exclusion, and opprobrium, creating a fear of cheap or free goods that may fall within our reach, but with which contact can be shown to be extremely dangerous.

These narratives constructed by intellectual property rights owners aim at modifying representations and changing behaviors while promoting legal reforms and securing the place of their products in the market. Their leitmotifs are the war against counterfeiting and piracy and stronger enforcement of intellectual property rights. This anti-A2K campaign is being developed to counter recent setbacks at the multilateral level, such as the Doha Declaration at the WTO, the Development Agenda at WIPO,[50] and adverse local initiatives, including the use of compulsory licensing to permit the production of certain generic drugs in Thailand and Brazil and national laws allowing the noncommercial downloading of music. The rhetoric used follows classic paths. One common trend is to appeal to fear of crime and concerns about safety, associating competitors' goods with the risk of death or imprisonment. From this viewpoint, the media storm that has risen during the past few years in the United States and Europe denigrating the quality of Chinese products, which burst in on Western markets at a time when their competitiveness is undeniable and their presence is significantly increasing, may call for some scrutiny. It offers a convincing example of the use of arguments and fears regarding quality and safety to influence people's opinion and serve commercial interests.

Already in the 1960s, when Asian countries where trying to secure the right to use intellectual work such as textbooks to meet the educational needs of their populations, they were branded by the United States as "pirates." In 1979, the call for an "anti-counterfeiting code" was used as an entry point by the United States to push the issue of an international agreement on intellectual property during the Tokyo Round of the GATT.[51] More recently, the link between counterfeiting and terrorism surfaced following the World Trade Center attack. In 2003, INTERPOL secretary general, Ronald K. Noble, warned governments that "there [was] *growing* evidence of a link between intellectual property crime and terrorist financing" (author's italics).[52] These kinds of association between terms became a staple of the political communication spins of the U.S. government over the past decade—associations between Iraq and 9/11 for example—and more generally of the neoliberal state, so much so that it is not surprising to hear participants at governmental

Let's Terminate Piracy, antipiracy advertisement featuring Arnold Schwarzenegger and Jackie Chan (available on-line at http://www.youtube.com/watch?v=Wf4pnY1wFiU&feature= related; California Commission for Jobs and Economic Growth and the Intellectual Property Department).

or international copyright meetings talk about terrorism as if it were a problem with which they were actually dealing, while the equation between "generic" and "counterfeit" has been unquestioningly assimilated into the thinking of many people in many countries. These propaganda techniques are connected to coercive measures and tactics aimed at controlling and regulating people's behaviors — which represents yet another component of neoliberal rationality.·

Yet another tendency of the neoliberal argument is to convince people that there is no alternative to what is offered and what is being done. This is where neoliberals have proven efficient at the performative practice of presenting their choices as the only realistic fictions available. It is a belief that expresses itself in developing countries' attitude toward free trade: For many leaders of developing countries, free-trade agreements are perceived as the only possible path to secure their country's position in the globalized economy: Not getting into the game would mean losing market access and competitiveness, while their neighbors are probably keen to launch negotiations with the United States, if they haven't already signed an agreement. This sense of inevitability also means that countries tend to accept U.S. demands on intellectual property issues once they have gotten involved in negotiations.

Associated with the power of beliefs is the use of statistics expected to reveal the economic truth underlying the neoliberal system. To support their assertions, advocates for intellectual property restrictions and neoliberal crusaders do not hesitate to invoke numbers and statistics, often showing few qualms regarding their veracity or the methods used to produce them. The estimated losses of corporate profits due to counterfeiting and piracy are figures of this sort, regularly fed to policy makers as well as to the public, whether in institutional debates or in commercials on TV. Numbers are conveyed from the companies to the lobby groups and from the industry associations to the government and the media. The media reports them, and governments use them in international negotiations. The ways these numbers are obtained often lack any semblance of scientific method or even a basis in reality, but they are nonetheless rarely questioned.

Drahos and Braithwaite mention a case in Italy in which the estimates of company losses because of video piracy were established based on the assumption that for every illegal cassette, there was an unsold movie-theater ticket: If you multiply the number of cassettes by the price of a movie ticket, you therefore supposedly get the total of the producer's loss on the movie.[53] Similar doubts regarding accuracy can be raised over the economic projections used by the U.S. government or industries promoting free-trade agreements. But as Drahos and Braithwaite note, skepticism and doubt do not prevent the U.S. trade representative from using the figures provided by the industry to promote free-trade agreements or to threaten countries with sanctions through the 301 process,[54] the most convincing argument for doing so being that "they [are], after all, the only figures that [are] available."

This practice of modifying the perception of reality, which occurs under the auspices of propaganda, indoctrination, and ideology, was routinely used in the process of establishing and maintaining neoliberal hegemony. It provides a good example of what Antonio Gramsci called "political questions... disguised as cultural ones," which "as such become insoluble."[55] As Harvey sees it, the utopian arguments of neoliberalism, such as freedom and individual responsibility, have served as a justification for and legitimization of the efforts of an economic elite to create or to restore its own power.[56]

Thomas Edsall describes how those who brought about the neoliberal revolution became organized:

> During the 1970s, business refined its ability to act as a class, submerging competitive instincts in favor of joint, cooperative action in the legislative arena. Rather than individual companies seeking only special favor... the dominant theme in the political strategy of business became a shared interest in the defeat of bills such as consumer protection and labor law reform, and in the enactment of favorable tax, regulatory and antitrust legislation.[57]

Instead of being a simple corporatist action, this movement became a political movement. An anecdote recalled by Jeff Faux in the introduction of *The Global Class War* offers a convincing example of how reliance on class solidarity operates within a worldwide elite. During a conversation he had with a corporate lobbyist from Washington, D.C., the latter, defending NAFTA, gave what she thought to be the ultimate argument: The president of Mexico is "one of us." Hence, his peers had to help him and support the agreement.[58] President Carlos Salinas had gone to the same type of school (Harvard, in this case) and was an important economic and political player, which, according to that lobbyist, made them all part of the same club.

Thus, there exists a loose network of individuals, not limited geographically, who, despite important differences in their origins, lives, experiences, cultures, and political views, share enough in terms of their position with regard to the market and political power to harbor a feeling of belonging and to act in a connected and even united way.[59] The members of this network, which can be seen as what Gramsci called a "fundamental social group," consider themselves to be the "organizer[s] of society," striving "to create the conditions most favourable to the expansion of their own class,"[60] and are often backed by those who are keen to compare themselves to this elite, yearn to be part of the club, and thus tend to support the elite and the policies it promotes, as if doing so were an elementary precondition of their acceptance.

This class phenomenon is mobilized and maintained according to functional rules similar to those described by Monique Pinçon-Charlot and Michel Pinçon in high social classes in France. Society life plays an essential role in the sharing of information and relations and in the building of strategies.[61] Members of this global governing class do favors for each other not only because they belong to the same club, but because doing so helps establish that belonging—and because each of the members will at one point or another need a favor in return, and knows it. A set of social techniques materializes class functioning and develops the collective consciousness of belonging among its members. It therefore creates the ties and the obligations to stand by each other, the solidarity summoned by the lobbyist talking with Faux and that Lloyd Bentsen expressed when he was asked why Congress should pass NAFTA: His response was, "One word. And it's spelled S-A-L-I-N-A-S."[62]

A new social group does not supplant an older group, but there is a reconfiguration of the vision of governing that allows alliances between a new economic power—those who got rich or richer through the privatization that occurred with the process of establishing the neoliberal hegemony—and the more traditional governing elite.

On the other end of the social spectrum, through free-trade agreements and other policies, neoliberalism has reproduced a population of poor individuals primarily used as producers deprived of control of the means of production and regarded simply as passive consumers of goods. This social group is composed of the traditional industrial working class, which, despite the closing of plants in developed countries, still exists and endures primitive working and living conditions in many parts of the world. According to estimates, the job loss in the United States from NAFTA has been between five hundred thousand and one million.[63] Meanwhile, car exports from Mexico to the United States doubled between 1993 and 1996. At the same time, productivity steadily increased, and wages decreased. Sherrod Brown estimates that Mexican workers have faced a 50 percent decline in their standard of living since the enactment of NAFTA. Most of the jobs that have disappeared in developed countries have not vanished from the surface of the Earth. The work is still done, the goods are still manufactured—only the places change. Free trade favors this low-cost labor, performed by workers living in free-trade zones, but mostly locked within national boundaries. A traditional form of exploitation that follows the rules of industrial capitalism under neoliberal governance persists at the peripheries of the knowledge economy, while productivity and wages are further dissociated from the costs of production and prices. As Brown notes, "when Nike moved all its production overseas, the price of its shoes did not decrease."[64]

These workers represent a classical form of proletariat that does not possess capital and is under the rules of free trade particularly vulnerable to exploitation. At the same time, by means of intellectual property protections, free-trade agreements help to exclude this population from the most liberating and creative dimensions of "cognitive capitalism," the capacity to use nonmaterial resources to produce freely, without restriction and for one's own interest.[65] A cognitive proletariat thus exists under the "information feudalism" described by Drahos and Braithwaite. It consists of women and men who manufacture jeans or DVDs, auto parts, electronic chips, or medicines and who, under the rules of intellectual property, are denied the access to knowledge that would allow them to be more than simple consumers with limited access to physical goods.[66] Indeed, the knowledge—the cognitive capital—necessary for them to develop their own products and to produce and sell them at cheap prices to their own advantage is instead confiscated by law.[67]

The valorization of privatized knowledge further dissociates the value that is attributed to a good on the market from the amount of social labor time necessary to produce it and from the simple cost of production. This affects immaterial goods (the price of a text message, for example), as well as material goods defined by immaterial qualities (the price for a pair of brand-name jeans, the cost of a patented drug). However, this is not the sort of liberation from the rule of valuation that puts abundance within the reach of the multitude.[68] The implementation of intellectual property rules compartmentalizes access both to knowledge and to what can be produced or invented with acquired knowledge. Intellectual property barriers reinforce the partition of society into categories of individuals—at least two for each type of goods: those who access the knowledge and the goods and those who do not or do not do so fully. Two different realities thus coexist: a market of abundance, where money, goods, and ideas flow, and a restricted market, where scarcity is the result of the limited capacity of those who constitute it to accumulate and use cognitive capital and to afford goods. We might draw an analogy with Marx's analysis of the class struggle and say that we see the information society "splitting up into two great hostile camps"[69] under the rule of cognitive capitalism and neoliberalism. However, this class structure cleaves the traditional social classes according to new divides that do not correspond to the habitual fault lines, even if they also highlight them occasionally, since what distinguishes people is not necessarily their ability to possess knowledge, but also their capacity to control nonrival knowledge or make use of it.

The expropriation of intellectual property does not affect only the poorest of the poor, those who work in sweatshops and have no access to education. One can see free-trade agreements as recreating and reinforcing social divides between a ruling class that continually strengthens its control and increases its capital accumulation and a portion of the world population with restricted access and choices and affected by increasing inequities. But there is also an important heterogeneity in those affected by the expropriation of intellectual property, a heterogeneity that leads to new fronts along which class confrontations can be seen. Many people do not belong to the most precarious social class. They may take part in the production of knowledge, but still remain captive consumers of proprietary goods, and because they are subject to enclosing and repressive intellectual property laws, they are deprived of influence over the conditions that determine and limit their access to knowledge and goods. For instance, they have no control over the types of medical innovation that are produced and that often do not necessarily meet their most urgent medical needs, and they do not control the pricing of the products and services that they do need, and hence, they can easily be excluded from access to them, a phenomenon that affects an increasing portion of the middle class in developed countries.

At the same time, all those who are among the cogs of production of the new form of capitalism that has flourished in the context of the knowledge economy, including those just discussed, are not just a workforce exploited to accumulate capital. Their knowledge, which defines them under this regime, is something that they contribute to production (that is, to their own exploitation), but unlike muscular force or other forms of physical effort, this input is not easily replaceable or always exchangeable.[70] Experience and the accumulation of knowledge and know-how give a unique character to the creative production of an individual. Part of what makes an individual's acts of creation possible—the skills, the thinking, and the experience of production itself—is capital for future creation and remains his or her own. Even if he or she is not the one who owns the property rights to what is produced and is not supposed to share it or use it without authorization, the skills, the thinking, and the experience involved in producing it are potentials that the individual can mobilize at any time and that others value.

Hence, the multitude of those who are, to one degree or another expropriated and exploited under intellectual property rules do not form a homogeneous class. (It is true that even during Marx's time, the "working class" was neither uniform nor harmonious, but rather a "mobilizing myth.")[71] While some of the people who belong to this category are kept away from the raw materials and tools of the knowledge economy, others, involved in the logic and the practices of cognitive capitalism, contribute to its production and have thus the ability to turn its "liberating power" against exclusion.[72]

Both the clever use of this liberating power and the formation of common fronts between the various categories of those who suffer exclusion offer interesting avenues, especially in the context of the clash created by free-trade agreements, to take advantage of a sort of recomposed class struggle that does not involve the classic confrontations, but that allows politicization of the conflicts that arise from appropriation and exclusion in the knowledge economy of the information society.

RESISTANCES

Mobilizations against free-trade agreements have taken on significant dimensions in many countries, bringing together people with HIV/AIDS, health defenders, students, farmers, workers, academics, parliament members, local generic producers, and, as in the Thai case, even bankers. In some countries, such as Thailand or South Korea, these mobilizations were very well organized. In Thailand, strong networks of collaboration were developed, sharing information and knowledge and building collaborations with civil servants from various institutions, including the Food and Drug Administration, the Health Ministry, and the Ministry of Commerce. Over the past few years, public protests and demonstrations have taken on increasingly sizeable proportions, and the issue of free-trade agreements has become central in national political debates. However, mobilizations and public opposition have failed to have a significant impact on the outcome of free-trade-agreement negotiations—TRIPS-plus provisions are still included, and agreements continued to be signed.[73]

What make the free-trade agreements such an efficient tool to promote neoliberal objectives is inherent in their structural characteristics and the conditions of their negotiations. Free-trade agreements are, as the U.S. trade representative puts it, "comprehensive": They affect many different domains, from tariffs and access to markets to sanitary measures, from trade remedies to the environment, from investments to electronic commerce, from intellectual property to government procurement. As I noted before, U.S. trade partners agree to negotiate particular aspects such as intellectual property—negotiations from which they have little to expect for their own benefit, in most cases—and to adopt the views of the United States because they are keen to obtain better tariffs and quotas for the goods they export. In this context, it is difficult to give issues such as health or education a high profile in negotiations when these are understood by political leaders to be secondary issues. And since the negotiations are kept secret, it is also difficult for concerned actors to make their arguments heard about details of agreements to which they have no access. Contrary to multilateral negotiations such as the ones carried out at the WTO, countries cannot create alliances and groups to confront the United States. The one-to-one power relation leaves the country negotiating with the United States at a disadvantage, not only because of the political or economic power imbalance, but also because of the inequalities in terms of human and technical resources (expertise, knowledge, and experience) that can be mobilized for the negotiations. As Drahos notes, "In bilateral trade negotiations between States involving a strong and weak State, generally speaking the strong state comes along with a prepared draft text which acts as a starting point for the negotiations. . . . In order to lower the transaction costs of bilateralism the United States has developed models or prototypes of the kind of bilateral treaties it wishes to have with other countries."[74] Free-trade-agreement talks also constitute an environment within which the use of fear, intimidation, make-believe, and threats play a determinant role in the negotiation dynamic. In this context, governments are all the more reluctant to take into account the opinions and requests of representatives of civil society.

The majority of the movements that contest the increase in intellectual property protection, especially in the context of free-trade agreements, try to draw attention to what they consider to be the abuses and dysfunctions of the intellectual property system. In most cases, they encourage countries to renounce provisions that increase protections above WTO standards or they try to promote compromises, such as the use of balancing mechanisms and softer provisions that would not fatally impede access to knowledge and to knowledge goods, that would limit the negative effects of what the United States requires. However, while they focus on the issue of access, rarely do these movements openly question the principle of intellectual property or the concept of property itself. Their strategy consists mostly in demonstrating that, in a number of cases, the intellectual property system fails to fulfill its role of promoting and fostering innovation, while with its strengthening of protections it increasingly generates social costs and limitations on development. Hence, they avoid taking stands that could be seen as ideologically stained—as associated with attacks on property as such—which does not prevent their detractors from calling them Communists.[75] Apart from exceptions such as popular political movements in opposition to free-trade agreements in Latin America,[76] the vast majority of those who have become mobilized on the issue of intellectual property and free-trade agreements tend to stay away from political and ideological rhetoric. While these movements strive to organize mass mobilization in order to affect the negotiation process and to impose resistance, their objective is predominantly to make reasonable and constructive criticisms and propositions about legal provisions and regulations in order to limit inequities, injustices, and inefficiencies, rather than to announce the advent of a politico-economical alternative model. Although free-trade agreements are obvious neoliberal vehicles, these movements seem to avoid even using the term, as if by doing so they would engage in an ideological battle that would cast them as dogmatic and make them lose credibility. They seek inclusionary politics, request their own participation into the negotiations, and use a rhetoric that articulates demands for social justice, egalitarian access, or plain decency, but that is predominantly anchored in a technical register that allows them to assume the identity of experts on intellectual property issues. One of the characteristics of these movements is to manifest a very detailed understanding of the legal issues at stake and the legal instruments involved. They can speak the language of copyright and patent lawyers. Their strategy is based on democratizing this technical knowledge, and in some cases, a large public becomes familiar with rather obscure legal concepts, as for example was the case in Thailand, where activists made the slogan "CL [compulsory licensing] = life" the message on a sticker spread widely throughout the country. Ironically, the pressure exerted to increase intellectual property protection thus ultimately contributes to the increasing knowledge of members of civil society about intellectual property matters and motivates their involvement with them.

While it is the case that most movements do not openly and directly contest the political ideology that underlies intellectual property regimes such as the TRIPS Agreement or specifically denounce market ethics, and while resistance to the escalation of intellectual property restrictions is rarely accompanied by formal rejection of the neoliberal economic order or the institutions that favor it, ideology, as a "narrative about a particular social order,"[77] is at the bottom of this system and of its efficacy. First, factually, the context of international negotiations over intellectual property rights is deeply ideologically coded and interlinked to the rise of neoliberalism. But moreover, "the extraordinary power of ideology itself," as Wendy Brown puts it, plays an important role: Ideology "does not simply (mis)represent the world, but is itself productive of the world, and particularly of the subject."[78] Thus, any attempt to fight avatars of neoliberal power such as the increase of intellectual property rights and the privatization of knowledge requires recognition of the way that ideology is used by the intellectual property movement. Building on the same impetus, it requires us to analyze and assess the modes of legitimacy through which this movement naturalizes its domination and the place and role that it ascribes to each of us.

Pointing to inadequacies and tensions between the proclaimed values and the concrete policies of neoliberalism may be a way to reach this goal and to undermine its credibility. Neoliberalism encourages state intervention when the market itself generates chaos, a phenomenon regularly observed when financial crises threaten financial institutions and losses are socialized as companies or banks are bailed out with public money. However, state interventions and pragmatic adjustments created to "rescue" firms tend to widen the gap between neoliberal theory, as commonly espoused, and practice. One might expect that the disjunction between what is understood as neoliberalism in the conventional wisdom and disseminated through propaganda and the reality of the policies that neoliberals have implemented might, by hint of repetition and because of the increasing proportion of the consequences, create rifts in representations of neoliberal dogma and undermine neoliberal fictions. Nevertheless, despite many indicators that neoliberal doctrine is not fulfilling its promises to secure economic efficiency,[79] the lessons taught by such empirical realities do not appear to be learned, nor is the enthusiasm or zeal of neoliberals undermined. A further effort is required.

Such an effort might suggest that critics of increasing intellectual property protection should try to oppose one ideology with another and develop a counterideology to substitute for the dominant one. However, besides the reluctance to promote ideologies that is common to many social movements—inherited from the critical disenchantment with Communism and socialism that spread since the end of the Cold War and that was influenced and encouraged by the denunciations led by neoliberals against these ideologies—alternative doctrines have not in fact emerged that could directly compete with neoliberalism.[80] If opposition movements do not want to rely on ideologies, other framing techniques and strategies need to be employed to produce cultural changes that could undermine neoliberal legitimacy and supremacy in people's minds.

Neoliberal discursive productions borrow terms from classical liberalism such as the central place of the notion of freedom while simultaneously both feeding and concealing existing tensions between liberalism and neoliberalism, thus sustaining confusion and making it difficult to delegitimize neoliberalism on the basis of internal contradictions and gaps between theory and practice. This is one of the particularities and one of the sources of the efficacy of neoliberalism: to be able to offer a fiction that can seem coherent and that at the same time integrates contradictions and discrepancies that give it the malleability that allows its strength and durability.

One way to break the spell of neoliberalism lies in questioning its use of the values and concepts that underlie it and in trying to reinvest these values and concept with other meanings. Such an approach seems consistent with those employed by the movements opposing high levels of protection for intellectual property rights that, beyond ideology, try to promote basic principles such as freedom, equality, justice or organizing concepts such as the commons and the public domain and that, for the most part, instead of contesting the principle of intellectual property itself, use the fulcrum of access strategically to reframe the terms of the debate. Moreover, the context of the debates and conflicts over intellectual property appears to be favorable. It is large enough, since it affects so many aspects of peoples lives, well-being, and destiny, yet specific enough, since it implicates a multitude of communities, identities, and particular interests from every corners of society, to support an attempt to initiate a process that explores a new balance between freedom and property or the relation between freedom and equality—to imagine a new blueprint for society.

The inadequacies and tensions between the proclaimed values and the concrete policies of neoliberalism are at least a part of this favorable context. Provoking the emergence of global crises whose dimensions are unprecedented and whose causes and consequences are more difficult to control, neoliberalism is increasingly showing signs of its failures to guarantee economic activity and sustain the market. Neoliberal legitimacy may finally be ripe to be called into question at a time when "the widening gap between rhetoric (for the benefit of all) and realization (the benefit of a small ruling class) [becomes] all too visible."[81] It may increasingly be possible to make effective use of the paradoxes and contradictions between classical liberal orthodoxy and neoliberalism. And when neoliberal policies depart so obviously from the obligation to respect neutrality, equity, and inclusion that are supposed to characterize a liberal government—not to influence competition and to provide to each one the opportunity to take part in it—using classical liberal principles against neoliberalism can be an effective way to undermine neoliberalism's pervasive representations and apparent coherence in the conventional wisdom.[82]

Liberalism in its classical form and freedom are two concepts that had a seminal and organizing importance in political debates and choices in the United States and are key notions publicized by neoliberalism. As Wendy Brown emphasizes, "'freedom' has shown itself to be easily appropriated in liberal regimes for the most cynical and unemancipatory political ends." Certainly, "the conservative political culture ascendant in the United States in the 1980s . . . further narrowed the meaning of freedom within liberalism's already narrow account."[83] Semantic shifts and reinterpretations, depending on the historical context and the use made of these terms, maintain ambiguities and confusions. Consequently, there is a risk that movements trying to reappropriate these concepts will fail to make clear the difference between what they promote and what their opponents invoke, and, contrary to their attempts to offer a counterpoint to neoliberal presentations, may reinforce the power of the neoliberal discourse by giving additional legitimacy to the concepts upon which it is based. Moreover, reclaiming the classical liberal inheritance without adopting a clear position critical of neoliberalism risks maintaining the given conditions of neoliberal ascendance insofar that their theoretical bases remain grounded in the same ethos.

Freedom of expression and consumer choice, both classical liberal values, are seriously limited in the current world of intellectual property protection.

Contesting those limits call for enlightening the public about the confusions that neoliberalism has perpetuated concerning the classical liberal conception of freedom. It may also require us to explore other frames of reference, questioning the definition and the meaning of the freedoms one wants to promote. If we follow Michel Foucault's advice, freedom can be understood as a (political) *practice* for individuals, a mode of relation between individuals, that empowers them.[84] Contrary to the tendency that Brown observes among thinkers who came to the conclusion that free enterprise is the most valid and worthy option of freedom, movements opposed to neoliberalism can adopt freedom as "a social and political practice," rather than "an individual good," and thus move away from its rendering in classical liberal thought.[85]

In the context of the knowledge economy, understood this way, freedom looks like what pirates exercise when using, creating, or spreading material or immaterial goods.[86] Put into practice, this freedom represents a form of resistance that may stand "in a position of exteriority to power,"[87] interfering to some extent with the political economy of neoliberal domination by promoting uncontrolled access, neither absolving the state of its responsibility to ensure the well-being of its citizens and to provide justice nor supplanting the need for fair administrative and legal systems to encourage innovation, but able to open breaches on some of the fronts of neoliberal rationality. Meanwhile, while obviously interfering with the logic of distribution, such a practice moves the issue of inequality to the level of production, questioning and disrupting old models—including the production of the self—and injecting anarchically the production of individuals into the public space.[88] Like the mechanical arts as described by Jacques Rancière, digital technologies, when "put into practice and recognized as something other than techniques of reproduction or transmission," hold the power to "confer visibility on the masses, or rather on anonymous individuals." What emerges then are "political subjects that challenge the given distribution of the sensible."[89]

But what are the conditions required to exercise this freedom? Who is in a position to do so? Or in other words, is the exercise of this freedom possible for those—who may be a majority—who are subjected to great social and economical inequalities, or would it represent too many risks and increase their vulnerability? Those are valid questions. Meanwhile, one may also note that for many reasons (from economic conditions to social conditions), the pirate attitude is found more frequently among those who reckon they have nothing to lose and everything to gain from being pirates, those who, by obligation and in some cases by choice, are at the peripheries and even the extreme peripheries of power spheres, rather than those who are close to their cores.[90] To this extent, the sociology of this practice may contradict the sociology of the current A2K movement. One may wonder if that creates a real antagonism between the practice of the pirate and the efforts of the movement.

As Lawrence Liang notes, the figure of the pirate provokes serious tensions within the A2K movement and within movements that oppose stricter exclusive rights protections.[91] To adopt or at least to consider as legitimate a practice of freedom inspired by the pirate, that is, as not taken as a pure matter of consumption, can certainly pose some paradoxes for these movements. Notably it may eventually call for clarifications as far as their relations with the neoliberal regime are concerned. However, without focusing on such tensions and conflicts, it is certainly possible for the movements to take the pirate for what he or she can provoke, for the "discontinuity in our experience" that pirates introduce in our encounters and relations with them. In short, it is possible to welcome the original reorganization of our conceptions, spaces, and relations that may emerge from this experience. As Maurizio Lazzarato points out, "one falls in love less with the person than with the possible world she expresses."[92] Likewise, without necessarily falling for the pirate's personality as a whole or the totality of the existing forms of piracy, without considering his or her modus vivendi or modus operandi as a solution in itself, we can consider the options that the pirate offers and that borrowing from the pirate's practices could generate.[93] From the practices it inspires could come new possibilities, "virtuals," Gilles Deleuze would call them; allowing recruitment at the frontiers of classes, where inequalities are made obvious by deprivation, where opposites confront each other (worker/capitalist, owners/ exploited) at the "locus of the fight," and where conflicts crystallize and injustices are denounced. It is there where an "awakening" may take place to the possibility of refusing the roles and models that are assigned to everyone, the functions required from each, and where instead it may be possible to develop, reinforce, and exercise the power of innovation that individuals possess and that can challenge institutionalized powers.

Without romanticizing or overrating the subversive power of piracy, and while asserting that piracy cannot be the only strategy adopted to address extremist regimes of intellectual property rights protection and repression, the different forms of political action and personal input that exist in piracy should not be dismissed (as they regularly are) because they remain in the shadow, or because they are too sporadic, or because they are done for self-gain. They instead can be seen as an opportunity to introduce useful disruptions from both a tactical and a theoretical point of view—even as a necessary strategy for doing so. This may look particularly appealing if one considers that protests against increasing and extremist intellectual property rights protections often lead to bureaucratic and heavily regulated adjustments to the prevailing regime. Think, for example, of the mechanism adopted at the WTO to allow generic export of patented medicines under compulsory licensing or the prequalification procedure established at the WHO to guarantee generic quality and thus trivialize the heavy propaganda against generic drugs conducted by the multinational pharmaceutical corporations. Practices inspired by the pirates may also help make obvious the realities of the new era of digital production, the conflicts it has caused, and the inequalities institutionalized by the current intellectual property system and by the tendency to increase those protections. Moreover, the resulting forms of resistance can combine with creative processes that can be directly invested by individuals and that can favor invention outside of the walls of the increasingly closed intellectual property regime. In doing so, individuals also generate the power that the movement needs to shape and conceptualize the freedom they try to exercise. For their part, the intellectual property moguls are not fooled and seem aware of these "virtuals." Hence their increasing pressure to enforce intellectual property rights and to criminalize those who infringe them.[94]

What could thus emerge in the context of the fight against free-trade agreements is the constitution of a common front against intellectual property protection policies, allying forms of activism such as the practices of piracy that have developed under the pressure of exploitation and expropriation typical of the neoliberal order or of cognitive capitalism with others inherited from the industrial era. Mobilizations against free-trade agreements such as those that took place in Thailand, South Korea, and some Latin American countries already show interesting alliances between worker-based movements and more classically liberal forms of activism. Traditional worker-based movements are often presented as declining, and these forms of mobilization have indeed been deeply affected by neoliberal offensives since the end of the 1970s, particularly in developed countries. However, similar movements arose during the 1980s in countries such as South Korea or South Africa and are particularly active in Latin America, to the point where they sometimes have taken power. In South Korea, Thailand, and Latin America, they have also been an important component of mobilizations against free-trade agreements in alliance with various components of civil society that rely on other forms of mobilization and contestation. While worker-based mobilizations follow the models of traditional political structure and organization, which allow them to provoke massive mobilizations across countries, other forms of mobilization show an ability simultaneously to address both technical and legal issues and the needs of a large spectrum of different social groups and particular constituencies not identified with or defined by their work.

Traditional forms reactivated in the context of the neoliberal restructuring of classes thus can combine with forms of protest that originate in sections of the society involved in knowledge production (and more imbued with the classical liberal political tradition). As I said before, some of these actors are not always replaceable because of their creative potential, and as a consequence, their recognized position as key elements in the creation of knowledge goods confers on them a special—legitimized—position as critics of its distribution, but, above all, as efficient creators who can fuel new movements. Although they develop arguments against the terms of the commodification, privatization, and financialization of knowledge goods that do not necessarily place them outside of the realm of the dominant political system, and although they do not reject as a whole the modes of consumption and trade that go with capitalism, by engaging in reinterpretation or redefinition of notions that are at the heart of the system that justifies neoliberalism, they may institute new forms of adverse interaction with it. By engaging in a routine practice of freedom, these alliances may help unleash practices that could override the boundaries of the enclosures imposed by intellectual property protection,[95] favoring the learning of knowledge, the empowerment and a better equalization of access to opportunities, while they also contribute to opening breaches to a counterculture.

Without considering strategies based on class struggle as obsolete or simply renouncing them, the alliance of those who directly challenge competition and private property and condemn the resulting alienations, those who promote new forms of collaboration between individuals, new forms of work and creation using knowledge technologies, and those who, inspired by the practices of pirates, engage in practices of freedom that disrupt neoliberal rules of appropriation and the control of access could result in the increase tenfold of the power of an A2K movement. Neoliberalism by virtue of the inequalities and discriminations it produces will thus have favored the emergence of a contestation specific to it,[96] and an A2K mobilization using every available tactic and strategy could organize itself as a creative force challenging knowledge capitalism and contributing to its reorganization. Reactions to the escalation of protections on intellectual property rights and to the intensification of the repression of intellectual property infringements, whether they are formally gathered under the A2K umbrella or not (or not yet), hold the seeds that could realize this efflorescence and thus help show a way out to the neoliberal order, which is certainly as rigid as it is unstable.

NOTES

The author is grateful to Laura Davis and Harriet Hirshorn for their assistance in translating this piece into English.

1   Peter Drahos, with John Braithwaite, *Information Feudalism: Who Owns the Knowledge Economy?* (London: Earthscan: 2002), p. 74.

2   For example, Indonesia adopted intellectual property laws while subjected to the colonization of the Netherlands, while the Philippines passed laws to protect intellectual property while being under the domination of Spain. See Jakkrit Kuanpoth, "The Political Economy of the TRIPS Agreement: Lessons from Asian Countries," in Christophe Bellmann, Graham Dutfield, and Ricardo Meléndez-Ortiz (eds.), *Trading in Knowledge: Development Perspectives on TRIPS, Trade, and Sustainability* (London: Earthscan, 2003), p. 47.

3   The Paris Convention for the Protection of Industrial Property is available on-line at http://www.wipo.int/export/sites/www/treaties/en/ip/paris/pdf/trtdocs_wo020.pdf (last accessed May 4, 2009). D. M. Mills, "Patents and Exploitation of Technology Transferred to Developing Countries (in Particular those of Africa)," *Industrial Property* 24 (1985): p. 120, cited in Kumariah Balasubramaniam, "Access to Medicines and Public Policy: Safeguards under TRIPS," in Bellmann, Dutfield, and Meléndez-Ortiz (eds.), *Trading in Knowledge*, p. 140.

4   Drahos and Braithwaite, *Information Feudalism*, p. 75.

5   *Ibid.*, p. 81.

6   *Ibid.*, p. 133.

7   David Harvey, *A Brief History of Neoliberalism* (Oxford: Oxford University Press, 2005).

8   Drahos and Braithwaite, *Information Feudalism*, p. 195.

9   Gaëlle Krikorian and Dorota Szymkowiak, "Intellectual Property Rights in the Making: The Evolution of Intellectual Property Provisions in US Free Trade Agreements and Access to Medicine," *Journal of World Intellectual Property* 10, no. 5 (September 2007).

10 Free-trade agreements are not the only bilateral vehicle used by the United States to increase intellectual property protection in other countries. Beside bilateral agreements on intellectual property, bilateral investment treaties that the United States negotiated with developing countries during the 1980s, including intellectual property as an investment activity, are other negotiating contexts in which provisions were included protecting intellectual property. However, for reasons that I will develop later here, free-trade agreements are probably one of the most efficient ways to obtain other countries' agreement on very detailed legal dispositions.

11 Drahos and Braithwaite, *Information Feudalism*, p. 83.

12 *Ibid.*, p. 87.

13 U.S. Patent and Trademark Office, "Special 301," available on-line at http://www.uspto.gov/web/offices/dcom/olia/ir_trade_special301.htm (last accessed May 4, 2009).

14 The Generalized System of Preferences, which has existed since 1976, is a program through which, since the Trade and Tariff Act of 1984, designated countries can gain duty-free export to the United States for some of their products in return for incorporating protection of U.S. intellectual property as a criterion for eligibility. Drahos and Braithwaite, *Information Feudalism*, pp. 86–87. Being placed under Section 306 monitoring is for countries subject to a section 301 investigation, for which measures have been agreed on with the United States to resolve the matter. If the U.S. trade representative concludes that measures are not satisfactorily undertaken, sanctions can be adopted.

15   19 U.S.C. § 3802 (b) (4) (A). The Trade Act of 2002 is available on-line at http://frwebgate. access.gpo.gov/cgi-bin/getdoc.cgi?dbname=107_cong_public_laws&docid=f:publ210.107 (last accessed May 4, 2009).

16   The United States signed many bilateral agreements on intellectual property during this period: Korea, 1986; Poland, 1990; Mongolia and Sri Lanka, 1991; Albania, Armenia, Czechoslovakia, China, Romania, Russia, and Taiwan, 1992; Azerbaijan and Tajikistan, 1993; the Philippines, Cambodia, Jamaica, Latvia, Lithuania, Thailand, and Trinidad and Tobago, 1994; China, 1995; Peru, 1997; and Nicaragua, 1998. See http://www.cptech.org/ip/health/c/agreements (last accessed May 5, 2009).

17   The Caribbean Basin Economic Recovery Act concluded two years earlier was only a one-way, duty-free trade agreement.

18   Fast-track authority for international trade agreements before Congress makes them subject to an up-or-down vote, but not to amendment.

19   The Southern African Customs Union includes Botswana, Lesotho, Namibia, South Africa, and Swaziland.

20   Jeffrey J. Schott, "Assessing US FTA Policy: Free Trade Agreements," in Jeffrey J. Schott (ed.), *Free Trade Agreements: US Strategies and Priorities* (Washington, D.C. Institute for International Economics, 2004), p. 359.

21   Frederick M. Abbott, "The Doha Declaration on the TRIPS Agreement and Public Health and the Contradictory Trend in Bilateral and Regional Free Trade Agreements," Occasional Paper 14 (April 2004), Friends World Committee for Consultation (Quakers), available on-line at http://www.quno.org/geneva/pdf/economic/Occassional/TRIPS-Public-Health-FTAs.pdf; Jean-Frédéric Morin, "Tripping Up TRIPS Debates: IP and Health in Bilateral Agreements," *International Journal of Intellectual Property Management* 1, nos. 1–2 (2006), available on-line at http://www.atrip.org/upload/files/activities/montreal2005/Morin%20IP-health-FTA.doc (both last accessed March 3, 2010); Krikorian and Szymkowiak, "Intellectual Property Rights in the Making."

22   David Vivas-Eugui, "Regional and Bilateral Agreements and a TRIPS-Plus World: The Free Trade Area of the Americas (FTAA)," TRIPS Issue Papers 1, Quaker United Nations Office (QUNO), Quaker International Affairs Program (QIAP), International Centre for Trade and Sustainable Development (ICTSD) (2003), available on-line at http://www.quno.org/geneva/pdf/economic/Issues/FTAs-TRIPS-plus-English.pdf; Oxfam, "Undermining Access to Medicines: A Comparison of 5 US FTAs," technical briefing note (2004), available on-line at http://www.twnside.org.sg/title2/FTAs/Intellectual_Property/IP_and_Access_to_Medicines/UnderminingAccessToMedicines.pdf (both last accessed March 3, 2010); Abbott, "The Doha Declaration on the TRIPS Agreement and Public Health and the Contradictory Trend in Bilateral and Regional Free Trade Agreements."

23   Ha-Joon Chang, *Bad Samaritans: The Myth of Free Trade and the Secret History of Capitalism* (New York: Bloomsbury Press, 2008), p. 127.

24   Under the previous Republican-controlled Congress, trade-related legislation passed by extremely thin majorities. Congressional votes on the Central America Free Trade Agreement, on the agreement with Oman, and on other free-trade agreements in 2006 indicated an increasingly negative attitude toward free-trade agreements among Democrats.

25   Drahos and Braithwaite, *Information Feudalism*; Susan K. Sell, *Private Power, Public Law: The Globalization of Intellectual Property Rights* (Cambridge: Cambridge University Press, 2003).

26　Drahos and Braithwaite have shown the complexity of this relationship and the fact that, for example, in the context of the 301 process, the U.S. administration is not ready to become totally dependent on or indentured to industry. It still has, to some extent, to take into account its image abroad and to preserve relations with other countries from seeming too confrontational. Diplomatic, economic, and political issues explain, for instance, the fact that the U.S. trade representative cannot punish every lapse from what the companies consider as proper intellectual property rights protection and sufficient intellectual property rights implementation. Drahos and Braithwaite, *Information Feudalism*, pp. 99–100.

27　*Ibid.*, pp. 115–19.

28　*Ibid.*, pp. 125 and 215.

29　Chang, *Bad Samaritans*, p. 97.

30　Jeff Faux, *The Global Class War: How America's Bipartisan Elite Lost Our Future—and What It Will Take to Win It Back* (Hoboken, NJ: John Wiley and Sons, 2006).

31　Wendy Brown, *Les habits neufs de la politique mondiale: Néolibéralisme et néo-conservatisme* (Paris: Les Prairies Ordinaires, 2007), p. 51.

32　*Ibid.*, p. 98.

33　*Ibid.*, pp. 52 and 97.

34　Michel Foucault, *Naissance de la biopolitique: Cours au Collège de France, 1978–1979* (Paris: Gallimard, Seuil, 2004) pp. 142–43, available in English as *The Birth of Biopolitics: Lectures at the College De France, 1978–1979*, trans. Graham Burchell (New York: Palgrave Macmillan, 2008).

35　See Marshall A. Leaffer, "Protecting United States Intellectual Property Abroad: Toward a New Multilateralism," *Iowa Law Review* 76, no. 273 (1991): p. 277.

36　Drahos and Braithwaite, *Information Feudalism*, p. 63.

37　Sell, *Private Power, Public Law*, pp. 67–74.

38　Drahos and Braithwaite, *Information Feudalism*, p. 87.

39　*Ibid.*, p. 166.

40　Sell, *Private Power, Public Law*, pp. 67–72.

41　Drahos and Braithwaite, *Information Feudalism*, p. 87.

42　*Ibid.*, p. 89.

43　*Ibid.*, p. 69.

44　President Ronald Reagan, "Message to the Congress on America's Agenda for the Future," available on-line at http://www.aegis.com/topics/timeline/RReagan-020686.html (last accessed May 6, 2009).

45　Harvey, *A Brief History of Neoliberalism*, p. 62.

46　White House Office of the Press Secretary, "Remarks by President Clinton, President Bush, President Carter, President Ford, and Vice President Gore in Signing of NAFTA Side Agreements," September 14, 1993, available on-line at http://www.historycentral.com/Documents/Clinton/SigningNaFTA.html (last accessed May 15, 2009).

47　See Lawrence Liang's essay "Beyond Representation: The Figure of the Pirate" in this volume.

48　See Peter Drahos's essay, "'IP World'—Made by TNC Inc.," in this volume.

49　Antipiracy advertisements offer different types of narrative. One invites you to be on the good side—the one with Arnold Schwarzenegger and Jackie Chan fighting piracy. See "Let's Terminate Piracy," available on-line at http://www.youtube.com/watch?v=Wf4pnYiwFiU&feature =related. Others discourage you from joining the bad side, the side of the criminals and those who help, encourage, and invite them: "You wouldn't steal a car.... You wouldn't steal a hand-bag.... You wouldn't steal a television.... You wouldn't steal a DVD.... Dowloading pirated films is stealing.... Stealing is against the law.... Piracy. It's a crime." See "Piracy: It's a Crime, available on-line at http://www.youtube.com/watch?v=iPcHhOBd-hI). "Buy a pirate DVD and you are inviting crime into your neighborhood," warns a British antipiracy ad, with vivid illustrations of other kinds of crime with which it equates DVD piracy. See "The Market," available on-line at http://www.youtube.com/watch?v=yXiHlY6iHqk (all last accessed May 6, 2009).

50 Susan K. Sell, "The Global IP Upward Ratchet, Anti-Counterfeiting and Piracy Enforcement Efforts: The State of Play," IQsensato Occasional Papers No. 1 (June 2008), p. 4, available on-line at http://www.twnside.org.sg/title2/intellectual_property/development.research/ SusanSellfinalversion.pdf (last accessed May 6, 2009).

51 Drahos and Braithwaite, *Information Feudalism*, p. 84.

52 Interpol media release, July 16, 2003, "INTERPOL Warns of Link between Counterfeiting and Terrorism. Cites Evidence That Terrorists Fund Operations from Proceeds," available on-line at http://www.interpol.int/public/icpo/pressreleases/pr2003/pr200319.asp (last accessed May 6, 2009).

53 Drahos and Braithwaite, *Information Feudalism*, p. 97.

54 *Ibid.*, p. 98.

55 Antonio Gramsci, *Selections from the Prison Notebooks* (New York: International Publishers, 1971), p. 149.

56 Harvey, *A Brief History of Neoliberalism*, p. 19. The view that, as a reaction to Keynesian policies, neoliberalism can be seen as aiming at the restoration of the political and economic power of an elite was first developed by Gérard Duménil and Dominique Lévy in *Crise et sortie de crise: Ordre et désordres néolibéraux* (Paris: Presses Universitaires de France, 2000).

57 Thomas B. Edsall, *The New Politics of Inequality* (New York: W. W. Norton, 1984), p. 128, quoted in Harvey, *A Brief History of Neoliberalism*, p. 48.

58 Faux, *The Global Class War*, p. 1.

59 Rick Fantasia, "Quand l'université reproduit les elites," *Manière de Voir: Le Monde Diplomatique* 99 (June–July 2008): pp. 48–52.

60 Gramsci, *Selections from the Prison Notebooks*, p. 5.

61 Michel Pinçon et Monique Pinçon-Charlot, *Les ghettos du Gotha: Comment la bourgeoisie défend ses espaces* (Paris: Seuil, 2007).

62 Sherrod Brown, *Myths of Free Trade: Why American Trade Policy Has Failed* (New York: New Press, 2004), p. 144.

63 Sherrod Brown, p. 151. Brown uses analyses from the U.S. Department of Labor and the Economic Policy Institute.

64 *Ibid.*, pp. 72, 152, 153, 157.

65 The notion of "cognitive capitalism" was developed by Yann Moulier Boutang in a series of articles published in the journal *Multitudes* and later gathered and elaborated in *Le capitalisme cognitif* (Paris: Éditions Amsterdam, 2007). Cognitive capitalism is based on the accumulation of immaterial capital. The management of knowledge and of information technologies to control all forms of production and accumulation, whether it is selling wheat on the market or stock exchange transactions, is at the heart of cognitive capitalism. One of the elements that manifests the emergence of this new form of capitalism is the importance that issues of intellectual property protection have taken in the past decades, together with the tensions

that came along with the increase in these protections (p. 73). Industrial capitalism is not disappearing, but coexists with this new form and is reconfigured by it—relocated, reorganized, become subaltern (p. 74). For many people, including workers whose job is still to transform material resources into material goods, the accumulation of immaterial capital essential to cognitive capitalism is often prohibited by interdiction of access to and free use of knowledge through the enforcement of intellectual property protections and the criminalization of violations of intellectual property rights.

66 Sean M. Flynn shows how in highly unequal society, "the legal right creates rational economic incentives to price the vast majority of consumers out of access." See his essay, "Using Competition Law to Promote Access to Knowledge," in this volume.

67 In the best-case scenario, under license, some are allowed to make a controlled use of the knowledge needed to reproduce goods.

68 Free and unrestricted access by the multitude to a product de facto leads to the dismantling of the rules of valuation and even of the empire of property, since it takes away from property the power that it confers on the owner. See the interesting discussions by Pierre Zaoui in "À propos de quelques paradoxes...," *Mouvements* 11 (October 2007), available on-line at http://www.mouvements.info/A-propos-de-quelques-paradoxes.html (last accessed March 3, 2010).

69 Karl Marx and Friedrich Engels, *The Communist Manifesto*, ed. Samuel H. Beer (Arlington Heights, IL: Harlan Davidson, 1955), p. 10.

70 Moulier Boutang, *Le capitalisme cognitif*, p. 131.

71 *Ibid.*, p. 164.

72 *Ibid.*, p. 152.

73 At the time this article was written, however, U.S. trade negotiations were seriously slowed down. Indeed, the fast-track procedure, which was in effect from 1975 to 1994 and which was restored in 2002, had expired on July 1, 2007. The U.S. trade representative needs to wait for its restoration before starting intensive bilateral negotiations again.

74 Peter Drahos, "BITs and BIPs—Bilateralism in Intellectual Property," *Journal of World Intellectual Property* 4, no. 1 (2001): p. 794, also available on-line at http://www.anu.edu.au/fellows/pdrahos/articles/pdfs/2001bitsandbips.pdf (last accessed May 7, 2009).

75 This is not a new tactic. Drahos recalls that at the time when the developing countries had became the majority within WIPO and were making efforts to promote reforms that would have favored them more than the existing international intellectual property rules, representatives of industry, namely the chairman and president of Pfizer International, did not hesitate to define the organization "as the representative of 'international socialism' when it came to intellectual property rights." Drahos and Braithwaite, *Information Feudalism*, p. 61.

76 Whether it was against the North American Free Trade Agreement (NAFTA), the Free Trade Area of the Americas (FTAA), CAFTA or the FTA between the Andean countries and the US, Latin America has been the scene of important popular political mobilizations against trade agreements and denunciations of neoliberal policies. The Zapatistas, the landless peasant movement in Brazil, the indigenous movements of Bolivia and Ecuador, the unemployed workers' activists in Argentina (*piqueteros*), and many others rallied against FTAs. The issue of the FTAs became a polarizing issue in national political debates, including during presidential elections (Costa Rica, Peru, Ecuador). Demonstrations, massive in some cases, were organized in many countries (Costa Rica, Peru, El Salvador, Guatemala, Bolivia, Colombia,

Ecuador, Honduras). Mobilizations even led to popular referendum in some countries (Colombia, Ecuador, Costa Rica).

77   Brown, *Les habits neufs de la politique mondiale*, p. 142.

78   *Ibid.*

79   Crises that can be attributed directly or indirectly to neoliberal reforms forced on developing countries have been numerous, but of course, these examples pale since the subprime disaster and the subsequent international financial and economic crisis.

80   Latin American movements opposing free-trade agreements are in some cases close to center and left coalitions that are in power and govern. They are critical of neoliberalism and draw their political agendas from the traditional left. Some rely on Communist and socialist ideologies, some try to promote a new left and alternatives to the existing political chessboard, creating in some cases new political parties, for example, the Citizens' Action Party—the Partido Acción Ciudadana—in Costa Rica. However no new alternative ideology presented as such seems to be on the horizon.

81   Harvey, *A Brief History of Neoliberalism*, p. 203.

82   Michel Feher, *S'apprécier: Pourquoi et comment épouser la condition néo-libérale* (Paris: La Découverte, forthcoming).

83   Wendy Brown, *States of Injury: Power and Freedom in Late Modernity* (Princeton, NJ: Princeton University Press, 1995), pp. 5 and 9.

84   Michel Foucault, "Space, Knowledge, and Power," interview by Paul Rabinow, in Paul Rabinow (ed.), *The Foucault Reader* (New York: Pantheon, 1984), p. 245.

85   Brown, *States of Injury*, p. 13.

86   Freedom looks this way insofar as pirates do not reinforce monopoly hegemonies by spreading the standards of those hegemonies and cultivating public taste for them.

87   Brown, *States of Injury*, p. 22.

88   See Lawrence Liang's essay "The Man Who Mistook His Wife for a Book" in this volume.

89   Jacques Rancière, *The Politics of Aesthetics: The Distribution of the Sensible*, trans. Gabriel Rockhill (New York: Continuum, 2004), pp. 32 and 40.

90   I may not be rich enough to buy the software I want in my computer, but because I belong to a university and because my research unit already has bought Microsoft Office, I can get through it a license for only sixty euros. It is more than tempting then to acquire it that way instead of hacking it or using open-source alternatives.

91   See Liang, "Beyond Representation."

92   "On tombe amoureux moins de la personne que du monde possible qu'elle exprime." Maurizio Lazzarato, *Les revolutions du capitalisme* (Paris: Les Empêcheurs de Penser en Rond, 2004), p. 17.

93   In "Beyond Representation: The Figure of the Pirate," Lawrence Liang suggests that we ask "not what piracy is, but what piracy does."

94   As Peter Drahos notes in "'IP World'—Made by TNC Inc.," "Increasingly, criminal-law-enforcement agencies have begun to play a much greater role in enforcement as states have moved down the path of criminalizing the infringement of intellectual property."

95   After all, some of the most innovative forms of production and consumption lately have come more from Indian pirates, the Nigerian movie industry, free-software movements, and Brazilian DJs, rather than from the intellectual property establishment.

96   Feher, *S'apprécier*.

Information/Knowledge
in the Global Society of Control:
A2K Theory and the Postcolonial Commons

Jeffrey Atteberry

The international intellectual property regime has approached a historical, politi-cal, and legal conjuncture whose negotiation will crucially affect the distribution of both knowledge and surplus value in the emerging globalized system of informa-tional capitalism.[1] The access to knowledge movement (A2K) has emerged in this context as a force advocating fundamental changes in the international intellectual property regime. An important, if not central component of A2K theory has been its support for an information commons. As the A2K movement has become global-ized, it has increasingly identified its promotion of an information commons with the interests of the developing world. Many critics, however, have begun to reflect upon various tensions between an A2K model of development and the proposals of other reform movements within international intellectual property circles. The most notable, perhaps, are the latent tensions between A2K and the movement to protect so-called "traditional knowledge."[2] In short, while the concept of the commons implies a certain curtailment of property rights, many proposals for the protection of traditional knowledge imply the creation of new property rights.

These tensions might begin to be resolved, I would suggest, by a critical assess-ment of the concept of the commons. Despite the importance of the information commons to A2K's discourse, however, there has been scant theoretical work on the material and ideological importance of the commons to the historical devel-opment of colonialism. Such a perspective will become ever more important as the A2K movement progressively networks with a number of other movements that have emerged from the legacy of colonialism and that continue the struggle against it in its various new guises. Without a theoretical awareness of how the A2K movement's own discourse is rooted in this history, the movement runs the risk—despite its best intentions—of becoming ideologically conscripted by the forces of informational imperialism, rather than resisting them.

At present, a latent tension lurks between the A2K movement's theoretical and political practices. This tension results, I suggest, at least in part from an insufficiently critical relationship to the historical and social context of the movement's own formation. Socially, the bulk of A2K theory has been articulated from within the Global North, where the transformations attending the emergence of the information economy have reached practically all levels of society. The A2K movement's theoretical resources have, in significant part, been drawn from the liberal traditions of law and political economy. Moreover, the development of A2K theory has been informed by the emergence of open-source software and the changes in the social relations of production wrought by peer-to-peer networking. In short, the production of digital goods presents the possibility of an alternative to the traditional commodity production that has characterized industrial capitalism. At present, however, these transformations also have been centered on the whole in the Global North. As A2K groups have sought to coordinate and consolidate their critiques in a global context, voices emanating from other critical traditions have begun to be heard within the A2K movement, and the movement has politically aligned itself on numerous fronts with various groups from the developing world. In order to forge truly progressive alliances as it enters this global terrain, the A2K movement needs to develop a critical understanding of the ideological roles its own theoretical resources have played in the historical struggle that the movement is now entering.

The A2K movement has come to use "the commons"—a theoretical construct with a long ideological history rooted in the development of capitalist imperialism—as a figure for representing the interests of the developing world. In such situations, Gayatri Spivak tells us that two "irreducibly discontinuous" senses of "representation" must be rigorously tracked: "representation as 'speaking for,' as in politics, and representation as 're-presentation,' as in art or philosophy."[3] In speaking for the subaltern, the First World intellectual necessarily re-presents her as an object of discourse. The subaltern appears in the form of a figure. A danger then lurks in ignoring "this double session of representations," the danger of failing to recognize the figures as "subject formations that micrologically and often erratically operate the interests that congeal the macrologies."[4] Spivak's basic point, despite its philosophically dense expression, is one that is familiar to us in other guises. For example, relationships of representation (such as that between a lawyer and a client) are necessarily fraught with ethical perils. Those dangers are only magnified when (as with class-action lawsuits or certain public-interest legal practices) the representative is in effect self-appointed. In these situations, a panoply of possible subterfuges may subvert the process, despite the best intentions of everyone involved.

The basic lesson is the same when one leaves the confines of the law for the realm of global politics, even if the mechanics of the problem are more complex and the play of forces more subtle. The subterfuge lurks in the figures of representation themselves, that is, the discourse that is used by the self-appointed representative to represent the represented. Without a critical reflection on the ideological character of our own discourse, we risk not understanding how our figures of representation are historically produced within the global structures of power and therefore not understanding how they are potentially recuperable within them. In other words, we run the risk of confusing, in Louis Althusser's terms, the "object of knowledge" with the "real object."[5] These figures are the products of a social and historical reality that stamps them with the ideological imprimatur of their production. As a result, conceptual figures that would appear to be full of critical potential are continually, if not immediately, reappropriated by the dominant ideology and deployed to restore the existing hegemony within a historically new social formation.[6] This process is inevitable, because it is nothing less than the material history of ideologies as such. In the case of the A2K movement, the concern then becomes how the figure of the commons may ideologically operate to reinscribe the movement's efforts to represent the developing world within the ongoing historical development of capitalist imperialism.

The successful negotiation of the current moment therefore requires a historicized discursive analysis of the conceptual terms shaping the present struggle over the international intellectual property regime. In "Can the Subaltern Speak?" Spivak's reading of the *sati* (the ritually self-sacrificing widow) demonstrates how, in legally abolishing the practice, the British produced a figure of the subaltern woman who needed the protection of the British state apparatus. The appearance of the *sati* within the British cultural and legal imagination, Spivak contends, "has a clear and complex relationship with the changeover from a mercantile and commercial to a territorial and administrative British presence" in India.[7] The figure of the *sati* emerged at a moment when the basic structures of the British colonial project were changing. Widow sacrifice was abolished in 1827.[8] At the same moment, the East India Company was in the process of losing its monopoly control over India.[9]

The transition from the colonial rule of the East India Company to the establishment of the British Empire was accompanied by a corresponding shift in the discursive strategies of legitimation, a shift from a rhetoric of profit to one of civilizing humanitarianism and enlightened legal order.[10] The figure of the *sati* was one discursive fulcrum advancing this transition. Through the figure of *sati*, "Imperialism's image as the establisher of the good society is marked by the espousal of the woman as *object* of protection from her own kind."[11] The ideological persistence of this figure (*sati*) and the object of knowledge/protection that it created (the "Third

English colonies (The Granger Collection).

World woman") continues to shape relations between the Global North and South as patriarchal in character. Similarly, in the current moment, the A2K movement represents the object of its protection as the "public domain," which is theoretically conceptualized through the figure of "the commons." The "commons," however, has a complex historical relationship with capitalist imperialism. The critical issue, therefore, is how the persistence of the commons within the A2K movement's theoretical discourse may unintentionally continue the ideological work of historically reproducing the colonizer-colonized relationship in a manner that is consistent with the relations of production that characterize our current globalized order or information imperialism.

HISTORICIZING ENCLOSURE

One of the defining tropes of the A2K movement has come to be "the second enclosure of the commons." Expressing a concern over intellectual property rights encroaching upon territory that had previously been considered safely a part of the public domain, a number of scholars began to employ the figure of an "enclosure of the commons" to discuss this trend.[12] The analogy serves an important critical function in revealing the fundamental role that intellectual property plays in the development of informational capital. At the same time, however, the trope of a new "enclosure of the commons" works a potentially mystifying ideological effect on the A2K movement when it engages the political economy of globalized informational capital. As James Boyle explains, the first enclosure movement, which took place in stages in England from the fifteenth to the nineteenth centuries, was the long historical "process of fencing off common land and turning it into private property."[13] The analogy from the current moment to the historical enclosure of the commons rests upon the observation that "once again things that were formerly thought of as either common property or uncommodifiable are being covered with new, or newly extended, property rights."[14] As a general proposition, this comparison is true enough, but when pursued with a historically informed rigor, the analogy of the "second enclosure movement" reveals deeper ambiguities within the current historical moment with which A2K theory must contend.

A rigorous treatment of the analogy reveals the central role of the current intellectual property regime in the development of the informationalist mode of capitalist production. The English enclosure movement is the subject of some of the most dramatic pages of Marx's *Capital*. Marx describes in great detail the violent processes, both legal and extralegal, through which "the proletariat [was] created by the breaking-up of the bands of feudal retainers and by the forcible expropriation of the people from the soil."[15] In the structure of Marx's argument, the English enclosure movement serves as the paradigmatic example of a more general process that Marx calls "primitive accumulation [*ursprüngliche Akkumulation*]."[16] This process creates the necessary preconditions for the establishment of capitalist relations of production. European society transitioned from feudalism to capitalism through the process of primitive accumulation. As Marx explains, " The capital-relation presupposes a complete separation between the workers and the ownership of the conditions for the realization of their labor. As soon as capitalist production stands on its own feet, it not only maintains this separation, but reproduces it on a constantly extending scale.... So-called primitive accumulation, therefore, is nothing else than the historical process of divorcing the producer from the means of production."[17]

Through the enclosure movement, the feudal peasantry was transformed into the wage-earning proletariat by being dispossessed of their communal property rights in the commons. As such, primitive accumulation creates the basic social relations that form the basis of capitalist production. Marx stresses, however, that the term *ursprüngliche*, which is typically translated not as "primitive" but "original," should not be understood as designating a process that is simply relegated to the prehistory of capitalism. Rather, it is a process that, once begun, capitalism must continually reproduce and repeat on more levels as capitalism territorializes ever new terrain.

In this respect, the "second enclosure movement" is more than an analogy to the first; it is the reproduction of the same historical process on an extended scale as capitalism transitions into an informationalist mode of development and submits the "intangible commons of the mind" to its inexorable logic.[18] Just as the historical enclosure movement transferred real property, which provided the raw materials for production, into the hands of the emerging bourgeoisie, the current expansion of intellectual property protections is having the similar effect of concentrating control over information, which is the raw material for production in the informational age. From patents on the human genome,[19] and from business methods to the Digital Millennium Copyright Act and the Trademark Dilution Revision Act,[20] the scope and strength of intellectual property keeps expanding in ways that separate the means of knowledge production from the greater portion of the people.[21] In deploying the figure of a "second enclosure movement" as a rhetorical way of deploring the expansion of intellectual property rights, therefore, those who use the figure imagine themselves, at least implicitly, on the side of "Marx" against the violent and unjust of effects of subjecting the realm of knowledge production to a capitalist regime of private property. The danger, however, is that this self-image, as long as it remains uncritical, may not end up corresponding in material fact to the position that the A2K movement's theory may otherwise prescribe for it. This risk is nothing other than the ever-present problem of ideological inscription in the strictest Althusserian sense.[22]

Precisely because primitive accumulation is an ongoing process at the heart of the continual reproduction of the capitalist mode of production, Marx does not limit his exposition of primitive accumulation to a historical depiction of the enclosure movement in England. Whereas Part VIII of *Capital*, titled "The So-Called Primitive Accumulation," begins with the enclosure movement, it ends with the chapter titled "The Modern Theory of Colonization." The historical relationship between primitive accumulation and colonialism has been most clearly expressed, however, by Rosa Luxemburg in *The Accumulation of Capital*. Luxemburg stresses that primitive accumulation is continued in "modern colonial policy," where "each new colonial expansion is accompanied, as a matter of course, by a relentless battle of capital against the social and economic ties of the natives, who are also forcibly robbed of their means of production and labor power."[23] Colonialism is linked to enclosure, therefore, as part of a continuous material-historical process endemic to capitalist development. From the perspective of political economy, colonialism is the primitive accumulation of enclosure operating on a globalized scale.

Colonialism is, in fact, doubly articulated with enclosure, for they are related on the ideological level, as well.[24] It is important to remark equally here the history of the ideological role played by the figure of "the commons" in the colonialist form of primitive accumulation. The figure of the commons appeared as early as John Locke's *Second Treatise of Government*, where the "commons" functions as an ideological object of knowledge, rather than as the real object of history analyzed by Marx. For Locke, the commons is a necessary figure for his famous labor theory of value, which posits that a man has property in anything that is "by him removed from the common state Nature placed it in, it hath by this *labour* something annexed to it, that excludes the common right of other men."[25] Moreover, Locke posits that God "hath given the World to Men in common" and confines property-entitling labor to labor that assumes the form of Western agricultural practice.[26] In this way, the ideological and legal apparatus is established for disappropriating land from colonized peoples who were previously at home on it. In short, the native peoples encountered by colonial expansion were denied any property interest in their own lands on the basis of the idea that the Earth was given to humanity in common to be appropriated in the form of private property by those whose social modes of production resembled those of the colonizing Europeans.

With Locke then, the idea of the "commons" provided an ideological basis and justification for colonialist appropriation and enclosure of non-European territory. To this extent, Locke's theory provides the ideological groundwork for the first stage of capitalist imperialism.[27] Once this difference between Marx's material history of the commons and Locke's ideological conception of the commons comes into focus, the A2K movement's invocation of the commons opens two possible paths of theoretical development. On the one hand, A2K theory may be rigorously informed by Marx's material history of the commons and articulate the critical resources necessary to forestall the reproduction of old colonial relations within the new informationalism. Or, on the other hand, A2K may become theoretically beholden to a political economy grounded in Locke's ideological conception of the commons. To the extent that it hews to the former path, the A2K movement promises to be a powerful force in the decolonization of the current imperialism of informational capitalism. To the extent that it falls into the latter rut, however, the A2K movement risks disappointing that promise.

THE (POST)COLONIAL COMMONS

Thus far, A2K theory has primarily found its conceptual resources in the foundational figures of liberal capitalism. When Boyle's essay "The Second Enclosure

Movement," for example, turns from "criticisms of the logic of enclosure," which he says are not enough, to developing "the vocabulary and the analytic tools necessary to turn the tide of enclosure," the conceptual resources are found in the writings of Thomas Babington Macaulay and Adam Smith.[28] That is, the resources that are held out as offering a way toward turning the tide of enclosure are precisely the conceptual resources that are at the very heart of the political economy of capitalist imperialism, which was nothing other than a global regime of disciplinary enclosure. It is unclear how conscious this theoretical choice has been. In any case, whether it is symptomatic or strategic, this theoretical reliance on classical liberalism reflects the fact that A2K theory is materially and historically situated within liberal capitalism. Nevertheless, lest A2K's own development be swept away by that material history, a theoretically critical relationship to A2K's positionality needs to be undertaken.

Thomas Babington Macaulay has a certain fame among intellectual property scholars for pithy comments concerning copyright that he made as a member of Parliament.[29] Macaulay's more enduring historical legacy, however, is as one of the chief architects of the British imperial project in India.[30] From 1834 to 1838, he served in India on the Supreme Council. During this time, he wrote his "Minute on Indian Education," which set the educational policy objective of the British Empire: "We must at present do our best to form a class who may be interpreters between us and the millions whom we govern; a class of persons, Indian in blood and colour, but English in taste, in opinions, in morals, and in intellect."[31] During the same period, he also drafted the Indian Penal Code, which would be adopted and put into force only in 1861, when the British Crown took control of India away from the East India Company after the Sepoy Mutiny.

Given his central role in the formation of the British Raj, therefore, it should come as no surprise that Macaulay's discourse, quoted at length by Boyle, is punctuated with references to the East India Company. In arguing against expanding the term of copyright protection, Macaulay asks his opponents for "any reason why a monopoly of books should produce an effect directly the reverse of that which was produced by the East India Company's monopoly of tea."[32] This rhetorical gesture is posed against the backdrop of a century-long struggle between Whigs and Tories, both within the Parliament and in the East India Company itself, over the role of the company within the political economy of the empire.[33] Consequently, Boyle is right to observe that Macaulay's "free-trade skepticism about intellectual property" was essentially a concern with monopolies in general.[34] Nevertheless, this concern with monopoly was historically rooted in the politics surrounding the East India Company and therefore was ultimately a concern with the political economy of empire.

The terms of this historical debate were largely shaped by the publication of Adam Smith's *The Wealth of Nations* in 1776. As is commonly recognized, one of the central arguments of this historically influential text is that free markets are naturally the most efficient way to structure a market economy and that monopolies, insofar as they are state-enforced limits on free trade, should be viewed with great skepticism. What is less commonly recognized, however, is the extent to which Smith formulates this argument with the policy objective of finding the most efficient way to administer the colonial project politically. Thus, for example, Smith repeatedly makes such statements as "If the colony trade, however, even as it is carried on at present, is advantageous to Great Britain, it is not by means of the monopoly, but in spite of the monopoly."[35] This policy objective is written all over Smith's text, so it is not surprising that it, rather than a concern with intellectual property, should form the heart of the lengthy passage from Smith quoted by Boyle in "The Second Enclosure Movement." The long passage quoted by Boyle comes from an extended meditation upon the limited utility of monopolies in the early stages of a colonizing effort. Smith writes, and Boyle quotes:

> A temporary monopoly of this kind may be vindicated upon the same principles upon which a like monopoly of a new machine is granted to its inventor, and that of a new book to its author. But upon the expiration of the term, the monopoly ought certainly to determine; the forts and garrisons, if it was found necessary to establish any, to be taken into the hands of the government, their value to be paid to the company, and the trade to be laid open to all the subjects of the state.[36]

The reference to what we now call intellectual property rights is ancillary to the thrust of Smith's argument, whose objective is to explain the limited utility of monopolies in the initial stages of colonization. After the monopoly's initial purpose has been served—that is, after the colony has been firmly established—the trade should be "laid open to all the subjects of the state." If one is to take the A2K discourse seriously, then, one must ask to what extent arguments in favor of an unfettered information commons, in opposition to stronger intellectual property rights, reflect the fact that global informationalist capital may be beginning to conceive of itself as exiting the initial mercantile stage of its neocolonial enterprise. In the current transitional moment, intellectual property rights may have already secured informational capital an international foothold in the form of the TRIPS Agreement. Now that "the forts and garrisons" are in place, it may be time to enter into a neoliberal phase of free trade. Of course, it may be objected that the difference between Smith's argument in 1776 and the A2K argument in 2009 is that the public domain is and should be open to exploitation by all, not just "the subjects of the state." Research suggests, however, that this is simply not the case—that the public domain is and will be "exploited asymmetrically" by Western corporate interests.[37] The concern is that A2K's informational commons, in the absence of additional changes in the international intellectual property regime that are yet to be theoretically elaborated, will work to effectuate this transition.

The shift from the mercantilist phase to a free-trade phase is, as Smith and the A2K movement both argue, necessitated by reasons of efficiency. The question remains: "Efficient" at what, exactly? In keeping with the classical liberalism inherited from Smith, the current neoliberal school of thought would answer the question with the reply "efficient at allocating resources between market actors by making them internalize their externalities." Such an answer merely begs the question insofar as participatory membership within the market is precisely what is at issue in the globalized struggle. The classic economic response takes the market and membership in it for granted. It does not reach the real issue in at least two structurally related ways. First, global capital has never seriously contemplated internalizing the externalities of colonialism, which is to say the tremendous burdens that the history of colonialism has imposed on the "developing world." Second, it fails to contend with the extent to which the market comes into being at all through the creation of a more radical externality — call it "subalternity" — that cannot be represented within the market at all.[38] The logic of the market, like all logical systems, is founded upon a constitutive exclusion that cannot be represented within the system, but the exclusion does, nevertheless, manifest effects within the system in the form of a trace that marks the system's own internal limit. The subaltern, as "the absolute limit of the place where history is narrativized into logic," traces the limit where capitalist economic logic fails to grasp the necessary historical condition of its own development and continued survival.[39] In the context of today's informational capitalism, TRIPS, for example, purports to integrate the developing world into the system of global capital, yet it is able to do so only to the extent that the intellectual property regime imagines the developing world as a subject of property rights. Similarly, A2K theory aims to address this particular problem by proposing an economic model of the commons in which the developing world would have greater access to informational goods. However, such an integration takes place again only to the extent that the developing world is imagined as having access to the finance capital, social capital, and human capital that is required to make such access economically productive. In both instances, both economic theory and the legal regime necessarily fail to account for the foundational act of expropriation that is required for the capitalist production of surplus value. This process of constructing the capitalist free market through the production of subalternity is, once again, what Marx calls primitive accumulation.

Smith's antimonopolistic argument is aimed, therefore, not only at achieving efficiency in wealth distribution within the functioning of the capitalist market, but at achieving efficiency in the production of those markets through primitive accumulation. This objective is starkly manifest, for example, when Smith writes:

> In Africa and the East Indies, therefore, it was more difficult to displace the natives, and to extend the European plantations over the greater part of the lands of the original inhabitants. The genius of exclusive companies, besides, is unfavourable, it has already been observed, to the growth of new colonies, and has probably been the principal cause of the little progress which they have made in the East Indies.[40]

Progress here is plainly defined as clearing "the natives" from their land. Where these "natives" are well entrenched, however, monopolies do not do the best job of it. As those who have lived through colonialism know, being denied monopoly privileges is not the only way of being excluded from the market. There are, indeed, more efficient means of exclusion, more thorough forms of material alienation. Even within the developed world, monopolies are not necessary to keep the bulk of the working class alienated from the prevailing means of production. Economic barriers to market entry do just fine. This has been all the more true for those on the periphery of the world system, where the regime of liberal free trade ephemerally appeared in the global transition to state monopoly capitalism.[41] The arguments for a global commons, therefore, may in effect clear the way for a more efficient capitalist territorialization of the informational terrain.

CONTROL IN THE INFORMATION SOCIETY

In response to this historicized argument, a rigorous line of reasoning in A2K theory would likely insist that important differences between informational goods and more traditional industrial goods are being overlooked. There are two major differences. First, informational goods are "nonrival," which means that their consumption by one user does not preclude their use by another. As a result, the marginal cost of producing informational goods approaches zero.[42] Second, informational goods are both inputs and outputs of their own production. Since innovation is cumulative, the reduction in the productive consumption of informational goods as a consequence of intellectual property protection results in externalized social costs in the form of reduced innovation.[43] Ignoring these differences conflates intangible and tangible goods, the argument would claim, and results in conflating the different historical moments of enclosure. Informational resources are different from more tangible resources, so a global information commons would operate differently than classic free-market liberalism under colonial rule because the market of informational capitalism differs so fundamentally from the market of industrial capitalism.

By focusing in this way on the microeconomics of informational goods, A2K theory has shown that the current distribution of both information and the benefits it brings cannot be rectified until the iron grip of maximalist intellectual property protection is broken. The formation of an information commons may prove to be a necessary condition for unleashing the productive potential of informational capitalism. Current A2K theories demonstrate not only why this is so, but why this is an economic imperative that may well be irresistible as the social conditions of production become increasingly networked. "Information wants to be free," as the slogan goes.[44] To the extent, however, that A2K discourse has tended to adopt an uncritical discourse of freedom, its theoretical project risks replacing a material conception of the commons with an ideological one. It must not be forgotten that capitalist liberal ideology has always advanced under the sign of "freedom." The process of primitive accumulation, for example, frees the peasant from the land, transforming him into "a free worker...free in the double sense that as a free individual he can dispose of his labor-power as his own commodity, and that on the other hand, he has no other commodity for sale."[45] Similarly, British imperialism—like all the others, past and present—ideologically justified itself by asserting that "under the charm of this beneficent spirit the chief colonial establishments of Great Britain have already achieved substantial freedom."[46] Indeed, the danger for the A2K movement grows only stronger as it increasingly engages with the contemporary iterations of the historical struggle against imperialism.

The microeconomic differences between tangible and intangible goods can be overstated to the extent that existing A2K theory does not adequately contend with the fact that the emerging market of informational capitalism is situated within a world system whose basic structures remain the same as they were under the imperialism of industrial capitalism. The unique microeconomic characteristics of informational goods, which A2K has so systematically developed on the micro level, may also necessitate the macrological development of a new and different apparatus for capturing the wealth that informational capitalism produces. The commons clearly plays an integral role in this new apparatus, just as it did in the old one. The role of the commons in both cases is to clear a space of raw goods open to expropriation by the creative labor of man, which is in fact not determined by either work or creativity, but by access to the means of production. The problem, therefore, becomes how an information commons can best be constructed in order to distribute the flow of wealth in a way that does not reproduce the existing neocolonial order. The information commons may be a necessary condition for unleashing the productive potential of informational capitalism, but it may not be a sufficient condition for realizing the liberating potential of that increased productivity.

Perfectly aware of the potentially duplicitous character of bourgeois freedom, Gilles Deleuze and Félix Guattari have described the process of capitalism's development in terms of a "generalized decoding of flows" and a dynamic of "deterritorialization."[47] The current informationalist regime and the discourses attending to it—including that of the A2K movement—fit nicely within Deleuze and Guattari's paradigm. The informationalist mode of production represents a new order of decoded flows. The freeing of information promises to restructure the relations of production, replacing vertically integrated structures of production with horizontally networked ones.

The terrain of social production is being reterritorialized. The information that is struggling to be free continues, however, to circulate within a very material geography. While the circuits of production have become networked, the nodal points of the network continue to be places such as New York, London, and Tokyo.[48] The creation of a global information commons may even render the networks of production slightly more flexible, facilitating the integration of places such as São Paulo, Bangalore, Mumbai, and Shanghai. The deterritorialization that results from the networked production of globalized informational capitalism is necessarily accompanied, however, by a corresponding reterritorialization.[49] While a few postcolonial metropolises may enter the network, billions of impoverished people around the world will remain off the grid. "These neoterritorialities," Deleuze and Guattari write, "are often artificial, residual, archaic; but they are archaisms having a perfectly current function."[50] In the case of the emerging territorialization of the Earth under informational capitalism, the archaisms of the colonialist world order threaten to reassert themselves with a vengeance. The solution to this problem of the continued colonialist distribution of wealth, therefore, will not be found simply in an information commons, although an information commons will surely have an important role to play. When faced with the dynamic of deterritorialization, Deleuze and Guattari ask, could it be that the revolutionary path is to "go further still, that is, in the movement of the market, of decoding and deterritorialization? For perhaps the flows are not yet deterritorialized enough."[51] The questions for the A2K movement then become what function would the information commons serve in the globalized economy, and how might we accelerate the process that it promises by finding ways to resist its potentially neocolonial reterritorializations?

While the specific economic behavior of the information commons may differ from any kind of commons we have previously seen, the question remains whether its basic function within a capitalist world system would be fundamentally any different. A rigorous approach to the question must situate the A2K movement's information commons not only within informational capitalism, but also within the emerging society of control.

In "Postscript on the Societies of Control," Gilles Deleuze argued that society was transitioning from a disciplinary society, which had been the object of Michel Foucault's classic works, to a different mode of social organization. According to Deleuze, disciplinary society organizes society in and through "spaces of enclosure." The paradigmatic spaces of enclosure would include prisons, hospitals, factories, schools.[52] Within these spaces of enclosure, the exercise of power is static, discontinuous, and rigid. In contrast, societies of control construct open spaces where power is exercised in a dynamic, continuous, and flexible manner.[53] In a further elaboration of his terms, Deleuze schematically links each type of society with a particular kind of machine that exemplifies the respective social relations of production: The disciplinary societies function through "machines involving energy" such as the steam engine or nuclear reactors; societies of control operate with "computers." Furthermore, the technological shift that characterizes the transition to the society of control is itself the function of a "mutation of capitalism."[54] This mutation is nothing other than the emergence of informational capitalism.

The so-called "new enclosures" operate within the logic of disciplinary societies. The open access of the information commons, on the other hand, exhibits an organizing structure that typifies the society of control. The regulation of information production under a commons regime functions according to a principle of open access, which reflects changing social relations of production that have themselves become more elastic and variable. As such, a commons-oriented regime, in contrast to the current intellectual property regime, would be more suited to the changing conditions of social production under informational capitalism.

Our current historical moment, then, is one of transition. In the realm of political economy, the transition appears as the passage from industrial to informational capitalism. At the level of the social organization of power, it takes the form of the passage from disciplinary societies to societies of control.

The current legal debate between maximalist protection and access to knowledge emerges as a symptom of this transition from discipline to control. "If our law is hesitant, is itself in crisis," Deleuze writes, "it's because we are leaving one in order to enter into the other."[55] The transitions from industrial to informational capitalism and from discipline to control appear as parallel historical changes, and the discourse of the A2K movement is situated within these passages.

Despite the rhetoric of manifest liberation that attends each of these transitions, however, they should not be narrativized as movements of either liberation or oppression. Given its tendency to adopt such a binary rhetoric, the A2K movement needs to come to the broader recognition that "there is no need to ask which is the toughest or most tolerable regime, for it's within each of them that liberating and enslaving forces confront one another."[56] The theoretical debate within the

A2K movement has sometimes been framed as if legal policy makers were faced with a choice between an enslaving regime of enclosure and a liberating regime of the commons. The revolutionary zeal shown by some within the A2K movement must be accompanied by the sober realization that informational capitalism is no less capitalism than industrial capitalism, with all the corresponding hopes and fears. When approached from within this framework, the information commons is not necessarily a force of either liberation or enslavement. Rather, the terrain of information that the information commons territorializes becomes, like land and labor before it, a site of struggle.

INFORMATION/KNOWLEDGE AS AN APPARATUS OF CAPTURE

In a variation on Marx's "trinity formula,"[57] Deleuze and Guattari develop a model of what they call the "apparatus of capture."[58] The trinity formula consists of land/rent, labor/wages, and capital/profit. This formula encapsulates the transformation of basic material elements into the "historically specific social forms" of capitalist relations of production that generated wealth in the mid-nineteenth century.[59] For example, under the social relations of private property, land is transformed into a source of wealth in the form of rent.[60] Deleuze and Guattari characterize each of the constitutive parts of the trinity formula as an apparatus of capture insofar as each functions as a means of generating and capturing flows of wealth.[61] Unfortunately, Deleuze and Guattari do not remark the historical specificity of these apparatuses, which Marx on the other hand was always careful to do. When properly historicized, however, Marx's trinity formula displays the basic features of disciplinary modes of social ordering. In particular, they all proceed by means of enclosure. Land is enclosed within private property to produce rent, labor is enclosed within the factory to produce wage labor, and capital is enclosed within circuits of production and consumption to produce profits.

As capitalism transitions into its current informationalist stage, however, new social relations of production are bound to emerge. Not only are these new relations of production themselves produced by informational capitalism, but they are also constitutive of the emerging society of control. Moreover, because information exhibits economic characteristics that differ significantly from traditional material goods, informational capitalism will necessarily develop new means of seizing, controlling, and distributing the wealth that it creates through these new relations of production. Informational capitalism will have developed a new apparatus of capture commensurate with both the economic characteristics of informational goods and the social characteristics of control. Alongside land/rent, labor/wages, capital/profit, a new apparatus, of information/knowledge, appears. The information commons will undoubtedly play an integral part in the operation of this information/knowledge apparatus.

An information commons in the context of a world system organized under a regime of control forms an integral component in information/knowledge as a new apparatus of capture. Every apparatus of capture operates by creating a "general space of comparison" that serves as the necessary condition for "a mobile center of appropriation."[62] This general space of comparison is created through a formal analytic category that abstracts away all particularities in the objects of comparison and thus creates a baseline against which differences can be measured. In the case of labor/wages, for example, the space of comparison is created by the category of "labor-power."[63] In Marx's labor theory of value, the price of labor ("wages") is determined not by the value produced by the laborer, but by the value of the laborer's labor power, which is nothing more than the capacity of the laborer to work, irrespective of any differences in the relative productivity of that labor. The fair-market value of labor power, like all other commodities, is determined by the costs of its production. The cost of producing labor power therefore is simply "the average amount of the means of subsistence necessary" to reproduce the kind of labor in question in that particular society at that particular historical moment.[64] The category of labor power thus creates a space of comparison that enables the capitalist who purchases labor power to appropriate the difference between the value of the labor power purchased and the value produced by its exertion. Thus, while labor in itself is without "value," it appears as the source of value by virtue of being valorized through the social relations of production in which the laborer offers his labor power for sale in the labor market.[65] The laborer must sell his labor power, rather than productively consume it himself, precisely because he has been separated from the means of production that are the necessary preconditions for its productive consumption. At root, then, the difference in value between labor power and its exertion is predicated upon the relative difference between the capitalist and the worker in their respective capacities to consume labor power productively.[66]

The information commons would likely operate as an abstract space of comparison for capitalizing on differences in the relative capacities of the Global North and South to consume the information commons productively. Just as labor or the Earth are in themselves without value prior to their valorization by their respective apparatus of capture, so too is information. Information in abstract isolation from informational capitalism has no "value" in the market-economy sense. The information commons, however, creates a space of comparison for information—as labor power does for labor—wherein information becomes valorized as a source of capital by exploiting the difference in the capacity to consume information productively in the creation of knowledge goods. These differences are the legacy of the uneven development resulting from the historical process of imperialist expansion of monopoly capital, and the global "digital divide" is the material manifestation of these differences in the capacity for productive consumption in the age of informational capitalism.[67]

Under these circumstances and absent any measures to redress them, the existing colonialist relations of production characteristic of the world system are likely to remain largely intact as informational capitalism expands globally. That is, information as one of the factors of production will appear to be free in the commons. Nevertheless, the other factors required for the productive consumption of information—such as computers and communication networks, not to mention social and cultural capital—will remain unevenly distributed between the core and the periphery. Consequently, the nations currently at the periphery of the world system will continue to provide raw materials to the information commons in such forms as genetic resources and "traditional knowledge," while the core nations will enjoy a relative monopoly on the productive consumption of information.[68]

Differences in capacity to consume information productively may also be framed, in the context of finance capital, as differences in so-called human and social capital.[69] The capacity to consume information productively requires not only a material infrastructure in the form of computerized networks, but also resources of both social capital, such as having access to the research networks in which the information circulates, and human capital, such as educational know-how. By defining the conditions that determine which kinds of social and human capital matter, colonial discipline has historically and routinely set up these kinds of disparities and then exploited the resulting differences. This technique was an integral component of the process of colonial primitive accumulation. In India, for example, the British passed the Indian Forest Act of 1865, which established a legal mechanism for claiming commonly held forest land as governmental land.[70] Under the act, Forest Settlement Officers were required to publish notices in official gazettes of the government's intention to claim forest land. Individuals who wished to make a property claim on the land were required to do so within three months of notice publication. This method of legal disappropriation established a procedure that capitalized on differences in social and human capital. First, by taking advantage of the fact that colonized peoples living in the affected forests were not likely readers of the official gazettes, it exploited a difference in social capital. Second, the affected individuals were not likely readers of the gazette in part because they are unlikely to be educated in English. Thus, the procedure also exploited a difference in the types of human capital that the social conditions of production under British imperialism deemed to be significant.

Insofar as the end result of this process was a property claim in the forest held by the colonial government, this historical example of the colonial construction and exploitation of differences in human and social capital analogizes nicely with the contemporary role of such differences in the current international intellectual property regime. In both this historical example and the current intellectual property regime, the entire process of disappropriation followed by enclosure is consistent with social forms of discipline that are static and discontinuous. Under these conditions, the procedure for disappropriation need operate only once, because the result is legally consolidated by the property enclosure. Simply scaling back the extent of enclosure, however, only cuts the process short without reaching the initial means of disappropriation. Indeed, in the present regime of social control, the moment of enclosure is no longer needed, for the process of disappropriation is now dynamic and continuous. The process of decolonization itself may be read as a transition on the geopolitical terrain from the disciplinary enclosures of colonialism to a neocolonialism of social control.[71]

A similar moment may have arrived in the evolution of informational capitalism, when the disciplinary logic of enclosure can give way without necessarily altering the fundamental relations of power. The fact that the information commons would not be subject to enclosure does not necessarily mean that it would not function as a key component of the information/knowledge apparatus of capture in a new regime of information imperialism.

CONCLUSION

A2K theory must abandon the comfort of tidy binary oppositions where a theoretical stance against enclosure and in favor of the commons necessarily implies that one is similarly positioned against emerging information imperialism. As the A2K movement continues to unfold its theoretical and political practice, its underlying critical theory must incorporate this fundamental realization. Positionality is not determinable by a single theoretical position; rather, positionality is determined by a network of historical and material relations. If A2K is to become an effectively globalized movement, it will necessarily have to accommodate a number of different voices, each with their own positionality. The positionality of the law professor in the United States is not the same as that of the local grassroots organizer in India, and the positionality of the governmental agent in Uganda is different still. Many have come to perceive in the A2K movement an important element in the ongoing struggle against the legacies of colonialism, and the movement has found itself propelled toward alliances with other movements from the formerly colonized world in large part because of a shared opposition to what has been successfully characterized as a new historical moment of enclosure. Nevertheless, if effective alliances are to be made, the A2K movement must collectively develop a critical theory with the flexibility and sophistication necessary to articulate a theoretical and political practice that can account for the multiple valences that are operative within any single position.

A critical interrogation of the limits of the commons doctrine is a good place to start. Given, as we have seen, that an uncritical version of the commons may well end up serving the long-term interests of informational imperialism, the A2K movement's commitment to the commons should not, for example, necessarily imply a dogmatic opposition to any attempt on the part of the developing world to withdraw some informational goods from the commons, whether in the form of "traditional knowledge" or otherwise. With a rigorous critical theory guiding its theoretical and political practice, the A2K movement may well then become an indispensable component of the ongoing global struggle against imperialism.

NOTES

1   On the notion of "conjuncture," see Louis Althusser, *For Marx*, trans. Ben Brewster (New York: Verso, 2005). "Informational capitalism" is Manuel Castells's term for the mode of production, based upon technological innovation and the production of knowledge as surplus value, that has developed since the restructuration of global capital since the 1980s. See Manuel Castells, *The Rise of the Network Society*, 2nd ed. (Oxford: Blackwell, 2000), pp. 18–21. The varied attempts to trace the contours of this epochal change in the character of capitalist production have resulted in many different names for this new mode of production. In addition to Castells's "informational capitalism," Michael Hardt and Antonio Negri's notion of "immaterial production" provides a constructive model for conceptualizing the issues at stake in this paper. See Michael Hardt and Antonio Negri, *Multitude* (New York: Penguin, 2004), pp. 140–57 and Michael Hardt and Antonio Negri, *Empire* (Cambridge, MA: Harvard University Press, 2000), pp. 22–24. Another theoretical formulation along the same lines may be found in Yann Moulier-Boutang's "cognitive capitalism." See Yann Moulier-Boutang, *Le capitalisme cognitif: La nouvelle grande transformation* (Paris: Éditions Amsterdam, 2007). Finally, the notion of the postindustrial society was pioneered by Alain Touraine. See, for example, Alain Touraine, *The Post-Industrial Society* (New York: Random House, 1971).

2   The work of Anupam Chander and Madhavi Sunder has been highly instructive on this issue. See in particular, Anupam Chander and Madhavi Sunder, "The Romance of the Public Domain," *California Law Review* 92 (2004): pp. 1331–73.

3   Gayatri Spivak, "Can the Subaltern Speak?" in Cary Nelson and Lawrence Grossberg (eds.), *Marxism and the Interpretation of Culture* (Urbana: University of Illinois Press, 1988), p. 275.

4   *Ibid.*, p. 279.

5   Louis Althusser, *Reading Capital*, trans. Ben Brewster (London: Verso, 1997), pp. 40–45.

6   "Ideology changes therefore, but imperceptibly, conserving its ideological form; it moves, but with an immobile motion which maintains it where it is, in its place and its ideological role." *Ibid.*, p. 142. Michael Hardt and Antonio Negri provide a vivid example of this process when they discuss how "hybridity," which would appear to serve a critical function within a "disciplinary society," ultimately provides the basis for a subjectivity that is commensurate with the current "society of control" that characterizes the flexible capitalism of a networked society. See Hardt and Negri, *Empire*, pp. 142–46. For a discussion of disciplinary and control societies, see Gilles Deleuze, "Postscript on the Societies of Control," *October* 59 (1992): pp. 3–7. For an analysis of why the claims for inscribing traditional knowledge within the public domain works as a further disappropriation of local populations, see Carlos M. Correa's essay, "Access to Knowledge: The Case of Indigenous and Traditional Knowledge," in this volume.

7   Spivak, "Can the Subaltern Speak?" p. 298.

8   *Ibid.*, p. 297.

9   This process ran from 1813 to 1833. See C. H. Philips, *The East India Company: 1784–1834* (Manchester, UK: Manchester University Press, 1961); Karl Marx, "The East India Company — Its History and Results," in *Collected Works*, vol. 12 (New York: International Publishers 1979), p. 153.

10 See, for example, Antony Anghie, *Imperialism, Sovereignty, and the Making of International Law* (New York: Cambridge University Press, 2004), p. 69.

11 Spivak, "Can the Subaltern Speak?" p. 299.

12 The use of the figure can be traced back to David Lange's seminal work. See David Lange, "Recognizing the Public Domain," *Law and Contemporary Problems* 44, no. 4 (Autumn 1981): pp. 147–78. The figure began to gain wide circulation, however, in the 1990s. See Keith Aoki, "Authors, Inventors and Trademark Owners: Private Intellectual Property and the Public Domain," *Columbia Journal of Law and Arts* 18 (1993): pp. 191–267. See also Yochai Benkler, "Free as the Air to Common Use: First Amendment Constraints on Enclosure of the Public Domain," *New York University Law Review* 74, no. 2 (May 1999): pp. 354–446. In the years since, the notion of "enclosure" has become commonplace in the literature. A recent use of the figure in an international context has been made by Peter Yu. See Peter Yu, "The International Enclosure Movement," *Indiana Law Journal* 82, no. 4 (Winter 2007): pp. 827–907.

13 See James Boyle, "The Second Enclosure Movement and the Construction of the Public Domain," *Law and Contemporary Problems* 66, nos. 1/2 (Winter/Spring 2003): pp. 33–34, available on-line at http://www.duke.edu/pd/ papers/boyle.pdf (last accessed May 20, 2009).

14 *Ibid.*, p. 37.

15 Karl Marx, *Capital*, vol. 1, trans. Ben Fowkes (New York: Penguin, 1990 p. 896.

16 *Ibid.*, pp. 873–76.

17 *Ibid.*, pp. 874–75.

18 Moulier-Boutang asserts that "cognitive capitalism is in its phase of primitive accumulation in the sense that the collection of property rights put in place between the seventeenth and nineteenth centuries, which form the rational basis of classical political economy . . . now constitute an insurmountable barrier to the inscription of the potential development of the productive forces of human activity in a way that follows a trajectory of regular growth and that effects an institutional compromise with the forces of the old economy." Yann Moulier-Boutang, "Richesse, propriété, liberté et revenu dans le 'capitalisme cognitif,'" *Multitudes* 5 (2001): pp. 20–21 (my translation).

19 See Rebecca S. Eisenberg, "Patenting the Human Genome," *Emory Law Journal* 39, no. 3 (Summer 1990): pp. 721–45.

20 See *State St. Bank & Trust Co. v. Signature Fin. Group*, 149 F.3d 1368 (D.C. Cir. 1998); see also John R. Thomas, "The Patenting of the Liberal Professions," *Boston College Law Review* 40, no. 5 (September 1999): pp. 1139–85.

21 Boyle, "The Second Enclosure Movement," pp. 37–41.

22 As Althusser understands it, "what is represented in ideology is therefore not the system of real relations which govern the existence of individuals, but the imaginary relation of those individuals to the real relations in which they live." Louis Althusser, *Lenin and Philosophy*, trans. Ben Brewster (New York: Verso, 1971), p. 165. Ideology works a second-order distortion in our representation of the real state of affairs. Ideology does not simply misrepresent the real relations that structure a given moment, but misrepresents our relation to those relations. In ideology, one misperceives where one stands in the real macrological struggle.

23 Rosa Luxemburg, *The Accumulation of Capital*, trans. Agnes Schwarzchild (New York: Routledge, 2003), p. 350.

24 For a reading of Daniel Defoe's *Robinson Crusoe* as a seminal text in the creation of an ideological fantasy of the commons and its constitutive role in the creation of colonialist culture, see Robert P. Marzec, "Enclosures, Colonization, and the *Robinson Crusoe* Syndrome: A Genealogy of Land in a Global Context," *boundary 2* 29, no. 2 (Summer 2002): pp. 129–56.

25 John Locke, *Two Treatises of Government* (New York: Cambridge University Press, 1988), p. 288.

26 *Ibid.*, pp. 286 and 290.

27 See Herman Lebovic, "John Locke, Imperialism, and the First Stage of Capitalism," in *Imperialism and the Corruption of Democracies* (Durham, NC: Duke University Press, 2006).

28 Boyle, "The Second Enclosure Movement," p. 52.

29 See, for example, Benkler, "Free as the Air," p. 355, quoting Macaulay's quip that copyright is "a tax on readers for the purpose of giving a bounty to writers."

30 See Eric Stokes, *The English Utilitarians and India* (Oxford: Oxford University Press, 1959), pp. 184–219.

31 Thomas Babington Macaulay, "Minute on Indian Education," in Bill Ashcroft, Gareth Griffiths, and Helen Tiffin (eds.), *The Post-Colonial Studies Reader* (New York: Routledge, 1995). This text has become a key reference for postcolonial studies. See, for example, Gauri Viswanathan, *The Masks of Conquest: Literary Study and British Rule in India* (New York: Columbia University Press, 1989). See also, Homi Bhabha, *The Location of Culture* (New York: Routledge, 1994), pp. 93–101.

32 Boyle, "The Second Enclosure Movement," p. 54, quoting Thomas Babington Macaulay, "A Speech Delivered in the House of Commons (Feb. 5, 1841)," in *The Life and Works of Lord Macaulay*, vol. 7 (London: Longmans, Green, and Co. 1897), p. 201.

33 See Bruce G. Carruthers, *City of Capital: Politics and Markets in the English Financial Revolution* (Princeton, NJ: Princeton University Press, 1999). See also, Timothy L. Alborn, *Conceiving Companies: Joint-Stock Politics in Victorian England* (New York: Routledge, 1998).

34 Boyle, "The Second Enclosure Movement," p. 55.

35 Adam Smith, *The Wealth of Nations* (New York: Modern Library, 1965), p. 575.

36 Boyle, "The Second Enclosure Movement," p. 55 n.5, quoting Smith, *Wealth of Nations*, p. 712.

37 This is a central argument of Anupam Chander and Madhavi Sunder in "The Romance of the Public Domain."

38   The concept of "externality" invoked here, on the other hand, comes from the literature of law and economics. The term "externality" designates a cost or burden that results from a particular use of a resource that the user fails to account for because it falls upon others. The famous Coase Theorem argues that externalities tend to lead to the inefficient use of resources. See Ronald Coase, "The Problem of Social Costs," *Journal of Law and Economics* 3, no. 1 (October 1960): pp. 1–44. Consequently, the best way to facilitate efficient outcomes is to create regimes in which the cost associated with a particular use of a resource reflects the burdens that such a use imposes upon others. In other words, efficiency requires the "internalization of externalities." See *Ibid*. See also Harold Demsetz, "Toward a Theory of Property Rights," *American Economic Review* 57, no. 2 (May 1967): pp. 347–59 and Garrett Hardin, "The Tragedy of the Commons," *Science* 162 (December 1968): pp. 1243–48. On the macroeconomic level, the capitalist world system remains highly inefficient—and unjust—insofar as its use of resources creates tremendous burdens upon the colonized Global South. The status of the subaltern then marks the constitutive limit of the capitalist system in the sense that it represents the point at which the externalities of capitalism as a world system cannot be internalized while still remaining capitalism. The subaltern thus represents the constitutive externality of the capitalist world system.

39   Gayatri Chakravorty Spivak, *In Other Worlds: Essays in Cultural Politics* (New York: Routledge, 1988), p. 207.

40   Smith, *Wealth of Nations*, p. 599.

41   I use the term "world system" in a manner consistent with its development by Immanuel Wallerstein. See, for example, Immanuel Wallerstein, *World-Systems Analysis: An Introduction* (Raleigh, NC: Duke University Press, 2006). For a history of the ephemeral era of free-trade liberalism, see Eric Hobsbawm, *The Age of Capital* (New York: Vintage 1996).

42   The classic exposition of these market dynamics is Kenneth Arrow's "Economic Welfare and the Allocation of Resources for Invention," in *The Rate and Direction of Inventive Activity* (Princeton, NJ: Princeton University Press, 1962), pp. 609–26. For their exposition within A2K theory, see Yochai Benkler, *The Wealth of Networks: How Social Production Transforms Markets and Freedom* (2006), available on-line at http://cyber.law.harvard.edu/wealth_of_networks/Main_Page (last accessed May 23, 2009).

43   The bibliography here is extensive. See Suzanne Scotchmer, "Standing on the Shoulders of Giants: Cumulative Research and the Patent Law," *Journal of Economic Perspectives* 5, no. 1 (Winter 1991): pp. 29–41; Robert P. Merges and Richard Nelson, "On the Complex Economics of Patent Scope," *Columbia Law Review* 90, no. 4 (May 1990): pp. 839–916; Michael A. Heller and Rebecca S. Eisenberg, "Can Patents Deter Innovation?: The Anticommons in Biomedical Research," *Science* 280, no. 5364 (May 1, 1998): pp. 698–701; Michael A. Heller, "The Tragedy of the Anticommons: Property in the Transition from Marx to Markets," *Harvard Law Review* 111, no. 3 (January 1998): pp. 621–88; and Benkler, *The Wealth of Networks*.

44   See, for example, John Perry Barlow, "Selling Wine without Bottles: The Economy of the Mind on the Global Net," *Wired* 2.03 (1993), p. 86, available on-line at http://homes.eff.org/~barlow/EconomyOfIdeas.html (last accessed May 23, 2009). See also James Boyle, "Foucault in Cyberspace: Surveillance, Sovereignty, and Hardwired Censors," *University of Cincinnati Law Review* 66, no. 1 (Fall 1997): pp. 177–206.

45   Marx, *Capital*, vol. 1, p. 272; see also, Karl Marx, *Grundrisse*, trans. Martin Nicolaus (New York: Penguin, 1993), pp. 239–50.

46 J. A. Hobson, *Imperialism: A Study* (London: James Nisbet, 1902), p. 119, quoting Henry Crittenden Morris, *The History of Colonization from the Earliest Times to the Present Day*, vol. 2 (Norwood, MA: Norwoood Press, 1900), p. 80.

47 Gilles Deleuze and Félix Guattari, *Anti-Oedipus*, trans. Robert Hurley, Mark Seem, and Helen R. Lane (Minneapolis: University of Minnesota Press, 1983), p. 224.

48 See Saskia Sassen, *The Global City: New York, London, and Tokyo* (Princeton, NJ: Princeton University Press, 2001).

49 See Deleuze and Guattari, *Anti-Oedipus*, p. 258 n.47, where they state that "it may be all but impossible to distinguish deterritorialization from reterritorialization, since they are mutually enmeshed, or like opposite faces of one and the same process."

50 *Ibid.*, p. 257.

51 *Ibid.*, p. 239.

52 Deleuze, "Postscript," pp. 3 and 4. These examples have their origins in Foucault's work, which focuses on the modes of social organization that take place within and beneath the level of the nation-state. See, for example, Michel Foucault, *Discipline and Punish: The Birth of the Prison*, trans. Alan Sheridan (New York: Vintage Books, 1995); and Michel Foucault, *The Birth of the Clinic: An Archaeology of Medical Perception*, trans. A. M. Sheridan Smith (New York: Routledge, 2003).

53 Deleuze, "Postscript," p. 4.

54 *Ibid.*, p. 6

55 *Ibid.*, p. 5.

56 *Ibid.*, p. 4.

57 Karl Marx, *Capital*, vol. 3, trans. David Fernbach (New York: Penguin, 1991), pp. 953–70.

58 Gilles Deleuze and Félix Guattari, *A Thousand Plateaus*, trans. Brian Massumi (Minneapolis: University of Minnesota Press, 1987), pp. 437–48.

59 Adam Smith had already identified rent, profit, and wages as the "original sources of revenue." See Smith, *Wealth of Nations* p. 52 n.36. The difference between Smith and Marx here is that Smith fails to perceive that these original sources of revenue are not originally productive at all, but social structures of capitalist production that merely capture and distribute the wealth produced by living labor. See Karl Marx, *Capital*, vol. 2, trans. David Fernbach (New York: Penguin Books, 1992), pp. 438–65.

60 Marx, *Capital*, vol. 3, pp. 953–57.

61 Deleuze and Guattari, *A Thousand Plateaus*, pp. 440–44.

62 *Ibid.*, p. 444.

63 See Marx, *Capital*, vol. 1, pp. 270–80 and 675–82; Marx, *Grundrisse*, pp. 359–64.

64 Marx, *Capital*, vol. 1, p. 275.

65 *Ibid.*, p. 677.

66 *Ibid.*, p. 717. A similar structure characterizes land/rent as an apparatus of capture. The category of landed property creates a space of comparison against which the relative productivity of land is measured, which in turn serves as the basis of differential rent. See Deleuze and Guattari, *A Thousand Plateaus*, p. 441; Marx, *Capital*, vol. 3, pp. 779–823.

67 See, for example, Pippa Norris, *Digital Divide: Civic Engagement, Information Poverty, and the Internet Worldwide* (New York: Cambridge University Press, 2001).

68 See Wallerstein, *World-Systems Analysis*, pp. 25–26. This geopolitical division of productive consumption and passive consumption was itself the basis of the mercantile system. See

Marx, *Capital*, vol. 3, p. 139.

69 It is not incidental that these notions have moved to the center of the World Bank's developmental policies. See, for example, Ben Fine, Costas Lapavitsas, and Jonathan Pincus (eds.), *Development Policy in the Twenty-first Century: Beyond the Post-Washington Consensus* (New York: Routledge, 2001).

70 See Chhatrapati Singh, *Common Property and Common Poverty: India's Forests, Forest Dwellers and the Law* (New Delhi: Oxford University Press, 1986), p. 11.

71 See Kwame Nkrumah, *Neo-Colonialism, The Last Stage of Imperialism* (New York: International Publishers, 1966).

Beyond Representation: The Figure of the Pirate

Lawrence Liang

> In civilizations without boats, dreams dry up, espionage takes the place of adventure, and the police take the place of the pirate.
> —Michel Foucault, "Of Other Spaces"

> The English live with the turmoil of two incompatible passions: a strange appetite for adventure and a strange appetite for legality.
> —Jorge Luis Borges, "Chesterton and the Labyrinths of the Detective Story"

> Whoever enters into or upon property in the possession of another with intent to commit an offence or to intimidate, insult or annoy any person in possession of such property, or having lawfully entered into or upon such property, unlawfully remains there with intent thereby to intimidate, insult or annoy any such person, or with intent to commit an offence, is said to commit "criminal trespass."
> —Section 441, Indian Penal Code

> Thus our first step has been to remember the proletariat body; we have tried to translate it out of the idiom of monstrosity.
> —Peter Linebaugh and Marcus Rediker, *The Many-Headed Hydra: Sailors, Slaves, Commoners, and the Hidden History of the Revolutionary Atlantic*

The transformation of intellectual property law from an esoteric legal subject to a topic of daily conversation and debate has occurred in a relatively short span of time. Over the past few years, the aggressive expansion of property claims into every domain of knowledge and cultural practice has interpellated almost everyone, from the academic to the musician, into the heart of the debate. No account of the contemporary moment would be complete without an examination of the dominance of the copyright sign or the effect of the small print of the trademark notice on our lives. In many ways, the mere act of looking at, reading, listening

Air Pirates cover (The Walt Disney Company). The Air Pirates were a group of underground cartoonists (Dan O'Neill, Ted Richards, Bobby London, and Gary Hallgren). They were sued by Disney in 1972 because they drew Mickey Mouse and other Disney characters dealing illicit drugs and having sex with each other. Hallgren and Richards settled with Disney. O'Neill and London decided to go to court. After years of legal battles, Disney agreed in 1980 to drop the case in exchange for the Pirates' agreeing not to infringe their copyrights.

to, making, understanding, or communicating any objects that embody thought, knowledge, or feeling is as fraught with danger and anxiety today as the appropriation of material wealth or trespassing onto private property were through much of human history.[1]

The anxiety and conflict are certainly not restricted to a set of geographical locations, but the nature of conflict gets configured differently as we move from the United States and Europe to parts of Asia and Africa. In the United States, the crisis is represented in terms of the shrinking of the public domain and of the commons by the extension of copyright, the linking of file sharing and peer-to-peer activities with the global war on terror, and the emergence of a new breed of criminals in the form of students sued by music companies for downloading MP3s online. In South Africa, the government is bulldozed by pharmaceutical corporations who have attempted to prevent it from declaring statutory licenses that will make AIDS drugs more accessible, and in many parts of Asia, the proliferation of cheap technologies of media reproduction creates a parallel economy that threatens the monopoly of old media players.[2]

The concern over the expansionist tendency of intellectual property has also motivated a rearticulation of the importance of the commons of knowledge and cultural production. This is exemplified by various phenomena, among the increasing popularity of nonproprietary modes such as free software and open content. A number of these concerns historically have emerged from the experience of Europe and the United States. But when one attempts to translate the terms of the intellectual property debate into the contemporary experience of countries in Asia, Latin America, and Africa, it is difficult to locate any easy indexical reference to ideas such as "the digital commons."

In a similar vein, scholarship on the concept of the public domain has opened out the debate on intellectual property and has forced us to pay closer attention to the political economy of information and the cultural politics of copyright. It has also sought to foreground public-interest considerations within international intellectual property policy. The terms established by work on the public domain enable the articulation of alternative normative claims to contest stricter intellectual property standards and the reintroduction of the public interest into intellectual property policy. They have also been very useful in challenging moves toward the greater criminalization of infringements on intellectual property rights. However, here again, while the scholarship on intellectual property and the public domain has been highly inspiring and influential for work in South Asia, it offers no easy fit with the concerns of daily life in that region and the role that intellectual property and the conflicts surrounding it play there.

The concept of intellectual property in many of these countries has been unfolded through the dual tropes of the triumphalist fantasy of harnessing intellectual property "to catch up with the West" and an account of paralyzing fear and images of the ruin, destruction, and violence that surround the reality of intellectual property infringement. The latter is best exemplified by the sharp conflicts and anxieties over the prevailing mediascape (from nonlegal software to cheap DVDs) that are a part of the contemporary urban experience in most countries. The dominant account of the unfolding of the new-media experience in these countries is also marked by the hyperprofiling of the act of piracy and the emergence of the figure of the pirate.[3]

It would seem almost paradoxical to suggest, as the title of this paper does, that there is a representational problem that emerges with respect to the figure of the pirate in contemporary discourse. If accounts in the mainstream media are anything to go by, it would seem that the figure of the media pirate is everywhere, and the problem would seem to be one of overrepresentation. However, we are not concerned with the way in which the pirate is narrated as a figure of illegality by the usual suspects, such as Jack Valenti (the longtime president of the Motion Picture Association of America), or the RIAA (the Recording Industry Association of America), or, closer home, the Indian Performers Rights Society, all of whom have argued for a more stringent enforcement of copyright. My focus instead is on the role of the pirate in the debate on intellectual property and the public domain that has emerged over the past few years to challenge the hegemonic account of intellectual property.

While the critical scholarship on intellectual property has been vital in the framing of an alternative paradigm, a quick survey of the range of debates also reveals the relative absence of any serious engagement with the world of quotidian nonlegal media consumption and circulation—or media piracy. This is surprising, given that intellectual property plays itself out in everyday life through an extraordinary focus on the pirate. What is it about the nature of piracy that creates this uncomfortable silence around it? Or is it possible that there is instead something about the way in which the critical responses to intellectual property have been framed that makes it impossible for them to deal with piracy or for piracy to redeem itself? Perhaps we will have to start asking different kinds of questions if we are to understand the status of the pirate in contemporary intellectual property debates and move beyond it.

Let's first look at the various ways in which the figure of the pirate enters the contemporary discourse of intellectual property. In the predominant logic of intellectual property enforcers, the pirate is demonized, seen as the ultimate embodiment of evil. That evil takes a variety of forms, from terrorism and the criminal underworld to causing the decline of the entertainment industry and evading of taxes. The figure of the pirate as criminal invites the legal attention of the state and of private enforcers. In recent times, the criminalized figure of the pirate has also become the subject of media attention, and rarely does a day go by without some sensational account of a raid.[4]

At the other end of the spectrum, that is, among those who work on limiting the expansion of intellectual property rights and on defending the public domain, the figure of the pirate is treated with embarrassed silence or outright disavowal. In Richard Stallman's work for instance, it is very clear that piracy is as unacceptable to the free-software movement as it is to copyright enforcers. The significant difference is that they would not argue for more criminalization or stronger enforcement and would have a more charitable understanding of the phenomenon, based on their reading of political economy.[5]

Scholars such as Lawrence Lessig and others have responded to the debate on intellectual property by looking beyond the binaries of legality/illegality that are set up by traditional copyright law, but when it comes to piracy, there still has been no effort to accommodate the concept of piracy within the accepted discursive

PIRACY HELPS STOP GLOBAL WARMING
Instead of driving to get CDs, buying pre-packaged software from megacorporations and wasting energy, you can help fight Global Warming by using P2P. Help save resources, fight pollution and save the environment. – It's what Jesus would do.

Piracy and global warming (© ipostr.com).

parameters of the debate. What, then, is the exact problem of piracy and why can it not be accommodated within the terms of public-domain theorists? Surely, it cannot be just the fact that it is tainted by illegality, since many other acts, including downloading music, are also tainted by illegality. Yet there are ways in which these acts find redemption, while the pirate cannot. Is the problem peculiar to the nature of this particular illegal act, the domain within which it operates, and the subjectivities that it interpellates?

The resistance to the concept and practice of piracy seems to be affected by several factors. First, it is seen as compromised because it is a commercial enterprise. Since piracy operates within the logic of profit and within the terms of commerce, it cannot claim the sort of moral ground that other nonlegal media practices can. For critics of the copyright regime dominated by media conglomerates, it would be an embarrassment to admit that they are supporting a nonlegal commercial enterprise. Their stance against piracy may therefore stem from either a strategic or an ethical position. The strategic stance against piracy may for instance be adopted by people who do not per se have any serious objections to piracy, but who recognize that it would be counterproductive, in their struggle against stricter intellectual property regimes, for them to be seen as espousing commercial piracy. On the other hand, there are a number of advocates for the free-software movement, including Stallman and Lessig, who would argue that even if a certain law exists and we do not agree with it, either we have to reform the law or create an alternative legal paradigm. However, if the law exists, we cannot encourage the violation of such a law.

Another reason for the suspicion of commercial piracy, in this case in relation to entertainment, stems from the fact that what is pirated often pertains to the domain of pleasure. Unlike access to affordable medicines and access to learning materials, piracy that provides people with low-cost DVDs, MP3s, and other copyrighted content seems to lack pragmatic justification and simply fulfils consumers' desires. We will examine this in some detail later.

Yet another critique of commercial piracy is that unlike young musicians who illegally download, then remix the music to produce new music, those who undertake piracy for purely commercial ends are unable to redeem their actions by claiming that they encourage and support further acts of creativity. Instead, in the case of commercial piracy, there is a slavish making of copies without any transformative redemption.

Finally, any justification of piracy is seen to fall within larger accounts of the collapse of the rule of law. Scholars working on understanding the phenomenon of piracy are accused of romanticizing illegality, and a sympathetic look at piracy is equated with support for anarchy and lawlessness.

Because piracy thus has not been able to be accommodated within the terms of public-domain theory, we need to understand how the terms of representation that public domain scholarship sets for itself operate to effect this exclusion. Although the public domain has emerged as the most viable alternative to the expansion of intellectual property, the question is whether the public domain is the only way by which we can understand both the contemporary conflicts around intellectual property and the limits of the approach with regard to accounting for the status of piracy. Can the world of the public domain and the world of the pirate be narrated as though there is a seamless web that should necessarily tie the two?

In many ways, advocates for the public domain deploy classical terms of representation that they borrow from either political or cultural theory. These terms include the classical categories of citizenship, resistance, and creativity.[6] One of the problems that we have when we try to understand piracy is that it often does not fit within any of these existing categories, and there is a positivity or excess in the body of the pirate that cannot be disavowed. As we have noted, the only manner in which the copyright infringer is rescued from the accusation of being an illegal pirate is through an act of redemption, for instance by showing that his or her acts of infringement actually result in an increase in creativity, and this redemption is formalized in doctrines such as the idea of "transformative authorship." But what happens to entire realms of nontransformative authorship or "Asian piracy," which does not necessarily transform anything, but merely reproduces ceaselessly using cheap technologies?

The high priest of open content and the founder of the Creative Commons movement has this to say:

> All across the world, but especially in Asia and Eastern Europe, there are businesses that do nothing but take others people's copyrighted content, copy it, and sell it— all without the permission of a copyright owner. The recording industry estimates that it loses about $4.6 billion every year to physical piracy (that works out to one in three CDs sold worldwide). The MPAA estimates that it loses $3 billion annually worldwide to piracy. This is piracy plain and simple. Nothing in the argument of this book, nor in the argument that most people make when talking about the subject of this book, should draw into doubt this simple point:
>
> This piracy is wrong....
>
> The copy shops in Asia, by contrast, are violating Asian law. Asian law does protect foreign copyrights, and the actions of the copy shops violate that law. So the wrong of piracy that they engage in is not just a moral wrong, but a legal wrong, and not just an internationally legal wrong, but a locally legal wrong as well.[7]

How do we read this as part of an account of the public domain? While one can understand that Lessig would have to be careful about the ways in which he pitches a reform of copyright law within the context of the United States, it is also difficult not to miss the linkages in this paragraph to older accounts of illegality in Asia. In many such accounts, the urban experience in Asia—and in Latin America— has been narrated in terms of its preponderant criminality and illegality. This is particularly true not merely in the context of the colonial imagination, but also in the ways that cities and everyday life in Asia are understood. The United States has always narrated itself through the tropes of constitutionalism and the rule of law, but with the arrival of Internet, all of a sudden, the language of criminality and illegality that was used to account by contrast for much of the world arrives home in the ordinary form of the criminalization of students downloading music. Clearly, one cannot have an account of such pervasive illegality in a country that prides itself on its constitutional tradition and its emphasis on the rule of law.

Consequently, one narrative strategy is to redeem the acts of ordinary American citizens through the discursive construction of an other—in this case an Asian other. The categories of the public domain serve as the neutral ground on which the two kinds of pirates are pitted, and the terms of reference of this public domain are the received notions of creativity and innovation.

Underlying much of copyright's mythology are the modernist ideas of creativity, innovation, and progress. The narrative conjunction of these ideas is represented as universal, and indeed, it is shared by both advocates of stronger copyright and advocates of the public domain.[8] By offering themselves as alternative accounts of the idea of progress and creativity, arguments for the public domain merely seek to provide a counterfactual: While copyright aspires to promote creativity, it actually fails to do this, and excessive protection has actually resulted in a decrease of creativity or a threat to creativity.

The difference between scholars who advocate for the public domain and copyright advocates lies in their understanding and interpretation of the idea of the creative. Lessig insists that we should protect some illegal works, based on the criterion of "transformativity," but the creative subject invoked here is in fact a very particular kind of subject—a disembodied classical liberal subject. The public domain is represented as a space in which everyone can participate as citizens bearing equal rights. The linking of public-domain theories to the freedom of speech and expression is not accidental, and the model of the public domain as the sphere of rational communication borrows from existing accounts of the public/private divide.[9]

Many postcolonial scholars have seriously contested the category of the citizen as the universal bearer of rights, and the representative capacity of the citizen to participate in the public sphere as an unmarked individual remains mythical, at best. In India, for instance, the creation of the category of the citizen subject demanded a move away from the oversignified body of the individual marked by religion, gender, caste, and so on to an unmarked subject position, "the citizen," a category based on equality and access and guaranteed rights within the constitutional framework. But the majority of the people in India are only precarious citizens who often do not have the ability to claim rights in the same manner as the Indian elite do. Instead, the manner in which they access the institutions of democracy and "welfare" is often through complex negotiations and networks and often is marked by their illegal status.[10]

In their work on "rowdy sheeters"—individuals with a criminal record, or "rap sheet," as it's called in the United States—Vivek Dhareshwar and R. Srivatsan suggest that "some bodies—like the 'rowdy' or the 'lumpen'—will not be disincorporated," that is, made to "speak and act as a citizen, "so tied are [sic] their shameful positivity to their bodies." Thus, the project of disincorporation into citizenship almost immediately creates a discursive other, the illegal citizen who refuses to shed his or her social excesses or who just cannot do so.[11] Thus, while citizenship and modernity are normatively constructed as highly desirable and the grand project wills everyone into a state of modernity, there arises from the start a clear lack or inability in the bulk of the population to occupy this space. So what happens when people fall off these official maps and plans? How do they find their way back into official memory and create for themselves avenues of participation? There is a great deal of work to be done on engaging with how people create vibrant spaces outside of official plans and spaces, and more often than not, these spaces are marked by their high degree of illegality.

Pirates are among those unable to shed these illegal excesses and play a role in or become a part of a reconstituted public domain. Pirates cannot play a role there, because they cannot claim the representative status given to the transforming creator within the productive public domain. There are very few possibilities for the pirate to occupy the normative terms established in the public domain for the creative citizen. And yet, despite this, a look at both history and the present indicates that there is a certain stubbornness on the part of those who do not find a representative space in the public domain—those who refuse to disappear and instead coexist at the margins of civil society and the law and at the margins of the narrative dominated by the creative, innovative citizen. Historically, for instance, there is an entire realm that is inhabited by figures such as the trickster, the copier, the thief, and the pirate, figures who inhabit a marginal site of production and circulation.[12]

If we move away from the normative account of the creator citizen and engage with an entire set of practices that renders any straightforward representation impossible or difficult, what intellectual horizons open out? As with any journey into unfamiliar terrain, it might be useful to have a few maps charted during previous moments of anxiety to help guide us. As with any maps, these are only tentative and provisional guides.

The simplistic opposition between legality and illegality that divides pirates from others renders almost impossible any serious understanding or engagement with the phenomenon of piracy. Following Nietzsche, we should perhaps advocate the virtues of slow reading. The dizzying speed with which one is forced to respond to issues in the era of globalization can sometimes hinder any reasoned response. The first task for us is to avoid the Enlightenment blackmail, a variant of which in recent times has been the blackmail of "You are either for terrorism or support the war on terror." In other words, before we jump into making normative policy interventions, which often draws black-and-white distinctions, we need to explore the various shades and depths of gray. We would only ask for patience from the scholars of the public domain and ask the same careful attention that they pay to understanding the larger political and cultural politics of copyright when they look at the phenomenon of piracy.

Let us reformulate our object of enquiry. Let's take for granted the illegal status of piracy, but let's not stop there. Instead, it might be more useful for us to ask not what piracy is, but what piracy does. The shift in focus from the discursive and moral representation of the illegal deed to the wider social world in which the deed is located allows us to bring to light the nature of the law that names a particular act as an illegal one.

And the naming of the deed as an illegal act indeed prevents us from reflecting on the nature of the act. When we look for instance at the act of sharing, it is an act immediately invested with a sense of virtue. But the same act when rendered through the prism of private property becomes an act of infringement and a crime. The debate between morality and ethics is now a familiar one, and indeed, it might even be argued that the law's monopoly over official definitions of morality does not render obsolete the question of whether an act can still be considered in terms of ethics.

The shift away from what piracy is to what piracy does enables us to consider on the same plane its linkages to the normative considerations for which public-domain advocates argue and that they are often unable to achieve. The best example is in the area of cheap books. While public-domain advocates try to reform copyright law to enable more educational exceptions, pirated books and the unauthorized photocopying that is the order of the day accomplishes what they cannot. Rather than looking at the neat spaces created by the opposition between the "legal" and the "illegal," it might be more fruitful to consider the spaces in which piracy plays itself out, the transforming urban landscapes and the specific histories of the nooks and crannies that render this space an illegal one, along with the accumulated histories of regulation, tactics, and negotiations that render this topography intelligible.

Definitions of legality do not exist in a vacuum, and they are constituted through specificities and relationships, even as they attempt to define constitutive legal and social relations.[13] Similarly stories of law and legality have to find a space in which they resonate, and often they exist as abstract, unintelligible murmurs. For instance, when the story of copyright piracy is narrated, it is usually through the language of statistics and figures and the narrative strategy of excess, designed to induce a "shock and awe" response at the alarming rate of piracy and illegality that exists, especially in non-Western countries, and it rarely succeeds in its desired effect.

To understand why these stories don't work in some contexts, we will have to travel to distant cities such as Delhi and Sao Paulo and perhaps even walk through the more unfamiliar byways of familiar cities such as New York. The discipline of urban studies has made the idea of "the illegal city" familiar to us. One reads, for instance, that an average of 40 percent and in some cases 70 percent of the population of major cities lives in illegal conditions. Furthermore, 70 to 95 percent of all new housing is built illegally.[14] How do we understand this older idea of illegality alongside the new illegality of the mediatized city? The task will be to pose the question of how the older form and the newer form integrate and intertwine—to interrogate our classical liberal assumptions of legality and highlight the limitations of any study based on a strictly legal understanding of contemporary urban practices.

Writing about the modernist project of planning, James Holstrom and Arjun Appadurai note:

> modernist planning does not admit or develop productively the paradoxes of its imagined futures. Instead it attempts to be a plan without contradictions or conflict. It assumes a rational domination of the future in which its total and totalizing plan dissolves any conflict between the imagined and existing society in the enforced coherence of its order. This assumption is false and arrogant as it fails to include as its constituent element, the conflict, ambiguity and indeterminacy characteristic of actual social life.[15]

The information era props up a master plan similar to that of modernist planning. The institutional imagination of the era relies on the World Trade Organization (WTO) as the chief architect and planner and copyright lawyers as the executive managers of this new plan, while the only people who retain their jobs from the old city are the executors of the old plan, the police force and the demolition squad. Just as one cannot understand land tenure in terms of the classical liberal concept of legality alone, any attempt to understand the complex networks of economic and social relations that underlie the phenomenon of piracy will have to engage with the conflict over control of the means of technological and cultural production in the contemporary moment of globalization. The ways in which the illegal media city emerges and coexists alongside the vibrant, innovative, and productive debris of the older city and the schizoid relationships between legality and illegality in postcolonial cities suggest that we may need to turn the gaze of the law from the usual suspects of legality to legality itself and to the relations that underlie its existence.[16]

The transformation of the urban experience in the past few years and the proliferation of the labyrinth experience of media forms have made pirate cultures a significant part of the experience of our contemporary era. What is perhaps different about the media experience in non-Western countries is the fact that there are no clear lines between the old and the new media, between physical and virtual experience, and often, the virtual extends from high-end shopping malls,

Drawing on the Wall (Tobias Vemmenby, http://twilightshadows.wordpress.com).

to low-end cybercafés, to pirate markets. This comfortable moving to and fro between different mediatized spaces creates a sensorial experience in which different classes actualize the global experience differently.

Piracy transforms the technological experience, which traditionally has been rooted either in monumentalist visions of development (the discourse on information and communication technologies for development) or in the aspirational imagination of the elite in India (Bangalore's aspirations to be Singapore), and it provides an entry point for a much wider array of people to experience on their own terms the "information era." The cheap CD or DVD supplements the experience of cyberspace while at the same time being rooted within diverse spaces in the city. Even as the urban landscape is being transformed and older media spaces such as movie theaters give way to high-rise malls with multiplexes, and even as the spaces of traditional mass media begin to shrink because of their prohibitive prices, you see the emergence of a widely distributed chain of the circulation of media commodities that challenges the regime of intellectual property. The crisis of intellectual property is narrated into the crisis of South Asian cities in general, and interventions in implementing property rules sit alongside lamentful pleas for reworking urban imaginations. The critical difference between this world of everyday media and the celebratory approach of radical new-media activists or scholars of the public domain is that the world of a quotidian media experience does not articulate itself in the terms of resistance or appropriation. Piracy obviously does not stake a claim in the world of official creativity, either. It remains what it is: a culture of the copy that exists alongside livelihood and labor, profit and pornography.

RETHINKING CREATIVITY: PIRATE INFRASTRUCTURES

A world of everyday media that transforms our contemporary experience and yet paradoxically does not make a claim to creativity as it is commonly understood invites us to revisit our ideas of creativity's relation to the copy.[17] The reproducible work that brings into play a network of circulation also inaugurates a series of cultural possibilities and readings.

Roland Barthes and Michel Foucault have already enabled us to shift our understanding of the locus of originality and creativity from the text and look for it instead in the process of consumption. What would happen if we also extended the search for creativity into the domain of circulation? The production and circulation of the ubiquitous pirate DVD, that prized commodity of pirate aesthetics, helps us understand the possibility of creative acts outside the domain of what is traditionally considered "creative."

To do so, we need to consider the conditions under which DVDs, these new products of digital reproduction, are pirated and circulate. Brian Larkin's work on piracy in Nigeria, for example, forces us not merely to look at and listen to the onscreen content of videos, but also to focus on those conditions of appropriation and circulation. Larkin demonstrates the critical importance of paying attention to the infrastructures of production in developing countries, where the process of cultural production is tied to the relative lack of infrastructure and becomes the basis for the transformation of the conditions of production by generating a parallel economy of low-cost infrastructure. He says that "a cycle of breakdown, repair, and breakdown again is the condition of existence for many technologies in Nigeria. As a consequence, Nigeria employs a vast army of people who specialize in repairing and reconditioning broken technological goods, since the need for repair is frequent and the cost of it cheap."[18]

This economy of recycling, which Ravi Sundaram describes as "pirate modern,"[19] becomes the arena for all sorts of technological innovation and extends further to experiments with cultural forms such as parodies, remixes, cover versions, and so on. In a sense, Larkin's invocation of the importance of infrastructure contrasts with the obsessive fixation with content that one sees in most Western accounts of creativity, although in fact, on a metaphorical level, infrastructure frequently gets invoked in Western discourses as a way to understand the public domain of ideas, with references to "the well of ideas," "bridging the information gap," "the information superhighway," and so on. In piracy, however, the content also has to be filtered through the regime of its own production. Piracy imposes particular conditions on the recording, transmission, and retrieval of data. Constant copying erodes data storage, degrades image and sound, and overwhelms the signal of the media content with the noise produced by the means of reproduction. Larkin says that since pirated videos are often characterized by blurred images and distorted sound, they create a kind of material space "that filters audiences' engagement with media technologies and their senses of time, speed, space, and contemporaneity. In this way, piracy creates an aesthetic, a set of formal qualities that generates a particular sensorial experience of media marked by poor transmission, interference, and noise."[20] Larkin uses the question of pirate infrastructure to open out the debate on intellectual property and to foreground the importance of addressing the question of content while looking at the legal aspects of culture. If infrastructures represent attempts to order, regulate, and rationalize society, then breakdowns in their operation and the rise of provisional and informal infrastructures highlight both the failure of that ordering and the recoding that takes its place.

When we subject the material operation of piracy and its social consequences to scrutiny, it becomes clear that pirate infrastructure is a powerful mediating force that produces new modes of organizing sensory perception, time, space, and economic networks.[21] Doing so also forces us to acknowledge the material linkages between content and infrastructure. One of the significant approaches used by scholars of the public domain is an emphasis on the ability to create new content building on existing works. This overemphasis on the creation of new content raises the question of who uses the new content and what the relationship is between such content and the democratization of infrastructures. In most cases, the reason for the fall in price of computers and other electronic goods and the increase in access to materials via the increase in photocopiers and the general infrastructure of information flows is not caused by any radical revolution such as free software or open content, but by the easier availability of standard, mainstream commodities such as those produced by Microsoft and Hollywood. When Stallman and others castigate people for pirating Hollywood's productions, it is only because they are in the position of being able to disavow the global economy. But for many people, finding their place within the global economy includes engaging with a world of counterfeit commodities, replicating the global economy's output.

We can play the game of seizing the higher moral ground and speak of the real information needs of these people, or we can provide crude theories of how they are trapped by false consciousness. Or better yet, we can move away from these judgmental perspectives and look at other aspects of globalization, such as the impact that the expansion of the market for these gray-market goods has on the general pricing of goods, on the spread of computer/Internet-technology culture, on lowering the price of consumables such as blank CDs and DVDs, on the popularity of CD writers, and so on. I find it a little strange and messianic that people who preach about access also preach about the kinds of access that should be allowed.

PLEASURABLE TRANSGRESSIONS

Such prohibitions take many forms. As I noted before, one of the objections to piracy seems to lie in the fact that it is associated more with the world of pleasure and desire than with meeting "pure needs." Let me begin to discuss this objection in greater depth with an interesting story about the intersection between the world of desire, subjectivity and the experience of piracy. It is a typical example of interventions in the field of the digital divide. An NGO in Bangalore that works in the field of information and communication technologies for development was conducting a workshop on accessing the Internet for the information needs of rural women working to empower other poor rural women in India. The facilitator guided the women through the basics of the Internet, including how to access information relevant to their work, which ranges from providing access to credit to promoting women's health. The training was highly appreciated, and all the women volunteers seemed to be enjoying themselves while fiddling with the computers and exploring the Internet. At the end of the training, when the NGO started cleaning up the computers, including the browsing histories and the cached copies of the sites accessed, they were a little aghast to find that most of the women volunteers had been surfing pornography—and a range of pornography, at that. So while the trainers were holding forth eloquently about the real information needs of the poor, the poor were quite happy to access their real information needs.

The links between pleasure, desire, aspiration, and trespass have always been complicated, and the closer that the transgressive act is to the domain of pleasure, the more difficult it seems for it to be redeemed socially. Thus, while one finds easier justifications for transgressions that deal with questions of livelihood and survival, and in the case of intellectual property, easier justifications for transgressions that appeal to claims to free speech and access to information, when the matter involved is about new subjectivities and pleasurable transgressions, the issue gets very differently framed. In particular, the terms set up by existing scholarship on the public domain end up excluding the ability to engage with practices guided not as much by necessity as by curiosity. The rhetoric of inclusiveness that is implicit in discourse on the issue of the public domain is necessarily accompanied by the prospect of exclusion, an exclusion that relies on either piety or pedagogy. What happens when we move toward the realm of nonlegal media practices, where all of a sudden the transgression is highly pleasurable, but not in any way connected to the essential character of what Gayatri Spivak calls the "subaltern subject"?[22] The sheer proliferation of these practices, both within the elite and also by the traditional subaltern classes, forces us to question our own assumptions about the terms in which people engage with the global economy of information and go about finding their place in the global economy. What critical conceptual resources can we draw on to address the question of pleasurable transgressions and subjectivities that resist easy framing?

Jacques Rancière paves the way for us to start thinking seriously about the hidden domain of aspiration and desire of the subaltern subject while at the same time thinking about the politics of our own aspirations and desires. Rancière examines an unexplored aspect of the labor archive of nineteenth-century France: small, obscure, and short-lived journals brought out by workers in which they were writing about their own lives. But they were not necessarily writing about their work, and if they were, they were not writing about it in glorified terms, but with immense dissatisfaction. For the most part, however, they were interested in writing poetry, writing about philosophy, and indulging in other pleasures in which nonworkers or intellectuals were entitled to indulge. Of course, from the other side of the class divide, intellectuals have been fascinated with the world of work and the romance of working-class identity. Rancière asks "what new forms of misreading will affect this contradiction when the discourse of labourers in love with the intellectual nights of the intellectuals encounters the discourse of intellectuals in love with the toilsome and glorious days of the labouring people?"[23]

Rancière's motley cast of characters include Jerome Gillard, an ironsmith tired of hammering iron, and Pierre Vincard, a metal worker who aspires to be a painter—in other words, people who refused to obey the role sketched out for them by history and who wanted to step across the line and perform the truly radical act of breaking down the time-honored barrier separating those who carry out useful labor from those who ponder aesthetics. Rancière says:

> A worker who has never learned how to write and yet tried to compose verses to suit the taste of his times was perhaps more of a danger to the prevailing ideological order than a worker who performed revolutionary songs. . . . Perhaps the truly dangerous classes are not so much the uncivilized ones thought to undermine society from below, but rather the migrants who move at the borders between classes,

individuals and groups who develop capabilities within themselves which are useless for the improvement of their material lives and which in fact are liable to make them despise material concerns.[24]

The moral dictates that govern the lives of the poor are not imposed only by the state ("Don't steal," "Don't beg") but equally by those who theorize the lives of the poor ("Be aware of your class," "Don't get trapped by false consciousness"). And when people start moving out of the frame of representation that has been so carefully and almost lovingly crafted for them, they either have to be shown their true essence or their transgressions have to be brought within the terms of their representative class. Thus, when Victor Hugo was shown a poem written by a worker, his embarrassed and patronizing response was, "In your fine verse there is something more than fine verse. There is a strong soul, a lofty heart, a noble and robust spirit. Carry on. Always be what you are: poet and worker. That is to say, thinker and worker."[25] This is a classic instance of what Rancière would term an "exclusion by homage."[26] Thus, the aspiration and desires of the poor have to be "something more than fine verse," and the information needs of the poor have to be something more than wanting to watch a film or even dreaming of becoming a filmmaker. These injunctions certainly tell us more about the fantasies of the state and of the intellectuals than they do about people engaging in the fulfillment of their aspirations and desires, and we may do well to start rethinking the terms in which the scholars of intellectual property engage the language of access.

REVISITING THE HISTORY OF THE COMMONS AND DISPOSSESSION

Prominent among the terms employed in recent scholarship on intellectual property and the public domain has been the metaphor of the modern commons and the threat that it faces from this limitless expansion of intellectual property. More often than not, the commons is allegorized as a mythical ideal governed by principles of sharing, access, and collaboration that was lost after the first enclosure movement. The argument proceeds to caution against a similar enclosure, a second enclosure movement in the realm of information ecology that threatens to privatize every aspect of information, thereby threatening creativity. The invocation of the commons is indeed a useful starting point in discussions of intellectual property regimes, but it would be incomplete if we did not acknowledge the histories of contestation, conflict, and violence that accompanied the first enclosure movement and its subsequent history.

Social historians of crime, for instance, have rigorously alerted us to the intertwined histories of property and criminalization. It may therefore be insufficient for us to invoke the commons only in allegorical terms, and it may be more fruitful to look at current conflicts as part of a wider historical continuum in a way that interrogates the nature of contestation over the definition, the contours, and the enforcement of what constitutes "property." The history of the commons is also a history of criminalization and of the definition of the ideas of trespass and encroachment.

In *The Many-Headed Hydra*, as a way of thinking about the challenges faced by the world of capital, Peter Linebaugh and Marcus Rediker begin with an invocation of the twin myths of the Hydra and Hercules' task of slaying it.[27] Confronted with the monstrous, many-headed water snake, the Hydra, Hercules found that as soon as he cut off one head, two grew in its place. With the help of his nephew Iolaus, he used a firebrand to cauterize the stump of the beast's neck. Thus they killed the Hydra. Hercules then dipped his arrows in the blood of the slain beast, whose venom gave his arrows a fatal power.

Using the allegory of the Hydra to characterize the various obstacles that capital has faced and, like Hercules, overcome from the eighteenth century to the present, Linebaugh and Rediker start with the material organization of many thousands of workers into transatlantic circuits of commodity exchange and capital accumulation and then proceed to look at the ways in which they translated their cooperation into anticapitalist projects of their own. The first enclosure movement resulted in the expropriation of the commons, freed large territories for capitalist agriculture, logging, mining, and speculation in land, and at the same time created a vast army of the dispossessed, who were then freed to become wage earners in new industrializing areas at home or abroad or who were criminalized by harsh laws that imposed penal servitude in the colonies. Those dispossessed from the land also became the bulk of the workforce for the new engine that transported commodities across continents, the ship. Sailors and ships linked the modes of production and expanded the international capitalist economy. The ship was also the site of the coming together of diverse forms of labor and of diverse laborers from different ethnicities, bound together by a pidgin tongue. The solidarity of this motley crew, like many others in the era, was forged by their shared situation of dispossession and their shared labor.

Linebaugh and Rediker document in detail the very difficult conditions under which these sailors worked and the dangers to which they were constantly exposed, which at the same time created the conditions for solidarity among those who would challenge the smooth flow of capital: pirates. The first pirates in this sense were often "the outcasts of the land" who mutinied against the conditions of their work and created an alternative order challenging the division of labor and capital. In fashioning what Linebaugh and Rediker call their "hydrachy," these buccaneers often drew from the memory of utopias created by theoreticians in which work had been abolished, property redistributed, social distinctions leveled, health restored, and food made abundant. By expropriating a merchant ship (after a mutiny or a capture), pirates seized the means of maritime production and declared it to be the common property of those who did its work. Rather than working for wages using the tools and larger machinery owned by a merchant capitalist, pirates abolished the wage and commanded the ship as their property, sharing equally in the risks of common adventure.

Piracy's redistribution of wealth was considered to be a massive international problem, and pirates were declared to belong to no nation. In fact, piracy emerged as one of the earliest crimes of universal global jurisdiction in a time when nation-states were still carving out their own local absolute sovereignties. But piracy was not merely a problem of the failure of the implementation or enforcement of the laws of property. Piracy also established an alternative ethic and an alternate mode of being. Piracy was democratic in an undemocratic age and egalitarian in a highly unequal age. Linebaugh and Rediker provide various accounts of instances in which the pirate ship inverted all rules of social hierarchy and in which, for brief spells, the laws of private property were suspended to allow for experimentation with alternative social imaginaries, even if only very briefly.

Summarizing the characteristics of this hydra of the era of early capitalism, Linebaugh says:

It was landless, exploited. It lost the integument of the commons to cover and protect its needs. It was poor, lacking property, money, or material riches of any kind. It was often unwaged, forced to perform the paid labors of capitalism. It was often hungry, with uncertain means of survival. It was mobile, *transatlantic*. It powered industries of worldwide transportation. It left the land, migrating from country to town, from region to region, across the oceans, and from one island to another. It was *terrorized, subject to coersion*. Its hide was calloused by indentured labor, gallery slavery, plantation slavery, convict transportation, the workhouse, the house of correction. Its origins were often traumatic: enclosure, capture, and imprisonment left lasting marks. It was *female* and *male, of all ages*. (Indeed, the very term *proletarian* originally referred to poor women who served the state by bearing children.) It included everyone from youth to old folks, from ship's boys to old salts, from apprentices to savvy old masters, from young prostitutes to old "witches." It was *multitudinous, numerous*, and *growing*. Whether in a square, at a market, on a common, in a regiment, or on a man-of-war with banners flying and drums beating, its gatherings were wondrous to contemporaries. It was *numbered, weighed*, and *measured*. Unknown as individuals or by name, it was objectified and counted for purposes of taxation, production, and reproduction. It was *cooperative* and *laboring*. The collective power of the many, rather than the skilled labor of the one produced its most forceful energy. It moved burdens, shifted earth, and transformed the land-scape. It was *motley*, both dressed in rags and multiethnic in appearance. Like Cali-ban, it originated in Europe, Africa, and America. It included clowns, or cloons (i.e., country people). It was without genealogical unity. It was *vulgar*. It spoke its own speech, with a distinctive pronunciation, lexicon, and grammar made up of slang, cant, jargon, and pidgin—talk from work, the street, the prison, the gang, and the dock. It was *planetary*, in its origins, its motions, and its consciousness. Finally, the proletariat was *self-active, creative*; it was—and is—alive; it is onamove.[28]

It is in the struggles of these multitudes that Linebaugh and Rediker see the hidden history of revolutionary ideas of freedom, entitlement, dignity, and every-thing else claimed in the name of rights and citizenship. The multitude was limited neither by the narrow allegiances of ethnicity nor by the vulgar claims of nation-hood, and yet ironically, the moment of the formal institutionalization of a number of these rights was also the moment that resulted in the exclusion of the very class that had suffered to gain them.

Linebaugh and Rediker say that "the new revolts created breakthroughs in human praxis: the Rights of Mankind, the strike, the higher-law doctrine, that would eventually help to abolish impressments and plantation slavery. They helped more immediately to produce the American Revolution, which ended in reaction as the Founding Fathers used race, nation, and citizenship to discipline, divide, and exclude the very sailors and slaves who had initiated and propelled the revolutionary movement."[29]

There is perhaps a lesson to be learned here for those of us interested in look-ing at the linkages between the multitudinous experience of living through the consolidation of intellectual property. Intellectual property is also created through transnational networks of new forms of capital and labor, made in virtual vessels that pass each other in the global night on the high seas of data. The tall ships of our times fly many flags of convenience. They are the software sweatshops, the media networks, the vast armadas of the culture industries and the lifestyle factories. They produce high-value primary commodities, stars, stories, sagas, software, idols, lifestyles, and other ways of ordering meaning in an increasingly chaotic world. Typically, even though they sell the fantasies of place and identity in an increasingly enmeshed world, they are produced in a global everywhere and delivered through electronic pipelines everywhere, when necessary, more or less instantaneously, through telecommunication networks.

Their ubiquity and their global reach are also hallmarks of their greatest vul-nerability, for like their precursors, the tall ships of the new economy are freighted with cargo that is just as vulnerable to attacks of piracy. The new electronic pirates are located in the interstices of the global culture economy, which are the nodes that make the network viable in the first place. We cannot imagine a global media industry without the technology that made possible the phenomenon known as peer-to-peer networking on intranets, but it is precisely the same technology on the Internet that renders any attempt to police the distribution channels of media content in the interests of proprietary agencies almost impossible. Just as the piracy of the past disturbed the equilibrium composed of slavery, indentured labor, the expropriation of the commons, the factory system, and penal servitude, the electronic piracy of the present is destined to wreck the culture industry, either by making the economic and social costs of policing content prohibitive or by ushering in a diversity of new protocols for the use, distribution, and reproduction of cultural and intellectual content that will make the whole enterprise of making vast sums of money out of the nothing of data and culture a difficult business.[30]

CONCLUSION

Any account of the conflicts over access to knowledge and culture in the contemporary world will have to be aware of the complicated terrain that knowledge occupies. Our examination of the figure of the pirate has been an attempt to chart out the ways in which familiar issues of political economy, inequity, and reform meet with aspirations, desires, and creativity in unlikely encounters in unexpected spaces. As scholars and activists interested in a more just information order, it might well be the case that we need to abandon any simple, one-size-fits-all approach to reforming the public domain. We need instead to be aware of the fact that there can be no accounts of access that are not simultaneously accounts of exclusion, and it is in the awareness of this productive tension that we may be able to engage with a wider set of practices through which people can access knowledge and culture.

NOTES

1   *Contested Commons, Trespassing Publics: A Public Record* (New Delhi: Sarai/CSDS, 2005), p. vi, available on-line at http://www.sarai.net/publications/occasional/contested-commons-trespassing-publics-a-public-record (last accessed May 11, 2009).

2   See Philippe Cullet, "Patents Bill, TRIPS and Access to Health," *Economic and Political Weekly* 43, October 27, 2001, also available on-line at http://www.ielrc.org/content/n0107.htm (last accessed May 11, 2009).

3   *Ibid.*

4   A statement by the U.S. Department of Transportation states that "they run computer manu-
    facturing plants and noodle shops, sell 'designer clothes' and 'bargain basement' CDs. They
    invest, pay taxes, give to charity, and fly like trapeze artists between one international venue
    and another. The end game, however, is not to buy a bigger house or send the kids to an
    Ivy League school—it's to blow up a building, to hijack a jet, to release a plague, and to kill
    thousands of innocent civilians." U.S. Department of Transportation, Office of Safety and
    Security, *Transit Security Newsletter* 36 (May 2003), p. 2. For a scathing critique, se also Nitin
    Govil, "War in the Age of Pirate Reproduction," in *Sarai Reader 04: Crisis/Media*, pp. 378–83,
    available on-line at http://www.sarai.net/publications/readers/04-crisis-media/50nitin.pdf
    (last accessed May 11, 2009). This statement has been similarly followed up by the Indian
    copyright enforcers, led by the former commissioner of police, Julio Rebiero, who have
    claimed that music piracy funds jihadist terrorists. See R. Rangaraj, "Music Piracy and Terror-
    ism," available on-line at http://www.chennaionline.com/musicnew/films/09musicpiracy.
    asp (last accessed May 11, 2009).

5   See chapter 4 of Lawrence Lessig, *Free Culture: How Big Media Uses Technology and the Law
    to Lock Down Culture and Control Creativity* (New York: Penguin, 2004).

6   Rosemary Coombe, *The Cultural Life of Intellectual Properties: Authorship, Appropriation,
    and the Law* (Durham, NC: Duke University Press, 1998); Rosemary Coombe, "Tenth Anni-
    versary Symposium: New Direction: Critical Cultural Legal Studies," *Yale Journal of Law and
    the Humanities* 10 (Summer 1998); Yochai Benkler, "Through the Looking Glass: Alice and the
    Constitutional Foundations of the Public Domain," available on-line at http://www.law.duke.
    edu/pd/papers/benkler.pdf (last accessed May 11, 2009).

7   Lessig, *Free Culture*, pp. 63–64.

8   Michael D. Birnhack, "The Idea of Progress in Copyright Law," *Buffalo Intellectual Property
    Law Journal* 1, no. 3 (Summer 2001): p. 3.

9   James Boyle, *Shamans, Software and Spleens: Law and the Construction of the Information
    Society* (Cambridge, MA: Harvard University Press, 1996); Diane Leenheer Zimmerman,
    "Information as Speech, Information as Goods: Some Thoughts on Marketplaces and the Bill
    of Rights," *William and Mary Law Review* 33 (1992); Yochai Benkler, "Siren Songs and Amish
    Children: Autonomy, Information, and Law," *New York University Law Review* 76 (2001).

10 Partha Chatterjee, "On Civil and Political Societies in Postcolonial Democracies," in Sudipta
    Kaviraj and Sunil Khilnani (eds.), *Civil Society: History and Possibilities* (Cambridge: Cam-
    bridge University Press, 2001), pp. 165–78.

11 Vivek Dhareshwar and R. Srivatsan, "'Rowdy-sheeters': An essay on Subalternity and Politics,"
    in Shahid Amin and Dipesh Chakrabarty (eds.), *Subaltern Studies IX* (New Delhi: Oxford Uni-
    versity Press, 1996), p. 223. Also available on-line at www.cscs.res.in/dataarchive/textfiles/
    textfile.2007-08-16.5404440406/file (last accessed May 13, 2009).

12 John Hope Mason, *The Value of Creativity: The Origins and Emergence of a Modern Belief*
    (Aldershot, UK: Ashgate, 2003).

13 For an account of the everyday life of law and social relations, see Susan Silbey and Patricia
    Ewick, *The Common Place of the Law: Stories from Everyday Life* (Chicago: University of Chi-
    cago Press, 1998).

14 Alain Durand-Lasserve and Lauren Royston (eds.), *Holding Their Ground: Secure Land Tenure
    for the Urban Poor in Developing Countries* (London: Earthscan, 2002); Arthur J. Jacobson,

"The Informal Economy: The Other Path of the Law," *Yale Law Journal* 103 (1994).

15 James Holsten and Arjun Appadurai, "Cities and Citizenship," *Public Culture* 8, no. 2 (1996).

16 Lawrence Liang, "Porous Legalities and Avenues of Participation," in *Sarai Reader 05: Bare Acts* (New Delhi: Sarai/CSDS, 2005).

17 Ravi Sundaram has suggested that it might be fruitful for us to revisit the histories of the copy, from early print culture to the forger in art history through the crisis in aesthetic experience precipitated by the "age of mechanical reproduction" as a way of understanding the current transitions and conflicts. It is also a useful way in which to understand the general anxiety about the consumption and circulation of cheaply reproduced media commodities. See Ravi Sundaram, "Other Networks: Media Urbanism and the Culture of the Copy in South Asia," in Joe Karaganis (ed.), *Structures of Participation in Digital Culture* (New York: SSRC Books, 2007).

18 Brian Larkin, "Degrading Images, Distorted Sounds: Nigerian Video and the Infrastructure of Piracy," *Public Culture* 16, no. 2 (Spring 2004).

19 Ravi Sundaram, "Recycling Modernity: Pirate Electronic Cultures in India," in Ackbar Abbas and John Nguyet Erni (eds.) *Internationalizing Cultural Studies: An Anthology* (Malden, MA: Blackwell, 2005), p. 47.

20 Brian Larkin, *Signal and Noise: Media, Infrastructure, and Urban Culture in Nigeria* (Durham, NC: Duke University Press, 2008), pp. 218–19.

21 Larkin, "Degrading Images, Distorted Sounds."

22 Gayatri Chakravorty Spivak, "Can the Subaltern Speak?" in Cary Nelson and Larry Grossberg (eds.), *Marxism and the Interpretation of Culture* (Chicago: University of Illinois Press, 1988), p. 284.

23 Jacques Rancière, *The Nights of Labor: The Workers Dream in Nineteenth-Century France*, trans. John Drury (Philadelphia: Temple University Press, 1989), pp. x–xi.

24 Jacques Rancière, "Good Times, or Pleasures at the Barriers," in Adrian Rifkin and Roger Thomas (eds.), *Voices of the People* (London: Routledge and Kegan Paul, 1988), p. 50, quoted in Donald Reid, "Introduction," *The Nights of Labor*, p. xxix.

25 Rancière, *The Nights of Labor*, p. 13.

26 Jacques Rancière, "A Personal Itinerary," in Andrew Parker (ed.), *The Philosopher and His Poor*, trans. John Drury, Corinne Oster, and Andrew Parker (Durham, NC: Duke University Press, 2004), p. xxvi.

27 Peter Linebaugh and Marcus Rediker, *The Many-Headed Hydra: Sailors, Slaves, Commoners, and the Hidden History of the Revolutionary Atlantic* (Boston: Beacon Press, 2002).

28 *Ibid.*, p. 332.

29 *Ibid.*, p. 328.

30 Raqs Media Collective, "Value and its Other in Electronic Culture: Slave Ships and Private Galleons," DIVE, a CD-ROM and text collection compiled for Kingdom of Piracy show, Armin Medosch (ed.), FACT Liverpool, 2003, available on-line at http://www.raqsmediacollective.net/texts6.html (last accessed May 14, 2009).

The Pirate Party demonstrates outside the Swedish parliament building in central Stockholm (Frederick Andersson).

Virtual Roundtable on A2K Politics

Amy Kapczynski and Gaëlle Krikorian

with Onno Purbo, Jo Walsh, Anil Gupta, and Rick Falkvinge

The A2K movement today appears as a coalition of groups united by opposition to a particular configuration of intellectual property law. These groups have very diverse relationships with more traditional or "foundational" political ideologies. Some work within them. Some, however, aspire to create their own ideology of A2K in an effort to theorize the conditions of freedom and justice in a networked and informational age. The following "virtual roundtable" was put together via e-mail to create space for some who wish to explore what a theory of A2K might look like and to provide a sense of the stakes and divergences within the A2K movement. Questionnaires were sent to four individuals who represent different approaches and involvements in the movement. They were free to respond to all five questions or only to some. Their responses were gathered and organized in a single document to facilitate the perception of their various opinions. What follows is not intended to give an exhaustive account of the variety of positions and thoughts that exist in the movement. Instead, it gives a sample of the different perspectives within it.

The four perspectives represented here are those of Onno Purbo, who is an IT engineer working with Penulis IT Independent in Indonesia; Jo Walsh, who is an open-source programmer in the United Kingdom working with the Open Knowledge Foundation and with the Free Software Foundation; Anil Gupta, who is the executive vice chair of the National Innovation Foundation in India and the editor of the *Honeybee* newsletter; and Rick Falkvinge, who is the founder of the Swedish Pirate Party and of the international politicized pirate movement. We asked them to respond to the following questions.

QUESTION 1 Is the A2K movement developing its own form of critique or ideology, or new theories of justice, equality, or freedom? Should it? What role, if any, do the following ideas play in the emergence of an ideology of A2K: theories of the generativity of pervasively connected digital networks or peer-to-peer systems, an ethics of sharing and nonrivalry, and the blurring of the distinction between producers and consumers of information?

QUESTION 2 What role do or should ideologies such as the following play in the A2K movement: antiglobalization, anitcapitalism, anticolonialism, libertarianism, classical liberalism, free-market competition, and the public/private opposition?

QUESTION 3 What are the most important historical conditions of the emergence of the A2K mobilization? What shape does or should this mobilization take?

QUESTION 4 Do you think it possible or desirable to include the various components of the A2K movement in a common ethic? If so, what might that ethic be? Where does it come from, and who has had the most influence on its shape?

QUESTION 5 The A2K movement could be viewed as quite creative in its strategies and modes of collaboration. Can the movement be seen as contributing to the creation of a blueprint for a certain social order, or should it instead be thought of as seeking particular changes in discrete areas (for example, "copyleft" or other such methods for making a computer program or other work free, or access to medicines)?

ONNO PURBO

QUESTION 1 From my point of view as a person in a non-English-speaking and developing country Indonesia, our practical barriers to A2K are language (most knowledge is written in English), a high-cost infrastructure, and the education level of our society (less then 10 percent of the society receives a university education and thus it is difficult for many to digest written knowledge). These three together create a major barrier to A2K. It is not so much a question of ideology or theory. What is needed is more practical action and engagement to reduce the barrier. I personally would like to see more pressure on the Indonesian government to increase literacy and education among the Indonesian people. The government is more interested in getting money and getting investors to come, and they don't really care about the people's well-being.

However, if the A2K movement is to have an ideological basis, perhaps a new ideology of justice and equality will be sufficient. The GNU General Public License [a free, "copyleft" license for software and other kinds of works for the GNU operating system] and the Creative Common licenses created in the West would be representations of such an ideology at work. At this moment in Indonesia, only the Ministry of Research and Technology recognizes these licenses. Others are keen for other kinds of license.

But most Indonesians are practical in nature, and they basically don't care about ideology. Indonesians like to benefit themselves, and they'll buy anything that's cheap. Unfortunately, this means that Indonesians will buy pirated stuff such as pirated Microsoft CDs and MP3 songs. They don't care about licenses.

Since 2005, the Indonesian police have been actively sweeping cybercafés and CD shops to find pirated software and MP3s. In 2007–8, the police brutally raided private offices to find pirated software and MP3s on computers. They even checked laptops at Wi-Fi hotspots and airports. They regularly seize computers with pirated software, and normally, the owner of the computer or laptop does not get it back.

Such actions create a huge demand for legal software and materials, especially open-source and open-license materials. For example, the Indonesian Ministry of Education is now buying the rights from the authors of Indonesian education books. They publish the soft copy of the e-books, and Indonesians can download the e-books for free. Software mirroring efforts are currently being undertaken by the open-source community at http://opensource.telkomspeedy.com/bse and http://kambing.ui.edu. The open-source community is very active now—it uses about 10 percent of total local Indonesian bandwidth from several major servers. Several Web sites are devoted to it: http://opensource.telkomspeedy.com, http://kambing.ui.edu, and http://www.vlsm.org.

QUESTION 2 To be honest, for a less educated society such as Indonesia, we will be placed on an unequal playing field by free-market competition. I personally would prefer to emphasize anticolonialism and anticapitalism in the A2K process. I would like to see the localization of knowledge (the translation of knowledge into local languages) and focus on the education of society at the local level. . . .

QUESTION 3 Overcoming the sense of inequality is one of the conditions for the emergence of the A2K mobilization. We normally think that people in the West and people in the East are equal. In reality, most Indonesians see Westerners as superior, if not as "gods." Unfortunately, this leads most Indonesians to assume that Indonesians are inferior and will not be able to acquire knowledge as well as their Western partners. As a result, most Indonesians tend to be passive and less willing to learn and to access knowledge. In my experience, inviting a person from Indonesia or from other developing countries to A2K events provides a good example to other Indonesians that someone from a developing country is equal to others in the West. Such examples will motivate others in Indonesia to learn, to gain more knowledge, and feel they are equal. Learning by seeing a real-life example is important to motivate others, especially in a developing country such as Indonesia.

QUESTION 4 Hmm...to be honest, at least in Indonesia, a concern for ethics is very rare...corruption, cheating during the national exam, and bribing are common in Indonesia. The bureaucracy filters any movement of information from the outside world to local people. Acknowledgment from the West of those who work hard to empower local people would certainly help to break these unhealthy ethical practices in Indonesia.

QUESTION 5 I think that the A2K movement should seek particular changes. One of the most strategic moves would be to support any development and writing of a Creative Commons–licensed e-book handbook for Indonesian school students in the Indonesian language and hosting the materials on the Web. Such an action would significantly reduce the barrier to knowledge and provide most people with access to knowledge.

JO WALSH

QUESTION 1 Regarding the blurring of the distinction between producers and consumers of information, in fact, the distinction between "producer" and "consumer" isn't really there. Information isn't consumed, just passed around (or not). Some have the power to withhold and direct knowledge, others don't.

Regarding the role of ideology, one can learn more from the theory of "network effects" than the broader (and vaguer) theory of "generativity." As a network grows and the benefit in belonging to it increases, the costs (not necessarily monetary) of not participating also increase. So a "network effect" leads selfish actors to behave generously for merely economic reasons. Competitors find they profit most from transporting things among themselves on a free or not-for-profit collective basis. Big banks don't charge for transactions between ATM networks; big Internet service providers don't charge each other to exchange data at the Tier 1 level. These are pragmatic business decisions, not ideological ones. A2K should be seen in the same way to gain serious momentum.

QUESTION 2 I don't see a place for being anti anything or much to be gained from a dialectical approach. When I got interested in A2K, I held a perspective of "information anarchy," yet ended up collaborating with a conservative Cambridge economist.

QUESTION 3 An important historical condition for the emergence of the A2K movement is "the university in ruins."[1] Academic institutions lose their status as a repository of knowledge and of wisdom. Increasingly, their work is subservient to trends in industry and politics. Young academics chafe against the restraints,

connect with pioneers such as Peter Murray-Rust, and start to effect a culture change—which their institutions need in order to evolve and to regain credibility. So some of the most developed "open data" work is in academic science as it meets the commercial sphere. "Network effect data merging" is a phrase coined by Dan Brickley. It pretty well sums up the technosocial conditions that are enabling more open data. But as the network has spread, our ability to mobilize ourselves in real-world ways seems to be stalling.

QUESTION 4 One can see a "hacker ethic" of free software development as a model for open knowledge development. Common ground for open access in different domains has been expressed by "definitions" explicitly modeled on their open-source/free-software counterparts—the Open Knowledge Definition, and the Definition of Free Cultural Works.[2]

These definitions propose a very pragmatic ethic. Attempts to define "ethical terms of use" don't have the same traction or reusability—there are so many different reasons why people produce open knowledge. It is important to remember that the hacker ethic is primarily one of "enlightened self-interest"; the collective benefit of open-source software is a side effect of one's own benefit. "Open-source values" are often oversold in theorization. There's a continuum of "open data, open standards, open source" reinforcing one another, and most A2K activists I have met are involved in at least two of these efforts.

QUESTION 5 This text accompanied the World Summit on Free Information Infrastructures in 2005, an event bringing together free networks, science, and civic data, free maps, open money, and open hardware: "Every kind of research effort can be re-oriented, from protectionist bubbles formed by the pressure to Be First, towards a space into which we can all arrive together, by working on the things that we enjoy most, having the means to contribute to enhancing the things that we like, and getting together to talk about it." The most interesting possibilities were in the intersections and the gaps between the different topics. In viewing subjects as "discrete," we cut off simplifying connections. We see more in common the more we view at once.

ANIL GUPTA

QUESTION 1 The blurring of the distinction between the producer and the consumer of information and knowledge is one of the most important trends in the emergence of an ideology of A2K. We need to pay attention to an ethics of sharing, and the nonrival consumption of innovative ideas and technologies has led us to develop the concept of a technology of the commons. There should be no restrictions on copying or improvisation in people-to-people exchanges, but people-to-firm exchanges must be governed by licensing protocols. The Honeybee Network, for example,[3] is a social movement created to recognize, respect, and reward creativity, innovation, and traditional knowledge in local communities and individuals and to link these people, who are rich with creative knowledge, though economically poor, with each other and with formal science and technical institutions in order to help them develop socially and economically productive businesses. It is essentially a knowledge-based approach to the alleviation of poverty and the conservation of the environment. The underlying concept of the poor as provider is contrary to the vision that treats them only as consumers. Not all disadvantaged people are creative or innovative so as to be able to solve their problems on their own, optimally or even suboptimally. But some indeed are very creative. And for many of them, innovation is imperative.

QUESTION 2 I think the major roles in A2K ideology should be played by the concept of the commons and support for institutions of sharing and caring, by support for the means of reducing transaction costs for knowledge-rich, economically poor people, and by emphasis on the possibilities of overcoming the asymmetry in the flow of information and knowledge between developed and less developed social and regional groups.

QUESTION 3 The philosophy of the Honeybee Network indicates one direction that A2K mobilization should take. Support for the Creative Commons movement is another. The use of shared patents is still another. I think the historical exploitation of traditional knowledge has contributed to a considerable concern in the WTO as well as WIPO about the ethics of such a situation.

QUESTION 4 I think, that yes, there is a common ethic: Those who share most, most liberally, and without often asking anything in return have got to contribute the most—that is the local communities and grassroots innovators and those who possess traditional knowledge.

QUESTION 5 A2K should both promote the creation of a blueprint for a new social order and seek particular changes. At least, it can do both if there is collaborative research and discussion among contending thinkers. This is not happening enough as yet.

RICK FALKVINGE

QUESTION 1 After having talked about A2K issues from a political standpoint for several years, we have definitely seen a new ideology emerge. Some summarize it as "information ethics," with the key question being "What information is public, and what information is private, and how is the line between the two safeguarded?"

The Pirate Party is not only about access to knowledge and culture; it is also about the right to privacy. This can be seen as the direct opposite of access to knowledge: I do not have access to all knowledge about my fellow citizen, nor should I. But these two classes of information are not unrelated; they are interdependent on one another. The political movement is therefore much more about raising awareness about the border between public and private information.

As it turns out, any attempt by the government to force information to cross this line—private to public, or public to private—requires repressive governmental action that infringes on the civil rights of real people to varying degrees. The most obvious example would be how your tax records become public records. Anybody in Sweden (or in the world, for that matter) can easily check how much money I make. This is an example of private information that is forced into the public domain by the government. More egregious examples would include wiretapping my phone calls, confiscating my computers and their content, and so on.

Reminder from the Recording Industry Association of America (Patrick Broderick/ModernHumorist.com).

An important axiom here is data leakage: It has been proven again and again, and must be assumed so in policy making, that all databases leak. They leak and are converted to uses not originally intended, uses that go against the will of the people concerned. Any information that is to remain private must therefore simply never be collected by an agent of the government.

The opposite — public information — includes any published information that can be digitized, including movies and music, which are usually the target of file sharing. One key question in the file-sharing debate is "How many repressive measures are the government prepared to undertake in order to prevent citizens from communicating already-published information to one another in private?" If I am sending you a piece of music in an e-mail, that is illegal. If we're in a private video chat, and I drop a copyrighted video clip there, that is illegal. But in order to enforce today's copyright laws, where breaches happen in such private communications, those private communications must be monitored — by law enforcement, and by private interest groups. Copyright has reached a point where it can no longer coexist as a concept with private communications.

Realizing this, private communications is a much more important aspect of society than the commercial monopoly of copyright. Copyright advocates therefore need to take several steps back: Information that has once been made public, as in a published book or movie, cannot be forced back into being privately controlled without severe repression.

So everything boils down to awareness of the line between private and public information and that repressive measures are needed in order to force information across this line involuntarily.

Thus, it is not so much the ethics of sharing or the blurring of line between consumers and producers as it is vital civil liberties that are at stake in this debate. In a yet larger view, for a society to advance culturally and technologically, its inhabitants require the right to privacy and private communications. The copyright lobby threatens that notion.

QUESTION 2 We find that our supporters come from all previous political ideologies, evenly distributed. The pirate ideology can be supported fully from at least capitalist, communist/socialist, liberal/libertarian, technological, and civil rights standpoints.

Once you rephrase copyrights and patents as "government-sanctioned private monopolies," you have most of the capitalist, liberal, and socialist people on your side.

The antiglobalization movement is not very well represented. You see some anticolonialism, because the United States is the driving force behind today's copyright and patent maximalism. The vast majority of key players, however, are seeing this not in the light of what they are against, but in the light of what they want to build.

I have a vision of a society where all of humanity's knowledge and culture is available 24/7 to anybody with access to the Internet. And you know what? We already have the technology for this! We can do this already! The only thing standing in the way of this vision—which would be a huge leap ahead, much larger than when public libraries arrived a hundred and fifty years ago—are a couple of anachronistic monopolies.

QUESTION 3 Without a doubt, the emergence of the A2K movement has been a result of the power of everybody to self-publish ideas. What we are seeing now has enormous parallels with the emergence of the printing press, which had similar catalyzing effects on the public discourse.

Where previously the power to publish ideas was limited to a publishing elite— the newspaper houses—this power now sits with every citizen. There is no longer any control of ideas, no control of public discourse. This is an immensely positive development for democracy.

QUESTION 4 I don't think it is possible to be normative to the degree that a common A2K ethic is the result. Whereas one might try to describe the various

Richard Stallman, founder of the Free Software Foundation and the GNU Project, as Che Guevara (http://geekz.co.uk/shop/store/show/che-stallman-sticker).

components in a common umbrella ideology, there will always be people who disagree with that ideology and state their own motives for pursuing the same goals. This has been very apparent in the pirate movement, where people have vastly different motives for exactly the same goals.

Yes, it is true that sharing is ethical—"sharing is caring," as we say—but some people do it because they can, others because it is politically correct, and yet others do it without thinking about it.

It is equally true that access to knowledge and culture has been a driving force behind the ongoing development of our civilization, but there are plenty of people who couldn't care less.

So I see the movement more as the power of a focal point of many different viewpoints that converge on the idea of access to knowledge and culture.

QUESTION 5 The mere fact that we have a political party on the topic of A2K, which has been successful beyond everybody's expectations (though yet without representation), shows that these indeed are high-order issues with general implications for the shape of society. I have elaborated somewhat on the fundamental questions above. Overall, the changes that are caused by the mass democratization of the published word go well beyond any discrete area. The events unfolding now will not just crumble today's power structures, but put them in the kitchen blender and set it to "Disintegrate," happily leaning against the kitchen counter with one hand on the blender lid while leisurely whistling folk songs.

I wouldn't go so far as to call it a "blueprint," though. That implies that there is some sort of plan, or at least beforehand knowledge of the effects. We have no idea what is going to emerge on the other side, only that knowledge, ideas, and culture can no longer be controlled by an elite few.

And that, just in itself, is a great development.

NOTES

Cecilia Oh translated Onno Purbo's contribution to this roundtable discussion.

1   Bill Readings, *The University in Ruins* (Cambridge, MA: Harvard University Press, 1996).

2   For the Open Knowledge Definition, see http://freedomdefined.org/Definition; for the Definition of Free Cultural Works, see http://www.opendefinition.org (both last accessed May 23, 2009).

3   The Honeybee Network encourages volunteers to join walking journeys throughout rural India, where they scout for innovations and outstanding traditional knowledge, recognize, respect, and reward the innovators, and try to spread these innovations to other parts of the country and the world. They are registered with National Innovation Foundation, which aims to build the value chain by adding value, providing risk capital under the Micro Venture Innovation Fund, the first of its kind set up in the country, and helping to protecting intellectual property so that an entrepreneur can easily and efficiently license the rights to bring the product to market. At the same time, the Honeybee Network also helps to disseminate a large number of technologies as open-source knowledge for the greater common good around the world for sustainable resource management. See http://knownetgrin.honeybee.org/innovation_database.asp (last accessed March 4, 2010).

PART FOUR

STRATEGIES AND TACTICS OF A2K

top: On July 31, 2000, for the opening day of the Republican National Convention in Philadelphia, members of ACT UP Philadelphia dropped a giant 30- by 75-foot banner on a downtown billboard (ACT UP Philadelphia).

bottom: Women of Via Campesina protest so-called terminator technology (which would render seeds sterile so that farmers would have to purchase them each season) inside the COP8 Meeting (http://fr.banterminator.org/content/view/full/414).

A Comparison of A2K Movements:
From Medicines to Farmers

Susan K. Sell

In recent years, developing countries, NGO activists, multinational corporations, and governments increasingly have clashed over intellectual property policies. But not all intellectual property is alike, and neither are all A2K campaigns. A2K campaigners come from diverse backgrounds and emphasize different things. Some campaigners champion creative expression and equate creativity with freedom. Others focus more on human rights, health, nutrition, and economic development. Still others worry about the disappearance of the scientific commons and the stifling of innovation. All share a skeptical view of the contemporary intellectual property rights regime and agree that the proliferation of intellectual property rights is having a negative effect on many aspects of the human experience. This essay focuses on A2K campaigns for access to books, drugs, and seeds.

It does so because access to educational materials (including software, books, digital media, and scientific and other scholarly journals), life-saving medicines, and seeds for sustainable and nutritious agriculture is the sine qua non of economic development. First, access to these resources has the potential to spur lasting development and to build capacity for innovation. Yet the strong trend toward transforming these into private commodities for sale at premium prices by making them subject to higher levels of intellectual property protection has made them less available to those who need them most. Second, increasing economic concentration over the past several decades characterizes each of these sectors. Fostered by relaxed antitrust enforcement and expanded intellectual property rights in the life sciences and academic publishing industries, economic concentration has increased these industries' political power in U.S. trade policy. This has led to an expansion of intellectual property rights multilaterally, regionally, and bilaterally. Third, medicine and agriculture have become increasingly commercialized.

For example, agricultural research once was a primarily public-sector activity; the United States established land-grant universities to develop and give away seeds to farmers. Now private investment in agricultural research has far outpaced declining public-sector investment. As for-profit enterprises, agricultural firms and brand-name pharmaceutical firms do not focus on the needs of the poor. Private-sector incentives to develop crops for subsistence, smallholder agriculture, and medicines for tropical diseases are lacking. Brand-name pharmaceutical firms do not invest in research on tropical diseases that mainly affect the Global South. The needs of a vast majority of the world's population continue to go unmet.

While NGO activists and developing countries have worked together and achieved notable victories, their potential to provide an economic and political counterweight to the aggressive expansion of intellectual property rights remains unfulfilled. In addition to facing opponents that have far greater resources, until quite recently, the lack of mobilization across issue areas has held the access campaigns back. The global corporations that lobbied for the 1994 World Trade Organization Agreement on Trade-Related Intellectual Property Rights (TRIPS) represented diverse sectors such as software, pharmaceuticals, and entertainment. They succeeded because these actors set aside their differences and joined forces to achieve common goals.[1] Today, however, it appears that the once invincible intellectual property juggernaut is beginning to slow because of sharp divisions between patent-dependent industries and information technology industries, on the one hand, and outside pressures, including the A2K movement, on the other. All of the access campaigns share the goals of rebalancing the scales from private reward to public interest and introducing a norm-setting process that includes multiple stakeholders—not just property holders.

This essay presents some of the core issues animating the campaign for access to medicines, the push for access to educational materials, and the various agricultural campaigns and considers how they might be taken further under the broader A2K banner. The survey is not meant to be exhaustive, but rather aims to give the reader a general sense of compatible elements and similar strategies across campaigns. It also seeks to identify some of the tensions within the campaigns and the differences between them with an eye to both the obstacles and the opportunities involved in going forward. The essay begins by identifying shared strategies, then examines both the similarities and the differences between the campaigns in terms of the ways that each successfully employed advocacy for the different issues is involved. It concludes with a discussion of the obstacles and opportunities for A2K exemplified by the similarities and differences between these campaigns.

DISCURSIVE AND INSTITUTIONAL STRATEGIES: REFRAMING ISSUES AND SHIFTING FORUMS

A2K campaigners have employed both discursive strategies and institutional strategies in the effort to combat the expansion of intellectual property regimes. The discursive strategies that they have employed have sought to reframe the issues, redefining them in ways that will activate political support. As James Boyle suggests, "one must convince people that one's arguments are good, one's institutional innovations necessary, and one's horror stories disturbing."[2] In the early 1980s, the TRIPS campaigners reframed intellectual property as a trade issue and thus were able to activate and mobilize new constituencies (exporters of intellectual property–laden goods and services and trade ministries) and institutions (the office of the U.S. trade representative and the General Agreement on Tariffs and Trade, GATT). As part of the post-TRIPS backlash, the access-to-medicines campaign redefined intellectual property as a public-health issue. This, too, effectively activated new constituencies, including developing countries facing the HIV/AIDS pandemic, NGOs such as ACT UP, the Treatment Access Campaign, Health Action International, Médecins Sans Frontières (MSF), and the Consumer Project on Technology, generic drug producers in India such as Cipla and Ranbaxy, health ministers, supportive academics, and legal experts. The public-health frame also made new institutions such as the World Health Organization (WHO) and national health ministries available as forums for deliberations on intellectual property. Words matter if they increase the political costs of supporting the status quo. In the case of access to medicines, the language of "profits before people" and "patents before patients" increased those costs.

The institutional strategies that campaigners employed on both sides of intellectual property issues, that is, those who wish to expand access and those who wish to ration access, have included strategic forum shifting.[3] Governments, private actors, and NGOs can shift forums horizontally across institutions, that is, from the World Intellectual Property Organization (WIPO) to the World Trade Organization (WTO) and from the WHO to the WTO. They can also engage in vertical forum shifting, as the United States has done with its bilateral and regional free-trade agreements to secure TRIPS-plus protection in developing countries.[4] According to Lawrence Helfer, there are four main reasons why actors choose to shift forums: to achieve a desired policy goal, to relieve political pressure for new laws by offloading an issue to another forum, to generate feasible alternatives or "counter-regime norms," and to integrate those norms into institutions such as the WTO and WIPO. Participants pursue the second, the "safety valve" strategy, in order to preserve the status quo. Here, participants confine an issue to a forum that is unlikely to effect meaningful change.[5]

Choosing institutions that afford actors better access or those whose philosophies reflect their own goals can provide opportunities to propose and experiment with policy approaches. Forum shifting can provide governments with a safe space in which to exchange information, develop "soft law" (quasi-legal instruments that do not have legally binding force), and craft viable policy alternatives that address their concerns.[6] Soft-law forums such as the WHO and the Convention on Biological Diversity have become significant incubators of alternative approaches to intellectual property protection.

This strategy also can generate competing discourses that can change the way parties read TRIPS and are willing to apply it.[7] Furthermore, these competing discourses can challenge various domestic political bargains, integrate a broader range of viewpoints and parties into the issues, and raise the political costs of defending the status quo. The campaigns for access to medicines, to books (educational material, digital media), and to seeds (farmer's rights, food security, access to agricultural biotechnology, and the defense of traditional knowledge and biodiversity) employed both discursive and institutional strategies. Highlighting these two strategies helps to illuminate the various strategic choices and outcomes that differentiate the campaigns for access to medicines, educational materials, and seeds.

ACCESS TO MEDICINES

Initial American concern about access to medicines focused on high prices. In October 1995, U.S. consumer advocate Ralph Nader and his colleague James Love wrote to then U.S. Trade Representative Mickey Kantor suggesting that U.S. trade policy had been too narrowly focused on protecting the interests of the Pharmaceutical Research and Manufacturers of America abroad. They pointed out that significant amounts of taxpayer money had underwritten drug development, yet the drugs thus developed had become the exclusive and very lucrative property of private firms. They began an Internet newsletter, *Pharm-Policy*, that became an important tool for mobilizing interest in drug-pricing and patent policy. These initial moves set the stage for advocates for access to medicines to insert the issue into debates over trade policy into international forums.

In achieving this success, the discursive strategy that these advocates employed was to identify the links between intellectual property and public health. The institutional strategy that they employed was to propose an alternative in a venue more congenial to the issue of access to medicines than the WTO—notably, the WHO.[8]

In October 1996, Health Action International (HAI) organized an NGO meeting in Bielefeld, Germany, on health care and TRIPS in which Ellen 't Hoen (then of HAI) and James Love of the Consumer Project on Technology (CPTech, now

Knowledge Ecology International) participated. HAI and CPTech became active in negotiations on the Free Trade Agreement of the Americas, criticizing TRIPS-plus provisions. This nascent campaign also began working on the World Health Organization's Revised Drug Strategy in preparation for the World Health Assembly in January 1998.[9] In this way, the campaign's discursive framing of intellectual property issues in terms of public health also helped to activate new institutional allies for the cause.

As 't Hoen notes regarding the reframing strategy that the access-to-medicines campaign employed, "The public-health advocates coordinated by Health Action International first raised concerns about the consequences of globalization and international trade agreements for drug access during the 1996 World Health Assembly. They sought to get the WHO to intervene in intellectual property issues because it became apparent that the GATT negotiators had drawn up the rules without any consideration for health issues."[10]

As 't Hoen also notes here, they successfully shifted the debate on the issue of access to medicines from a forum dedicated to world trade issues, the WTO, to a forum friendlier to their argument. When they did, the WHO's Executive Board approved the recommended Revised Drug Strategy, endorsing compulsory licensing and parallel importing and underscoring the priority of health concerns over commercial interests. The U.S. trade representative and the Pharmaceutical Research and Manufacturers of America were upset because the WHO Revised Drug Strategy, meant to offer guidance to developing countries, endorsed the same practices against which they were vehemently fighting via trade policy in the WTO. Thanks to this forum shift, the WHO became an important incubator of an alternative approach to access to medicines.

Campaigners supplemented institutional strategies with public demonstrations that further bolstered their efforts. Another big push in the access-to-medicines campaign came in reaction to very aggressive U.S. efforts to support and defend the Pharmaceutical Research and Manufacturers of America's interests in countries in the grip of the HIV/AIDS pandemic. In 1998, the United States applied extensive trade pressure on Thailand and South Africa, two countries that were trying to use TRIPS flexibilities to provide affordable HIV/AIDS drugs. The United States threatened these countries with trade sanctions. The U.S. trade representative also supported the Pharmaceutical Research and Manufacturers of America's lawsuit against South Africa to try to get South Africa to repeal its 1997 law incorporating TRIPS flexibilities. In the United States, the advocacy group ACT UP visibly and vocally disrupted Al Gore's presidential campaign launch in the summer of 1999 in an effort to get the Clinton administration to back off, and within a week, the Clinton administration reversed two years' worth of objections to South Africa's law. In the face of mounting public criticism and effective activism by groups such as the Treatment Access Campaign, MSF, and CPTech, the pharmaceutical firms dropped the lawsuit.

A second strategy involved exploiting the general disapprobation, even within classical liberal economics, of monopolies and their deleterious effects. The international campaign against the South Africa lawsuit gained traction partly because NGOs, activists, and the media recognized the problems that intellectual property monopolies were causing. Publicizing this monopoly issue in reaction to pressure by the United States and by Big Pharma and its South African affiliates was a turning point.

In 1998, there was a dramatic decrease in the costs of producing generic HIV/AIDS drugs. Brazil's successful program of providing antiretroviral drugs free to all those with HIV/AIDS had altered the market for bulk fine chemicals. Brazil had purchased $150 to $200 million worth of these chemicals, and in three years, the cost of the chemicals dropped from $10,000 to $750 per kilo. Dr. Yusuf Hamied, CEO of the Indian manufacturer of generic drugs, Cipla, offered to produce the drugs for one dollar a day per dose. This rendered an intractable problem suddenly tractable because the solution was plausibly affordable. Furthermore, it effectively isolated the impact of intellectual property protection on the affordability of these drugs. The connection between intellectual property and high prices was stark. The high prices looked more like monopoly rents than anything else. Some thus argued that the Pharmaceutical Research and Manufacturers of America valued patents and profits over patients. The raw emotional context of the HIV/AIDS pandemic vividly animated the urgency behind the political and technical issues around intellectual property, and distrust of monopoly power proved an effective way to mobilize that emotion.

Aligning the access-to-medicines movement with the obvious interest of developing countries in the issue also increasingly became a major factor in mobilizing effective support, and a group of African nations pressed for a special TRIPS Council meeting on the issue. Developing countries formed an effective coalition on TRIPS and public health and aggressively negotiated for the Doha Declaration on TRIPS and Public Health.[11] This was a significant application of the forum-shifting strategy, but this time employed in the opposite direction, using an approach devised and eventually accepted in terms of health-care issues in the WHO to press for changes in trade policy in the WTO.

Finally, in August 2003, the negotiating parties agreed to amend TRIPS to make it possible for countries without generic production capabilities to import generics manufactured elsewhere. Member countries agreed to waive temporarily Article 31(f), which restricts compulsory licensing only to supply one's domestic market, and to include procedural safeguards to prevent the diversion of cheap medicines to rich-country markets.[12] At the Hong Kong WTO ministerial meeting in December 2005, member states agreed to amend TRIPS permanently. Member states agreed to retain the waiver of 31(f) so that countries could export medicines produced under compulsory license.

However, these successes were not achieved without generating some dissension within the movement. While some African delegates expressed relief that the amendment resolved some of the uncertainty that the temporary waiver had generated,[13] many observers criticized the amendment as being cumbersome and difficult to use.[14] Also troubling was the so-called "Chairman's Statement" that reportedly had been crafted by Karl Rove and Henry McKinnell, then CEO of Pfizer. This underscored campaigners' suspicions about the close relationship between the Pharmaceutical Research and Manufacturers of America and the U.S. government.[15] To the extent that the amendment can increase the availability of generic drugs, this will be a positive outcome, but the jury is still out.

ACCESS TO BOOKS AND EDUCATIONAL MATERIALS

The campaign for access to books and educational materials has focused on copyright, software patents, and technological protection measures as barriers to access to books and educational materials, including software and digital media. Ironically, as the Internet and the digital revolution have made so many resources available, they have also engendered a persistent rights holders' push to find new ways to ration access and control the dissemination of copyrighted material. Many analysts, educators, librarians, museum curators, and activists have bemoaned the erosion of fair use, the uncompensated use of copyrighted material for noncommercial purposes.[16] This campaign emerged out of three related but distinct movements: the emergence of the free and open-source software movements and the creation of the GNU General Public License, both of which posed a direct challenge to Microsoft's proprietary model for software; Creative Commons licensing practices; and the political battles over the 1996 digital treaties negotiated at WIPO and over the sharp restrictions on the use of copyrighted material in the U.S. Digital Millennium Copyright Act of 1998.

Richard Stallman founded the Free Software Foundation in 1984 and developed the GNU ("GNU Not Unix") General Public License.[17] Stallman has been an ardent champion of free software. "Free" signifies a commitment to four freedoms: freedom to run the program for any purpose, freedom to study how the program works and adapt it to your needs, freedom to redistribute copies so you can help your neighbor, and freedom to improve the program and release your improvements to the public so that the whole community can benefit.[18] The GNU General Public License embodies the notion of "copyleft," in which one uses copyright law to require users to share.

The movement for free and open-source software holds promise for development in two ways. First, it provides users with access to source or programming code. Second, access to source code can transform the relationship between the user and the product. Rather than being a passive consumer of proprietary object code, such as is found in Windows, the user becomes an active and potential producer of source code. This stimulates adaptive and innovative possibilities for users to alter and experiment with programs and tailor them to their specific needs.[19] It is empowering in a way that proprietary object code is not.

However, tensions that have divided the movement for access to books and educational materials have led to different policy emphases and different strategies for advocating them. Lawrence Lessig, a cofounder of Creative Commons licensing, has promoted practices that aim to expand, rather than reduce access. However, in recent years, Richard Stallman has broken ranks with the Creative Commons licensing project on principle. For example, some Creative Commons licenses bar commercial use. By definition, this violates the first of Stallman's four freedoms and thus is not free software.[20] As Stallman states, "I cannot endorse Creative Commons as a whole, because some of its licenses are unacceptable."[21] In particular, he unsuccessfully has urged Creative Commons to drop or disassociate itself from two licenses, the Sampling Licenses and the Developing Countries Licenses, a.k.a. Recombo and DevNat, because they prohibit noncommercial verbatim copying.[22]

However, the vigor of proponents of greater intellectual property protection has produced solidarity between sometimes unlikely allies in opposition to this agenda. In the United States, opposition to the Clinton administration's vision of copyright in the digital age generated a potent domestic political constituency. In the early 1990s, Bruce Lehman, then head of the U.S. Patent and Trademark Office, issued a "white paper" outlining this vision, which highlighted the economic and competitive potential of the "information superhighway." Lehman, like others in the administration, wanted to harness and develop this technology in a way that would strengthen American economic competitiveness. This promising web of communications networks had the potential to make astounding amounts of information available to all sorts of disparate users. Yet Lehman's policy recommendations included a dramatic expansion of copyright holders' rights and a sharp contraction of users' rights. Lehman argued that without giving content providers expansive rights, they would refuse to make their content available, and the "highway" would be empty.[23] Particularly pernicious aspects of the recommended policy included provisions to restrict the scope of fair use, to prohibit the production or distribution of any devices or services designed to circumvent technological protection measures, and to prohibit circumvention of such measures.

Opponents of this policy mobilized against it by building a coalition of the many diverse stakeholders who would be adversely affected by it. Alarmed law professors and librarians, led by Peter Jaszi of American University and Adam Eisgrau of the American Library Association, united in a commitment to preserve fair use. Jaszi and Eisgrau then brought together multiple stakeholders, including library organizations, on-line service providers, phone companies, and civil rights and consumer-protection organizations, including CPTech.[24] They called themselves the Digital Futures Coalition (DFC) and received significant support from the Home Recording Rights Coalition and Sun Microsystems. The DFC and such diverse allies as Cory Doctorow's Electronic Frontier Foundation, the Electronic Privacy Information Center, and the National Education Association successfully stalled rapid passage of the white paper's version of copyright law.

Facing domestic legislative deadlock, in 1996, Bruce Lehman pursued his preferred version of copyright's future in WIPO copyright negotiations. Using vertical forum shifting, he hoped that international acceptance of the U.S. vision would facilitate domestic implementation of these rules. The coalition of opponents who had mobilized domestically likewise shifted their attention to the WIPO proceedings. Executives from Sun Microsystems and Netscape and lobbyists from the various mobilized NGOs worked to highlight concerns over this high-protectionist agenda and participated informally as observers and lobbyists in the negotiations. Their core focus in terms of discursive reframing was on the principle of fair use. None of the original U.S.-sponsored digital proposals passed, and the proposed database treaty did not emerge at all.[25]

Yet coalition building and the strategies of discursive reframing and forum shifting have not been an unqualified success for the campaign for access to books and educational materials. Although these strategies defeated Lehman's attempt to prevail by shifting forums to WIPO, the U.S. Digital Millennium Copyright Act of 1998 incorporated even more expansive protections for copyright holders, narrower exceptions to copyright, and anticircumvention provisions. As Jessica Litman documents, a complex combination of negotiating fatigue, Senate and House committee rivalries, convoluted concessions, intense lobbying, and a Christmas-tree approach (each party hanging its own desired policy ornament), led to the act's ultimate passage.[26] As she states, "when the groups involved in the DFC [the Digital Futures Coalition] agreed to withdraw their opposition to the DMCA in return for modest concessions . . . they believed the deal they had made was the best deal they could get. By that point in the process, it probably was."[27] The Digital

Millennium Copyright Act granted copyright holders extensive new rights. It also incorporated preemptive measures that made producing, importing, distributing, or offering any technology or service designed to circumvent access protection or copy protection illegal. In addition, it incorporated nonnegotiable contracts such as click-through end-user license agreements that make the traditional defenses and statutory immunities of copyright law unavailable.[28] However, the coalition that tried to fight the Digital Millennium Copyright Act is still active, and many of those who had mobilized against the WIPO copyright treaties of the 1990s were able to defeat the proposed, highly protectionist WIPO Broadcast Treaty in June 2007.[29]

FOOD SECURITY, FARMERS' RIGHTS, AND ACCESS TO AGRICULTURAL BIOTECHNOLOGY

The 1980s seed wars over farmers' rights, the campaign for access to agricultural biotechnology, including genetically modified organisms, and the campaigns that emerged around the issues of biodiversity, biopiracy, and traditional knowledge have not offered as coherent an approach and as unified a front as the campaigns for access to medicines and to educational materials.

Of the two principal strategies that access campaigns have shared, discursive reframing and forum shifting, discursive reframing has been the most problematic in this area, because agriculture for development and access to seeds are complex topics, and some discursive frames are particularly confusing. A term such as "food security," for example, means different things to different people. To some, it means reliance upon landraces (domesticated animals or plants related to their local natural and cultural environments), local cultivars, organic-farming methods, protection from biopiracy, and the preservation of biodiversity. To others, it means higher yields and pest-resistant, drought-tolerant crops produced via genetic modification. Similarly, "farmers' rights" evokes different images for different people. Some think of farmers' rights as the right to save seed and the right to resist globalization, while others think of them as rights to compensation for the appropriation of landraces. This lexical confusion makes it harder for parties to come together and makes it more difficult for discursive strategies to mobilize potential activists. Depending upon who says it, "food security" can be shorthand for what an opponent of multinational agribusiness advocates or for what advocates for genetically modified organisms such as Monsanto claim as their goal.

Forum shifting therefore is a more reliable strategy in this area. NGOs and developing-country governments have shifted to forums such as those offered by the Convention on Biodiversity and the UN Food and Agriculture Organization (FAO) to try to achieve outcomes that are more favorable than what they would expect in forums more oriented toward the support of property rights, such as the WTO and WIPO. The Convention on Biodiversity and the Food and Agriculture Organization have served as incubators for norms that support the A2K movement. While these other forums have less power as norm setters in the global economy, the contest is ongoing.

Forum shifting proved to be the most effective strategy in the so-called "seed wars." In the early 1980s, Cary Fowler, a political activist opposed to the extension of intellectual property rights to life forms, adopted just such a strategy. According to Fowler, the progressive extension of property rights with no commitment to conserving genetic diversity led activists such as Fowler himself and Pat Mooney to "develop a strategy and set it to work in a new but potentially friendlier arena."[30] The 1981 biennial conference of the FAO also marked a shift in the discursive strategy concerning plant genetic resources from patenting plant varieties to the broader connections between patenting, genetic conservation, and development.[31] Fowler and Mooney had consulted with the Mexican ambassador to the FAO prior to the meeting and had provided him with reports on genetic resources. The ambassador arrived armed with proposals for the establishment of an international gene bank and an international legal convention for the exchange of genetic resources.[32]

In pursuing these strategies, the main concern for Fowler and Mooney was preventing the one-way flow of germplasm from the genetically rich Global South to the North.[33] Developing countries sought to address the inequities arising from a system in which they lacked open access to improved varieties bred by seed companies, especially when developing countries' "raw" germplasm was the base for some of those "worked" resources.[34] And in this, the advocates for the Global South were partially successful. In 1983, negotiators agreed to the International Undertaking on Plant Genetic Resources against the adamant opposition of the United States, the United Kingdom, Australia, and major seed companies. The nonbinding undertaking was designed to ensure conservation and unrestricted availability and sustainable utilization of plant genetic resources for future generations by providing a flexible benefit and a burden-sharing framework.[35] Negotiators also established a new commission on Plant Genetic Resources at the FAO where states could discuss and monitor the nonbinding undertaking.

Beginning in 1989, the International Undertaking on Plant Genetic Resources also incorporated Pat Mooney's notion of farmers' rights "to acknowledge the contribution that farmers have made to conserving and developing plant genetic resources."[36] The idea was to safeguard the rights of farmers to "work with, and live from, farming systems based on diversity, in the face of expanding monocultures and uniform seeds."[37] The 1989 agreement asserted the principle of unrestricted access and common heritage of both raw and worked germplasm. However, the industrialized countries rejected the principle of open access for worked plant genetic resources.[38]

Along with such setbacks, tensions within the farmers' rights movement emerged, and with them, differences over goals and the strategies for reaching them. In the late 1980s, the developing countries became interested in property rights in raw germplasm—"green gold"—as a way to acquire wealth.[39] The developing countries began to assert their sovereign rights to the exploitation of their biological resources. Gene-rich countries could deny access to their biological riches and cut deals with companies that wanted to engage in bioprospecting. Asserting property rights in biological resources has been controversial for both worked and raw varieties. As Jack Kloppenburg has noted, it can be questioned "whether the whole farmers' rights orientation was the proper way to go and whether there simply weren't too many contradictions embedded in trying to use the master's tools to dismantle the master's house."[40] These tensions and differences are still palpable today and have been particularly acute in debates over so-called "traditional knowledge." Some oppose property rights per se, whereas others advocate property rights and compensation for collective endeavors as a matter of social justice and to prevent biopiracy. The prospect of bestowing more property rights is anathema to many A2K campaigners, who bemoan a world with too many property rights.

In line with this newfound interest in "green gold," the 1991 annex to the 1983 FAO undertaking incorporated the principle of states' sovereignty over their plant genetic resources. Critics have argued that the 2001 FAO Treaty that came into force in 2004 is a bonanza for plant breeders and a disaster for farmers' rights. On the plus side, the FAO Treaty recognizes that access itself is the main benefit to be shared, yet it clearly enshrines the notion of property rights over genetic resources. In the negotiations, developed countries resisted any measures that would prevent their patenting of genetic resources, whereas developing countries wanted to limit the scope of the treaty to protect any business opportunities that might result from providing individual genes on the global market.[41]

As in the situation we have just examined, in dealing with issues of agricultural biotechnology, including the problem of genetically modified organisms, discursive reframing proved to be a difficult strategy for A2K to employ effectively. In this area, opponents and supporters of the development and spread of genetically modified organisms have employed three different discursive frames. Opponents maintain that genetically modified organisms are a curse for rich and poor alike. They argue that they damage ecosystems and reduce biodiversity. Windborne seeds can pollute land used for natural crops. The technology also promotes monocropping and encourages the large-scale farming of cash crops for export in lieu of growing nutritious staple crops for local consumption. Activists opposed to the development and spread of genetically modified organisms, such as Greenpeace and Oxfam, support organic farming, low-input-oriented farming, the use of landraces, on-farm saved seeds, and biological insecticides.[42] Others, especially in Europe, highlight the potential health risks of using genetically modified organisms. Still other arguments focus on the dangers of adopting genetic modification technologies as ensnaring poor farmers in inextricable webs of dependency on multinational corporations for expensive inputs. Indian activist Vandana Shiva has dramatized this antiglobalization position with her accounts of farmer suicides in Andhra Pradesh and Maharashtra.[43] On political-economy grounds, many African countries are appreciably wary of pollution of their export crops by genetically modified organisms. Those countries exporting to Europe would lose those markets because Europe does not accept genetically modified agricultural products.[44]

On the other hand, proponents see genetically modified organisms as a blessing. Agrifood corporations promote genetic modification as a technology that can increase yields and productivity and solve the problem of hunger. To the extent that genetically modified organisms increase food production for smallholder subsistence farmers, create income opportunities for poor farmers, increase the demand for labor, and lower staple food prices, they could help alleviate poverty and malnutrition.[45] Genetically modified organisms such as Bt cotton (cotton modified with the insecticidal genes found in the bacterium *Bacillus thuringiensis*) can render plants toxic to prominent pests, reduce crop losses, and reduce the need for toxic pesticides. They can also contribute to a country's competitiveness in globally integrated markets in agricultural products.[46]

A third discursive frame valorizes the use of genetically modified organisms as the "farmers' choice."[47] Based on the marketing and adoption of illegal Bt cotton seeds in Gujarat, India, proponents recast the debate over the use of genetically modified organisms by emphasizing that when given the choice, farmers have embraced the technology. They have claimed that campaigns against the use of genetically modified organisms such as Shiva's were out of touch with the actual farmers whom they purported to represent. When Monsanto sought to pressure the government of Gujarat to burn the illegal plantations, farmers who had planted the illegal Bt seeds protested so vehemently that the government backed off.[48] The "farmers' choice" rationale depicts farmers not as victims of globalization, but rather "as decision makers and voters for the technology."[49] The increasing use of genetically modified organisms on a grand scale by China, Argentina, and Brazil is said to underscore this element of choice.[50] China is moving full-speed ahead in public-sector agricultural biotechnology, and one can hardly depict China as a "victim" of Monsanto.[51]

Note that these discourses, whether in opposition to the use of genetically modified organisms or in support of their wider dissemination, do not strictly fall along the fault lines between industrialized countries and the Global South or between industry and the public sector. There are reasonable, well-meaning people on all sides of the issue. Because these discursive frames represent substantively different positions on the issues that are based on wholly conflicting assumptions, it is difficult to reframe the issue of the use of genetically modified organisms in a coherent and consistent way that illuminates a common ground on which a sufficient number of interested parties can agree. These differences create real problems when considering NGO representation, for example. Who is Vandana Shiva really representing when poor Indian farmers scramble to obtain genetically modified seeds? How can Monsanto insist that genetically modified organisms promote food security when windborne genetically modified seeds may pollute an organic farmer's plot and undermine her way of living? What happens if an African farmer seeking to export her crops to Europe cannot guarantee that they are free of genetic modification?

The obverse of the issue of the dissemination and increased use of genetically modified organisms is the appropriation of traditional knowledge. The most effective forum for the defenders of traditional knowledge from appropriation by patents and claims to intellectual property rights that industries in developed countries have staked has been the Convention on Biological Diversity. The Convention on Biological Diversity emerged out of the 1992 United Nations Environment Program's Earth Summit. It was a response to growing alarm about species loss, the erosion of genetic diversity, and the accelerating destruction of the rainforests. Unlike TRIPS and WIPO, the Convention on Biological Diversity more explicitly incorporates intellectual property provisions that developing countries favor. In particular, Article 8 (j) recognizes communal or traditional knowledge, challenging the TRIPS endorsement of the Western, individualistic conception of knowledge ownership. The convention stresses that biological resources are sovereign resources of states, whereas TRIPS enforces private property rights over them. The convention seeks to promote the wider application of traditional knowledge with the approval and involvement of the holders of such knowledge. Many developing countries and NGOs endorse the convention as a way of combating biopiracy.

Once TRIPS entered into force, the Convention on Biological Diversity offered an opportunity for forum shifting on the issue of protecting traditional knowledge. Developing states, including China and India, and environmental NGOs raised questions about TRIPS's compatibility with the convention's access and benefit-sharing rules.[52] In 2002, the Conference of the Parties, which is the governing body of the convention, adopted what are known as the Bonn Guidelines, which stipulated that applicants for intellectual property rights should disclose the origin of any genetic resources or related knowledge relevant to the subject matter. Such disclosures are meant to facilitate monitoring whether applicants have received the prior informed consent of the country of origin and have complied with the country's conditions of access.[53]

Despite the difficulty of employing discursive reframing in this area, industry has attempted to reframe these agriculture issues as investment issues. Since the 1980s, when it linked intellectual property to trade, it has underscored a strong relationship between high levels of intellectual property protection and foreign investment. Many developing countries seek foreign investment, and industry leverages this vulnerability to press its agenda.[54] The life-sciences conglomerates have resisted any attempt to force them to include disclosure of origin in patent applications. They argue that it would inject unacceptable levels of uncertainty into the process of commercializing biotechnology. They argue that this uncertainty would have a chilling effect on investment in bioprospecting.[55]

On the larger question of access to biotechnology, which is where the particular concerns in this set of issues intersect most clearly with each other and with the A2K movement, the strategic burden lies in devising policies and ways of advocating them so that intellectual property rights do not prevent the public sector from delivering essential goods. Clearly, actors in the public and nonprofit sector in developing countries will bear the lion's share of the task of agricultural innovation for staple crops, and they will need access to biotechnology tools. Current intellectual property obstacles to access include patent thickets (overlapping patents created to frustrate competition), the patenting of research tools, and the sharply reduced realm of open science.[56] This has made it more difficult to bring useful plant genetic resources to market and has hampered the efforts of public-sector agronomists.

BACK TO WIPO: THE GENEVA DECLARATION, A2K, AND THE DEVELOPMENT AGENDA

By the late 1990s, intellectual property issues had received extensive critical attention in the WTO. In 1999, the United States and the European Union shifted forums to take their unmet concerns to WIPO and to pursue a Substantive Patent Law Treaty.[57] Many suspected that WIPO's work on patent harmonization in this treaty aimed to universalize TRIPS-plus standards.

The developing countries, too, seized opportunities to press their agendas within WIPO, although they had successfully developed these more fully in other forums. They sought to link biodiversity issues to the negotiations on the 1999 WIPO Substantive Patent Law Treaty by proposing the incorporation of the

Convention on Biological Diversity principles of prior informed consent and access and benefit-sharing rules. This reflects the deliberate discursive and institutional strategies facilitating issue migration from a friendlier forum to a tougher one. In September 2004 on the eve of the WIPO General Assembly meeting, James Love of CPTech organized a workshop in Geneva on the future of WIPO. This was a significant moment for the A2K movement, because it united a diverse range of stakeholders and brought numerous campaigns and issues under a common umbrella. The workshop resulted in the Geneva Declaration on the Future of WIPO, which a broad range of stakeholders and activists endorsed.

At the October 2004 WIPO General Assembly, a group of developing countries (the Group of Friends for Development) proposed a development agenda for WIPO.[58] The United States objected and sought to confine discussions of what they framed as "development" issues, such as the disclosure of origin and genetic resources to the WIPO Intergovernmental Committee on Intellectual Property and Genetic Resources, Traditional Knowledge and Folklore, presumably a forum where "words don't matter." Yet in June 2005, the "Friends of Development," led by Brazil, issued an ultimatum, refusing to discuss the Substantive Patent Law Treaty without forward movement on the WIPO Development Agenda.[59] The group succeeded in halting progress on the treaty, holding it hostage to meaningful, substantive progress on a development agenda.

In the past several years, WIPO has moved much further on its development agenda, and in 2007, its membership approved numerous items on it. Negotiations are ongoing. This has been remarkable progress, especially in light of the fact that WIPO gets about 90 percent of its operating budget from the Patent Cooperation Treaty, whose users are large global firms that hold extensive intellectual property portfolios. Many of these firms have objected to the development agenda.

If I were to place the various access campaigns surveyed here on a spectrum from the most successful from an A2K perspective and the most cohesive to least successful and least cohesive, I'd say that the access-to-medicines campaign has been the most successful and cohesive. Its issues appear to be the most clear cut. The campaigns for access to educational materials and software, as undertaken, for example, by the Free Software Foundation, the free and open-source software movement, and the Creative Commons campaign, would lie somewhere in the middle. Some profound philosophical and/or pragmatic fissures have evolved in that movement over time. Perhaps this dissention is analogous to the rift in the environmental movement between Earth First and Conservation International over who is the most pure and who is achieving the most. Finally, the campaigns involving issues of agricultural biotechnology are probably the least successful and least cohesive. Yet while the access-to-medicines movement has enjoyed the most negotiating success, the campaigns for access to "books" and "seeds" both have benefited from differing applications of the strategies of discursive reframing and institutional forum-shifting strategies to press their cases, along with other strategies appropriate to their differing situations. Perhaps most astonishing is the extent to which the relatively underresourced A2K campaigners have promoted skepticism about contemporary intellectual property protection.[60]

A2K: OBSTACLES AND OPPORTUNITIES

As this survey of these campaigns and the often successful strategies they have employed shows, the A2K movement holds much promise. The conditions that pertained when countries agreed to TRIPS have changed markedly. The original pro-TRIPS coalition of intellectual property–rich firms is disintegrating. Major information technology firms are questioning patent protection, because so-called "patent trolls" have attacked them in expensive litigation. They no longer support the pharmaceutical and agrichemical patent agenda. In the United States, the Internet technology firms are gaining the upper hand. In the summer of 2007, the U.S. House of Representatives passed the Patent Reform Act, which tightens the criteria for issuing patents. Important cases in the U.S. Supreme Court and the Circuit Court of Appeals in the spring of 2007 reasserted the importance of novelty and invention in intellectual property rights cases. In the United States, many policy makers now recognize that U.S. intellectual property policy may be impeding innovation. Congressional Democrats succeeded in altering provisions in U.S. bilateral and regional trade deals to remove many TRIPS-plus provisions. Representatives such as Henry Waxman have protested U.S. policies toward Thailand that impede that country's ability to address its HIV/AIDS crisis. While TRIPS-plus provisions remain in earlier trade pacts, it seems unlikely that the U.S. trade representative will be keen to enforce them, given this political shift. At the very least, efforts to do so would smack of blatant hypocrisy. In 2008, Democratic presidential candidates, responding to American concerns over health care, advocated new approaches to medical innovation, such as prizes (John Edwards), and more vigorous antitrust enforcement to protect consumers from abuses of monopoly power (Barack Obama). Meanwhile, the A2K movement, beginning in 2004 with the WIPO Development Agenda and the Geneva Declaration, has mobilized a far broader coalition in an effort to unite disparate access campaigns.

Furthermore, many practical alternatives to intellectual property have emerged. Increasing numbers of developing countries are using Linux, an open-source operating system, instead of Microsoft Windows. Computational biomedical research holds promise for an open-source approach to medical innovation. If successful, Google's efforts to put five research libraries' contents on-line might help to overcome some barriers to access to educational materials, such as high prices, technological protection measures, and abuses of monopoly power in the publishing industry. Creative Commons licenses from the "copyleft" movement have made more resources available. Proposals for a global research-and-development treaty for medical research and the WHO's strong endorsement of TRIPS flexibilities offer hope for delivering lifesaving drugs to the poor. Innovative legislation for agriculture that avoids the strictures of the Union for Plant Variety Protection helps to promote sustainable agriculture, biodiversity, and traditional farming practices. The Public Intellectual Property Resource for Agriculture and the African Agricultural Technology Fund enable smallholder farmers and public-sector agricultural services to get access to biotechnology tools. Both Duke and Yale Universities are promoting open licensing of core technologies for the Global South. Land-grant universities such as the University of Minnesota are finding ways to make their biotechnology available to the Global South. MIT has provided free on-line educational materials with its OpenCourseWare initiative.[61] These are just a few examples of promising approaches to access.

Perhaps the biggest challenge facing the A2K movement will be to cultivate and maintain unity. Just below the surface of solidarity, campaigners will have to address and manage many cracks and fissures. There are philosophical fissures between the free-software advocates and Creative Commons licensing practices. There is palpable tension between environmentalists supporting biodiversity and advocates for access to genetically modified organisms. Traditional knowledge divides campaigners. Some are antiproperty — period. Others feel that the intellectual property system needs to protect traditional knowledge so that its holders and developers may benefit and receive compensation.

However, just as the pro-TRIPS campaigners of the 1980s and early 1990s managed to put aside fundamental differences in order to present a united front and achieve their goals, A2K campaigners must do the same. While the cracks and fissures of the original TRIPS coalition are finally becoming problems for those who advocate increased intellectual property protections, until recently, they had achieved a high standard global intellectual property agreement. While not everyone endorses a convergence approach,[62] the template of the pro-TRIPS industry coalition is instructive. Advocates for TRIPS's intellectual property protections mounted a united effort, one that required bracketing some deep differences in approach and philosophy, especially between patent and copyright interests. These were privileged advocates with significant resources and enviable access to elite decision-making circles. The A2K movement, on its face, is underresourced and relatively weaker. It needs to recruit powerful industry partners, but also needs to construct unity across disparate sectors. Just as the environmental movement became possible only after "tree huggers" and hunters joined forces, A2K campaigners must discursively frame their preferences and values in such a way as to activate their most eclectic and ardent supporters. Furthermore, the Global South and its allies and advocates must act cohesively in favorable forums if they are to make continued progress in obtaining access to knowledge.

A2K campaigners already have astonished observers because, without obvious economic or political clout, they have generated notable skepticism and prompted a fresh conversation about contemporary intellectual property rights. A2K campaigners must find ways to maintain solidarity and to mobilize the broadest public support if they are to succeed in achieving their goals.

NOTES

1 See Peter Drahos's essay, "'IP World'—Made by TNC Inc.," in this volume.

2 James Boyle, "Cultural Environmentalism and Beyond," *Law and Contemporary Problems* 70, no. 3 (Spring 2007): pp. 5–21. Some scholars argue that capabilities and interests drive politics. See John Mearsheimer, "The False Promise of International Institutions," *International Security* 19, no. 3 (Winter 1994–95): pp. 5–49. Others, however, assign significant explanatory weight to language, discourse, and the construction of meaning. See Neta Crawford, *Argument and Change in World Politics* (Cambridge: Cambridge University Press, 2002) and Edward Comor, "The Role of Communication in Global Civil Society," *International Studies Quarterly* 45, no. 3 (2001): pp. 389–408.

3 Scholars differ in their assessments of the role of institutions. Some argue that institutions merely reflect the underlying distribution of capabilities and that the powerful prevail. See Stephen Krasner, "Global Communications and National Power: Life on the Pareto Frontier," *World Politics* 43, no. 3 (April 1991): pp. 336–66 and Mearsheimer, "The False Promise of International Institutions." Others see the function of institutions as reducing transaction costs, increasing information exchange, and solving contracting problems. See Robert Keohane and Lisa Martin, "The Promise of Institutionalist Theory," *International Security* 20, no. 1 (Summer 1995): pp. 39–51. Still others assert a more transformative role for institutions as sites that can change actors' understandings, interests, and preferences. See Ernst B. Haas, *When Knowledge is Power: Three Models of Change in International Organizations* (Berkeley: University of California Press, 1990) and Crawford, *Argument and Change in World Politics*.

4 See Gaëlle Krikorian's essay, "Free-Trade Agreements and Neoliberalism: How to Derail the Political Rationales that Impose Strong Intellectual Property Protection," in this volume.

5 Lawrence Helfer, "Regime Shifting: The TRIPS Agreement and New Dynamics of International Intellectual Property Lawmaking," *Yale University International Law Journal* 29, no. 1 (Winter 2004): p. 56, available on-line at http://papers.ssrn.com/sol3/papers.

cfm?abstract_id=459740 (last accessed May 26, 2009). The United States initially pursued this approach in discussions of a development agenda at WIPO, recommending that the discussions be sidelined to a committee with a mandate only to *discuss* the issue. Industry representatives have complained of WIPO's "mission creep" into development issues. Notably, however, they did not complain about GATT's "mission creep" into intellectual property issues during the Uruguay Round of trade negotiations.

6　*Ibid.*, pp. 55 and 58.

7　E-mail from Gregory C. Shaffer, Professor of Law, University of Wisconsin Law School, to Susan K. Sell, November 23, 2003 (on file with author).

8　See Ellen 't Hoen's essay, "The Revised Drug Strategy: Access to Essential Medicines, Intellectual Property, and the World Health Organization," in this volume. The WTO's 2001 Doha Declaration on TRIPS and Public Health owes much to this earlier work.

9　*Ibid.*

10　*Ibid.*, p. 132.

11　John Odell and Susan K. Sell, "Reframing the Issue: the WTO Coalition on Intellectual Property and Public Health, 2001," in John Odell (ed.), *Negotiating Trade: Developing Countries in the WTO and NAFTA* (Cambridge: Cambridge University Press, 2006), pp. 85–114.

12　Duncan Matthews, "WTO Decision on Implementation of Paragraph 6 of the Doha Declaration on the TRIPS Agreement and Public Health: A Solution to the Access to Essential Medicines Problem?" *Journal of International Economic Law* 7, no. 1 (2004): p. 73.

13　Intellectual Property Watch, "African Countries Ready to Accept TRIPS and Public Health Deal," December 6, 2005, available on-line at http://www.ip-watch.org/weblog/2005/12/06/african-countries-ready-to-accept-trips-and-public-health-deal (last accessed May 25, 2009).

14　Duncan Matthews, "Is History Repeating Itself?: Outcome of the Negotiations on Access to Medicines, the HIV/AIDS Pandemic and Intellectual Property Rights in the World Trade Organisation," *Electronic Law Journal LGD* 2004 (1), p. 11, available on-line at http://www2.warwick.ac.uk/fac/soc/law/elj/lgd/2004_1/matthews (last accessed May 25, 2009).

15　James Love, "No Gift to the Poor: Strategies used by the US and EC to Protect Big Pharma in WTO TRIPS Negotiations," *Working Agenda*, December 1, 2005, available on-line at http://workingagenda.blogspot.com/2005/12/no-gift-to-poor-strategies-used-by-us.html (last accessed May 25, 2009).

16　Ruth Okediji, *The International Copyright System: Limitations, Exceptions and Public Interest Considerations for Developing Countries in the Digital Environment* (Geneva: International Centre for Trade and Sustainable Development, 2005), also available on-line at http://www.iprsonline.org/unctadictsd/docs/Okediji_Copyright_2005.pdf (last accessed May 25, 2009).

17　Steven Weber, *The Success of Open Source* (Cambridge, MA: Harvard University Press, 2004), pp. 47–48.

18　Benjamin Mako Hill, "Towards a Standard of Freedom: Creative Commons and the Free Software Movement" (2005) available on-line at: http://lists.ibiblio.org/pipermail/cc-community/2005-July/000639.html (last accessed May 25, 2009).

19　Alan Story, "Intellectual Property and Computer Software: A Battle of Competing Use and Access Visions for Countries of the South" (2004), available on-line at http://www.iprsonline.org/unctadictsd/docs/CS_Story.pdf (last accessed May 25, 2009).

20　Mako Hill, "Towards a Standard of Freedom."

21　Richard Stallman, quoted in Boyle, "Cultural Environmentalism and Beyond," p. 20 n.70.

22 Lawrence Lessig, quoted in *ibid*. While they may not meet Stallman's test, Creative Commons licensing practices have been successful in increasing access to educational materials and, broadly, knowledge.

23 Jessica Litman, *Digital Copyright* (New York: Prometheus Books, 2001), p. 93.

24 *Ibid.*, p. 123.

25 Pamela Samuelson, "The U.S. Digital Agenda at WIPO," *Virginia Journal of International Law* 37, no. 2 (Winter 1997): pp. 374–75. See also Spring Gombe and James Love's essay, "New Medicines and Vaccines: Access, Incentives to Investment, and Freedom to Innovate," in this volume.

26 Litman, *Digital Copyright*, pp. 122–45.

27 *Ibid.*, p. 145.

28 Jerome Reichman and Paul Uhlir, "A Contractually Reconstructed Research Commons for Scientific Data in a Highly Protectionist Intellectual Property Environment," *Law and Contemporary Problems* 66, nos. 1–2 (2003): p. 378, available on-line at http://dlc.dlib.indiana.edu/archive/00002601/01/A_Contractually_Reconstructed_Research_Commons.pdf (last accessed May 25, 2009).

29 See Viviana Muñoz Tellez and Sisule Musungu's essay, "A2K at WIPO: The Development Agenda and the Debate on the Proposed Broadcasting Treaty," in this volume.

30 Cary Fowler, *Unnatural Selection: Technology, Politics and Plant Evolution* (Yverdon, Switzerland: Gordon and Breach: 1994), p. 180.

31 *Ibid.* pp. 181–82.

32 *Ibid.*, p. 181.

33 Jack Kloppenburg, "Interview with GRAIN," *Seedling*, October 2005, available on-line at http://www.grain.org/seedling/?id=414 (last accessed May 25, 2009).

34 Kal Raustiala and David G. Victor, "The Regime Complex for Plant Genetic Resources," Working Paper #14, Stanford University Institute for International Studies, Center for Environmental Science and Policy (May 2003): p. 12, available on-line at http://iis-db.stanford.edu/pubs/20190/pgr_regime_complex.pdf (last accessed March 4, 2010).

35 Graham Dutfield, *Intellectual Property Rights and the Life Sciences* (Aldershot, UK: Ashgate, 2003), p. 216.

36 *Ibid.*

37 GRAIN, "The FAO Seed Treaty: From Farmers' Rights to Breeders' Privileges," *Seedling*, October 2005, available on-line at http://www.grain.org/seedling/?id=411 (last accessed May 25, 2009).

38 Raustiala and Victor, "The Regime Complex for Plant Genetic Resources," p. 13.

39 Jakkrit Kuanpoth, "Closing in on Biopiracy: Legal Dilemmas and Opportunities for the South," in Ricardo Melendez-Ortiz and Vicente Sanchez (eds.), *Trading in Genes: Development Perspectives on Biotechnology, Trade and Sustainability* (London: Earthscan: 2006), pp. 139–52.

40 Kloppenburg, "Interview with GRAIN."

41 GRAIN, "The FAO Seed Treaty: From Farmers' Rights to Breeders' Privileges."

42 Robin Pistorius and Jeroen van Wijk, *The Exploitation of Plant Genetic Information: Political Strategies in Crop Development* (New York: CABI Publishing, 1999).

43 Vandana Shiva, *Protect or Plunder?: Understanding Intellectual Property Rights* (London: Zed Books, 2001). But see Ronald Herring, "Miracle Seeds, Suicide Seeds, and the Poor: GMOs, NGOs, Farmers, and the State," in Raka Ray (ed.), *Social Movements in India: Poverty, Power,*

and Politics (Rowman and Littlefield, 2005) and Anitha Ramanna, "Farmers' Rights in India: A Case Study," Background Study 4, The Farmers' Rights Project (Lysaker, Norway: The Fridtjof Nansen Institute: 2006) available on-line at http://www.gtz.de/de/dokumente/en-biodiv-fni-background-study4-farmers-rights-India-2006.pdf (last accessed June 10, 2009).

44 Jennifer Clapp, "Unplanned Exposure to Genetically Modified Organisms: Divergent Responses in the Global South," Journal of Environment & Development 15, no. 1 (March 2006): pp. 3–21. But see Robert Paarlberg, The Politics of Precaution: Genetically Modified Crops in Developing Countries (Washington, D.C.: International Food Policy Research Institute, 2001).

45 Sakiko Fukuda-Parr, "Introduction: Genetically Modified Crops and National Development Priorities," in Sakiko Fukuda-Parr (ed.), The Gene Revolution: GM Crops and Unequal Development (London: Earthscan, 2007), pp. 6–9.

46 Ibid. p. 8.

47 Ramanna, "Farmers' Rights in India," p. 10.

48 Ibid.; Ronald Herring, "Why Did 'Operation Cremate Monsanto' Fail?: Science and Class in India's Great Terminator Technology," Critical Asian Studies 38, no. 4 (December 2006).

49 Ramanna, "Farmers' Rights in India," p. 11.

50 Sakiko Fukuda-Parr, "Institutional Changes in Argentina, Brazil, China, India and South Africa," in Fukuda-Parr (ed.), The Gene Revolution, pp. 202–203 and 218.

51 However, the case of Argentina is complicated by the fact that, unlike China, it is utterly dependent on foreign technology and adopted it in the context of International Monetary Fund structural-adjustment lending in the early 1990s. See Daniel Chudnovsky, "Argentina: Adopting RR Soy, Economic Liberalization, Global Markets and Socio-economic Consequences," in Fukuda-Parr (ed.), The Gene Revolution, p. 85.

52 Helfer, "Regime Shifting," p. 33.

53 Ibid., p. 29.

54 Carlos M. Correa, "Investment Protection in Bilateral and Free Trade Agreements: Implications for the Granting of Compulsory Licenses," Michigan Journal of International Law 26 (2004): pp. 331–53.

55 Jacques Gorlin, "Additional Patent Disclosure Requirements for Biotech Inventions in Play in WTO, WIPO and CBD," Stockholm Network: Know IP—Stockholm Network Monthly Bulletin on IPRs 2, no. 1 (February 2006): p. 4.

56 Brian Wright and Philip Pardey, "Changing Intellectual Property Regimes: Implications for Developing Country Agriculture," International Journal of Technology and Globalization 2, nos. 1/2 (2006): pp. 93–114.

57 Jerome Reichman and Rochelle Cooper Dreyfuss, "Harmonization without Consensus: Critical Reflections on Drafting a Substantive Patent Law Treaty," Duke Law Journal 57, no. 1 (October 2007): pp. 89–90.

58 Argentina and Brazil presented the proposal to the WIPO General Assembly. The proposal was cosponsored by Bolivia, Cuba, the Dominican Republic, Ecuador, Egypt, Iran, Kenya, Peru, Sierra Leone, South Africa, Tanzania, and Venezuela. WIPO doc. no. WO/GA/31/11 (September–October 2004), available on-line at http://www.wipo.int/edocs/mdocs/mdocs/en/iim_1/iim_1_4.pdf (last accessed April 5, 2009).

59 International Centre for Trade and Sustainable Development, "Moving Forward the 'Development Agenda' in WIPO" (October 6, 2004), available on-line at http://ictsd.net/i/ip/39690/ (last accessed June 11, 2009).

60 Mark Schultz and David Walker, "How Intellectual Property Became Controversial: NGOs and the New International IP Agenda," *Engage* 6, no. 1 (October 2005): pp. 82–98, available on-line at http://www.fed-soc.org/doclib/20080313_IPSchultz.pdf (last accessed May 26, 2009).

61 Okediji, *The International Copyright System*.

62 GRAIN, "The FAO Seed Treaty: From Farmers' Rights to Breeders' Privileges."

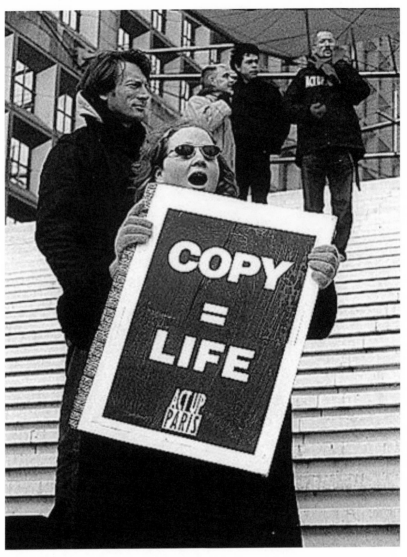

Act Up–Paris demonstration on March 5, 2001 in Paris, in support of people with HIV/AIDS facing the pharmaceutical industry in court in South Africa.

TRIPS Flexibilities: The Scope of Patentability and Oppositions to Patents in India

Chan Park and Leena Menghaney

India was one of the last World Trade Organization–member developing countries to come into compliance with the Trade-Related Aspects of Intellectual Property Rights Agreement (the TRIPS Agreement), the WTO's treaty on intellectual property. With the spread of the global AIDS pandemic, the global outrage at the inequity in access to lifesaving HIV/AIDS treatment, and the concomitant rise of the access-to-medicines movement, the global landscape had changed quite dramatically between 1995, when TRIPS came into existence, and 2005, when India finally acceded to its demands.[1]

In the interim came the WTO's Doha Declaration on the TRIPS Agreement and Public Health.[2] That statement recognized that patent protection should not interfere with a country's sovereign right and responsibility to protect the health of its citizens and that TRIPS included inherent flexibilities that allowed countries to override patents in order to protect public health.

Also in the interim came the announcement in 2001 by Cipla, an Indian manufacturer of generic drugs, that it was willing to provide a generic version of a triple-combination AIDS cocktail for a dollar a day, as compared with the (US) $10,000 a year then being charged by the Western multinational pharmaceutical companies. And with this drop in price came the drive toward "universal access," which resulted in an unprecedented global scale-up of AIDS treatment to over 1.3 million people living with HIV/AIDS in developing countries by the end of 2005.[3]

The world in 2005, then, was a very different one from what existed ten years prior. Indian generic companies had become the key supplier of affordable AIDS and other essential drugs throughout much of the developing world. A host of latent TRIPS flexibilities had been explicitly recognized by the Doha Declaration and had been expounded upon by experts in the field. And the global community of treatment activists had found their voice, refusing to be ignored or brushed aside.

These and other factors led India to include several unique features in its TRIPS-mandated patent law, features that, if rigorously and exactingly utilized, have the potential to lessen dramatically the likelihood that patent monopolies will place essential medicines out of reach. In addition to focusing on the much-discussed flexibilities with respect to the compulsory licensing of patented medicines (a subject on which Indian law frankly still leaves much to be desired), India also turned its attention to the criteria on which patents are granted in the first place.

These patent criteria, which are markedly more stringent than prevailing standards in effect just about anywhere in the world, potentially provide a complementary solution that may be able to overcome at least partially the inherent limitations of relying on the compulsory licensing mechanism alone. By establishing stricter criteria for obtaining a patent, fewer patent monopolies are created, thus creating more space for generic competition to enter the market with lower-cost alternatives. And where there are no patents on an essential drug, there is no need to issue a compulsory license.

This essay examines some of the unique features of Indian patent law and the experiences of civil society in utilizing them to help ensure that the Indian domestic industry can continue to provide affordable lifesaving medicines throughout the developing world.

COMPULSORY LICENSING: BENEFITS AND LIMITATIONS

Since Doha, the TRIPS flexibility that has received by far the most attention is the freedom that countries have to issue so-called "compulsory licenses" on patented drugs. By issuing a compulsory license, a country can override an existing patent (or patents) on a drug. In exchange for "adequate remuneration" paid to the patent holder, the country is then free to purchase generic versions of the medicine, often at a fraction of the price of the branded version. The Doha Declaration explicitly recognized that countries have the "right to grant compulsory licenses" and "the freedom to determine the grounds upon which such licenses are granted." But the Doha Declaration has a much broader significance than clarifying the rules concerning compulsory licensing. It serves as both an interpretive guide and a source of obligation on WTO member countries in implementing *all* aspects of the TRIPS Agreement: "We agree that the TRIPS Agreement does not and should not prevent Members from taking measures to protect public health. Accordingly, while reiterating our commitment to the TRIPS Agreement, we affirm that the Agreement can and should be interpreted and implemented in a manner supportive of WTO Members' right to protect public health and, in particular, to promote access to medicines for all." As the use of the obligatory "should" here implies, the Doha Declaration recognized the positive obligation on member states to ensure that intellectual property protection did not come at the cost of safeguarding public health.

For many developing countries, compulsory licenses are the most effective and immediate way to lower the prices of patented drugs. One such instance is Thailand. In late 2006 and early 2007, the Thai government issued a series of compulsory licenses on the AIDS drugs efavirenz and lopinavir/ritonavir and on the heart medication clopidogrel. The immediate cost savings were significant.[4]

The reaction to the Thai compulsory licenses gives us some insight as to why more countries have not dared come forth to use the compulsory licensing mechanism. The Thai compulsory licenses were issued in perfect accordance with both Thailand's domestic laws and the TRIPS requirements. The legality of it all notwithstanding, the U.S. trade representative promptly placed Thailand on its Special 301 Priority Watch List, a list that purports to identify the world's most egregious intellectual property "offenders" and that places countries at risk of retaliatory trade sanctions from the United States.

Abbott Laboratories, the owner of the patents on lopinavir/ritonavir, retaliated by withdrawing all of its pending drug-approval applications from Thailand, with the clear implication being that it would refuse to register its new drugs in any country that dared to issue a compulsory license on its patents. And suddenly, in editorial pieces popping up in newspapers all over the world, pundits supporting the pharmaceutical industry lambasted the actions of Thailand's "military dictatorship" in "stealing" the property of Western pharmaceutical companies.

As the Thailand example shows, issuing a compulsory license often comes with a price. The U.S. trade representative and the pharma lobby have made sure this message is heard, loud and clear. Understandably, not many governments are keen on angering the United States or any of their other rich and powerful Western trading partners. But in those relatively rare instances in which there is the political will within a developing country's government to stand up to these pressures, the benefits from issuing a compulsory license must be tangible and immediate. More often than not, these benefits will be sourced from India. For example, a handful of other countries, including Brazil, Indonesia, and Malaysia, have issued compulsory licenses on essential medicines in recent years. Upon issuing the compulsory licenses, they all promptly proceeded to import cheaper versions from Indian generic companies, at least until their domestic manufacturing capacity could be developed to make these drugs in-house.

This fact underscores a few critical issues. The first and most obvious is that the Indian pharmaceutical industry plays a central role in the supply of affordable generic drugs throughout the developing world. The importance of the Indian generic-drug industry is difficult to overstate. According to one study, Indian companies supply up to 85 percent of the generic AIDS drugs currently being used in sub-Saharan Africa.[5]

The second critical fact is that these countries were able immediately to realize significant cost savings because Indian generic companies had already been manufacturing these drugs in India for years. Because India did not recognize product patent protection on drugs prior to 1995, Indian companies were free to reverse engineer medicines that were under patent in other countries and sell them at a fraction of the cost. The result: a thriving generic industry some fifteen thousand entities strong, selling drugs at some of the lowest prices in the world. Consequently, once these countries issued compulsory licenses, they had a well-established and competitive market to turn toward in purchasing lower-cost alternatives, rather than having to create such a market from scratch.

Which brings us to the third and most critical point. Compulsory licenses have been effective in immediately and significantly lowering drug prices largely because there are countries—such as India—where these medicines are not patented at all. This is the case for at least two reasons. First, successfully reverse engineering a drug simply does not happen overnight. Historically, it has taken anywhere from one to six years after the original drug has been launched for an Indian company to come out with a generic alternative.[6] Additionally, the presence of just one or two generic competitors has been shown to be ineffective in significantly lowering drug prices. The cost of a drug is inversely proportional to the number of generic competitors on the market, and it has been observed that dramatic price reductions kick in only once the "rule of five"—that is, five or more competitors entering the market for a given drug—has been satisfied.[7]

Thus, at least in the area of AIDS drugs, where multinational pharmaceutical companies are under intense pressure and where they charge reduced (but still expensive) "access prices" in developing countries, it is not uncommon for the price of the first generic version of a drug to be similar to or even slightly more expensive than the prices being charged by the originator company (see Table 1).

As this table illustrates, there are several generic companies manufacturing the standard "first-line" AIDS drugs: lamivudine (3TC), nevirapine (NVP), stavudine (d4T) and zidovudine (AZT). Correspondingly, the prices for both originator and generics are very low. However, in the case of the newer AIDS drugs such as abacavir (ABC) and lopinavir/ritonavir (LPV/r), where there are fewer generic competitors, the prices remain alarmingly high.

The point of all of this: From the perspective of a government official faced with the politically risky decision of issuing a compulsory license for an important drug, the most compelling reason in favor of issuing the compulsory license will often be the immediate benefits that accrue from the significant cost savings that

compulsory licenses only effective if non-patented generics exist

can be achieved. But if there is no preexisting source of cheap drugs (such as India) where the drugs are already being manufactured and from whence such immediate cost savings can be realized, the political calculus becomes much murkier.

Does the government official issue a compulsory license in the hope that one to six years down the line, a generic company will be able to reverse engineer a cheaper version successfully? Does she issue multiple licenses to several generic manufacturers in the hope that competition will eventually drive the prices down? And does she, in the face of these uncertainties, issue the compulsory license(s) anyway, despite the virtual certainty of facing the rather odious forms of political and economic pressure that will be applied by the U.S. trade representative, the pharma lobby, and their cohorts?

And yet, with India now under a product patent regime and with the majority of the other developing and least-developed countries having already implemented the TRIPS Agreement, this is precisely the situation that the world is potentially facing: nearly universal patent protection for pharmaceutical products and no immediate relief available through the issuance of a compulsory license. Thus, despite the Doha Declarations's mandate that countries can and should take measures to promote access to medicines for all, the increasingly global reach of

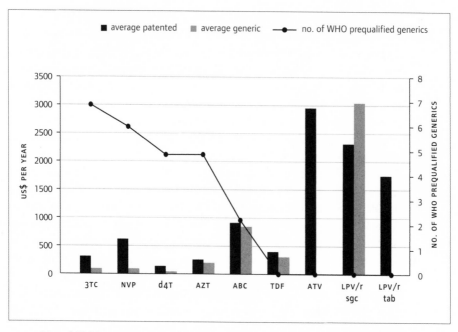

TABLE 1 Prices of AIDS drugs versus number of generic competitors (MSF, *Untangling the web of price reductions*, 2007).

parallel importation requires two separate compulsory licenses

product patent protection makes taking such measures more and more unrealistic. Given both the political and economic realities, the use of the compulsory licensing mechanism in a world of universal patent protection has the potential to be inadequate to produce the rapid and dramatic reduction in prices that is often necessary to make essential medicines truly accessible.

For the large number of developing and least-developed countries that, unlike a handful of developing countries such as India, China, Brazil, South Africa, and Thailand, lack the ability to manufacture generic medicines domestically, the situation is even more bleak. This is because the issuance of a compulsory license serves no function if the country is unable to manufacture its own generic drugs or to import drugs from countries that can. The Doha Declaration recognized this and directed the WTO to come up with an "expeditious solution" to this problem. However, the purported "solution" that the WTO came up with in August 2003 has proven to be fraught with difficulties,[9] and has thus far been used just once, with limited success.[10] Given the numerous difficulties with these rules, Médecins Sans Frontières (MSF) and others have declared the WTO's decision to be "unworkable" and "neither expeditious, nor a solution."[11]

PATENTS AND THE SCOPE OF TRIPS FLEXIBILITIES: A BRIEF INQUIRY

As we have seen, in order for compulsory licenses to be most effective in achieving immediate and significant cost savings, there should be a readily available source of cheaper generic alternatives. However, the global reach of patent protection that TRIPS entails threatens the very existence of such readily available alternatives. With India now under a product patent regime, it is an open question as to whether, for newer drugs that are eventually granted patent protection in India, issuing a compulsory license will result in the immediate and dramatic savings that countries have been able to take advantage of simply by switching to a preexisting Indian generic alternative.

Fortunately, it is by no means a foregone conclusion that simply because a drug has been patented in the United States or Europe, the same drug will be patented in India. Patent standards often vary from country to country, and what may be deemed to be sufficiently "new" and "inventive" in one country will not necessarily be deemed to be so in another. Indeed, what may be considered to constitute patentable subject matter, and thus an "invention" at all, varies quite significantly from country to country.

Countries may choose to adopt varying levels of protection for any number of reasons. Historically, many countries, shifted from having "weak" intellectual property regimes to "strong" ones as their economies evolved from being net users of intellectual property to net producers. This raises several questions. Which came first? Did the stronger levels of protection cause the innovation, or was it the other way around? And is this relationship (if there is one at all) linear, such that more protection always equals more innovation? Also, what sorts of innovation, exactly, are generated by the patent system? There are no definite answers to these questions, but there are some clues.

First, there is very little empirical evidence—particularly in the pharmaceutical context—to suggest that *stronger* levels of patent protection result in increased levels of innovative output. With the entry into force of the TRIPS Agreement in 1995, the world witnessed a dramatic strengthening of patent protection on a global scale. However, since 1995, the number of new drugs being released onto the market has, if anything, declined (see Table 2).

As Table 2 illustrates, after a peak of over fifty new drugs approved by the United States Food and Drug Administration in 1996, the numbers have steadily declined in the era of global patent protection ushered in by the TRIPS Agreement, dipping to fewer than twenty new drugs in 2002. The levels of innovation (as measured by new drug approvals) before TRIPS, when many countries did not recognize product patents on pharmaceuticals, are not appreciably lower than the levels

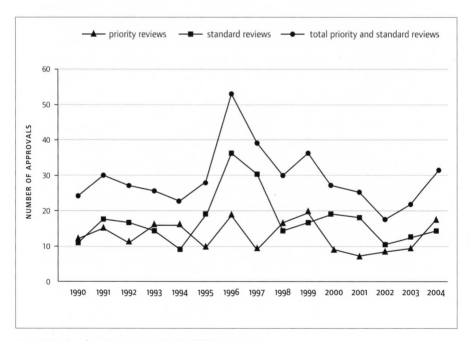

TABLE 2 Number of new drugs approved by the USFDA, 1990–2004.

of innovation after TRIPS. Thus, as far as the pharmaceutical sector is concerned, there has been no increase in the levels of innovation as a result of stronger levels of patent protection around the world.

Second, there is also very little evidence to suggest the other extreme: that in the absence of patent protection, no innovation will occur at all. In fact, there are numerous instances where this notion has been shown to be false—particularly with respect to what are called "incremental" innovations: advances that simply add on to and improve already existing technologies. Often, the incentives created by the market alone will be sufficient to allow such types of innovation to happen.

One example of an extremely useful incremental innovation occurring in the absence of patent protection comes from India. In 2001, Indian generic-drug companies, without the benefit of a product patent regime, were able to develop a single, three-in-one fixed-dose combination of AIDS medicines that revolutionized AIDS treatment in the developing world. Because three different drugs were combined into one easy-to-swallow pill, treatment regimens were dramatically simplified, patient adherence was improved, and this combination became the weapon of choice in the dramatic global scale-up of AIDS treatment. Market pressures, and not patent protection, were a sufficient incentive to allow this important innovation to occur.

Ironically, this is also a situation in which patent protection actually proved to be a *barrier* to further innovation. Because three different companies owned the patents on the three individual drugs, there was little incentive for these companies to enter into complicated cross-licensing arrangements in order to produce a single product that was useful in scaling up treatment in the developing world. Indian companies, unhindered by patent protection, were able to take advantage of the situation.

The ongoing debate about whether patent protection is necessary, and if so, what the "optimal" level of protection should be has not been confined to developing countries struggling to come to grips with its TRIPS obligations. In a recent landmark judgment, the United States Supreme Court recognized that:

> Granting patent protection to advances that would occur in the ordinary course without real innovation retards progress and may, in the case of patents combining previously known elements, deprive prior inventions of their value or utility.... When there is a design need or market pressure to solve a problem and there are a finite number of identified, predictable solutions, a person of ordinary skill has good reason to pursue the known options within his or her technical grasp. If this leads to the anticipated success, it is likely the product not of innovation but of ordinary skill and common sense."[12]

Implicit in the above quote is the recognition that market pressures will often be sufficient to produce many types of innovations and that providing patent protection to such improvements would in fact *hamper*, not encourage, real innovation.

Finally, while it is unclear what kinds of additional incentives are created by stronger levels of patent protection, it is fairly clear what kinds of incentives are created by the patent system that exists today in many developed countries. Because the patent system depends upon the pharmaceutical company being able to recoup its research-and-development (as well as marketing) costs by charging high drug prices, the pharmaceutical companies will naturally have an incentive to develop drugs for markets that can afford to pay for these drugs.

Thus, drug companies will have little incentive to conduct research and development into the special health needs of the Indian market, which accounts for about 1.2 percent of the global pharmaceutical market, or for the African market, which accounts for about 1.1 percent.[13] Rather, drug companies focus on the diseases and conditions that affect the developed world, where consumers are able to afford the high costs that the patent system entails. As a result, the vast majority of the new drugs that are developed are tailored for diseases that predominantly affect the developed world, such as cancer or heart disease, or for nonessential "lifestyle" drugs, such as treatments for erectile dysfunction or obesity. Thus, while there are no fewer than three separate drugs for erectile dysfunction on the market, there are as yet no effective treatments developed for potentially fatal diseases such as kala azar (leishmaniasis, a chronic and potentially fatal disease of the viscera, caused by protozoan parasites) or sleeping sickness.

As a result of the skewed incentives created by the patent system, the WHO Commission on Intellectual Property, Innovation and Public Health recently concluded: "There is no evidence that the implementation of TRIPS agreement in developing countries will significantly boost R&D in pharmaceuticals on Type II [diseases substantially affecting the developing world] and particularly Type III diseases [those almost exclusively affecting the developing world]. Insufficient market incentives are the decisive factor."[14]

As all of these complexities suggest, the debate surrounding patent protection on pharmaceuticals cannot be easily reduced, as some would have it, to a zero-sum game that pits access to medicines on one side against innovation on the other. The two concepts are neither mutually exclusive nor even necessarily inherently hostile to each other. Faced with the myriad complexities of the debate, individual countries can and often do choose to adopt varying standards for what qualifies as a patentable invention within the confines of TRIPS.

The TRIPS Agreement implicitly recognizes this variance and provides that patents granted or revoked in one country are completely independent of patents in other countries.[15] As such, TRIPS makes no attempt to define what an "invention" is, nor does it define the basic criteria for patentability, namely, "novelty," being an "inventive step," and possessing an "industrial application." This leaves countries with a significant degree of space to define these concepts according to their own policy priorities.

To take just one example, most countries make a general distinction between "discoveries" and "inventions" and provide patent protection only for the latter. For example, the European Patent Convention, like the patent law of many countries, includes a provision that "discoveries, scientific theories and mathematical methods" are not to be regarded as inventions.[16] Thus, for instance, "discovering" a hitherto unknown element on the periodic table or a new law of nature would not give rise to a patentable claim in most countries, no matter how useful the new element or this new law of nature may be or how much human ingenuity or investment was required to make such a discovery. In practice, however, the distinction between "invention" and "discovery" is not so clear.

For instance, U.S. patent law defines an "invention" as an "invention or discovery" and provides that "whoever invents or discovers any new and useful process" or "machine" is entitled to a patent.[17] Thus, under U.S. law, even if someone discovers a substance that already exists in nature, it can nevertheless qualify as an invention and be awarded a patent. For instance, the U.S. Food and Drug Administration recently approved a new drug for type 2 diabetes called exenetide. One of the several patents covering this drug (U.S. 5,424,286) discloses that exenetide was not created in a lab, but discovered in the venomous saliva of the Gila monster, a poisonous lizard found in the deserts of North America. Nevertheless, the patent covers not only the "method of treating" diabetes by using a compound found in lizard spit, but the naturally occurring compound itself.

To take another example, despite its broad exclusion of "discoveries" from patentability, the European Patent Convention carves out a specific exception to this rule and provides that "an element isolated from the human body . . . including the sequence or partial sequence to a gene, may constitute a patentable invention, even if the structure of that element is identical to that of a natural element."[18]

But why carve out an exception to the general rule only for elements isolated from the *human* body? Could not the Europeans just as easily, if they chose, have drafted this provision as exempting any element isolated from the human or *lizard* body? Would doing so have done violence to some ideal Platonic form of "inventionness"?

Or, perhaps, the Europeans simply made a policy determination that the costs of providing patent protection for gene sequences and the like would eventually be outweighed by the fruits of the research and development into this potentially significant area. Whether or not tweaking the definition of invention in this

manner is a smart policy decision may be open to debate, but Europe's prerogative to make this decision certainly is not. And mere disagreement with this policy
choice most certainly does not provide the basis for challenging its legal validity,
under TRIPS or otherwise.

But as we will see, this is essentially what formed the motivation behind an
unsuccessful legal challenge brought against India's definition of "invention" to
suit its own (vastly different) context: that India's exercise of its own policy prerogatives in narrowly defining "invention" was illegal because it failed to provide
sufficient incentives for research and development. But TRIPS is a legal document,
not a policy document. As such, it does not require countries to adopt and embrace
any particular policy priority (save, of course, the policy of promoting "access
to medicines for all," as stated in the Doha Declaration), as long as its minimum
requirements are satisfied.

If there is no generally agreed-on concept of what an "invention" is, then,
do not countries have the freedom correspondingly to narrow this definition to
address countervailing policy priorities, such as promoting access to medicines?
The TRIPS Agreement has little to say in this regard. It merely provides that
countries may, but are not obliged to, provide more extensive protection than
is required under the agreement. But this simply poses the question of what *is*
required under the agreement. To say that TRIPS requires countries to provide patent protection to all "inventions," but not necessarily to some or all "discoveries,"
is redundant, because it presupposes that there is a generally accepted definition
of both and a clear delineation between them. But as we have seen, there is not.

The point here is that there is sufficient variance in practices from around the
world to give countries a large degree of flexibility in defining what shall and shall
not constitute patentable subject matter to suit their particular policy needs. Promoting research and development into new drugs undoubtedly is one policy goal,
but merely one among many competing goals and one that enjoys no particular
pride of place under the TRIPS Agreement. Indeed, after Doha, the policy goal of
promoting easy access to affordable medicines could be seen as paramount.

INDIA'S USE OF TRIPS FLEXIBILITIES IN PATENT MATTERS: A CASE STUDY

With the benefit of the Doha Declaration as an interpretive guide and with the
global community of treatment activists breathing down its collective neck, the
Indian Parliament in 2005 passed a unique set of provisions that collectively define
"invention" in what is likely to be the most stringent manner in the world. Section
3 of the Indian Patents Act is entitled "What are not inventions" and lists fifteen
broad categories as "not inventions within the meaning of this Act."

Certainly the most contentious of these provisions was section 3(d), which was amended to state that the "mere discovery" of a new form of an existing drug would not be considered an invention unless this new form made the drug significantly more "effective." As the parliamentary debates surrounding this section made clear, this provision was designed to prevent a practice known as "evergreening" (or "product life-cycle management," a euphemism coined by those within the industry), whereby drug companies artificially extend the monopoly on an existing drug by filing for and obtaining patents on what are often trivial or minor changes to that drug.

Just two years after its birth, this provision was made the subject of a legal challenge in the Indian courts by the Swiss multinational drug company Novartis. The story of this litigation, how it came about, and how it was decided is illustrative of how civil society, if provided with the opportunity to participate in the patenting process, can potentially serve as sentinel against the granting of harmful patents.

This opportunity was provided by way of a broad pregrant opposition provision in the Indian Patents Act, which allowed for "any person" (as opposed to "any person interested" as provided in the old law), to oppose the granting of a patent at any time before grant. Seizing on this opportunity, the Cancer Patients Aid Association (CPAA), an Indian organization providing care and support services to those living with cancer, sought out the assistance of an HIV/AIDS legal organization—the Lawyers Collective HIV/AIDS Unit—in filing a pregrant opposition against Novartis's patent application for imatinib mesylate, trademarked and marketed as Glivec/Gleevec by Novartis, a vital treatment for people with chronic myeloid leukemia.

Imatinib mesylate was developed by Novartis with significant assistance from scientists at the Oregon Health Sciences University, which was funded largely with public funding from the U. S. National Cancer Institute. The drug was hailed widely as a "miracle drug," transforming what had been an invariably fatal condition into a treatable one—that is, if one was lucky enough to afford it.

Novartis was selling imatinib mesylate in India at its global price of about $2,600 per month—this in a country where the per-capita income is about $820 per year—and had obtained court orders preventing several manufacturers of generic drugs from selling their versions at less than one-tenth of Novartis's price. When the patent application for imatinib mesylate came up for examination, the Cancer Patients Aid Association, along with several generic-drug companies, filed pregrant oppositions to the application, claiming that, among other things, imatinib mesylate did not qualify as an invention under section 3(d). The Indian Patent Office agreed and denied Novartis a patent.

Novartis then filed suit against the Cancer Patients Aid Association, the Indian generic-drug companies, and the government of India in the Madras High Court, challenging not only the denial of the imatinib mesylate patent, but also the validity of section 3(d) itself. Novartis claimed, among other things, that section 3(d) was in violation of the TRIPS Agreement and that it was inconsistent with the Indian Constitution.

On the TRIPS-compliance issue, Novartis claimed, predictably, that the flexibilities contemplated by the Doha Declaration were limited only to compulsory licensing and that TRIPS very narrowly limited what countries may and may not define as patentable subject matter. In a press release to assuage the growing public outrage at Novartis's actions, it stated, rather paradoxically, "Novartis is not challenging any provisions of the Indian patent law that were put in place to promote access. We are challenging parts of the Indian patent law that led to the rejection of the Glivec/Gleevec patent."

During oral arguments that dragged on and off from January to April 2007, Novartis's lawyers repeatedly based their objection to section 3(d) as providing inadequate incentive to engage in much-needed research and development into new drugs and warned the court of the dire consequences of allowing section 3(d) to stand. At one heated moment, the counsel for Novartis opined that it would be better for the poor to wait twenty years for cheaper medicines than to retain section 3(d), under which no new drugs would be developed at all.

Several weeks after the close of the dramatic (and at times melodramatic) oral arguments on the matter, the Madras High Court dismissed Novartis's challenge to section 3(d). On the TRIPS-compliance issue, it rightly noted that an Indian court was the inappropriate forum to resolve such a dispute and that the appropriate place for resolving such a dispute between Switzerland and India was before the WTO dispute-resolution panel. On the constitutional issue, however, the court in rather glowing terms praised section 3(d), stating that in upholding this provision it had "borne in mind the object of [section 3(d)], namely . . . to provide easy access to the citizens of this country to life saving drugs and to discharge the Constitutional obligation of providing good health care to its citizens."[19]

In addition to its participation in the courtroom, civil society played an enormous role outside it. Throughout the streets of Chennai, Mumbai, and Delhi, hundreds of protestors shouted slogans decrying Novartis for attempting to "shut down the pharmacy of the developing world." In Washington, D.C., activists delivered to Novartis's offices a Golden Coffin award for Novartis CEO Daniel Vasella. In Basel, Switzerland, the Swiss NGO Berne Declaration awarded Novartis the 2007 "Public Eye Swiss Award," awarded annually to what it deemed to be the most irresponsible corporation. Despite a raging blizzard with forty-mile-per-hour winds, a handful of determined protestors gathered in front of Novartis's offices in Boston, Massachusetts, chanting "Novartis! Stop it! People over profit!" And international

On August 15, 2006, honoring India's Independence Day, Indian activists organized a demonstration during the XVI International AIDS Conference in Toronto. They marched through the conference center with posters in the colors of the Indian flag that read "Big Pharma Quit India," "Time to Deliver" and "Life Before Profits." In reference to the "Quit India Movement" of 1942 against the British colonial rule, they asked global pharmaceutical companies to "withdraw all patent applications." (Gaëlle Krikorian)

organizations such as MSF, Oxfam, Care, and others collected nearly half a million signatures from around the globe demanding that Novartis "drop the case."

Although Novartis never did drop the case, it was very much aware of the negative publicity it was receiving from all corners of the globe and announced shortly after the decision that it had no intention of appealing it. And the truly international attention this campaign garnered served to bring the issues around patents and access to medicines to the attention of millions, including, no doubt, the two judges who decided the matter, not to mention the Swiss trade minister, who announced shortly after the judgment that the ruling "does not concern the Swiss Confederation" and that Switzerland had no intention of filing a complaint at the WTO.

Following upon the Cancer Patients Aid Association's successful opposition to patenting imatinib mesylate, a host of other pregrant patent oppositions followed. Organizations of people living with HIV/AIDS opposed several applications for patents on critical AIDS drugs and have already achieved some success. Rather than risk a negative precedent that may cast doubt on the validity of its patents in other countries, the pharmaceutical giant GlaxoSmithKline chose to withdraw its pending patent applications for crucial components of anti-HIV therapy in developing countries, such as its fixed-dose combination of the AIDS drugs lamivudine and zidovudine, as well as its application for abacavir.

Indian civil-society groups are now slowly beginning to expand their attention to drugs beyond AIDS drugs, recently filing an opposition against Roche's application for PEGinterferon, a critical treatment for hepatitis C, a common and often fatal liver disease, that Roche sells in India for about $5000 for a six-month course.[20] Many of these patent oppositions are currently pending before the Indian Patent Office, and the coming months will reveal the extent to which India's unique patentability provisions, along with active civil-society participation in their application, can prevent harmful patents from being granted. A hopeful sign for opposition to patents in the future came in June 2008, when the Indian Patent Office rejected a patent application for a pediatric formulation of the AIDS drug nevirapine. In rejecting the formulation as a "new form" of a "known substance" and thus not patentable under section 3 (d), the Patent Office agreed with the opponents' contention of the need to "give a strict interpretation of patentability criteria, as decision . . . thereof shall affect the fate of people suffering from HIV/AIDS for want of essential medicine."[21]

Assuming that the various safeguards in the Indian Patents Act are rigorously enforced, there are indications that they will be extremely effective in controlling the number of patents granted on medicines. As mentioned, section 3 (d) is just one of several provisions that set stricter criteria in defining the scope of patentable subject matter. Others include an exclusion preventing "any living thing or non-living substance occurring in nature" from being patented (thus precluding things such as gene sequences and naturally occurring substances from patentability), a broad exclusion barring new uses of already-known substances from patentability (without regard to whether the new use is for a first or subsequent medical use), a provision that states that the "mere admixture" of substances is not patentable (thus preventing most patents on "formulations" of known drugs into specific dosage forms, such as tablets or capsules), and a broad exclusion that prevents the patenting of any medicinal treatment of human beings (thus precluding the common practice of drafting patent claims for known substances in the form of "the method of treating disease X by administering substance Y").

This last exclusion of "methods of treatment" is of particular significance. A surprising number of "new" drugs that are approved are not actually new drugs at all, but merely "discoveries" that a known substance is also (or only) good at treating another condition. A well-known example is sildenafil, better known to most

Viagra / Sildenafil

as Viagra as marketed by Pfizer. Sildenafil was originally developed and investigated as a drug for heart disease, but was ultimately approved for use for erectile dysfunction. But if a substance is already known, it cannot qualify for a patent— in any jurisdiction—for lack of novelty. The solution? Simply draft the patent claims to cover not the sildenafil itself, but for the "method of treating" erectile dysfunction by giving a patient sildenafil. But because Indian law includes a broad exclusion on methods of treatment, such a claim would be excluded from patentability. This same exclusion would similarly preclude the patenting of many other "new" drugs and in fact would have precluded the patenting of the first AIDS drug ever developed—AZT—which had been known since the 1960s and which was originally investigated for use as a cancer drug.

In addition to these broad substantive safeguards, there is one significant procedural bar to patentability that bears mentioning. Because India did not recognize product patent protection for pharmaceuticals prior to entering into the TRIPS Agreement on January 1, 1995, it was not bound to give retroactive effect to its TRIPS obligations. Thus, all new drugs that were "invented" before 1995 are ineligible, as a matter of law, for patent protection in India.

If these provisions are given robust interpretations, it is likely that the majority of new drugs that are coming out today would not qualify for patent protection in India. In fact, if one examines the patents covering the thirty-four new drugs approved by the U.S. Food and Drug Administration between January 2005 and March 2007, up to 70 percent of these new drugs would have no valid patent associated with them as a result of one or more of these substantive and procedural safeguards.

Of course, the true efficacy of these safeguards depends on how rigorously the Indian Patent Office applies them. Civil-society groups, with their limited resources, cannot possibly challenge the validity of every one of the thousands of patent applications that are pending in India to ensure that these standards are applied rigorously.

Indeed, there are already indications that some of these standards are not being applied by the Patent Office in a regular and systematic manner. For instance, the Patent Office recently granted a patent to Roche for an application covering valganciclovir, a substance used in the treatment of cytomegalovirus retinitis, a common and treatable AIDS coinfection that can cause blindness. Remarkably, despite the markedly tougher patentability standards that prevail in India, an investigation into the patent revealed that the Indian Patent Office had granted a number of patent claims that even the U.S. Patent and Trademark Office had rejected.[22] Indian civil-society groups have filed a suit opposing this patent, and it is currently pending.

SOME CONCLUDING OBSERVATIONS

A radical understanding of the scope of available TRIPS flexibilities is not only possible, but required in order to fulfill the Doha Declaration's mandate to promote access to medicines for all. India's experiments in defining the scope of patentable subject matter to suit its overriding policy objective of protecting public health should be seen as the start, not the end, of an effort to push at the very boundaries of what TRIPS allows. To do less is to resign ourselves to a world in which universal patent protection handcuffs the effective utilization of even the "accepted" flexibilities that are available.

As the Indian experience has shown, the utilization of these TRIPS flexibilities in a manner that gives civil society an opportunity to participate in a meaningful way is an essential component of the struggle for access to medicines, one that other developing countries can and should incorporate into their laws. But as essential as civil society's participation is, the limited funds and resources of civil-society groups will often constrain their level of participation to triage, selecting out of necessity only those cases that are of utmost importance.

In order to address these issues adequately on all levels, alternative models of promoting innovation without sacrificing access need to be explored. Whether the alternative models suggested by the Inter-Governmental Working Group on Public Health, Innovation and Intellectual Property prove to be workable remains to be seen. Our fingers are crossed.

NOTES

1   Although the TRIPS Agreement came into force in 1995, developing countries, including India, were given a five-year "transition period" to come into compliance with its obligations. Additionally, for countries that did not recognize product patent protection for certain areas of technology prior to 1995, there was an additional five-year transition period available to introduce product patent protection in these areas. Because India did not recognize product patent protection for pharmaceuticals prior to 1995, it had ten years—until 2005—to recognize product patents in this area. See TRIPS, Articles 65.2 and 65.4. The TRIPS Agreement is available on-line at http://www.wto.org/english/tratop_e/trips_e/t_agm0_e.htm (last accessed March 22, 2009).

2   See Sangeeta Shashikant's essay, "The Doha Declaration on TRIPS and Public Health: An Impetus for Access to Medicines," in this volume. The Doha Declaration is available on-line at http://www.wto.org/english/theWTO_e/minist_e/min01_e/mindecl_trips_e.htm (last accessed June 7, 2009).

3   World Health Organization, *Progress on Global Access to HIV Antiretroviral Therapy: A Report on "3 by 5" and Beyond* (Geneva: World Health Organization, March 2006), p. 7, available on-line at http://www.who.int/hiv/fullreport_en_highres.pdf (last accessed May 30, 2009).

4   See the essay by Jiraporn Limpananont and Kannikar Kijtiwatchakul, "TRIPS Flexibilities in Thailand: Between Law and Politics," in this volume.

5   Tenu Avafia, Jonathan Berger, and Trudi Hartzenberg, *The Ability of Select Sub-Saharan African Countries to Utilise TRIPs Flexibilities and Competition Law to Ensure a Sustainable Supply of Essential Medicines: A Study of Producing and Importing Countries*, Tralac Working Paper No. 12 (Stellenbosch, South Africa: Trade Law Centre for South Africa, 2006), p. 4, available on-line at http://www.iprsonline.org/unctadictsd/docs/Trade%20and%20Competition%20 30%203%2006%20final%20Edit1%20_2_%20_2_.pdf (last accessed May 30, 2009).

6   Biswajit Dhar and C. Niranjan Rao, *Transfer of Technology for Successful Integration into the Global Economy: A Case Study of the Pharmaceutical Industry in India* (New York and Geneva: United Nations Conference on Trade and Development, 2002), pp. 7–8.

7   Jonathan D. Quick, *Managing Drug Supply: The Selection, Procurement, Distribution, and Use of Pharmaceuticals* (West Hartford, CT: Kumarian Press, 1997).

8   Available on-line at http://www.msf.org/source/access/2007/Untangling_the_Web_10th_ ed_2007.pdf (last accessed May 30, 2009).

9   Under these rules, where a country with insufficient domestic manufacturing capacity to make its own drugs wants to import a drug produced under a compulsory license from a country that has the capacity, two separate compulsory licenses are required: one in the importing and another in the exporting country. In addition to this, a host of other reporting and "non-diversionary" measures need to be satisfied. See WTO General Council, "Implementation of Paragraph 6 of the Doha Declaration on the TRIPS Agreement and Public Health," Decision of the General Council of 30 August 2003, WTO doc. no. WT/L/540, available on-line at http://www.worldtradelaw.net/misc/dohapara6.pdf (last accessed May 30, 2009).

10   In 2008, after a four-year struggle, the Canadian generic company Apotex made shipments of a generic AIDS cocktail to Rwanda under the WTO "solution." Given the numerous delays and expenses incurred in making use of this mechanism, Apotex has stated that it would be unwilling to use the mechanism in the future.

11   Médecins Sans Frontières, "Neither Expeditious, Nor a Solution: The WTO August 30th Decision is Unworkable," (August 2006), available on-line at http://www.msfaccess.org/fileadmin/ user_upload/medinnov_accesspatents/WTOaugustreport.pdf (last accessed May 30, 2009).

12   *KSR International Co. v. Teleflex, Inc. et al.*, 550 U.S. 398 (2007), pp. 15 and 17 (slip opinion): 127 S.Ct. 1727 (2007), available on-line at http://www.supremecourtus.gov/opinions/ 06pdf/04-1350.pdf (last accessed June 2, 2009).

13   Commission on Intellectual Property Rights, Innovation and Public Health, *Public Health, Innovation and Property Rights* (Geneva: World Health Organization, 2006), p. 15, available on-line at http://www.who.int/intellectualproperty/documents/thereport/ENPublic HealthReport.pdf (last accessed June 2, 2009).

14   *Ibid.*, p. 85.

15   This is provided for in Article 4bis of the Paris Convention for the Protection of Industrial Property, which was incorporated by reference into TRIPS.

16   European Patent Convention, Article 52(2), available on-line at http://www.epo.org/patents/

law/legal-texts/html/epc/1973/e/ma1.html (last accessed June 2, 2009). But, as we will see, the convention carves out an important exception to this rule.

17 35 U.S.C. §§ 100 (a), 101.

18 European Patent Convention, Rule 23(e)(2).

19 *Novartis, AG v. Union of India and others* (2007) 4 MLJ 1153, para. 19 available on-line as "High Court Order Novartis Union of India," p. 34, at http://www.scribd.com/doc/456550/High-Court-order-Novartis-Union-of-India (last accessed March 5, 2010).

20 In April 2009, the Indian Patent Office rejected Indian civil society's opposition against PEGinterferon. Civil-society groups are reportedly considering appealing this decision.

21 *Boehringer Ingelheim v. Indian Network for People Living with HIV/AIDS*, Appln. no. 2485/DEL/1998.

22 C. H. Unnikrishnan, "Patent Denied in US, but Granted in India," *Mint*, January 15, 2008, available on-line at http://www.livemint.com/2008/01/14235738/Patent-denied-in-US-but-grant.html (last accessed March 5, 2010).

April 26, 2007—demonstration organized by students and supporters of the Stop AIDS Campaign to denounce Abbott´s retributive actions against Thailand after the government used compulsory licenses to provide generic medicine to Thai patients (Sarah Waldron).

TRIPS Flexibilities in Thailand: Between Law and Politics

Jiraporn Limpananont and Kannikar Kijtiwatchakul

The United States succeeded in forcing the members of the World Trade Organization to accept the Trade-Related Intellectual Property Rights (TRIPS) Agreement, but also had to accept that intellectual property monopolies must incorporate "breathing space" for developing countries in the form of "flexibilities" in the protection of intellectual property rights because of the problem that patents can cause. Compulsory licensing by a government for public, noncommercial use is one of these special mechanisms or "flexibilities" permissible under the TRIPS Agreement, allowing countries to override drug patents in order to save lives.[1]

The TRIPS Agreement has been in effect for over a decade. Although developing and least-developed countries were allowed to delay its implementation,[2] many of them amended their laws to comply with TRIPS more quickly than they were required to do so. For example, Thailand amended its law in 1992, eight years before the deadline. However, until recently, none of these nations made use of the provision for compulsory licensing, in part because one of the areas in which it made the most sense to employ this provision was that of access to medicines, but most of the drugs used in the past decade were old enough not to be patented.

After the outbreak of HIV/AIDS in Thailand around 1989, the mortality rate of AIDS patients there became very high, as did the price of antiretroviral drugs. Similar scenarios took place in most developing countries. When public-health crises over the inaccessibility of medicines peaked, since more recent drugs were now under patent, governments began to implement government-use compulsory licensing under TRIPS in the Association of Southeast Asian Nations (ASEAN). This move was led by Indonesia, followed by Malaysia and then, more recently, Thailand.

Between the end of 2006 and the beginning of 2007, the Thai minister of health issued three compulsory licenses for drugs targeting HIV/AIDS and heart disease. This initiative provoked explosive responses from the U.S. administration and

Congress, as well as from multinational drug companies. The issuance of the compulsory licenses in Thailand was one of the first examples of a developing country with significant manufacturing capacities using compulsory licensing of medicines since the Doha Declaration on the TRIPS Agreement and Public Health, with its support for the "right to protect public health and, in particular, to promote access to medicines for all" and its affirmation of the use of TRIPS flexibilities to achieve this end, was adopted in 2001. It is also the first serious political showdown over developing countries' use of compulsory licenses.

This article examines the use of compulsory licenses in Thailand and is intended as a case study of the ability of developing countries to make use of them. Reactions to the Thai initiative highlight the tensions between the law and politics as far as the TRIPS Agreement and public health are concerned. They raise questions concerning the meaning of the TRIPS flexibilities and the role of political pressures in intellectual property policy.

THE ISSUING OF COMPULSORY LICENSES IN THAILAND

The impact of pharmaceutical product patents on accessibility to medicines is well recognized. For instance, they cause high prices for patented drugs and delays in the introduction of generic drugs in the market.[3] The Declaration on TRIPS and Public Health adopted in Doha reflected that impact in Article 3.[4] In the 1990s, the prices of antiretroviral drugs in Thailand kept them out of reach of people with HIV. The daily cost of the medicine is about two to ten times the average daily wage. In reaction to this situation, on December 22 and 23, 1999, about one hundred people with HIV and activists from NGOs camped in the grass yard in front of the Ministry of Public Health. They requested that the government authorities apply a compulsory license to an antiretroviral drug called didanosine (ddI), manufactured by Bristol-Myers Squibb and Barr Laboratories and patented as Videx, so that cheap generic drugs could be produced.

The government refused. Since its reason for not issuing a compulsory license was fear of U.S. trade sanctions, the NGOs wrote to the U.S. president, Bill Clinton. The reply confirmed Thailand's right to implement compulsory licenses under the TRIPS Agreement. However, the letter was not enough to reassure the minister of Public Health, who still refused to issue a compulsory license for ddI. This reluctance to use compulsory licenses to solve health problems is often found in most developing and least-developed countries, whose governments fear economic and politic retaliation.

At the time, most of the academics, senior staff members in the Ministry of Public Health, and pharmacists in the Government Pharmaceutical Organization,[5]

Sticker printed by Thai activists in support of the use of compulsory licensing to increase access to medicine.

Inspired by the Ecologist

ซีแอล การใช้สิทธิเพื่อผู้ชีวิต

Right to CL = Right to Live

which had already done the research to produce ddI, encouraged the government to issue the compulsory license (CL). The "Royal Thai Government should now implement CL for one of the needed ARVs [antiretrovirals] to learn about the process," said a senior staff member. And the deputy director of the Research and Development Institute at the Government Pharmaceutical Organization said: "We are conducting the research and development of patented drugs such as efavirenz, lopinavir, and the ddI pellet. But to produce these drugs, it requires the decision of the Royal Thai Government to issue compulsory licensing."

In 2002, in response to the submission of fifty thousand signatures from Thai citizens to the parliament, Thailand passed the National Health Security Act. According to it, the government has the obligation to provide health services to every Thai citizen. However, because of the limited health budget, high-cost treatments such as the antiretroviral drugs for AIDS were excluded from the service.

In 2003, the Thai Government Pharmaceutical Organization (GPO) launched an antiretroviral cocktail drug in one pill named GPO-vir at the price of only one U.S. dollar per day. Despite the patent law, this was possible because none of the three drugs involved were patented in Thailand. This combination, prescribed as a first-line regimen, came to be used by the vast majority of the patients in Thailand, except when contraindicated by side effects or medical conditions banning the use of one of the drugs included in the combination. Because the cost of the treatment had dramatically fallen to only one dollar per day, a request from NGOs and people living with HIV/AIDS to include antiretrovirals in the pharmaceutical benefit scheme was accepted by the government.

However, two years after the implementation of this new system, the National Health Security Office faced the risk of running out of funds for antiretrovirals as a consequence of the high prices of second-line drugs needed by patients who developed drug resistance to GPO-vir or who suffered from important side effects.[6] The National Health Security Office and the government health authority then tried to negotiate with drug companies to reduce these prices, but failed.

In 2004, negotiations for a free-trade agreement between Thailand and the United States were initiated. Resistance from the public to the agreement was important and grew in part in reaction to U.S. requests to increase intellectual property protections. The level of protection sought by the United States, going beyond WTO standards—which is why they are called "TRIPS-plus"—represented an increase in the barriers preventing access to medicines. Two years later, this matter was one of the main issues raised against the administration of the Prime Minister Thaksin Shinawatra, who was finally overthrown in a military coup. The issue of free-trade agreements and the inaccessibility of medicines were raised by most speakers during the campaign against Taksin's administration, which mobilized a hundred thousand people and produced massive demonstrations every evening for two months before the military coup took place.

In December 2005, the UN Development Programme, in cooperation with the World Health Organization (WHO), the UN Programme on HIV/AIDS, the Ministry of Public Health, and Chulalongkorn University, held a conference entitled Free Trade Agreement-related Intellectual Property Rights: The Case of Drug Consumption. Academics from around the world were invited to share their experiences and research work relevant to free-trade agreements and access to medicines. The conference concluded that the excessive U.S. demands, exceeding the TRIPS standards, would undermine Thailand's access to essential drugs. The meeting presented its policy proposals in a press conference, where delegates urged the government to preserve its sovereignty over the use of compulsory licensing provided by the TRIPS Agreement. They also recommended that Thailand apply compulsory licensing to second-line antiretrovirals and refuse free-trade-agreement provisions on intellectual property exceeding the level of protection in TRIPS Agreement. The conclusions of the conference were put into an article written by Dr. William Aldis, the representative of the WHO in Thailand, and were published in the *Bangkok Post* on January 9, 2006, the day before the beginning of the sixth round of

Thai-U.S. free-trade agreement negotiations in Chiang Mai. The U.S. administration reacted by putting pressure on the WHO Secretariat, and in March 2006, Dr. Aldis was suddenly removed from his position in Thailand.

The sixth round of the Thai-U.S. free-trade-agreement negotiations was also marked by intense mobilizations from civil society. About ten thousand people representing eleven networks of nationwide people's organizations came together to protest the free-trade-agreement negotiations in Chiang Mai.[7] Although the negotiations did not break down completely, the negotiators frantically had to find a new meeting place. The power of the people, mobilized by the dissemination of knowledge and efforts to increase public awareness, made the Thai-U.S. free-trade deal very questionable, especially U.S, demands for a TRIPS-plus standard for drug patent protection.

Like the December 2005 conference, a World Bank report entitled "The Economics of Effective AIDS Treatment: Evaluating Policy in Thailand" sent a message to Thailand that it was time to brave it out and use compulsory licensing.[8] This report was an evaluation of the three-year-long attempt by the Ministry of Public Health to expand its antiretroviral treatment services. It warned that without a prompt decision to ensure access to newer drugs, the increasing costs of treatment would be too high for the government to continue providing national access to the antiretroviral program, which had been highly praised. Without any action taken to lower the prices of these drugs, the government's budget to run this program would have had to be increased fivefold within fifteen years. During the Thaksin administration, Dr. Sanguan Nitayarumphong, the secretary-general of the National Health Security Office, and Pinij Jarusombat, the minister of public health, agreed that compulsory licensing was a mechanism that could support the national health system. They decided that the matter be investigated so that a decision could be made. A subcommittee dedicated to implementing government use of patented drugs and medical supplies was established.

This subcommittee was set up on January 12, 2006 and consisted of representatives from a wide range of concerned parties, including the commerce and public health ministries, the Council of State, the Law Society of Thailand, hospital doctors, networks of people living with AIDS and cancer patients. As a result of its work, the subcommittee passed a resolution calling for a compulsory license to produce efavirenz, which was finally approved by the National Health Security Board in August 2006. The minister of public health asked the subcommittee to consider carefully the written compulsory license announcement decree. One week later, on September 19, the military coup took place.

When Dr. Mongkol Na Songkhla took charge of the Ministry of Public Health in

CLs & welfare state ?

October 2006, he surrounded himself with a team of experts who, over the years, had accumulated knowledge on intellectual property issues. The secretary-general of the National Health Security Office, who had served as the chair of the previous subcommittee, sent him the conclusions of its work and the recommendation to issue the compulsory license for efavirenz. Finally, at the end of a process of give and take between the ministry and the subcommittee, the first compulsory license announcement about efavirenz was made, followed a few months later by two others.

The first compulsory license was issued on November 29, 2006 on the first-line drug efavirenz, commercially known as Stocrin, made by Merck Sharp and Dohme, the UK subsidiary of the U.S. pharmaceutical company Merck. On January 24, 2007, a second license was issued on the second-line drug lopinavir/ritonavir, commercially known as Kaletra, made by Abbott. Another license was granted on January 25, 2007 on clopidogrel, an antiplatelet drug used in the treatment of coronary artery disease, peripheral vascular disease, and cerebrovascular disease, commercially known as Plavix and made by Sanofi-Aventis.

"Public Health interest and the life of the people must come before commercial interest." This sentence appeared at the beginning of the white paper on drug patents issued by the Ministry of Public Health.[9] This document cites legal rights and justifications for the use of compulsory licensing for the public interest: The use of the patent rights for the three patented drugs serves noncommercial purposes and would be limited to patients covered under three government welfare systems, the National Health Security System, Social Security System, and the Civil Servant Medical Benefit Scheme.

The Ministry of Public Health's white paper pointed out the reasons behind this decision: "The Thai Ministry of Public Health firmly believes in a moderate and public interest oriented approach to implement the intellectual property right. We are convinced and committed to the view that 'Public Health interest and the life of the people must come before commercial interest.'" The government expected to save 1,035 to 1,665 million baht (32 million to 51 million U.S. dollars, using June 2010 exchange rates) in its annual budget and to provide an increase in drug access for patients of sixfold to twelvefold.

Since the end of 1999, when the request for a compulsory license for ddI had been made, the problem of accessibility to antiretrovirals had become more and more serious, and since then, the campaign for access to medicines has been carried on continuously by Thai NGOs,[10] by academics, and by several civil-society groups, supported by international NGOs.[11]

As this account has shown, compulsory licenses under the TRIPS flexibilities provisions could be implemented in Thailand because three main factors occurred concurrently: public awareness of and knowledge about compulsory licensing, the continuous pressure of social movements, and the political will of public agents. As the white paper on the compulsory licensing of patented drugs put it, these three factors form "the so-called 'Triangle that moves the mountain.'" The white paper also states what the Thai experience shows about the effort to implement compulsory licenses under the TRIPS flexibilities in any country. Doing so requires an "educated and motivated society that will push for and support the political commitment to bring real and sustainable success to any social reform movement."[12]

REACTIONS TO THE THAI INITIATIVE

There was overwhelming support for Thailand's use of compulsory licensing from various international organizations and NGOs, for example the UN Programme on HIV/AIDS,[13] Médecins Sans Frontières (MSF),[14] the Third World Network,[15] the Consumer Project on Technology,[16] and the Clinton Foundation.[17] Thanphuying (Dame) Preeya Kasemsan, chair of the Public Health Committee of the National Legislative Assembly, sent a letter dated February 20, 2007, praising the government's announcement of the exercise of compulsory licensing. "Such implementation is beneficial to a great number of people and will increase the people's access to essential drugs. As the government has a limited budget, such enforcement is legitimate and in compliance with international principles currently adopted by the global community."[18]

A letter sent by MSF to Condoleezza Rice, the U.S. secretary of state, and to Susan Schwab, the United States trade representative demanded that the United States stop interfering in Thailand's compulsory licensing. It pointed out that:

> Thailand's decision will have important consequences, not only for Thailand, but for any developing country that needs to obtain low-cost generic products. If Thailand follows through and begins to buy from generic suppliers, it will create a larger global market for generic products, stimulate competition, and lower prices everywhere for the newer products. While the benefits of expanded generic competition are widely appreciated, many developing countries have been reluctant to issue compulsory licenses because of fears that the United States government will oppose such actions and exert pressure.[19]

While the Thai Ministry of Public Health's commitment to the view that the right to life of the people must come before commercial interests has been praised by most of Thai media, international NGOs, including MSF, and Oxfam, some U.S. NGOs, and the UN Joint Nations Programme on HIV/AIDS,[20] it has also been disapproved of, opposed, and lobbied against by pharmaceutical companies and the countries that support them. When Thailand announced its three compulsory licenses, the pharmaceutical companies, anxious to stem the tide of such practices among developing countries, retaliated aggressively, using every means available. One indication of their political and economic power is the threat by the U.S. government to use Special 301 measures against Thailand, taking Thai exports hostage to prevent Thailand from doing any business with which the U.S. government is not in agreement.

The pharmaceutical companies and their supporters made several arguments to try to discredit the action of the Thai Ministry of Health. These arguments are characteristic forms of criticism from these actors. To make them, the pharmaceutical industry launched an intense media and publicity campaign. Many foreign media, especially the *Wall Street Journal*, accused Thailand of violating intellectual property rules. Several others admitted at some point that the move was legal, but elsewhere employed the language of "patent breaking" or "patent overriding," thereby connoting that Thailand was breaking the law. Public-relations lobbyists publicized false and misleading information in the international and local press, as well as on the Internet.

The Pharmaceutical Research and Manufacturers Association, the Thai affiliate of the Pharmaceutical Research and Manufacturers of America, which is the most powerful association of the multinational drug industry, took a leading role in initially opposing Thailand's compulsory licensing by denouncing it as appropriating the private sector's property. It also threatened to suspend its investments in Thailand. In addition, the multinational pharmaceutical companies argued that Thailand's exercise of compulsory licensing was illegal under both the TRIPS Agreement and the Thai Patent Act.

These tactics did not work, in part because their claims were wrong. Numerous authorities, including the U.S. trade representative,[21] the director general of the WHO,[22] and many intellectual property rights lawyers,[23] have made statements confirming Thailand's compliance with the TRIPS Agreement.

The use of compulsory licensing is not the appropriation of private-sector property, but the lawful exercise of TRIPS flexibilities. Article 31 (b) of TRIPS states:

> such use may only be permitted if, prior to such use, the proposed user has made efforts to obtain authorization from the right holder on reasonable commercial terms and conditions and that such efforts have not been successful within a reasonable period of time. This requirement may be waived by a Member in the case of a national emergency or other circumstances of extreme urgency or in cases of public non-commercial use. . . . In the case of public non-commercial use, where the government or contractor, without making a patent search, knows or has demonstrable

grounds to know that a valid patent is or will be used by or for the government, the right holder shall be informed promptly.

Reasonable remuneration was offered in recognition of the patent holders' rights. According to the Thai government's proclamation in the "Public Use of Patents for Pharmaceutical Products," patent holders were allocated a 0.5 percent royalty fee and could also seek to change this remuneration rate through negotiations with the government or appeal to the Department of Intellectual Property. But no patent holders have entered into such negotiations. Indonesia and Malaysia had similar experiences when they announced their public use of compulsory licensing in 2004. The multinational pharmaceutical companies preferred to give up their royalties rather than recognize and legitimate developing countries' legal rights by negotiating the remuneration rate for government use of their patents. Neither did they want to set an example of accepting remuneration. The companies, of course, could still sell their drugs in Thailand—all that was being taken away was a government-granted monopoly that the companies were abusing through their pricing practices.

And Thailand in fact had complied with Thai law, as well as with international agreements. When the multinational pharmaceutical companies sought to block Thailand's exercise of compulsory licensing by arguing that it was illegal under Thai law, the newspapers reported that the Department of Intellectual Property was poised to ask for the Council of State's interpretation.[24] Many Thai experts on intellectual property law and the world's leading experts in international intellectual property law, including Professors Brook Baker of the Northeastern University School of Law, Sean Flynn of American University, and Dr. Carlos M. Correa of the University of Buenos Aires, expressed their objections to the move. According to them, there was no need for the interpretation. Article 51 of the Thai Patent Act states very clearly the government's right to issue licenses on patented drugs clearly without prior consent of patent holder.

Representatives of the Thai Network of People living with HIV/AIDS and of other AIDS NGOs asked Mrs. Puangrat Assawapisit, director general of the Department of Intellectual Property, for clarification on the matter. On February 21, 2007, the Department of Intellectual Property turned down the company's appeal and suggested that any complaint about the Ministry of Public Health's failure to follow Thai legal procedures should be filed with the Administrative Court. The Department of Intellectual Property stated that it had jurisdiction only over appeals regarding the amount of remuneration. The company gave up at this point.

These legal claims against the use of compulsory licensing thus failed, but at least they were made in legal venues. By far the most insufferable component of the attack was the smear campaign run by USA for Innovation, a group claiming to be a nonprofit organization, but in fact a thinly disguised shill for drug firms connected to Edelman Public Relations Worldwide, a firm whose clients include many multinational drug corporations and the ousted prime minister, Thaksin Shinawatra. It bought advertising space in the Thai and foreign print media and created its own Web site accusing the military-backed government of Thailand of turning the country into a dictatorship like Burma by illegally issuing compulsory licenses.

This was clearly false, and it is wrong to attribute the decision to the coup. Efforts to counter life-threatening monopolies over essential drugs had been going on for more than a decade and were spearheaded by civil society, not the military. As we noted above, a subcommittee to implement compulsory licensing had been set up under the government led by Prime Minister Thaksin, and before the military coup of September 2006 that deposed Thaksin, this committee already had resolved to impose compulsory licensing on the antiretroviral drug efavirenz. The minister of public health who took office after the coup merely carried out this legitimate and orderly process. The military budget had nothing to do with the decision, which would have come regardless of the change in government.

The organization also claimed that GPO-vir, produced by the Government Pharmaceutical Organization, was of low quality, and it submitted letters to the U.S. Congress and the Bush administration urging them to retaliate against Thailand, claiming that the drug was ineffective and not approved by the WHO. USA for Innovation cited a study from Mahidol University in 2005 that purportedly showed that GPO-vir caused much more drug resistance than other antiretrovirals, from 39.6 to 58 percent of all patients taking it. However, the study did not provide references, and drug resistance rate is not related to the quality of the medicine at all. The fact is that this drug, produced by the Government Pharmaceutical Organization under a Thai utility patent, has saved tens of thousands of lives. This drug is registered in Thailand, and its quality has been assured by the Thai Food and Drug Administration since 2003. The reduction in the costs of antiretroviral treatment through GPO-vir from 2,000 baht to 400 baht per day has yielded savings that have allowed the expansion of the universal health coverage program to antiretrovirals.

Such false accusations provide an example or what sort of reaction countries may have to face when making use of their right to issue compulsory licensing. There is a risk that they may consequently discourage the implementation of compulsory licenses in other developing or less developed countries, even though the pharmaceutical companies' assertions are pure lies and intimidations.

The Bangkok Post and *The Nation* newspapers each dedicated for free one full page to voices from Thai civil society to publish their responses to the allegations of USA for Innovation:

USA for Innovation is an organization set up to serve the interests of U.S. drug companies upset by Thailand's recent announcements of compulsory licenses permitted by the WTO Agreement. This organization is adept at manipulating and distorting the facts to achieve its purposes.

The accusations by USA for Innovation, which appeared in an advertisement in the English-language press a few days ago, distort the facts by denigrating the antiviral drugs produced by Thailand's Government Pharmaceutical Organization as being worthless. The truth is that GPO-vir has played a significant role in reducing the annual mortality rate among Thai AIDS patients from 8,246 in 2001 to 1,613 in 2006. More than tens of thousands of lives have been saved over the past 3–4 years.

The study by Mahidol University cited by the USA for Innovation to claim that GPO-vir had a high resistance of between 39.6 and 58 per cent is in fact research that attempted to study the resistance among long-term patients who had already failed the treatment. There are no research findings available to compare the rates of resistance to GPO-vir with the rates of resistance to equivalent originator products.

Thailand is internationally recognized for its efforts to provide health coverage for all and to ensure that HIV/AIDS patients have universal access to the appropriate drugs.

The decision to authorize the compulsory licensing of necessary drugs shows the praiseworthy courage of the Thai government and the Ministry of Public Health to put the lives of the Thai people before commercial benefits.

When confronted by the forceful refutation of their claims and the strength of both international and internal support for compulsory licensing in Thailand, the multinational pharmaceutical companies had other strategies in reserve. They engaged in aggressive local lobbying. For example, Nimit Tienudom, director of the AIDS ACCESS Foundation, revealed that Abbott had asked to meet with the Thai Network of People living with HIV/AIDS and to offer to reduce the price of Kaletra for the government in exchange for the network's lobbying with the Ministry of Public Health to cancel its use of compulsory licensing.

The companies also sought to exert pressure on the Thai government by leveraging diplomats from their home countries. At the same time that Abbott was approaching AIDS activists, the ambassadors of the United States, the European Union, France, and Switzerland met with the Thai ministers of public health, commerce, and foreign affairs many times, signaling their disapproval of Thailand's pursuit of compulsory licenses. They cited Thailand's failure to hold prior negotiations with the patent holder, despite acknowledging that such negotiations were not legally required and even though there were several negotiations on price reduction with the companies before issuance of compulsory licensing.

Meanwhile twenty-two members of the U.S. Congress, led by Henry Waxman, sent a letter urging the U.S. trade representative not to impede Thailand's exercise of its legal rights. In her reply, the U.S. trade representative, Susan Schwab, admitted that Thailand was authorized to issue compulsory licenses.[25]

Abbott Labs resorted to the tactic of sending a letter to the Drug Control Division of Thailand's Food and Drug Administration, instructing it to withdraw all applications to register its new drugs in Thailand. The Thai newspaper *Matichon Daily* and the *Wall Street Journal* reported that "Abbott will not apply for the registration of new drugs and will withdraw all applications to register new drugs in Thailand until the government takes heed of intellectual property, including the cancellation of compulsory licensing." The withdrawal of the drugs — Zemplar, for the treatment of chronic kidney disease, Simdax, for heart failure treatment, Humira, a medicine for treating autoimmune disease, and Aluvia tablets, a new formulation of the heat-stable second-line AIDS drug — was clear retaliation for the Thai government's use of compulsory licensing for Abbott's Kaletra (lopinavir/ritonavir).

As soon as Abbott's move was widely reported, the Thai Network of People Living with HIV/AIDS, the AIDS ACCESS Foundation, the Centre for AIDS Rights, the Thai NGO Coalition on AIDS, and the Foundation for Consumers condemned the company for maliciously putting pressure on Thailand. Its action, they contended, reflected the pharmaceutical industry's limitless greed and a total lack of concern for the Thai people. These opposing organizations urged the Thai people to boycott Abbott's products, for example by buying generic versions or alternatives to drugs made by Abbott.

The move to strike back at Abbott spread to the Network of Parents, the Rural Doctors Club, the Rural Pharmacists Club, and the Network of Patients with Kidney and Heart Diseases. These groups denounced Abbott's action as holding patients and consumers hostage in order to force the government to end its use of compulsory licensing, even though the Thai government had done nothing in contravention of domestic law and international rules.

Anger against Abbott soon spread globally. On April 26, 2007, one day before Abbott's annual shareholders meeting in Chicago, demonstrations were held in front of the company's offices in cities in France, the United States, the United Kingdom, Germany, India, South Africa, China, Brazil, Argentina, Australia, Canada, Indonesia, Japan, and Singapore. Activists around the world urged a boycott of Abbott's products. In France, the group of people living with HIV/AIDS known as Act Up-Paris organized a "netstrike," or denial of service attack, on Abbott's Web site. So many supporters visited Abbott's site that the servers became overloaded and the Web site collapsed.

A day later, on April 27, Jon Ungphakorn, a former senator of Thailand and secretary general of the AIDS ACCESS Foundation, and Wirat Purahong, chairperson of the Thai Network of People Living with HIV/AIDS, participated in Abbott's annual shareholders meeting in order to act as proxies for the shareholders from religious groups and to question the unethical actions of Abbott. They used the opportunity to speak directly to Abbott executives, demanding that drugs not be used as political leverage and patients not be used as political hostages.

The student group Universities Allied for Essential Medicines and the Student Global AIDS Campaign sought to add pressure from a new direction. One of the drugs that Abbott was using as a bargaining chip had in fact been developed at the University of Wisconsin at Madison. Outraged that a medicine developed by their institution to better the lives of patients around the world was being used in so manipulative a fashion, former and present students sent a letter demanding that the Wisconsin Alumni Research Foundation condemn Abbott's action.

Thereafter, the sentiment that Abbott's actions were unacceptable reverberated around the world. A critical turning point arose when religious groups holding Abbott's shares condemned the company and called on it to end their action and reregister the drugs in question.

Finally, Abbott sought to redeem itself or to reduce the pressure and improve its image by establishing a three-tier price structure for Kaletra worldwide. The price for middle-income developing countries would drop from (US) \$2,200 per patient per year to \$1,000 per patient per year. But this reduced price would be available in Thailand only if no compulsory license was imposed on the new heat-stable form of lopinavir/ritonavir, Aluvia.

Thai civil-society groups led by the Foundation for Consumers, the AIDS ACCESS Foundation, the Thai Network of People Living with HIV/AIDS, and the Thai NGO Coalition on AIDS tried a new tactic: invoking Thai law on commercial competition to attack Abbot's decision to withdraw its pending drug registrations. It appealed to the Thai Competition Commission, arguing that Abbott's actions violate Sections 25 (3) and 28 of the Competition Act, B.E. 2542 (1999), because the company was a business operator with market domination and was using that dominant position to act in an anticompetitive fashion by unreasonably cancelling imports, leading to a decrease in the available choice of drugs to the point where patients' needs could not be met. The Competition Commission was thus urged, under Section 31 of the act, to order Abbott Laboratories to apply for registration of its new drugs and resubmit applications for the ten withdrawn drug registrations in Thailand.

The effect of these successes in the effort to provide access to medicines through the use of compulsory licensing under the TRIPS Agreement in Thailand have not been limited to the national level. One positive side effect has been the emergence of solidarity in a global community. A couple of months after the Thai compulsory license on efavirenz was issued, the Brazilian government issued one for the same drug.

In November 2007, about two hundred experts, social activists, and representatives of patient networks from all over the world gathered in Bangkok to discuss compulsory licensing, innovation, and access to medicines for all. They launched the Bangkok Declaration on Compulsory Licensing, Innovation, and Access to Medicines for All. The declaration recognized and applauded Thailand's leadership in the use of compulsory licensing to overcome legal monopolies, praised decisions by Brazil and Indonesia to issue compulsory licenses, and encouraged other countries to exercise their rights to do so, too. They also proposed a new global network on compulsory licensing, innovation, and access for all that links together patients, NGOs, academic and public health experts, government officials, and manufacturers of generic drugs to find ways to ensure that patients have access to medicines with acceptable quality and price.

CONCLUSION

Every country should have the right to use compulsory licensing systematically and routinely, as well as other means available to them under the TRIPS flexibilities. Governments all over the world use compulsory licensing in a variety of contexts and in many different fields, and the right to use compulsory licensing is incorporated in international law and precedent, including the WTO's TRIPS Agreement and the Doha Declaration. But when developing countries such as Thailand try to exercise their right to use compulsory licensing for essential drugs in order to solve the health problems, they face significant obstacles from the multinational drug companies and politicians.

The mechanism of compulsory licensing is powerful, but as the example of Thailand shows, it can be hard to implement in developing countries. However, as we have seen, pressure from interested parties, such as patients groups, academics, and politicians, can make it possible to do so. The difficulties involved in that process and the obstacles that powerful opponents can throw in the way of its successful completion nevertheless make it important to consider alternatives to the increase of intellectual property protection as ways to promote and guarantee medical research and development without contravening access to medicines.

NOTES

1   Several compulsory-licensing options exist in the TRIPS Agreement. Compulsory licensing can be applied in order to remedy anticompetitive practices or to address public needs in a wide range of situations, including when the patent is not used or when products are overpriced. Prior negotiations with the patentee are required. Another form of compulsory licensing is government use, which can occur without the prior consent of the patentee in exchange for remuneration and with the obligation to inform the patentee immediately after issuance. There are three conditions under which the government-use provision can be employed: a national emergency, a situation of urgency, and for public, noncommercial use. Article 51 of the Thai Patent Act applies to public, noncommercial use, whereas Article 52 applies in cases of national emergency or situations of urgency.

2   Nonindustrial countries, depending on their level of development and on whether they were already protecting pharmaceutical products with patents, could wait until 2000, 2005, or 2006. At the WTO conference in Doha in 2001, it was decided that least-developed countries that were not protecting pharmaceutical products prior to the ratification of TRIPS could wait until 2016.

3   Jiraporn Limpananont, "Thailand: The Impact of Pressure from the US," in M. Foreman (ed.), *Patent, Pills and Public Health: Can TRIPS Deliver?* (London: The Panos Institute, 2002), pp. 41–43.

4   Doha Declaration, para. 3: "We recognize that intellectual property protection is important for the development of new medicines. We also recognize the concerns about its effects on prices." The Doha Declaration on the TRIPS Agreement and Public Health, November 14, 2001, WTO doc. no. WT/MIN(01)/DEC/2, is available on-line at http://www.wto.org/english/theWTO_e/minist_e/min01_e/mindecl_trips_e.pdf (last accessed March 24, 2009).

5   The Government Pharmaceutical Organization is a pharmaceutical manufacturer that is owned by the state, but whose administration is independent from the government.

6   The National Health Security Office is the institution that was established to manage the universal health coverage program after the National Health Security Act 2002 was passed.

7   The eleven networks included the Thai Network of People Living with HIV/AIDS, the Alternative Agriculture Network, the Federation of Consumer Organizations, the Four-Region Forest Network, the Federation of Northern Farmers, the Four-Region Slums Network, the Council of the People's Organizations of Thailand Network, the Confederation of State Enterprise Workers, the Student Federation of Thailand, and the Free-Trade Agreement Watch Group, a coalition of NGOs and academics working on different issues related to free-trade agreements in areas such as agriculture, labor, the interests of different patients groups, and so on.

8   Ana Revenga et al., "The Economics of Effective AIDS Treatment: Evaluating Policy in Thailand," (Washington, D.C.: The World Bank, 2006), available on-line at http://siteresources.worldbank.org/INTEAPREGTOPHIVAIDS/Resources/AIDS-Treatment-effectiveness-TOC.pdf (last accessed June 4, 2009).

9   Ministry of Public Health and National Health Security Office, "Preface," *Facts and Evidences on the 10 Burning Issues Related to the Government Use of Patents on Three Patented Essential Drugs in Thailand* (February 2007), available on-line at http://www.moph.go.th/hot/White%20Paper%20CL-EN.pdf (last accessed June 3, 2009).

10 These NGOs include the Thai Network of People Living with HIV/AIDS, the AIDS Access Foundation, the Foundation for Consumers, the Drug Study Group, the Health and Development Foundation, Free-Trade Agreement Watch, and the Pharmacy Network on Health Promotion.

11 The international NGOs supporting the effort are primarily MSF and Oxfam.

12 Ministry of Public Health and National Health Security Office, "Preface," *Facts and Evidences on the 10 Burning Issues Related to the Government Use of Patents on Three Patented Essential Drugs in Thailand.*

13 Letter from Dr. Peter Piot, executive director, UNAIDS, December 26, 2006, reprinted in Ministry of Public Health and National Health Security Office, *Facts and Evidences on the 10 Burning Issues Related to the Government Use of Patents on Three Patented Essential Drugs in Thailand,* p. 92.

14 Letter from Nicholas de Torrente, director of MSF USA, and Paul Cawthorne, MSF Thailand, December 29, 2006.

15 Letter from the Third World Network, February 23, 2007.

16 Letter from James Love, director of the Consumer Project on Technology, December 11, 2006.

17 Letter from the Clinton Foundation, February 16, 2007.

18 Kannikar Kittiwatchakul, *The Right to Life* (Bangkok, Thailand: Pimdee Co., Ltd, 2007), p. 26.

19 See http://doctorswithoutborders.org/publications/article.cfm?id=1917 (last accessed June 6, 2009).

20 Letter from Dr. Peter Piot, *Facts and Evidences on the 10 Burning Issues Related to the Government Use of Patents on Three Patented Essential Drugs in Thailand,* p. 92.

21 Letter from the Honorable Susan C. Schwab, the United States trade representative, to the Honorable Sander M. Levin, Member of Congress, January 17, 2007.

22 Letter from Dr. Margaret Chan, director general of the World Health Organization, to the Honorable Dr. Mongkol Na Songkhla, Thai minister of public health, February 7, 2007.

23 Sean Flynn, "Thailand's Lawful Compulsory Licensing and Abbott's Anticompetitive Response," American University Program on Information Justice and Intellectual Property, April 26, 2007, available on-line at http://www.wcl.american.edu/pijip_static/documents/Thailandreport426.2_001.pdf?rd=1 (last accessed June 7, 2009).

24 The Council of State is the government consultation organization on the interpretation of the law.

25 Ministry of Public Health and National Health Security Office, *Facts and Evidences on the 10 Burning Issues Related to the Government Use of Patents on Three Patented Essential Drugs in Thailand,* pp. 53–56.

Using Competition Law to Promote Access to Knowledge

Sean M. Flynn

One of the points of convergence among the many strands of the A2K movement is resistance to the one-size-fits-all ratcheting up of intellectual property provisions around the world. The resistance is grounded in analysis showing that intellectual property rules often create social costs that far outweigh their intended benefits. Much of the A2K movement's advocacy for limitations of intellectual property rights is located within the field of intellectual property law—promoting the inclusion and use of balancing mechanisms within the laws granting intellectual property rights. But intellectual property rights are also shaped and limited by their interaction with other fields of law, competition law being a prime example.

Competition laws, often referred to as "antimonopoly" or "antitrust" laws, regulate the conduct of firms that face insufficient competition and thereby have the power to raise prices charged to consumers. Intellectual property laws, on the other hand, grant rights to exclude competition from the subject matter of the intellectual property right to create incentives to invent and produce new products for consumers. The rights to exclude competition that are at the core of intellectual property rights may create the ability to engage in anticompetitive conduct of the kind normally regulated by competition laws. In such instances, policy makers and enforcement agencies are called upon to determine the extent of interaction between the two legal regimes. This determination plays a central role in defining the limit of intellectual property rights and thus is an important site of legal advocacy for access-to-knowledge movements.

The argument of what follows is that competition law can and should be used to promote the goals of access to knowledge. It is a common misconception that competition laws and intellectual property laws are in irreconcilable conflict, necessitating blanket exclusions of intellectual property from competition-law regulation. Competition and intellectual property laws serve similar ends: increasing

Anti-Monopoly game developed by Ralph Anspach (http://www.antimonopoly.com/).

economic productivity and access to new goods. It is only when intellectual property protections fail to serve the end of net increases in economic productivity, for example by creating more barriers to access to current technology and production than they create incentives for additional innovation and creation, that competition law justifiably limits intellectual property rights. In such cases, there is a long history in the United States and other countries of using competition law, including enforcing duties to share access with potential competitors, to limit consumer harm from excesses practiced by holders of intellectual property rights and other property holders. A key example is when intellectual property rights in developing countries prohibit competition in the supply of goods such as essential medicines or access to information essential to the social or economic development of a country. In such instances, profit-maximizing actions taken by the holders of exclusive intellectual property rights will cause far more consumer harm through restricted access than they benefit consumers in the form of incentives for future innovation.

A2K advocates are using competition law to limit the scope of intellectual property rights, including in cases that lay the groundwork for implementing a more robust theory of intellectual property. Their work is shifting the question of *whether* competition law can limit intellectual property rights to *when* competition law should regulate intellectual property, making it possible to recognize a new category of "essential intellectual property" for which open-licensing duties should be frequently required. After describing the theoretical and doctrinal underpinnings of this shift of A2K legal advocacy toward the use of completion law, this part surveys some of the strategic advantages of using competition norms to reframe political debates and shift struggles into new, potentially more hospitable, forums.

THE INTELLECTUAL PROPERTY/COMPETITION INTERFACE

On the surface, the misconception that intellectual property and competition laws are locked in interminable conflict seems plausible. Intellectual property laws grant rights to exclude competition with the rights holder to create incentives and rewards for innovation, including through higher prices to consumers. Competition laws prohibit anticompetitive conduct that creates or abuses a lack of competition in the market to the detriment of consumer interests. Taken at this superficial level, every act to obtain and profit from an intellectual property right could be construed as an act of monopolization in violation of competition law mandates. However, this view of competition and intellectual property laws has been rejected by courts and enforcement agencies in the United States, Europe, and other countries. In such jurisdictions, competition law is used, explicitly or implicitly, as a policy tool to restrict the scope of intellectual property rights without negating them.

Modern competition statutes regulating the use and abuse of monopoly power are of fairly recent vintage, the first such statute being the U.S. Sherman Act of 1890.[1] But the history of circumscribing grants of monopoly privileges with social duties that restrain excessive pricing and other abuses dates back much further.

When King Edward's *quo warranto* campaign in the thirteenth century first required a "letter patent" as proof of a valid exclusive marketing franchise,[2] a central purpose was to regulate those who "take outrageous Toll, contrary to the common Custom of the Realm."[3] Complaints of excessive pricing and other violations of a duty of "reasonable use" could be brought before the king's courts and were grounds for forfeiture of the franchise. Duties to serve public interests were included in the patent grants themselves, which "often required the patentee to produce goods of a certain quality and sell them within certain price limits."[4] The seventeenth-century Statute of Monopolies authorizing letters patent for new inventions, but banning most other franchises, stated the condition that "they be not contrary to the Laws nor mischievous to the State, by raising the prices of Commodities at home."[5]

In the late nineteenth and early twentieth centuries, duties for patent holders to serve the greater public interest took various forms. The Paris Convention for the Protection of Industrial Property of 1883, the first major international treaty on patent standards, instructed that "the patentee shall remain bound to work his patent in conformity with the laws of the country into which he introduces the patented objects."[6] Many countries included within their patent laws general public-interest grounds for compulsory licensing when a patentee has failed to meet the country's demand for the particular item on "reasonable terms." The United States was different in this regard. Unlike most other countries, it does not have a general public-interest compulsory license standard. Public-interest grounds for compulsory licensing and for otherwise limiting the scope of exclusive intellectual property rights instead have been developed in major part through the then uniquely American institution of competition law.

For the first decades after the U.S. Congress passed the Sherman Act of 1890, courts largely interpreted it as placing little restraint on the practices of intellectual property holders. It was commonly considered that the right to exclude others from use of a creation included a right to refuse to license the technology to others and a correlative right to impose unlimited restrictions on the licenses that the holder chose to issue. It was common, for example, for patent and copyright holders to impose minimum price requirements and resale restrictions in intellectual property licenses. Over time, this reasoning lost favor, and a large number of restrictive licensing practices, including minimum-pricing and resale restrictions, tying patent licenses to purchases of other products, and charging royalties not strictly related to the use of the patented technology, were deemed to be beyond the scope of the patent grant and prohibited by competition law.[7] The reach of antitrust law in this area is commonly said to have peaked in the 1970s with U.S. enforcement agencies' use of a "Nine No-No's" list of intellectual property licensing practices deemed to be per se illegal.[8]

Courts often frame the legal analysis as discovering a core of intellectual property rights immune from competition law scrutiny, with the right to exclude others through refusals to license often considered the heart of the core, surrounded by a periphery of restrictive licensing practices regulated by competition law. Courts thus answer the question of whether a particular restrictive practice is prohibited by examining whether it is within the "scope of the patent"[9] and contributes to the "reward which the patentee by the grant of the patent is entitled to secure"[10] or is outside the scope of the grant and by its "very nature illegal."[11]

The line between core and periphery has shifted dramatically over time, and several recent cases in the U.S. and Europe have discarded much of the core/periphery analysis altogether. Some recent cases use use instead competition doctrines to force the sharing of intellectual property in particular cases. These cases in effect transform the core of intellectual property rights from a right to exclude to a liability rule giving the rights holder compensation for use by others.

The genealogy of duties to license leads back to doctrines in real property law. The right to exclude others from the use of real property has long been considered the core of the property right. But competition law has nevertheless been used to invade that core in a number of special circumstances. In a series of cases dealing with the ownership of unique infrastructure resources that are necessary to enable competition, from the only bridge across a river,[12] to electricity and telephone wires needed to promote utility service competition,[13] U.S. courts have propounded what has become known as the "essential facilities doctrine." This doctrine orders companies to share "access to their unique facilities, even to competitors, on a nondiscriminatory basis where sharing is feasible and the competitors cannot obtain or create the facility on their own."[14] U.S. and European courts have begun applying this doctrine to refusals of intellectual property owners to license their rights in special cases.

In the United States, the application of competition law standards to force the licensing of important intellectual property is highly controversial. However, the former chairman of the Federal Trade Commission, a primary competition law enforcement body, has argued that U.S. antitrust law can and should impose antitrust liability for a monopolist's refusal to licence intellectual property, just "as with any other kind of property, tangible or intangible . . . shown to constitute an essential facility."[15] And a small number of courts have concluded that a refusal to license intellectual property may violate antitrust law where the refusal does not sufficiently serve the purposes of intellectual property law in promoting new creation or innovation.[16] Yet an important appellate court for patent law questions has held that such theories may not be pursued and that courts instead are restricted to regulating practices that lie outside of the core right to exclude granted by the patent.[17]

By contrast, legal limits on the refusal to license in intellectual property matters are relatively well established in Europe. In one lead case, referred to as *Magill*,[18] three television broadcasters held copyrights on their respective listings for broadcasts in Ireland and refused to give permission for any firm to produce a comprehensive weekly guide combining the listings. European courts struck down the refusal to license, holding that the exclusion justified compulsory licensing because it prevented "the appearance of a new product . . . which the appellants did not offer and for which there was a potential consumer demand."[19] In another lead case, European courts ordered compulsory licensing of a copyrighted data tool that had become an industry standard and that prevented any other firm from competing in the same market.[20]

APPLYING COMPETITION LAW TO INTELLECTUAL PROPERTY ISSUES

Although many courts and commentators continue to engage in core/periphery thinking and attempt to define sets of practices inside and outside the central scope of the patent, the shifting boundaries between core and periphery over time and between countries expose the policy-laden nature of the task. Competition law can be used to limit intellectual property rights, including invading the core of the rights and prohibiting refusals to license others. Thus, the key question for policy advocates and enforcement officials is when competition law should limit intellectual property rights.

Many modern experts seek to answer the question of when competition law should restrict intellectual property rights by means of an explicit balancing of the costs and benefits of a particular practice. An influential article by Harvard Law School professor Louis Kaplow, for example, argues that enforcement officials should determine whether to use competition law to regulate a particular practice by reference to the net benefit or harm to society that the restrictive practice is causing.[21] According to this line of thinking, the social benefit of allowing intellectual property holders to engage in restrictive practices is that by raising their ability to profit from the intellectual property, the restrictive practices may increase incentives to innovate and create new products for future consumers. The costs of a restrictive practice may be decreased access to the existing technology, as well as other dynamic costs from limiting the diffusion of information and erecting barriers to follow-on innovation. The leading treatise on intellectual property law and competition in the United States similarly describes the question as calling for "balancing the social benefit of providing economic incentives for creation and the costs of limiting diffusion of knowledge."[22]

In the period before implementation of the World Trade Organization (WTO) agreements, including the Trade-Related Aspects of Intellectual Property Rights (TRIPS) Agreement imposing minimum standards on intellectual property laws and terms, countries could respond to a situation where the costs of intellectual property restrictions far exceeded their benefits by shortening the term of years when the restrictions on intellectual property rights could be exploited or by doing away with intellectual property in the particular field of technology. Those options are no longer available to WTO members, who must grant intellectual property rights in all industries, without discrimination and for minimum terms. But under TRIPS, countries may accomplish the same effects by using competition law to reduce the scope of restrictive practices that may be engaged in during the period. By allowing the use of competition mandates, TRIPS offers countries considerable flexibility to adjust both the costs and the benefits of intellectual property restrictions.[23] The two policy tools are interrelated: "The amount of reward provided and the monopoly loss arising in each additional year in which exploitation is permitted (and thus the appropriate length of patent life) depend on what practices patentees may employ during that time period."[24]

Using cost-benefit balancing tests, the interface of competition and intellectual property law in a country generally or for a particular industry or practice can be charted along a spectrum. At one end, the most dominant competition laws invade the core of intellectual property rights with enforceable duties to license competitors under the "essential facility" and related doctrines. At the other end, the most dominant intellectual property rights grant owners nearly total exemption from competition mandates. If restrictive practices cause particularly egregious social harm while creating relatively slight increases in reward to the intellectual property rights holder, then an expansion of the realm of competition law in restricting intellectual property rights is a WTO-compliant policy measure justified by the purposes of each doctrine. As described below, economic analysis suggests that the application of intellectual property law to essential goods and services in developing countries with high income inequality is just such a situation.[25]

THE ECONOMICS OF EXCLUSION

Just as there are public and consumer interests served by the recognition in competition law of the existence of "essential facilities" and the attendant doctrine requiring the sharing of some real property among competitors, public interests can be served by the recognition of the existence of what may be called "essential intellectual property" for which enforcement of exclusionary rights demonstrably and predictably causes far more social harm that it creates benefits. The fact that intellectual property rights are granted for essential goods and services in a highly unequal society is a key example of intellectual property becoming essential in this regard. In such a situation, as described more fully below, the legal right creates rational economic incentives to price the vast majority of consumers out of access. This lack of access, in turn, creates demonstrable losses to society that far exceed the minimal incentives to innovate that the owner of intellectual property receives for engaging in the socially harmful activity. Using the basic policy balance between costs and benefits articulated by modern commentators, it becomes eminently justifiable to apply the strongest competition law and other duties to share such essential intellectual property.

Monopoly economics teaches that profit-maximizing monopolists will serve a smaller segment of the consumer population (limiting output) at a higher price than would be the case under a competitive market. Absent some form of government price regulation or threat of entry by competitors, the only restraint on the monopoly's pricing will be a function of the willingness and ability of consumers to pay higher prices. The rational monopolist will keep raising prices until so many people cannot or will not pay the price that the loss of consumers eats into the firm's aggregate profits.

Economists illustrate the effect of consumer demand choices on monopolist pricing behavior through the shape and slope of a demand curve. A flat horizontal demand curve (see figure 1) would indicate that the seller has no discretion to raise prices. A small price increase would lead to all consumers foregoing purchases. Take, for example, a commodity such as wheat being sold at an exchange with a nearly infinite number of buyers and sellers. In such a market, the demand curve will be essentially flat—a small raise in price by a seller would result in all buyers shifting to another seller. Nearly flat demand curves also result when there is very limited discretionary income in a market, so that a small price increase will exclude all buyers from the market.

A vertical demand curve would yield no restraint on prices at all, referred to as a perfectly inelastic market (see figure 2). Consumers will purchase the amount of goods they require, regardless of the price set. Necessities without substitutes, such as utilities, food staples, and fuel, tend to be very inelastic, at least in the short run. One real-world example is the behavior of some electricity markets in the U.S. during the so-called "Enron scandal." The problem with markets that Enron and other companies exploited was that at certain times when energy demand is particularly high (for example, hot days in the summer), demand reaches the limits of supply, but cannot decrease on a short-term basis. The electricity producers thus achieve nearly absolute market power—they can charge any price they want without demand decreasing. During some of these crises, there were reports of electricity sales that cost $35 per unit the day or hour before spiking up past $10,000 per unit.[26]

Demand curves have a shape, as well as a slope. The shape of the curve is affected by how different consumers react to a price increase. If there is a large group of consumers that is very price sensitive and another group that is very price insensitive, then the curve will have a convex character, with part of it approaching vertical and another part nearly horizontal. This, too, will affect pricing behavior. If the number of consumers in the steeper section of the curve is large enough and willing to pay

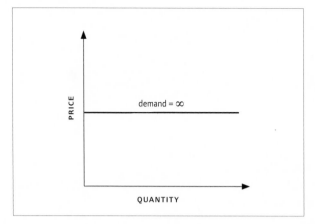

FIGURE 1 Perfectly elastic demand curve.

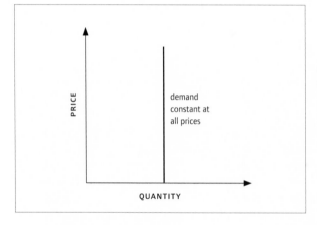

FIGURE 2 Perfectly inelastic demand curve.

high enough prices, then the monopolist may make more money by serving only that segment of the population than by serving all of the potential demand.

If a monopoly-provided good is an essential product with no substitutes, people will be willing to pay a very high portion of their income to enjoy access to it. Thus, the real restraint on pricing will be a function of ability to pay. This, in turn, means that the shape and slope of the demand curve will be a function of how income is distributed in a given society. In countries with very high income inequality, with a small number of superrich people living on First World incomes and a large number of superpoor people with very little discretionary income, the demand curve will be highly convex. And this will predictably lead to the monopolist pricing to the nearly vertical section of the curve, where large price increases can be implemented with very little additional loss of sales.

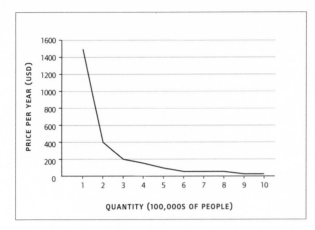

FIGURE 3 South Africa ARV demand if price = 5% income.

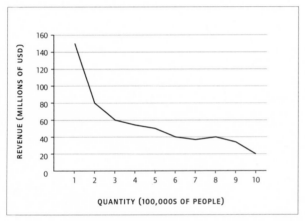

FIGURE 4 South Africa revenue per quantity sold.

Consider the case of South Africa, a country among the worst in terms of income inequality. Figure 3 is a demand curve constructed according to the assumption that people needing AIDS treatment in South Africa will purchase an antiretroviral if the cost is 5 percent of their income. If a firm prices its antiretroviral at $1,481 per patient per year, 100,000 people in the top income decile will buy it. In order to sell to a greater proportion of the population, the price must fall considerably—200,000 people with HIV/AIDS will buy the medicine if it is priced at $396, and half of the people in need of treatment can purchase an antiretroviral if it is priced at $92. In order to sell to all people with HIV/AIDS who need treatment, the price would have to be lowered to $18 per patient per year. [27]

Figure 4 shows the total sales revenue a firm will gain if it sells at each price on

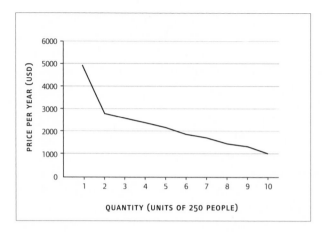

FIGURE 5 Norway ARV demand if price = 5% income.

FIGURE 6 Norway revenue per quantity sold.

the demand curve. The firm maximizes its sales in South Africa by selling at the price that only the top 10 percent can afford. If the firm lowers its price to what 20 percent can afford ($396), it will sell twice as many at a price far less than half the profit-maximizing price, earning substantially less ($79.2 million compared with $148.1 million in total revenue). If the monopoly firm continues to cut prices to raise sales volume, revenues fall further. In other words, at this level of wealth inequality in a society, the firm maximizes profit by setting a price that at least 90 percent of people in need cannot afford.

To understand the effect that inequality in income distribution has on the pricing and output decisions of a monopolist, compare the South African case with that of Norway, which has one of the most equitable income distributions.[28] Here, the demand curve would be much flatter, as shown in figure 5.

As shown in figure 6, under the same assumptions as the South Africa case above, that is, that a person will buy the essential product at 5 percent of income, the firm makes more money in Norway by selling more products at lower prices at each step along the income distribution until the firm serves between 80 percent and 90 percent of the population.

The lesson is this: The more unequal the distribution of income is in a country, the more people will be excluded from the market (what economists call "dead-weight loss") when a monopoly practices profit-maximizing pricing strategies for an essential good. At the same time, because sales in such a country are likely to be so few (making sales only to the very top income earners), the monopoly does not enjoy very high levels of overall profits. In other words, in countries with high income inequality, unrestrained monopoly pricing of essential goods is very likely to cause large social harms and comparatively small incentives to invest in innovative activities. In this context, it becomes an incredibly persuasive economic argument that whatever duties to promote competition and restrain monopoly pricing power exist, they should have their strongest application.

POST-TRIPS COMPETITION LAW ADVOCACY BY THE A2K MOVEMENT

In September 2002, the access-to-medicines movement took a decisive turn when, in South Africa, A2K advocates began to use competition law to advance their cause. South Africa's Treatment Action Campaign (TAC) shifted the focus of its advocacy for access to medicines to the South African Competition Commission. At the time, TAC and other access campaigners around the world were working to increase access to an important first-line AIDS drug regime commonly used in developing countries at that time. The cocktail—composed of the drugs AZT and 3TC, both patented by GlaxoSmithKline (GSK), and Nevirapine, patented by Boehringer Ingelheim (BI)—was being priced in the late 1990s in South Africa and around the world for over $10,000 per patient per year. That price was about three times the GDP per capita in South Africa. By the time of the complaint, prices in South Africa had fallen to about $3,000 a year for the same cocktail, compared with under $300 a year for generic versions from Indian firms. Indian companies also produced the medicines in a single-pill format, which was unavailable from the patent holders.

There had been previous requests for licenses by Indian pharmaceutical producer Cipla and from the medical services NGO Médicins Sans Frontières (MSF), both of which were rejected by the companies. BI admitted in documents filed in the South African Competition Commission that it had a general policy to refuse licenses for the generic supply of its products. The health minister had authority to issue authorizations for the use of generic versions of patented medicines for public-health purposes, but the Department of Health refused requests to use the law, leaving TAC searching for a new forum in which to bring its advocacy.

The complaint filed by TAC with the Competition Commission in 2002 alleged that the use of the patents by multinational pharmaceutical companies to demand prices that only a sliver of South Africa's population could afford violated section 8(a) of the South African Competition Act, which states that it is prohibited for a dominant firm to charge an "excessive price," defined as a price that is higher than the "reasonable economic value" of the good and that is to the detriment of consumers. In essence, this was a core/periphery complaint, asking the commission to leave the basic right of the company to refuse to license intact, but defining a periphery of pricing excess that is beyond the scope of the patent grant. A subsequent submission by the Consumer Project on Technology (CPTech) (now known as Knowledge Ecology International) encouraged the commission to adopt the concept of essential intellectual property. CPTech argued that whenever "(1) the number of people who need access to the medicines to prolong their life or improve their health significantly exceeds those with access to the drug, and (2) a substantial barrier to access is price," a legal burden "shift[s] to the pharmaceutical company to prove that it has promoted competitive pricing by issuing licenses of right to all qualified suppliers on reasonable terms."[29]

In October 2003, the commission announced that it found three abuses of dominance under Article 8 of the Act: excessive pricing, refusing to give a competitor access to an essential facility when it is economically feasible to do so, and engaging in exclusionary conduct if the anticompetitive effect of that act outweighs its technological, efficiency, or other procompetitive gains. Menzi Simelane, commissioner at the Competition Commission, explained:

> Our investigation revealed that each of the firms has refused to license their patents to generic manufacturers in return for a reasonable royalty. We believe that this is feasible and that consumers will benefit from cheaper generic versions of the drugs concerned. We further believe that granting licenses would provide for competition between firms and their generic competitors.
>
> We will request the Tribunal to make an order authorising any person to exploit the patents to market generic versions of the respondents patented medicines or fixed dose combinations that require these patents, in return for the payment of a reasonable royalty.[30]

Soon after the Commission's announcement, the two pharmaceutical companies settled the complaints through agreements requiring the issuance of multiple licenses to South African and Indian generic producers who now supply dramatically cheaper versions of the medications in South Africa, including in fixed dose combinations. The licenses also authorized exports to all of sub-Saharan Africa.

Other access campaigns have used competition law to achieve other objectives. In February 2007, Knowledge Ecology International filed a complaint in the U.S. Federal Trade Commission alleging that Gilead Science Incorporated was illegally using restrictive licensing policies, including banning licensees from serving some countries and charging royalties for countries where it did not have patents.[31] In Thailand, treatment activists filed a competition complaint against Abbott Laboratories for refusing to supply new drugs in the country to punish the government for issuing a compulsory license on the AIDS drug Kaletra.[32] In the West, competition-law cases were used successfully to open access to Microsoft's application programming interfaces for Internet browsers, and complaints have been filed to open access to the digital-rights-management software used to prevent iTunes music from being played on competing players.[33]

In these and other cases, the A2K movement has used the forum and language of competition law for strategic advantage. Some of these advantages exist regardless of the end result of the complaint. Rhetorically, the move allows A2K campaigns to shift the dominant frame for analyzing the issue from one of the protection of the intellectual property holder's "rights" to one of monopoly regulation. Institutionally, competition law strategies allow the A2K movement to shift into regimes with investigative resources and institutional cultures that are often wary of barriers to competition. Doing so also is a way to alter the terms of the political debate over access-to-knowledge issues.

REFRAMING POLITICAL DEBATE

Research from the cognitive and social sciences shows us that people interpret ideas and issues through existing frames and concepts that are culturally constructed and historically situated. This research suggests that it is important for policy advocates to focus on how people are thinking about a particular issue, rather than attempting only to change the amount of information they are using to reach conclusions. Social movements often engage in advocacy that reflects these teachings by reframing issues, that is, by "conscious, strategic efforts by groups of people to fashion shared understandings of the world and of themselves that legitimate and motivate collective action."[34]

The communications strategies of the dominant intellectual property industries show great attention to the importance of framing in their campaigns to ratchet up intellectual property protection around the globe. As discussed above, historically, patents and other intellectual property were viewed by the public and government officials as a form of monopoly that needed to be regulated to prevent abuse and to serve the greater public interest. The intellectual property industries very deliberatively and strategically shifted this frame to one of "property rights." Susan Sell explains:

> The language of rights weighs in favor of the person claiming the right. The language of privilege weighs in favor of the person granting the privilege. By wrapping themselves in the mantle of "property rights," they suggested that the rights they were claiming were somehow natural, unassailable and automatically deserved. They were able to deploy "rights talk" effectively in part because they were operating in a context in which property rights are revered. In that regard "rights talk" resonated with broader American culture. . . . The advocates of highly protectionist IP norms expressed indignation at those violating these "rights" and claimed that so-called violators were "pirates."[35]

Filing complaints in competition tribunals shifts the discourse back to the monopoly frame, where consumers hold an advantage. While the language of rights suggests deserving protection from state regulation, modern culture continues to be highly distrustful of monopolies. The term is synonymous with exploitation and abuse. Monopolies are entities to be regulated, not freed from state intervention.

The use of competition forums to shift the framing of intellectual property issues is evident in some of the A2K movement's advocacy documents and explanations of their strategies. In a statement on the day it filed complaints against GSK and BI, Action for Southern Africa (ACTSA) explained that the TAC submission showed that "largely as a result . . . of monopoly abuse, the pharmaceutical industry remains the most profitable industry in the world" and that "GSK continues to expand their profit margins by charging excessive prices for life-saving medicines in markets in which many people living with HIV/AIDS [have] little or no income."[36] Similarly, one of TAC's attorneys explained that the use of a competition law strategy was selected in part because of a perceived "need to revive the public debate about patent abuse and profiteering,"[37] which the competition forum enabled. Using competition law is thus important not only for the potential remedies one may achieve there, but because it helps a movement communicate to the greater public about an issue.

Another discursive advantage of using competition law is that it is punitive. Other TRIPS flexibilities, for example a general public-interest license, are often discretionary and do not necessarily brand the intellectual property holder as a bad actor. Using competition laws shifts the inquiry from whether the government should use its discretion to limit patent rents to whether the company deserves punishment for its abusive actions.

The importance of a punitive framework proved particularly evident in the Thai case against Abbott Laboratories. In that case, Abbott withdrew drug-registration applications for several drugs after the Thai government issued a compulsory license to authorize generic purchases of one of Abbott's AIDS drugs. Abbot claimed in the press that the Thai compulsory license was illegal. Health activists responded with a competition complaint against Abbott for the withdrawal of needed supplies for the Thai market, thus branding Abbott as the true illegal actor. Treatment campaigner and law professor Brook Baker explained in *The Nation* newspaper: "Instead of Thailand breaking the law, it is Abbott that has engaged in an unprecedented and probably illegal withdrawal from the Thai market, taking seven important medicines, including a heat-stable form of Kaletra, out of the drug registration process."[38]

Another framing advantage of competition proceedings is that they commonly provide a calendar of proceedings around which media and advocacy events can be staged. Unlike general public-interest licenses, which often lack set procedures or precedents, competition procedures are normally defined by regulations with set points for decision and input. The filing of a complaint, the filing of a response by the companies, a public hearing, the decision by the agency, a formal complaint or appeal to a tribunal, and so on, all become moments when public attention can be brought to bear on the complaint and focused on the story of illegal action and abuse told by activists. The proceedings may also produce documents and statements through the investigation that can be obtained through freedom of information laws and used in subsequent campaigns to explain industry dealings in the country.

REGIME SHIFTING INTO COMPETITION FORUMS

The concept of framing focuses on the strategic use of discourse to alter public perceptions of and reactions to an issue. By contrast, regime shifting, a concept from political science and international relations, is a strategy that attempts to alter the status quo ante by moving law-making initiatives and standard-setting activities from one venue into another.[39] Here, advocacy groups seek out forums that may be more hospitable to their cause. Regime shifting and framing often go hand in hand. One benefit of effectively reframing an issue is that it may open the possibilities of action in new forums. Reframing intellectual property issues as trade issues enabled the dominant intellectual property industries to shift the forum for international intellectual property law-making initiatives from the World Intellectual Property Organization (WIPO) into the WTO. Access campaigners responded by reframing pharmaceutical patents as a public-health issue, enabling the engagement of the World Health Organization (WHO) in intellectual property debates.

Framing intellectual property as a monopoly-regulation issue opens the potential for advocacy in competition law forums. These forums offer potential institutional advantages in developing countries, where laws are new and undefined.

The opportunities for competition law advocacy in the Global South arise from the flip side of economic liberalization that is exporting intellectual property restrictions and deregulation to much of the developing world. Based on the dominant Western model of liberalization, the "free markets" created by contract and property rights and by deregulated industries are supposed to be regulated to serve public interests primarily by competition law. Only a handful of developing countries had such laws before 1990. In the decade between 1990 and 2000, the same decade that witnessed the globalization of substantive intellectual property laws through the TRIPS Agreement (1994), fifty countries (most of them developing) added competition laws to their books. (See table 1.)

Although competition law is rapidly globalizing, it is not doing so in a uniform way. Unlike in intellectual property law, where binding minimum standards are established by the TRIPS Agreement, in competition law, countries remain largely free from any international obligation to draft, interpret, and enforce standards in any particular manner. Indeed, although U.S. and European Union laws are the obvious models for the substantive doctrines contained in most of the world's competition statutes, there are very noteworthy differences in the interpretive norms and policies that animate the laws of many developing countries.

Competition laws in developing countries often explicitly incorporate developmental objectives. For example, the South African Competition Act expresses the intent to create a competitive economic environment "focussed on development" in order to "advance the social and economic welfare," "to correct structural imbalances and past economic injustices," and "to reduce the uneven development, inequality and absolute poverty which is so prevalent in South Africa." The South African Competition Commission found these norms persuasive in determining the outcome of the complaint against GSK and BI, explaining: "Indeed the very goals of our Competition Act—promoting development, providing consumers with competitive prices and product choices, advancing social and economic welfare and correcting structural imbalances—have been made difficult in this context by the refusal of the respondents to license patents."[40]

Injunctions to consider equity objectives in the interpretation and enforcement of competition law may be heightened in countries that have adopted social and economic rights in their constitutions. To take South Africa as an example again, the constitution obligates the state to "promote the achievement of equality" and to "take reasonable legislative and other means" to realize the rights of everyone to access to health care. The constitution specifically delineates one key means of

1900–1969

Brazil (1945)
Brunei (1930)
Colombia (1962)
Chile (1963)
Germany (1909)
Haiti (1964)
Holland (1956)
India (1969)
Israel (1959)
Japan (1947)
Lebanon (1967)
Liechtenstein (1946)
Mexico (1934)

1970s

Australia (1974)
Austria (1972)
Bahrain (1970)
Cote D'Ivoire (1979)
El Salvador (1970)
France (1977)
Great Britain (1973)
Greece (1977)
Mauritius (1979)
Pakistan (1970)
Argentina (1980)

1980s

Canada (1986)
Gabon (1989)
Korea (1986)
Kuwait (1980)
Luxembourg (1986)
Malawi (1987)
Mali (1986)
New Zealand (1985)
Spain (1989)
Sri Lanka (1987)

1990s

Albania (1995)
Algeria (1995)
Belarus (1992)
British Virgin Islands (1990)
Bulgaria (1998)
Burkina Faso (1994)
Cameroon (1990)
China (1993)
Costa Rica (1994)
Croatia (1997)
Cyprus (1999)
Czech Republic (1991)
Denmark (1997)
Estonia (1998)
Finland (1992)
Hungary (1996)
Indonesia (1993)
Ireland (1991)
Jamaica (1993)
Kazakhstan (1991)
Kenya (1990)
Latvia (1997)
Lithuania (1999)
Malaysia (1991)
Malta (1994)
Mauritania (1991)
Norway (1993)
Oman (1990)
Panama (1996)
Peru (1992)
Poland (1993)
Romania (1996)
Russia (1991)
Slovak Republic (1991)
Slovenia (1993)
South Africa (1998)
St. Vincent and the Grenadines (1999)
Sweden (1993)
Switzerland (1995)
Taiwan (1992)
Thailand (1999)
Trinidad and Tobago (1996)
Tunisia (1991)
Turkey (1994)
Ukraine (1996)
Uzbekistan (1996)
Venezuela (1996)
Vietnam (1996)
Zambia (1994)

TABLE 1 Globalization of competition law.

promoting these rights, enjoining every court and agency to "promote the spirit, purport and objects of the Bill of Rights" whenever "interpreting any legislation, and when developing the common law." In the context of structural market problems that create incentives for providers of essential goods to exclude the majority of people in need from their products, promotion of the right to access to health care and the achievement of equality may counsel for interpretations of competition law favorable to access to intellectual property rights.

Making use of competition commissions is another regime-shifting strategy that is particularly well suited to promoting access to knowledge in developing countries. One of the very helpful attributes of Western competition laws that has been exported to many developing countries is the competition-advocacy agency. The role of these agencies is to receive complaints from competitors and consumers about potentially illegal practices, to investigate them using professional staffs and special legal authority (such as subpoena power), and to litigate complaints on behalf of the consumers or the state, often in a specialized tribunal.

In many countries, these agencies are relatively well funded. Aid programs from the United States and Europe support the institutional capacity of competition authorities as part of packages aimed at promoting the liberalization of economies. The agencies often have the capacity to hire top lawyers, economists, and other professionals. And because the laws themselves are often relatively new, these staffs may not be too overburdened with work to do a professional job in their investigations.

The availability of an advocacy agency may enable an access campaign to mount a highly technical legal campaign against a well-resourced intellectual property owner without the kind of legal war chest that such a battle would require on their own. Resources must still be spent on convincing the agency to act and on educating the agency about technical, medical, and intellectual property topics with which it may not be familiar. The mobilization of political resources may also be necessary to convince leaders with influence over the agency to prod it to act with sufficient determination. But where competition authorities are inclined to act in the greater public interest, their professional lawyers and staff can be extremely valuable additions to the resources of an access movement.

Finally, the decisions of competition courts and commissions may have a lasting precedential effect, altering the assumed background rules in the industry. This is evident in South Africa. Where BI once openly proclaimed a policy of not licensing generic companies to provide its products, industry lawyers now counsel that blanket refusals to license patents on AIDS drugs are legally suspect and open to challenge.[41] The precedent has similarly been relied upon by treatment activists, including a subsequent complaint against the pharmaceutical giant Merck, alleging that licenses granted for the AIDS drug efavirenz do not license the lowest-cost suppliers and do not allow new fixed-dose combinations.[42]

CONCLUSION

Courts and agencies can and do use competition law to help strike the balance between the aims of intellectual property laws to promote investment and innovation and competition law goals to maximize consumer welfare through competitive markets and lower prices. Where a developing country chooses to strike this balance may—and should—differ markedly from how the balance is struck in the Global North. Economic analysis suggests that rules should be drawn in developing countries to much more heavily favor open access to intellectual property on essential goods and services where the welfare implications of allowing exclusive dealing appear enormous.

No strategy is without risks, and there are significant risks to pursuing competition law strategies to open access to intellectual property. Many of the risks involved with competition law strategies are the flip side of the benefits. The fact that competition strategies can create legal precedents that will affect later cases means that losses in this forum can have lasting negative repercussions. The indeterminacy of law that provides opportunities for progressive legal movements also provides a fluid medium within which industry lawyers can work. The institutional structure of the dominant model of competition law, with a well-resourced advocacy agency as a gatekeeper to courts, may be a barrier to progressive use of the law if it is staffed with conservative bureaucrats. Finally, the opportunities for relying on Northern precedent should not be overstated: no Northern court has held that essential drug patents are subject to open-licensing duties.

The experiences in South Africa and other countries are showing that competition agencies can be valuable sites for political struggle over how intellectual property will be regulated. Such sites offer many advantages for access communities and should certainly be considered in any access campaign. But as with any site of struggle, the likelihood of success will depend on contextual circumstances: who will make the decision, what leverage movements have over the decision maker, and how successfully tactics are executed to leverage ideological and political power toward a favorable result.

NOTES

Mike Palmedo and Parva Fattahi provided valuable research assistance and helpful comments.

1   *Sherman Act*, 26 Stat. 209 (1890), current version at 15 *U.S. Code*, §§ 1–7 (2004).

2   See William Blackstone, *Commentaries on the Laws of England*, 4th ed., Robert M. Kerr (ed.) (1768; London: J. Murray, 1876), pp. 316–17, describing a letter patent as "so called because they are not sealed up, but exposed to open view, with the great seal at the bottom." See also Adam Mossoff, "Rethinking the Development of Patents: An Intellectual History, 1550–1800," *Hastings Law Journal* 52 (August 2001): pp. 1255, 1259.

3   The First Statute of Westminster, Chapter XXXI (1275), quoted in Charles M. Haar and Daniel William Fessler, *The Wrong Side of the Tracks: A Revolutionary Rediscovery of the Common Law Tradition of Fairness in the Struggle against Inequality* (New York: Simon and Schuster, 1986), pp. 63–64, explaining that charging an "outrageous Toll" was grounds for the king to "seize into his own hand the Franchise."

4   Oren Bracha, "Symposium: Intellectual Property at a Crossroads: The Use of the Past in Intellectual Property Jurisprudence: The Commodification of Patents 1600–1836: How Patents Became Rights and Why We Should Care," *Loyola of Los Angeles Law Review* 38 (2004), pp. 188–89, citing as "typical" examples: a clause voiding a manufacturing patent if the patentees "prove extortionate in their charges" and another requiring the patentee to "serve the same smalt as good and as cheap as the like brought from beyond seas."

5   An Act Concerning Monopolies, 21 Jac. I, c. 3 (1623) (Eng.), quoted in Mossoff, "Rethinking the Development of Patents," pp. 1272–73. See also Edward Coke, *Institutes of the Laws of England* (1644; London: E. and R. Brooke, Bell-Yard, near Temple-Bar, 1797), p. 184, where he describes the conditions for a valid monopoly patent as including "it must not result in the raising of prices."

6   International Convention for the Protection of Industrial Property, Article V, Paris, March 20, 1883, in Roger William Wallace, *The Law and Practice Relating to Letters Patent for Inventions*, appendix III (London: William Clows and Sons, Ltd., 1900).

7   See *Motion Picture Patents Co. v. Universal Film Mfg.*, 243 U.S. 502 (1917), banning tying the purchase of patented projectors to film purchases.

8   The Nine No-No's banned the following:
    1. Royalties not reasonably related to sales of the patented products.
    2. Restraints on licensees' commerce outside the scope of the patent (tie-outs).
    3. Requiring the licensee to purchase unpatented materials from the licensor (tie-ins).
    4. Mandatory package licensing.
    5. Requiring the licensee to assign to the patentee patents that may be issued to the licensee after the licensing arrangement is executed (exclusive grantbacks).
    6. Licensee veto power over grants of further licenses.
    7. Restraints on sales of unpatented products made with a patented process.
    8. Postsale restraints on resale.
    9. Setting minimum prices on resale of the patent products.

    Richard Gilbert and Carl Shapiro, "Antitrust Issues in the Licensing of Intellectual Property: The Nine No-No's Meet the Nineties," Brookings Papers on Economic Activity: Microeconomics (1997).

9   *Motion Picture Patents Co. v. Universal Film Mfg.*, 243 U.S. 502, 510 (1917).

10 *United States v. General Electric*, 272 U.S. 476 (1926).

11   *Bement v. Harrow Co.*, 186 U.S. 70, 91 (1902).

12   *United States v. Terminal Railroad Association*, 224 U.S. 383, 411 (1912).

13   *Otter Tail Power Co. v. United States*, 410 U.S. 366 (1973); *MCI Communications Corp. v. AT&T*, 708 F.2d 1081 (7th Cir. 1982).

14   Brett Frischmann and Spencer Weber Waller, "Revitalizing Essential Facilities," *Antitrust Law Journal* 75, no. 1 (2008): pp. 1–2.

15   Robert Pitofsky, Donna Patterson, and Jonathan Hooks, "The Essential Facilities Doctrine Under U.S. Antitrust Law," 70 *Antitrust Law Journal* 70 (2002): pp. 461–62. They cite the relevant components of the essential facility test as control of the essential facility by a monopolist, a competitor's inability practically or reasonably to duplicate the essential facility, the denial of the use of the facility to a competitor, and the feasibility of providing the facility.

16   See *Data General v. Grumman Sys*, 36 F.3d 1147, 1187 (1st.Cir. 1994).

17   *In re Independent Service Organizations Antitrust Litigation*, 203 F.3d 1322 (Fed. Cir. 2000).

18   *Radio Telefis Eireann v Commission of the E.C.* [1995] E.C.R. I-743.

19   *Ibid.*, para. 54.

20   *NDC Health and IMS Health* Comp D3/38.044 (July 3, 2001); *NDC Health and IMS Health Corp*, 2004 E.C.R. I-5039.

21   Louis Kaplow, "The Patent-Antitrust Intersection: A Reappraisal," *Harvard Law Review* 97 (1984). Kaplow suggests determining the bounds of competition law's interface with intellectual property by examining a "ratio between the reward the [intellectual property holder] receives when permitted to use a particular restrictive practice and the monopoly loss that results from such exploitation."

22   Herbert Hovenkamp, Mark D. Janis, and Mark A. Lemley, *IP and Antitrust: An Analysis of Antitrust Principles Applied to Intellectual Property Law*, supplement (New York: Aspen Publishers, 2007), pp. 1–10.

23   Section 8, Article 40 of the TRIPS Agreement specifically instructs that countries may take measures to regulate restrictive licensing practices for patents. Articles 40.1 and 40.2 state that countries can take "appropriate measures" to prevent or control "the abuse of intellectual property rights having an adverse effect on competition in the relevant market" by rights holders, practices that "restrain competition" or that have "adverse effects on trade," and practices that "may impede the international transfer of technology." Article 31 also provides special compulsory licensing flexibilities where the exploitation of a patent is found to be anticompetitive, including authorization of unlimited exports under the license. The TRIPS agreement is available on-line at http://www.wto.org/english/tratop_e/trips_e/t_agm0_e.htm (last accessed March 22, 2009).

24   Kaplow, "The Patent-Antitrust Intersection: A Reappraisal," p. 1822.

25   A longer version of the analysis that follows is presented in Sean Flynn, Aidan Hollis, and Mike Palmedo, "An Economic Justification for Open Access to Essential Medicine Patents in Developing Countries," *Journal of Law, Medicine and Ethics* 37, no. 2 (Summer 2009).

26   The scandal is that Enron created more of these conditions by strategically limiting supply. Nearly vertical demand curves still occur in some electricity markets today, but are frequently regulated with price caps to prevent the most extreme price spikes.

27   In this example, the price set on 5 percent of income for each decile is derived from real GDP and population figures from the World Bank's World Development Indicators Database for the year 2000, available on-line via DDP Quick Query http://ddp-ext.worldbank.org/ext/

DDPQQ/member.do?method=getMembers&userid=1&queryId=135, and the proportion of income held by each decile from the Income and Expenditure Survey (IES) 2000, Statistics South Africa, available on-line at http://www.statssa.gov.za/publications/earningspending2000/earningspending2000October2000.pdf (both last accessed November 9, 2009).

28 In this example, data for income distribution for the year 2000 is taken from the CIA World Factbook, archived at http://permanent.access.gpo.gov/lps35389/2000/sf.html#Econ. As in the example of South Africa, real GDP and population data for the same year are taken from the World Bank World Development Indicators Database.

29 Statement of Information by Consumer Project on Technology Concerning an Alleged Prohibited Practice in Terms of Section 49B(2)(a) of the Competition Act 89 of 1998, available on-line at http://www.cptech.org/ip/health/cl/cl-cases/rsa-tac/cptech-statement.doc (last accessed November 9, 2009).

30 Quoted in Consumer Project on Technology, "Competition Commission Finds Pharmaceutical Firms in Contravention of the Competition Act," October 16, 2003, available on-line at http://www.cptech.org/ip/health/sa/cc10162003.html (last accessed November 9, 2009).

31 The FTC has privately communicated to KEI that it is not pursuing the complaint.

32 The complaint was rejected by the Commission and is now under appeal.

33 Background and materials for many of these cases can be found on the Web site of American University's Washington College of Law Program on Information Justice and Intellectual Property Competition Project at http://www.wcl.american.edu/pijip/competitionpolicyproject.cfm.

34 Doug McAdam, John D. McCarthy, and Mayer N. Zald, "Introduction: Opportunities, Mobilizing Structures, and Framing Processes — Toward a Synthetic, Comparative Perspective on Social Movements," in Doug McAdam, John D. McCarthy, and Mayer N. Zald (eds.), *Comparative Perspectives on Social Movements: Political Opportunities, Mobilizing Structures and Cultural Framings* (Cambridge: Cambridge University Press, 1996), p. 6. See also "Topic: Framing Lessons from the Social Movements Literature," *Frameworks EZines*, no. 29, available on-line at http://frameworksinstitute.org/ezine29.html (last accessed November 9, 2009).

35 Susan Sell, "TRIPS and the Access to Medicines Campaign," *Wisconsin International Law Journal* 20, no. 3 (Summer 2002): pp. 489–90.

36 Information from Action for Southern Africa (ACTSA) Concerning Complaint from the Treatment Action Campaign (TAC) on 17 September 2002 Regarding Pharmaceutical Firms GlaxoSmithKline (GSK) South Africa (Pty) Ltd. and Others, available on-line at http://www.tac.org.za/Documents/DrugCompaniesCC/ACTSA_statement.doc. See also "Statement on Competition Commission Complaint Against GlaxoSmithKline and Boehringer Ingelheim," September 17, 2002, available on-line at http://www.tac.org.za/Documents/DrugCompaniesCC/DrugCompaniesCC.htm (last accessed November 9, 2009).

37 Jonathan Michael Berger, "Litigation Strategies to Gain Access to Treatment for HIV/AIDS: The Case of South Africa's Treatment Action Campaign," *Wisconsin International Law Journal* 20, no. 3 (Summer 2002): p. 609.

38 *The Nation*, "The Eight Deadly Lies of Big Pharma," April 23, 2007, available on-line at www.nationmultimedia.com/2007/04/21/opinion/opinion_30032324.php (last accessed November 9, 2009).

39 Laurence Helfer, "Regime Shifting: The TRIPs Agreement and New Dynamics of International Intellectual Property Lawmaking," *Yale Journal of International Law* 29, no. 1 (2004), p. 14.

40 Quoted in Consumer Project on Technology, "Competition Commission Finds Pharmaceutical Firms in Contravention of the Competition Act."

41 This was conveyed to me in a personal communication with a pharmaceutical industry lawyer from a prominent South African patent law firm.

42 See Sean M. Flynn, "Summary of the South African Competition Complaint against Merck," November 6, 2007, available on-line at http://www.wcl.american.edu/pijip/documents/pijip11062007.pdf?rd=1 (last accessed November 9, 2009).

Open-Access Publishing: From Principles to Practice

Manon A. Ress

The issues involved in providing access to knowledge are not new, and publishing historically has been an arena in which they have been most salient. Until the seventeenth century, scholarly findings were usually published in rare and expensive books. Scholars would share their findings by mailing their papers to selected individuals. However, beginning in 1665, the *Philosophical Transactions of the Royal Society of London*, first edited and published by the society's secretary, Henry Oldenburg, was published and in effect formalized and centralized the process of peer review and scholarly publication.[1] As is vividly described in Open and Shut?— Richard Poynder's blog on open access—the scholarly publishing system thus put into place was apparently efficient enough to last for more than three centuries.[2]

However, in the last thirty years, things have changed rapidly. There has been an enormous growth in research being conducted at the global level, but access to this knowledge increasingly has been restricted by the erection of substantial economic barriers. As a result of the need for new, global publications, commercial publishing organizations have become interested in the market and have promoted and launched more and more journals. These are for-profit organizations, unlike most traditional learned societies and scholarly publishers. Scholars and libraries have had no alternative—they have had to pay more and more for more subscriptions to these new journals. And since the 1990s, when features such as on-line access have been added, there has been additional cost for the publishers and their clients.

One response to this increasing restriction of access to published knowledge has been the rise of open-access electronic publishing. In what follows,using practical examples from the recent creation of the open-access electronic journal *Knowledge Ecology Studies*, I will examine both the value of open-access journal publishing for the broader A2K movement and some of the challenges that open-access journal publishing faces as it attempts to allow the unrestricted

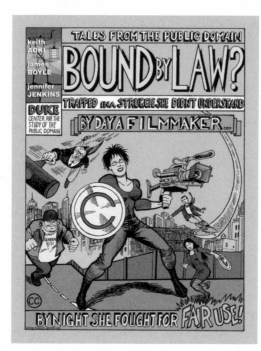

Cover of Keith Aoki, James Boyle, and Jennifer Jenkins, *Bound by Law? (Tales from the Public Domain)* (Durham, NC: Center for the Study of the Public Domain, 2006) (available at http://www.law.duke.edu/cspd comics/; License: http://creativecommons. org/licenses/by-nc-sa/2.5/).

participation of creative communities in electronic publishing. This essay is also meant to be of use to A2K activists in designing and implementing editorial policies and copyright policies and in dealing with technical questions when launching an open journal. My intention here is not to provide legal or technical details about open-access electronic publishing. Rather, I want to focus on the strategic significance for knowledge activists of advancing open access from principles to practice in developing and developed countries.

I begin by describing and explaining the essential principles of an open-access electronic journal. I also examine the recent history of the A2K movement in the context of such seminal open-access initiatives as the Budapest Open Access Initiative, the Bethesda Statement, and the Berlin Declaration on Open Access to Knowledge in the Sciences and the Humanities. After describing the rationale for the creation of an open-access journal dedicated to the multidisciplinary field of access to knowledge, I describe and explain the policy decisions and issues that, while informed by the principles of the movement, proved difficult to put into practice: intellectual property policy, quality and peer review, funding and sustainability, and technical and editorial decisions. Finally, although the ongoing experiment that is *Knowledge Ecology Studies* is far from over, I draw some conclusions about the role and nature of an open-access journal within the growing access to knowledge movement to gain a better understanding of the challenges ahead.

The A2K movement includes participants as diverse as scholars, activists, publishers, and government representatives. Academics—political scientists and sociologists, especially—may read the description of this experiment with the open-access journal from quite a different point of view from that of most activists, publishers, authors, and readers involved in this project today. Open access has been a way for scientific communities to regain some control over their work,[3] but as each new technical innovation has created the possibility of lower-cost distribution methods and new policies and practices concerning access and control, the proliferation of possibilities has redefined open-access publishing in different ways for different constituencies. One of my objectives here is to present the plurality and complexity of varying points of view as an essential component of the shift in the A2K movement from principles to practice.

PRINCIPLES

Over the last thirty years, authors publishing in academic journals have routinely been asked to assign their copyrights to publishers seeking exclusive distribution rights. Publishers, in turn, have been able to set the price of subscriptions to what the market (institutional libraries, mostly) can bear, and while there have been more and more journals, their prices have steadily increased. Libraries' budgets, however, have decreased or at best have remained static, and funding the cost of journals has eaten up an increasing percentage of them. University librarian David W. Lewis states: "Today the annual value of the peer-reviewed journal market is estimated at £25 billion [$50 billion], and consists of 23,700 journals, which between them publish 1.59 million articles a year." As he points out, if these journals are too expensive for the developed world, one can imagine the access problems faced by researchers and institutions in developing countries.[4]

Clearly, the prices of scholarly journals have again become an obstacle to broad access at a time when, paradoxically, new information technologies have expanded the possibilities for information sharing. As Leslie Chan describes the situation:

> The scholarly communication crisis encompasses two distinct though interrelated problems. On the one hand, serial-subscription costs, particularly for science and medical journals, have been increasing rapidly over the last two decades, often at rates far above the cost of inflation. At the same time, research-library budgets have been decreasing or are otherwise unable to keep pace with price increases. The result is that libraries are spending more, but they are in fact getting less, in terms

of journal titles and new monograph acquisition, as more of the budget is being consumed by serial subscriptions.[5]

This is what many describe as the "affordability" problem, which affects libraries and other research institutions in both developed and developing countries. Moreover, the cost to libraries is often passed on to the public, when it funds these institutions. Therefore, the public at large has a significant stake in seeing that essential scientific and scholarly research (often funded by governments) is made more widely available without unnecessary costs or delays. And along with educators, physicians, nurses, students, journalists, activists, and various investigators, one should consider also patients and their families as key stakeholders. For example, when patients or their families go on-line to find treatment options or updates on clinical trials, they need to get access to peer-reviewed research of the highest quality on a timely basis. For the public at large, access to knowledge and information are both essential elements for development, democracy, and social progress in general.

During the 1990s, new computer networks were full of promises, yet also were vexed by troubling paradoxes. There was a growing awareness that new technologies could both promote and hinder access to materials. Along with the price-inflation crisis, new technological regulations came quickly as new technologies were made available to the general public. Indeed, the new technologies came with "locks" called "technological protection measures" that made their ways into a new international digital copyright treaty and its national implementations.[6] Soon, on-line publishers adopted digital rights management (DRM) systems and technologies that could not be circumvented, publishing scholarly journals digitally and limiting their access. These new control mechanisms contributed to the perception that more access was possible than ever before, when in fact, the reverse was true.[7]

In many other areas, however, the Internet had laid the groundwork for a new culture of abundance in which creation is decentralized and often collective. From this positive potential and from threats to the new wealth of easily accessible information, a social movement promoting access to knowledge emerged.

One early and influential part of this movement began in 1990, when groups such as the American Library Association, the American Association of Law Libraries, and the Association of Research Libraries reached out to Ralph Nader and his Taxpayer Assets Project to address the issue of access to digital versions of government information. The Taxpayer Assets Project used Internet e-mail listservs to create a grassroots campaign demanding access to federally owned databases. Included in this effort was the Crown Jewels Campaign, which focused on access to the best-known and most valuable federal databases, including a database of corporate disclosure documents compiled by the U.S. Securities and Exchange Commission, the Medline database of biomedical articles maintained by the National Institutes of Health, the federal database of patent filings held by the U.S. Patent and Trademark Office, the full text of bills pending before the U.S. Congress, and databases of federal court opinions, to mention only a few. These efforts were considered the first use of the Internet to mobilize and organize individuals to exercise political power and were widely credited with a reversal of earlier policies created by the Reagan administration to privatize access to digital versions of government information. Moreover, the successful mass mobilization of stakeholders displayed the potential power of open access as an idea and as a social movement.

The movement for open access to scholarly publications has followed a different path than the movement for access to government documents, with the former being nurtured largely within academic institutions and informed by the open-access movement and the latter being driven by the activist community. Without listing all the conferences, events, and declarations that have led to the opening of scholarly access, it is important to note statements of principle drafted between 2001 and 2003 in the so-called "Three Bs," which helped to elevate public interest and policy debates over open access: the Budapest Open Access Initiative, The Bethesda Statement on Open Access Publishing, and the Berlin Declaration on Open Access to Knowledge in the Sciences and Humanities.

The Budapest Open Access Initiative came out of a conference convened by the Open Society Institute on December 1 and 2, 2001, and resulted in a clear definition of open access that permanently changed the field of scholarly communication. Imposing an old scholarly tradition on the use of new information technologies, the Budapest Open Access Initiative promoted scholarly communication as a public good. It defined open access as "the free availability of peer-reviewed literature on the public internet, permitting any user to read, download, copy, distribute, print, search, or link to the full texts of the articles."[8] The Bethesda Statement on Open Access Publishing and the Berlin Declaration on Open Access to Knowledge in the Sciences and Humanities[9] followed the Budapest Open Access Initiative with similar statements of principle in 2003.[10]

Also in 2003, during a meeting of consumer groups in Lisbon, Portugal, activists, government delegates to the World Intellectual Property Organization (WIPO), and academics from all over the world met to focus on a work program for WIPO. From that meeting emerged an agreement among a number of nongovernmental organizations and government negotiators to focus on the subject of access to educational and scholarly materials. In the aftermath of this meeting, another conference was organized by the Trans Atlantic Consumer Dialogue (TACD)

Working Group on Intellectual Property, on April 5, 2004, at the Ford Foundation in New York. The conference was dedicated to the subject of access to essential learning tools, including access to books, scholarly journals, databases, and software, as well as the delivery of education over the Internet (distance education). This event was greatly stimulating, as was a private meeting on April 6 at the Open Society Institute, where participants decided to broaden the overall theme of the effort to access to knowledge.[11]

As a historical footnote, this was where the decision was first made to use the term "A2K" as the "brand name" for a movement that included not only the learning-tools issues, but also the broader issues of access to knowledge, including access to medical and other inventions.

Subsequently, new projects were born to promote access to educational materials and access to knowledge and technologies, often focusing on the opportunities presented by new digital technologies and open-licensing strategies for copyright or patents. These multiplied rapidly, thanks to the efforts of a diverse set of actors. Many Web sites and discussion lists began to knit together the various strands of different access movements, and in 2005, two meetings were held to draft a proposal for a treaty on access to knowledge. These brought together a very diverse set of experts, interests, and constituencies, including librarians, free-software advocates, access-to-medicine activists, Internet service providers, software providers, development NGOs, and government representatives from the Global South. Collectively, the access to knowledge movement was becoming increasingly important and influential.

The public statements by the Three Bs have provided long-lasting and influential definitions of open access for publishing enterprises in this movement. While the definitions differ in some ways,[12] they are in accord on core principles, such as that open-access content must be free of charge for all users with an Internet connection and permission barriers must be removed for all legitimate uses.

The Budapest statement explicitly prescribes that open access must depend on authors' consent. In the Bethesda and Berlin statements, the owners of the rights to the content must consent to let all users copy, use, distribute, transmit, display, or make derivative works in any way and for any purpose with proper attribution.

The core principle is the removal of access barriers and permission—however, that principle does not require removing barriers to commercial reuse or imposing a policy on allowing or not allowing derivative works. Some open-access initiatives include mandatory measures for long-term preservation. Some, such as BioMed Central, specifies XML as a standardized format for open-access works.[13]

"How important is uniformity about the definition?" asks Peter Suber. "There is already uniformity on the core concept: removing price and permission barriers."[14]

For most proponents of open access, there is no need to create artificial factions in the movement. Rather, the definition of open access is an evolving and flexible concept with policy space to test new elements as they become necessary. The challenge is to keep it simple and not confusing, yet complex enough to accommodate diversity.[15] In order to realize the vision of a global and accessible representation of knowledge, the future Web has to be sustainable, interactive, and transparent. Content and software tools must be openly accessible and compatible.

PRACTICE

Though the publishers of open scholarly journals share many of the goals and principles of the A2K movement, including the open, unconstrained dissemination of information, publication in support of collaboration, and the advancement and spread of knowledge throughout the world, in practice, a troubling gap remains. Historically, scholarly and scientific journals existed in unison with a professional association, university, or university library. These journals generally had a narrow subject focus and a limited audience composed of the scholars/authors themselves and their peers.[16] These tended to be mostly U.S.-based scholars.

The field of scholarly publishing is still dominated by U.S.-based scholars, though foreign input of articles has risen since the 1990s.[17] However, U.S. dominance compounds access problems for scholars, researchers, and the public in developing countries. Many researchers cannot be published and are therefore never quoted. Many cannot even acquire journals published mostly in developed countries and are left without access to scholarly literature on scientific innovations. The price of academic journals, already historically high, has increased with the new licensing schemes of on-line publishing. For example, on-line publishers often license only large packages of journals, instead of the few that are really needed or wanted by an institution. In addition, on-line-only access (without hard copies) often results in empty archives when a subscription is dropped, often because it is too costly. Many libraries in developing countries cannot afford these licenses.

The practical implementation of open-access principles by open-access journals has solved many of these problems and in so doing has proved to be a fundamental piece of the A2K puzzle.[18] Not only do open-access journals expand access to scholarly works around the globe, they also provide opportunities for new voices and perspectives to enter the worldwide scholarly community.[19] Included in this expansion is an increased potential for authors and policy advocates in developing countries to affect the debate over the rationing of knowledge throughout the world.

While it is difficult to quantify the economic and social value of solving this problem, it is evident that access to knowledge has become one of the fundamental goals of developmental policies, even while developing countries face enormous trade pressures to increase enforcement of copyright laws, including the introduction of new legal and technological measures to restrict the copying of digital works.

The A2K movement, in many respects, exists to combat such harmful and entirely unnecessary policies. Because the movement is itself an affirmation or defense of the same policies on which open journals are based, it seems only fitting that the A2K movement should have an open scholarly journal of its own. That was the impetus behind the founding of the journal *Knowledge Ecology Studies*.

In 2006, as the A2K movement began to gain in influence and importance, it became increasingly necessary for the various actors involved to consolidate and disseminate not just information, but advocacy and policy writings touching on all of the diverse issues and disciplines that intersect in discussions of A2K. While there are so many quality on-line journals that it is almost impossible to keep track (SpringerLink alone publishes about fifteen hundred), they generally focus on traditional subject matters, and very few offer a multidisciplinary approach to issues and potential policies regarding A2K issues.

With no formal experience as publishers, but as longtime readers of and contributors of scholarly publications, our small team of eight decided that we could enhance our overall mission of promoting access to knowledge by creating our own on-line journal that would serve the A2K community. After consulting with open-journal and technology experts for about a year, the staff of Knowledge Ecology International (KEI) decided to come forward with a proposal to create a journal specifically designed and managed to provide opportunities to scholars and researchers who might not have access to existing on-line journals. An additional goal was to improve access to A2K issues and debates for the broader public. Developments in technologies, as well as in the policy space and history of open access, have fueled this ongoing experiment.[20]

For us, knowledge ecology is a multidisciplinary field of study recognizing the importance of relationships and links, the diversity of knowledge forms and types, and the need for community production, distribution, and use of knowledge. It is a broader and more inclusive theme than can be encompassed in terms such as "intellectual property" or "access," and it captures our own sense that it is useful to think about knowledge in a more holistic way, considering a wide range of economic, social, labor, distributional, security, moral, and political issues.

As a project of KEI, *Knowledge Ecology Studies* was designed to have a multidisciplinary and international scope and to aim for the most cost-effective approach to the dissemination and indexing of scholarly and scientific research and related information. Shifting away from knowledge and information "management" and its focus on assessing and profiting from knowledge, *KEStudies* looks at

The Intellectual Property Garden is an original 3D art installation created by Pavig Lok
for the opening of the International Justice Center in Second Life, and the Visions
of Global Justice art exhibition (Pavig Lok, courtesy of Not Possible in Real Life).

knowledge creation, distribution, and use as a system, an ecosystem, where human
intellect, innovation, and technology interrelate.

The following is an account of our experience in becoming an open-access journal publisher, including a discussion of the issues we had to address while designing the journal: the journal's basic policy intellectual property, quality and peer
review, and funding sources, as well as on technical issues such as software platforms and on editorial principles.

BASIC POLICY DECISIONS

The principal issues facing authors, publishers, and users of open-access journals
are related to the compatibility of open access with existing rules regarding copyright, peer review, revenues, printing, preservation, promotion, and indexing, all
of which are associated with conventional scholarly literature.

Clearly, open-access on-line publishing can function within the copyright system.
It is compatible with traditional exploitations of writings (printing, preservation,
professional promotion, and indexing), and it does not exclude the peer-review
process, but it also allows a more transparent and in some cases more efficient
use of that process. However, there is a wide variety of answers to the types of

questions faced by those looking to publish an open-access journal. Among the decisions that need to be made, three issues appear to be more difficult than the others: issues related to copyright and other existing rules, quality and peer-review issues, and funding issues.

One of the first questions facing an open access publisher involves what authors, publishers, and readers want and need from copyright in their works.[21] An additional question involves what kind of control and what kind of distribution authors and publishers need. Because we were born out of the A2K movement, we believe our journal has to provide free access to scholarly literature without undue copyright and licensing restrictions. To achieve this goal, the editorial board examined several possible licenses.

One alternative—and the one embraced by most open-access journals—was to apply the Creative Commons Attribution License (CCAL) to all works. Under the CCAL, authors retain ownership of the copyright for their article, but they allow anyone to download, reuse, reprint, modify, distribute, and/or copy the article as long as the original authors and source are credited. This broad license was developed to facilitate open access to, and free use of original works of all types. It ensures that works remain freely and openly available. Under the CCAL, readers are free to make derivative works, defined as "a work based upon one or more pre-existing works, such as a translation, musical arrangement, dramatization, fiction-alization, motion picture version, sound recording, art reproduction, abridgment, condensation, or any other form in which a work may be recast, transformed, or adapted. A work consisting of editorial revisions, annotations, elaborations, or other modifications which, as a whole, represent an original work of authorship."[22] Readers are also free to make commercial use of works on the condition that in doing so, the user gives credit to the original author and, in the case of reuse or distribution, makes clear the licensing terms of the derivative work. Any of those conditions can be waived with permission from the author.

However, whether authors should give everyone permission to make "derivative" works from their articles proved to be one of the most contentious issues we faced when establishing our policy on copyright issues. Copyright law usually prevents the creation of derivative works. What counts as a derivative in each different area of copyright is different, and there is enormous variety in the different kinds of works that one might license. For example, one author argued that his paper should not be used and transformed to misrepresent his policy suggestions. Another wanted to keep the right to control the quality of a possible translation. While many authors are not really interested in being involved in or in limiting what others can do with their work or even, on the contrary, may be happy to see others build on their work, some do want, for various reasons, to keep control of their productions. A multiplicity of factors contribute to differing attitudes toward permission to create derivative works: the type of work (a policy paper, an interview, a review, or a picture for example), the culture or origins of the author (academics versus activists, Europeans versus South Asians, for example), the author's expectations for the future of their work, and other factors, as well.

Consequently, despite the popularity of the CCAL license, the editorial team of *KEStudies* decided to rewrite the license in a publishing contract to reflect internal discussions regarding remix culture and, in general, transformative use. We decided to look at the needs of the three principal parties involved: the author, the publisher, and the public. We decided that copyright owners are within their rights to object to nontransformative, verbatim copying of their copyrighted materials on-line. This does not include instances where copyrighted materials are employed for purposes of comment, criticism, reporting, parody, satire, or scholarship or if within the bounds of fair use under U.S. law. Since this is not the case under every copyright law, specifying the public's right is necessary. In addition, we, as publishers, might want to make and distribute copies noncommercially or commercially without asking for permission. Therefore, we settled on the following terms:

1. Author's Rights. The Author retains: (i) the exclusive right to be credited as the Author; (ii) the right to reproduce, distribute, publicly perform, and publicly display the Work in any medium for any purpose; (iii) the right to prepare and distribute derivative works from the Work; and (iv) the right to authorize others to make any use of the Work, in all cases so long as the Journal is cited as the source of first publication of the Work.

   For example, the Author may make and distribute copies in the course of teaching, research or any other activities, may post the Work on personal or institutional web sites and in other open access digital repositories, or use the Work commercially.

2. Publisher's Rights. The Publisher obtains a royalty-free, worldwide: (i) right to publish, distribute, reproduce, display, publicly perform, prepare derivative works, edit as suitable for publication, translate to any language, and use the Work in any format—digital or in print or any other medium whether now known or hereafter devised—either separately or as a part of a collective work; (ii) right to receive compensation; (iii) exclusive right to receive proper attribution and citation as journal of first publication; (iv) right to include the Work in computerized archival or retrieval systems (such as SSRN, Westlaw, LexisNexis); and (v) right to assign, sublicense or transfer to third parties any rights obtained by this agreement, provided that third party fulfils the Publisher's obligations contained in this agreement towards the Author. In all cases these rights apply so long as the Author receives proper attribution as author.

For example, the Publisher may make and distribute copies in the course of teaching, research or any other activities, may post the Work on personal or institutional web sites and in other open access digital repositories, or use the Work commercially, without the Author's permission.

3. Public's Rights. Knowledge Ecology Studies is a journal dedicated to the open exchange of information; therefore the Author agrees that the Work published in the Journal be made available to the public under two licenses: A Creative Commons Attribution Non-commercial No Derivatives 3.0 License and a Knowledge Ecology Studies Developing Nations License.[23]

In addition, the editorial team decided to deal specifically with issues regarding developing countries by offering an expanded license to exploit works commercially, the terms of which are as follows:

The Knowledge Ecology Studies Developing Nations License will provide the members of the public accessing the Work from a developing country—any country not classified as high-income economy by the World Bank—with the option to obtain all the rights granted under the Creative Commons Attribution Non-commercial No Derivatives 3.0 License on a commercial basis (see: http://creativecommons. org/about/licenses/meet-the-licenses), as long as they implement reasonably effective mechanisms to prevent access to the Work by members of the public located in high-income economies as classified by the World Bank.[24]

The debate over not allowing the public to make derivative works by default and the special treatment clause for developing countries were the two most controversial issues both internally and with our advisory board. However, so far, no author has asked to redact the agreement.

Less controversial was our decision about how to attract and provide quality content. This followed from the decision to have an open system for peer reviewing. High-quality peer review is necessary, but some aspects of the traditional peer-review processes are being questioned. The traditional system has well-known flaws such as lack of transparency and issues of conflict of interest. With the new open access model, one can see the opportunity to reevaluate existing practices.

There are many examples of claims by traditional publishers that open access is a threat to peer review. However, Peter Suber, who makes a point of keeping track of such claims, notes that if some of the publishers clearly state that open access will undermine peer review, they are never specific in explaining why or how.[25] According to Stevan Harnad, a scientist and an active promoter of the Open Access philosophy,[26] there are no essential differences between the traditional journals that require a paid subscription and open-access journals when it comes to the objectives of peer review. However, the new platforms offer the possibility of distributing the tasks more equitably and more efficiently,[27] can make the process more transparent, and even can embrace a "utopian transformation" that has already taken place in some scientific fields, where live, preprint archives are available.[28] Even without using all the imaginative interactive processes that can be used for peer reviewing, most open-access journal editors are willing to experiment with new strategies. For example, many open-access journal editors are experimenting with unblinded review, postpublication review, and open submission. At *KEStudies*, we believe that anonymity is not a necessary condition. A reviewer may make herself known to the author if she so desires. One of the advantages of this practice is to encourage collaboration between authors and reviewers, which often results in a much better paper.

A wide variety of authors contribute to the journal, including some who do not necessarily find their way in more traditional publishing contexts and who are involved in different types of knowledge production. Authors may submit longer pieces for peer review, but they may also submit personal interviews, short research notes, book reviews, or opinion pieces, none of which are peer reviewed. The journal is also open to reports of events such as negotiations or meetings covered "in real time," and these do not require peer reviewing. This is one of the most attractive features of the journal: the possibility of quick publishing, particularly when articles are designed to shape and influence rapidly unfolding policy debates and negotiations. In this sense, peer review can become an impediment to quality.

Finally, funding *KEStudies* posed issues with which we had to deal. The fundamental difference between traditional journals and an open journal is that readers do not pay for access, which eliminates price as a barrier. "Artificial scarcity" is not the business plan for an open-access journal. With this in mind, the obvious question is how to pay for it. On the other hand, while readers do not pay for access for *KEStudies*, that doesn't mean that it's free. As a response in the Budapest Open Access Initiative Frequently Asked Questions notes,

> The term "free" is ambiguous. We mean free for readers, not free for producers. We know that open-access literature is not free (without cost) to produce. But that does not preclude the possibility of making it free of charge (without price) for readers and users. The costs of producing open-access literature are much lower than the costs of producing print literature or toll-access online literature. These low costs can be borne by any of a wide variety of potential funders, among which BOAI has no preferences.[29]

How can open-access journals become a more sustainable model for publishing? Jim Till, currently a member of the Executive Committee of Project Open

Source/Open Access at the University of Toronto, writes in his blog: "There's an ongoing debate about ways to pay the costs of OA, especially during the transition phase between traditional business models and newer OA-oriented ones. There are several potential sources of substantial revenues. One focal point for the debate has been on revenues from academic institutions (a major source of support for knowledge dissemination), relative to revenues from funding agencies (a major source of support for knowledge generation)."[30]

There are many funding scenarios that apply differently for different institutions. As long as the funding mechanism does not imply charging readers or their institutions for access, that is, as long as there are no subscription fees, no licensing fees, and no pay per view, the means of funding is consistent with the open-access philosophy.[31]

In our broader study of open journals, the editorial board considered a variety of funding mechanisms, including requiring authors to pay, seeking government funding, applying for funding from private donors or a consortium of institutions or libraries, funding through our own organization as a form of promoting our mission, and creating new, decentralized funding mechanisms based on prize systems.

For the purposes of this journal, the more ambitious methods that require broader acceptance by governments or other large institutions were not feasible. The editorial board most seriously considered only three possibilities: using the "author pays" approach, seeking support from charitable foundations, or making the journal an integrated part of Knowledge Ecology International's larger budget, connected to its mission and activities. However, the board ultimately rejected the "author pays" approach, since few authors in the field have independent funding, though this is not the situation for some journals in the sciences.[32] In the end, the board allocated some start-up funding from a 2006 MacArthur prize given to KEI, and the ongoing cost of the journal will be part of the basic KEI operating budget.

However, this is not a long-term solution, and we are actively working on proposals to have new funding mechanisms put into place. Options do exist. For example, Leslie Chan notes that we would all benefit if only 1 percent of any university library budget for journals was to be awarded to open-access publications. One can even imagine scholars and readers having a say in which journal should be awarded the 1 percent. One could imagine an allocation system as described by James Love in "Artists Want to Be Paid: The Blur/Banff Proposal."[33] One proposal could be that intermediaries such as the open-access publishers would compete against each other for the funding from an organizational budget. Other possibilities could include that all the money goes to journals in a specific field, or that a percentage would go to support new or rare fields of study, or that money would be allocated on the basis of readers' downloads. As Peter Suber states, open access is "not free to produce, but a very small subsidy will make possible a very large public good."[34]

TECHNICAL DECISIONS

When evaluating possible technical platforms on which to host the journal, the most important criterion was that the system had to deal with issues of management and publishing in an open fashion. In agreement with the open-access declarations, *KEStudies* required software tools to be openly accessible and compatible. Interoperability and long-term archiving were also main criteria. However, with few staff members qualified to write code, the journal also needed a system that did not require a full-time technical staff.

Through conferences and advice from more experienced colleagues (such as *First Monday*'s Nancy Jones), the *KEStudies* board decided to experiment with the Open Journal Systems, or OJS, a journal-management and publishing system that has been developed by the Public Knowledge Project.[35] Our technical advisor easily installed the system, and the editorial team learned how to set up and operate the system without much help from the technicians involved. Basically, OJS is open-source software offering a publishing template that is easy to use and modify. At the outset of the project, the software prompted us through a set-up menu, a series of templates requiring our editorial board to articulate the journal's policies on copyright, staff roles, and review and privacy guidelines. The structure was helpful in identifying the decisions necessary for the publication of an open journal.

OJS is installed on our local server and is fully controllable from in-house equipment, giving our editors the ability to configure, independently of the OJS system, requirements, sections, the review process, copyright policies, and so on. All content is submitted and managed on-line, from contacting and recording peer reviewers and their responses to approving galley proofs and working through the copyediting and formatting processes. Subscriptions are likewise handled by the system, with delayed open-access options, in which contents are made available a chosen time period after initial publication, and OJS manages archives and indexing. Readers are able to e-mail authors directly through the system.

EDITORIAL DECISIONS

In addition to the issues discussed above, several of the flexibilities inherent in the open, on-line publishing system are worth noting. Like any serious publisher, we hope that readers and authors will benefit from our policies, which allow for a great deal of openness and flexibility. The fact that the journal is published on a rolling basis, rather than as a weekly, monthly, or quarterly journal allows its readers and authors to engage in real time, as has been previously noted. Beyond that, the fact that the journal is very open in terms of the types of pieces it accepts and whether these pieces are peer reviewed allows for a variety of input from different disciplines and in a variety of languages. If an author or authors want to propose the publication of data or draft policy guidelines, for example, the editorial board has the ability to incorporate these ideas. In many ways, that makes *KEStudies* an evolving journal, responsive to user-generated ideas and decentralized in its decision-making processes, all with the goal of being as open a publication as possible without sacrificing the quality expected of a peer-reviewed, scholastic journal. The editorial board is now discussing improving our presubmission process, by which authors can presubmit an abstract that would benefit from early feedback from the editors. Since we are occasionally publishing first-time authors (and some nonnative English writers), we consider part of our mission to invest time and skills in presubmission dialogues.

CONCLUSION

From open-access principles to practice, the experiment of publishing an open-access journal, like the A2K movement itself, is ongoing. Technical decisions such as how tools must be accessible and compatible were probably the easiest to make, but not to implement. Editorial decisions are also still evolving and are still challenging. We are trying to find the right balance between being extremely inclusive and keeping a high level of expertise among our authors, as well as promoting balance in the scope and focus within their various fields.

The decisions related to intellectual property required a lot of time and seemed complex and political, but ended up being much less controversial than others. The process of moving from exclusive rights to nonexclusive rights obviously changes the dissemination of knowledge with respect to its legal aspects. To facilitate optimal use and access, we seek "to find solutions that support further development of the existing legal and financial frameworks," as the signatories of the Berlin Declaration on Open Access to Knowledge in the Sciences and Humanities state. We are still working on finding long-term solutions that implement this aim, especially when it comes to making our journal financially sustainable.

We are seeking to promote and support global participation in cultural, scientific, civic and educational affairs, and of course we recognize the opportunities arising from technological progress, such as the Internet. Providing more access to more knowledge is essential and should include both diversifying forms of publication and democratizing knowledge creation and access. At the same time, it also becomes more and more apparent that creative individuals and communities' interests must be recognized and protected without limiting access or creating scarcity when abundance can prevail.

Open-access journals do not by themselves revolutionize access to knowledge and the distribution of scholarly or scientific materials. As is well described by Jean Claude Guédon,

> Open Access is not an end in itself; it is merely a symptom of deeper processes linked to the growing role of digitization in our civilization. It is digitization that brings about opportunities for profound shifts in power. Open Access simply defines a battle front that refers to the challenges being thrown at the architectures of control supported by publishers. Like a litmus test, the quest for Open Access reveals an architecture of control on the wane.[36]

In early conversations about the Internet and its impact on publishing, what dominated was the potential "liberation" from the existing intermediaries. However, according to Michael Carroll, "after the revolutionary euphoria died down... many acknowledged that intermediaries are necessary to all kinds of transactions in commerce, culture, and news. Reintermediation soon follows from disintermediation, and the real question the Internet posed was not whether intermediaries are necessary but what kinds of intermediaries are necessary."[37]

Defining ourselves as publishers of scholarly and policy-relevant knowledge goods, we are, whether we like it or not, intermediaries for goods that can be accessed and copied fairly cheaply, sometimes at zero or close to zero cost. However, providing these knowledge goods also often requires high fixed costs and presents a number of policy dilemmas to the intermediaries that provide them.

As a provider of knowledge goods, the journal *Knowledge Ecology Studies* tries to base its policies on the fact that obviously knowledge goods are "nonrival in consumption," meaning that consumption by one does not preclude or diminish consumption by another. If one decides to ensure that they must also be nonexcludable and that it is neither necessary nor beneficial to restrict access to create traditional private incentives, it is important for the A2K movement to develop and create new mechanisms to support authors and their intermediaries. This is an issue that *KEStudies* is seeking to resolve by engaging in research on new methods and sources of funding, especially for developing countries' authors and intermediaries.

KEStudies has had to reinvent the role of an intermediary, a publisher, in the knowledge ecology in terms of funding, balancing creator and public rights, and distributing knowledge goods in keeping with the paradigm of access to knowledge. Through our experiences, we have come to believe that the complexities of A2K practice, including the many ways to structure an open publication, can all be resolved within the bounds of A2K principles.

NOTES

The views presented here are those of the author, with contributions and helpful suggestions from several others, including James Love and David Serafino, and do not necessarily reflect those of Knowledge Ecology International, its staff and board, or Knowledge Ecology Studies' staff and advisory board.

1   *The Philosophical Transactions of the Royal Society of London* first appeared on March 6, 1665 after the French *Journal des sçavans,* which first appeared as a twelve-page quarto pamphlet on January 5, 1665.

2   See, in particular, Richard Poynder, "The Open Access Interviews: Leslie Chan," June 20, 2008, available on-line at http://poynder.blogspot.com/2008/06/open-access-interviews-leslie-chan.html (last accessed November 19, 2009).

3   See Jean-Claude Guédon, *In Oldenburg's Long Shadow: Librarians, Research Scientists, Publishers, and the Control of Scientific Publishing,* Association of Research Libraries, 2001, available on-line at http://www.arl.org/resources/pubs/mmproceedings/138guedon~print.shtml (last accessed November 19, 2009); and John Willinsky, *The Access Principle: The Case for Open Access to Research and Scholarship* (Cambridge, MA: The MIT Press, 2009).

4   "We need to begin with a fundamental fact—the cost of scholarly journals has increased at 10 percent per year for the last three decades. This is over six times the rate of general inflation and over two-and-a-half times the rate of increase of the cost of health care. Between 1975 and 2005 the average cost of journals in chemistry and physics rose from $76.84 to $1,879.56. In the same period, the cost of a gallon of unleaded regular gasoline rose from 55 cents to $1.82. If the gallon of gas had increased in price at the same rate as chemistry and physics journals over this period it would have reached $12.43 in 2005, and would be over $14.50 today." David W. Lewis, "Library Budgets, Open Access, and the Future of Scholarly Communication: Transformations in Academic Publishing," *C&RL News* 69, no. 5 (May 2008), available on-line at http://sn.pronetos.com/documents/0000/0046/DLewis_Open_Access___Scholarly_Communication.pdf (last accessed November 18, 2009). http://www.ala.org/ala/acrl/acrlpubs/crlnews/backissues2008/may08/librarybudgetsscholcomm.cfm (last accessed June 26, 2008). Document ID: 471139.

5   Lesley Chan, "Supporting and Enhancing Scholarship in the Digital Age: The Role of Open Access Institutional Repository," *Canadian Journal of Communication* 29, no. 3 (2004), available on-line at http://www.cjc-online.ca/index.php/journal/article/viewArticle/1455/1579 (last accessed November 18, 2009).

6   WIPO Copyright Treaty, adopted in Geneva on December 20, 1996, available on-line at http://www.wipo.int/treaties/en/ip/wct/trtdocs_wo033.html (last accessed November 19, 2009). See Articles 11 and 12 and the treaty's U.S. implementation in 1998, the Digital Millennium Copyright Act, available on-line at http://thomas.loc.gov/cgi-bin/query/z?c105:H.R.2281. ENR:, last accessed November 19, 2009). Since 1998, the European Union, Japan, Australia, and Canada have also introduced similar legislative measures.

7   See Roberto Verzola's essay "Undermining Abundance: Counterproductive Uses of Technology and Law in Nature, Agriculture, and the Information Sector" in this volume.

8   See http://www.soros.org/openaccess and Melissa R. Hagemann, "Five Years On—The Impact of the Budapest Open Access Initiative," Information Program, Open Society Institute, available on-line at http://elpub.scix.net/data/works/att/168_elpub2007.content.pdf (both last accessed November 19, 2009).

9   The Berlin Declaration states that: "In order to realize the vision of a global and accessible representation of knowledge, the future Web has to be sustainable, interactive, and transparent. Content and software tools must be openly accessible and compatible," http://oa.mpg.de/openaccess-berlin/berlindeclaration.html (last accessed March 19, 2010).

10   See Peter Suber's timeline at: http://www.earlham.edu/~peters/fos/timeline.htm (last accessed November 19, 2009). The Bethesda Statement includes the following definition and notes:

An Open Access Publication[1] is one that meets the following two conditions:

1. The author(s) and copyright holder(s) grant(s) to all users a free, irrevocable, worldwide, perpetual right of access to, and a license to copy, use, distribute, transmit and display the work publicly and to make and distribute derivative works, in any digital medium for any responsible purpose, subject to proper attribution of authorship[2], as well as the right to make small numbers of printed copies for their personal use.

2. A complete version of the work and all supplemental materials, including a copy of the permission as stated above, in a suitable standard electronic format is deposited immediately upon initial publication in at least one online repository that is supported by an academic institution, scholarly society, government agency, or other well-established organization that seeks to enable open access, unrestricted distribution, interoperability, and long-term archiving (for the biomedical sciences, PubMed Central is such a repository).

[1] *Open access is a property of individual works, not necessarily journals or publishers.*

[2] *Community standards, rather than copyright law, will continue to provide the mechanism for enforcement of proper attribution and responsible use of the published work, as they do now.*

See http://www.earlham.edu/~peters/fos/bethesda.htm (last accessed November 19, 2009).

The Berlin Declaration states that "in order to realize the vision of a global and accessible representation of knowledge, the future Web has to be sustainable, interactive, and transparent. Content and software tools must be openly accessible and compatible." See http://oa.mpg.de/openaccess-berlin/berlindeclaration.html (last accessed November 19, 2009).

11   The varied set of actors included consumer/public interest representatives such as Rhoda Karpatkin (Consumers Union), Benedicte Federspiel (the Danish Consumer Council), Anna Fielder (Consumers International), James Love (CPTech), and Collette Caine (Consumer Institute South Africa); academics such as Jean-Claude Guédon (Université de Montréal), Peter Jaszi (Washington College of Law, American University), Raquel Xalabarder (the Open University of Catalonia, Barcelona), and Alan Story (Kent Law School); libraries' representatives Prue Adler (the Association of Research Libraries) and Bob Oakley (the American Association of Law Libraries); government representatives Luis Villaroel Villalon (Chile) and Ahmed Abdel Latif (Egypt); and foundation representatives Vera Franz (the Open Society Institute) and Becky Lentz (the Ford Foundation).

12   For more details on the differences between definitions, see Peter Suber, "How Should We Define 'Open Access'?" *SPARC Open Access Newsletter*, issue 64, August 4, 2003, available on-line at http://www.earlham.edu/~peters/fos/newsletter/08-04-03.htm (last accessed November 19, 2009).

13   BioMed Central Open Access Charter, originally issued July 1, 2002, available on-line at http://www.biomedcentral.com/info/about/charter (last accessed November 19, 2009).

14   Suber, "How Should We Define 'Open Access'?"

15   More details on definitions can be found on-line at http://www.earlham.edu/~peters/fos/

boaifaq.htm#defopenaccess and http://www.plos.org/about/openaccess.html (both last accessed November 19, 2009).

16 Disciplines typically represented include: architecture and design, astronomy, the biomedical sciences, business and management, chemistry, computer science, economics, education, engineering, the environmental sciences, geography, the geosciences, the humanities, law, the life sciences, linguistics, materials, mathematics, medicine, philosophy, physics, psychology, public health, the social sciences, and statistics.

17 Ali Uzun, "Assessing Internationality of Scholarly Journals through Foreign Authorship Patterns: The Case of Major Journals in Information Science, and Scientometrics," *Scientometrics* 61, no. 3 (November 2004).

18 See a review of recent trends in the open-access movement, as well as a discussion of the significance of those trends for information access in developing countries, by Leslie Chan and Sely Costa, "Participation in the Global Knowledge Commons: Challenges and Opportunities for Research Dissemination in Developing Countries," *New Library World* 106, nos. 3/4 (2005): pp. 141–63. These authors note that knowledge workers in developing countries now have access to scholarly and scientific publications at a historically unmatched level.

19 "Articles published as an immediate OA article on the journal site have higher impact than self-archived or otherwise openly accessible OA articles. We found strong evidence that, even in a journal that is widely available in research libraries, OA articles are more immediately recognized and cited by peers than non-OA articles published in the same journal. OA is likely to benefit science by accelerating dissemination and uptake of research findings." Gunther Eysenbach, "Citation Advantage of Open Access Articles," *PLoS Biology* 4, no. 5 (2006), available on-line at http://www.plosbiology.org/article/info:doi/10.1371/journal. pbio.0040157 (last accessed November 24, 2009).

20 We are also grateful for the information contained in *Gaining Independence: A Manual for Planning the Launch of a Nonprofit Electronic Publishing Venture*. This manual, published by SPARC, the Scholarly Publishing and Academic Resources Coalition, is designed to help universities, libraries, societies, and others implement alternatives to commercially published scholarly and scientific information. It is available on-line at: http://www.arl.org/sparc/ bm~doc/sparc_gi_manual_version1-0.pdf (last accessed November 24, 2009).

21 We are thankful to Michael W. Carroll, author of the "Addendum to Publication Agreement" on publication rights for authors (http://www.arl.org/sparc/bm~doc/Access-Reuse_Addendum. pdf), which started many of our internal discussions on the needs of open-access contributors, producers, and consumers. See the very useful and practical *Author Rights: Using the SPARC Author Addendum to Secure Your Rights,* SPARC (2006), available on-line at http:// www.arl.org/sparc/author/addendum.html (both last accessed November 24, 2009).

22 See the definition of a "derivative work" in 17 U.S.C. § 101 and in *U.S. Copyright Office Circular* 14: "A typical example of a derivative work received for registration in the Copyright Office is one that is primarily a new work but incorporates some previously published material. This previously published material makes the work a derivative work under the copyright law. To be copyrightable, a derivative work must be different enough from the original to be regarded as a 'new work' or must contain a substantial amount of new material. Making minor changes or additions of little substance to a preexisting work will not qualify the work as a new version for copyright purposes. The new material must be original and copyrightable in itself. Titles, short phrases, and format, for example, are not copyrightable."

Or as the Berne Convention for the Protection of Literary and Artistic Works says in Article 2: "(3) Translations, adaptations, arrangements of music and other alterations of a literary or artistic work shall be protected as original works without prejudice to the copyright in the original work."

23 For the Creative Commons Attribution–Noncommercial–No Derivative Works 3.0 License and the Knowledge Ecology Studies Developing Nations License, see http://creativecommons.org/licenses/by-nc-nd/3.0/us (last accessed November 24, 2009).

24 *Ibid.*

25 In May 2006, the Association of American Publishers (AAP), as part of its opposition to the Federal Research Public Access Act, declared: "If enacted, S.2695 could well have the unintended consequence of compromising or destroying the independent system of peer review that ensures the integrity of the very research the U.S. Government is trying to support and disseminate." See https://mx2.arl.org/Lists/SPARC-OAForum/Message/3006.html and http://www.earlham.edu/~peters/fos/2006_05_07_fosblogarchive.html#114726726169346460. And in February 2007, the DC Principles Coalition declared, in opposition to the same legislation: "Subscriptions to journals with a high percentage of federally funded research would decline rapidly. Subscription revenues support the quality control system known as peer review." See http://www.dcprinciples.org/press/2.htm and http://www.earlham.edu/~peters/fos/2007_02_18_fosblogarchive.html#117202895951705271 (all last accessed November 24, 2009). The Brussels Declaration of February 2007, organized by the Association of Learned and Professional Society Publishers and the International Association of Scientific, Technical, and Medical Publishers likewise opposed a proposed open-access mandate for the European Union by declaring: "Open deposit of accepted manuscripts risks destabilising subscription revenues and undermining peer review." See http://www.alpsp.org/ForceDownload.asp?id=304 and http://www.earlham.edu/~peters/fos/2007_02_11_fosblogarchive.html#117138731546079566. Finally, in August 2007, the new lobbying organization of the Association of American Publishers and Professional Scholarly Publishing, the Partnership for Research Integrity in Science and Medicine (PRISM), opposed government open-access mandates by declaring that open-access policies would "jeopardize the financial viability of the journals that conduct peer review, placing the entire scholarly communication process at risk." See https://mx2.arl.org/Lists/SPARC-OAForum/Message/3934.html and http://www.earlham.edu/~peters/fos/2007_08_19_fosblogarchive.html#365179758119288416. See also http://www.earlham.edu/~peters/fos/newsletter/09-02-07.htm (all last accessed November 24, 2009).

26 Stevan Harnad, Open Access Archivangelism, http://openaccess.eprints.org. See also http://en.wikipedia.org/wiki/Stevan_Harnad.

27 "Paper publishing's traditional quality-control mechanism, peer review, will have to be implemented on the Net, thereby re-creating the hierarchies of journals that allow authors, readers, and promotion committees to calibrate their judgments rationally—or as rationally as traditional peer review ever allowed them to do it. The Net also offers the possibility of implementing peer review more efficiently and equitably, and of supplementing it with what is the Net's real revolutionary dimension: interactive publication in the form of open peer commentary on published and ongoing work. Most of this 'scholarly skywriting' likewise needs to be constrained by peer review, but there is room on the Net for unrefereed discussion too, both in high-level peer discussion forums to which only qualified specialists in a given field have read/write access, and in the general electronic vanity press." Steven

Harnad, "Implementing Peer Review on the Net," in Robin P. Peek and Gregory B. Newby (eds.), *Scholarly Publishing: The Electronic Frontier* (Cambridge, MA: The MIT Press, 1996).

28 Steve Harnad mentions the example of high-energy physics (*ibid.*). Other examples of this "utopian revolution" include Ulrich Poschl's success story of the open-access journal *Atmospheric Chemistry and Physics*. The vision behind this interactive journal is to promote scientific knowledge through a multistage publication process that includes interactive peer review and public discussion. He writes: "Substantial improvement can be achieved by open access publishing with a two-stage publication process, public peer review, and interactive discussion (interactive open access journal concept). This approach enables rapid publication and dissemination of new scientific results in discussion papers followed by thorough and transparent peer review which is open for comments from the global scientific community (permanently archived and fully citable), and it leads to final revised papers with maximum quality assurance and information density." Ulrich Poschl, "Interactive Journal Concept for Improved Scientific Publishing and Quality Assurance," *Learned Publishing* 17, no. 2 (April 2004): pp. 105–13.

29 See http://www.earlham.edu/~peters/fos/boaifaq.htm#wishfulthinking (last accessed November 24, 2009).

30 Jim Till, "Scenarios for Paying for OA," available on-line at http://tillje.wordpress.com/2007/03/02/scenarios-about-paying-for-oa (last accessed November 24, 2009).

31 As the Budapest Open Access Initiative puts it, "there are many alternative sources of funds for this purpose, including the foundations and governments that fund research, the universities and laboratories that employ researchers, endowments set up by discipline or institution, friends of the cause of open access, profits from the sale of add-ons to the basic texts, funds freed up by the demise or cancellation of journals charging traditional subscription or access fees, or even contributions from the researchers themselves. There is no need to favor one of these solutions over the others for all disciplines or nations, and no need to stop looking for other, creative alternatives." The Budapest Open Access Initiative is available on-line at http://www.soros.org/openaccess/read.shtml (last accessed November 24, 2009).

32 In some cases, an institutional library may subsidize the author's fee as is described in Andy Gass and Helen Doyle, "The Reality of Open-Access Journal Articles," *Chronicle of Higher Education*, February 18, 2005, but this would not apply to our situation.

33 James Love, "Artists Want to Be Paid: The Blur/Banff Proposal," *Blur 02: Power at Play in Digital Art and Culture*, available on-line at http://www.nsu.newschool.edu/blur/blur02/user_love.html (last accessed November 24, 2009).

34 Peter Suber, "Removing the Barriers to Research: An Introduction to Open Access for Librarians," *College & Research Libraries News* 64, no. 2 (February 2003): p. 113, also available unabridged on-line at http://www.earlham.edu/~peters/writing/acrl.htm (last accessed November 24, 2009).

35 See http://pkp.sfu.ca/?q=ojs (last accessed March 19, 2010).

36 Jean-Claude Guédon, "A Take on Peter Suber's 'The Opening of Science and Scholarship,'" Publius Project Essays and Conversations about Constitutional Moments on the Net Collected by the Berkman Center, available on-line at http://publius.cc/take_peter_suber's_"_opening_science_and_scholarship" (last accessed November 24, 2009).

37 Michael W. Carroll, "Creative Commons and the New Intermediaries," *Michigan State Law Review* 45, no. 1 (2005): p. 1.

The Global Politics of Interoperability

Laura DeNardis

Technical standards are the least visible, but arguably most critical and least understood component of the Internet's technical and legal architecture. Information and communication technology (ICT) standards are not material products, like software or hardware, but exist at a much higher level of abstraction and control. They are literally blueprints for developing technologies that can communicate and exchange information with other technologies.[1] Most Internet users are familiar with well-known standards such as Bluetooth wireless, "Wi-Fi,"[2] the MP3 format for encoding and compressing audio files,[3] and HTTP,[4] which enables the standard exchange of information between Web browsers and Web servers. These are only a few examples of thousands of standards enabling the production, exchange, and use of information.

From a technological standpoint, standards are the agreed-upon rules structuring information in common formats and establishing communication interfaces that enable interoperability between diverse ICT environments.[5] From an economic standpoint, interoperability standards carry significant externalities, such as enabling competition and innovation in product areas based on common technical specifications.[6] Technical standards are not only technical design decisions that can carry significant economic externalities, they also make political decisions about global knowledge policy.[7]

These resources necessary for information production and exchange are examples of what Yochai Benkler calls "knowledge-embedded tools," similar to enabling technologies for medical and agricultural resources.[8] Knowledge-embedded tools such as open (versus proprietary) standards are necessary for enhancing welfare and enabling innovation itself. Internet standards such as TCP/IP (Transmission Control Protocol/Internet Protocol) and HTML (Hypertext Markup Language) have historically been openly available, enabling citizens and entrepreneurs to

Aaron Gustafson (http://
creativecommons.org/licenses/
by-sa/2.0/deed.en).

contribute to Internet innovation, cultural creativity, and electronic discursive spheres. For example, Web standards such as HTTP and HTML enable citizens both to create new software applications based on these standards and to contribute actual content to the Internet. However, the institutional processes, intellectual property arrangements, and technical approaches of some standards do not necessarily provide sufficient openness to enable this innovation and cultural production. Who can participate in the development of standards? Who controls standards-related intellectual property rights? Who can access and use the standard, once developed? Will the standard provide interoperability between competing products? The degree of openness in standards has effects on technical interoperability and cultural innovation and also has significant political implications.

Design decisions underlying standards sometimes also are decisions about individual privacy, property rights, and public access to government documents. The intellectual property arrangements underlying standards establish policy about the economic competitiveness of certain markets, how innovation should proceed, and what opportunities might exist for developing countries. As traditional barriers to trade have diminished, standards are also emerging as alternative global trade barriers. Incompatible standards can prevent interoperability and access to knowledge, as was the case when incompatible applications impeded communications during rescue and recovery efforts after the 2004 south Asian tsunami disaster.[9] As described in the Berkman Center's *Roadmap for ICT Ecosystems*, Thailand's government encountered difficulty obtaining vital information because various agencies and nongovernmental organizations used software products based on incompatible data and document formats. The incompatibility of standards that impeded disaster-response efforts helped to encourage the Thai government to make open-document standards a national priority.[10] Furthermore, in a globally distributed information society, standards create scarce resources, such as radio-frequency bands and IP addresses (Internet Protocol addresses) and make decisions about how these resources are structured and allocated. Standards, once adopted, are an enduring form of global public policy, but are developed primarily by private corporations, rather than by those representing publics directly affected by these policies.

This essay describes the challenges and opportunities that technical standards present to access to knowledge, emphasizing the pronounced implications of standards for developing countries. It also describes emerging responses from a variety of stakeholders, including a nascent multistakeholder "open standards movement" and government policies to procure technologies based on open standards.

THE A2K IMPLICATIONS OF STANDARDS

Jack Balkin has described the access to knowledge movement (A2K) as "a demand of justice," as "an issue of economic development and an issue of individual participation and liberty," and as "about intellectual property," but "also about far more than that."[11] This serves as a useful framework for forging connections between technical standards and the goals of access to knowledge. From a purely technical standpoint, common standards are designed to enable the exchange of information, while proprietary specifications are, in a correlative and pragmatic way, designed to prevent, limit, or control the exchange of information.

A completely proprietary standard is one that is developed and controlled by a single company, is not available for others to develop interoperable and competing products based on that standard, and results in single-vendor lock-in, making users dependent on a single source for a product. The closed specifications historically underlying Microsoft's traditional Office products are an example of a closed standard. The technical blueprints for formatting information are completely proprietary. Other classic examples of closed specifications were those underlying the proprietary on-line systems of the late 1980s and early 1990s—for example, Prodigy, CompuServe, and America Online—before on-line service providers expanded their services to more open Internet protocols. Analogously, the dominant social networking sites of the early twenty-first century are based on proprietary systems that prohibit the exchange of information between competing social networking services.

A completely open standard can be defined as one that is developed in an open membership forum, that is made freely available to anyone interested in developing products based on the standard with minimal intellectual property restrictions, and that results in multiple competing products, interoperable services, and innovation based on the standard. An example of an open standard is TCP/IP, a group of the Internet's underlying network protocols. In reality, most standards reside on a continuum between completely proprietary and completely open. Depending on the context, the extent of a standard's openness can have important political and economic implications.

This section describes three ways in which standards and the degree of openness in standards fit within an access to knowledge framework. Standards raise issues of distributive justice when they create scarce resources necessary for meaningful participation in the information society. They also have direct implications for individual political participation and liberty in the information society. Finally, they can create pronounced political and economic challenges for developing countries.

STANDARDS AND DISTRIBUTIVE JUSTICE

Many technical standards create and allocate the finite resources required for access to information networks. How these resources are distributed and by whom can raise issues of distributive justice. Some standards partition and allocate the radio-frequency spectrum among users—for example, broadcast standards, Wi-Fi, and cellular standards. Others prioritize the flow of information over a network based on the type of application being transmitted, such as prioritizing voice applications and decelerating peer-to-peer video. Other standards divide up orbital slots in satellite systems. Some assign rights of access to local broadband services. Some, such as digital subscriber lines, provide an asymmetrical distribution of bandwidth whereby downstream communications to users are privileged over upstream communications from users to the network.

The Internet Protocol (IP) standard is an exemplar of how standards create finite resources necessary for access to information networks. IP is the central protocol of the Internet. As Internet engineers have described it, if a device uses IP, it is "on the Internet." If it does not use IP, it is not on the Internet. One of the central functions defined by this standard is Internet addressing. Every device exchanging information over the Internet possesses a unique number (an IP address) identifying its virtual location, somewhat like a unique postal address identifying a home's unique physical location. The long-prevailing Internet address standard is called IPv4,[12] which originated in the early 1980s and specifies a unique 32-bit number for each Internet address.[13] This address length of 32 bits provides 232 or nearly 4.3 billion unique Internet addresses.

In 1990, the Internet standards community identified the potential depletion of these 4.3 billion addresses as a critical technical and political concern in the context of Internet globalization. U.S. institutions had received enormous IP address allocations when the Internet was primarily an American enterprise, raising concerns that the remaining addresses might not meet emerging requirements of rapid growth and new applications such as wireless Internet access and Internet telephony. The Internet standards community selected a new network protocol, IPv6,[14] to expand the Internet address space. The new standard ultimately expanded the address length from 32 to 128 bits for each address, supplying a staggering 2^{128} or 340 undecillion unique addresses.

Upgrading to a new standard technically can not happen all at once, like flipping a switch. In the case of IPv6, it requires incremental software and hardware upgrades within local Internet service provider (ISP) infrastructures, upgraded end-user software, and the allocation of new IPv6 addresses. Interest in IPv6 in the United States has been relatively limited, primarily because U.S. institutions received ample IP addresses prior to Internet globalization. In contrast, governments in India, China, the European Union, Japan, Korea, and elsewhere have considered IPv6 a national priority, both as a solution to projected address shortages and as an economic opportunity to develop new products and expertise in global Internet markets.

Even with the availability of the IPv6 standard to increase exponentially the number of devices potentially able to connect to the Internet, questions of distributive justice remain. The availability of a standard does not mean that the standard will be implemented in products, and the availability of products does not mean that products will be adopted in infrastructures. The long-anticipated upgrade to IPv6 has not occurred to any great extent. It is difficult for citizens in parts of the world with limited IPv4 addresses, such as Kenya and other parts of Africa, to benefit from new IPv6 addresses unless their local service providers upgrade network infrastructures to handle IPv6 traffic. A scarcity of IPv4 addresses already limits Internet access and slows the deployment of new Internet services requiring IP addresses. Until service providers have market or public incentives to upgrade to IPv6, this problem will remain. Additionally, the conservative momentum of IPv4 and associated lack of interest in IPv6 in dominant Internet markets such as the United States have an effect on developing markets relying on IPv6. Some software and hardware innovators have limited incentives to develop new products based on IPv6 unless a critical mass of Internet users adopts the new standard.

A related issue of distributive justice is the question of who should allocate the scarce resources created by standards. In the case of the IP standard, this question has been a central controversy in Internet governance. Centralized control has historically existed in the area of IP address allocation, in part to maintain the architectural principle of globally unique addresses. Currently, regional institutions such as the American Registry for Internet Numbers (ARIN), the Réseaux IP Européens–Network Coordination Centre (RIPE-NCC), the Asia Pacific Network Information Centre (APNIC), the Latin America and Caribbean Network Information Centre (LACNIC), and the African Network Information Centre (AfriNIC) allocate addresses. These organizations are not governmental bodies. They are paidmembership, nonprofit corporations that control addresses. Member institutions consist primarily of Internet service providers, telecommunications companies, and other large corporations in each respective region. One public policy question involves the implications of organizations ultimately controlled by private companies controlling the vital resources required for Internet participation.

Despite the regional distribution of IP addresses, definitive control of the entire address reserve, including the allocation of address resources to international registries, has remained centralized and is currently an administrative function within ICANN, the Internet Corporation for Assigned Names and Numbers, a private entity incorporated in California and traditionally overseen by the U.S. Commerce Department. International concerns have centered on questions about ICANN, retaining this Internet governance authority and control over global resources, as opposed to distributing control to an international entity.

The central function of the IP standard is the definition of the length and therefore the number of available IP addresses. The question of who controls the scarce resources created by such standards is a critical one. In a global Internet economy, control over standards and the scarce resources they can create increasingly determines wealth. The citizens in countries with greater control over the development and adoption of standards and the scarce resources created by standards have distinct advantages in opportunities to access knowledge-embedded tools, to innovate, and to produce and control information.

STANDARDS, INDIVIDUAL LIBERTY, AND PARTICIPATION

Standards also can have direct policy implications related to civil liberties and democratic processes.[15] The content of standards can determine the extent of individual civil liberties such as user privacy. For example, in developing the IPv6 standard, the Internet Engineering Task Force (IETF) faced a design decision about whether a virtual Internet address should incorporate a physical address, such as the number associated with a computer's network interface card. This linkage between software-defined and physical addresses would create an environment in which information transmitted over the Internet could potentially be associated with a citizen's computer and therefore with a citizen's identity and physical location. Internet engineers, rather than elected officials, faced policy questions about location privacy and anonymity. Ultimately, privacy options were built into the standard, but what are the implications of a standards-setting institution making decisions about the public's civil liberties? Similar decisions about privacy arise in the development of encryption standards and also in the development of electronic health-care-information standards that make decisions about how citizens' health-care records are electronically exchanged.

The degree of openness in document-format standards also has political implications. To be digitally stored or exchanged, information (images, words, numbers, video, and audio) must be converted into a digital format (encoded into 0s and 1s). Examples of these formats include JPEG for images, MPEG for video, and MP3 for audio. Office applications such as text documents, spreadsheets, and presentations are encoded using document-format standards. If the standard is openly published with minimal intellectual property restrictions, competing vendors can make software capable of creating or reading documents encoded in this common format. If the document format is proprietary, citizens can exchange and read these documents only if they use a specific software product.

One way the public interest is implicated is when governments make electronic public documents available in these proprietary formats. The archiving of documents is a fundamental responsibility of democratic governments, and public access to these documents is essential for government accountability and for deliberation over the efficacy of government institutions and policies. Using a proprietary standard does not meet this obligation of democratic governments, because it locks public documents in a format that requires citizens to use a particular vendor's software, raises concerns about backward incompatibility and lack of interoperability, and creates the possibility of public documents becoming inaccessible in the future because the proprietary format is no longer supported.

Technical standards have even greater democratic implications when involved in processes of political authorization and representation, such as standards related to electronic voting machines and electronic voter registration. Transparency in these formal democratic processes is necessary for legitimacy and civic trust in government. Vote tabulation processes have historically been available for public scrutiny, with volunteers gathering in a room and scrutinizing election ballots. The question of whether standards for electronic voting tabulations and information exchange are open for viewing, as well as in a format that can

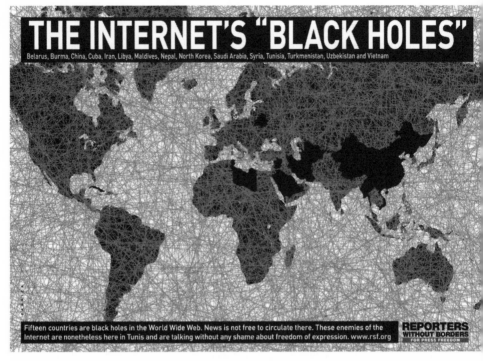

The Internet's "black holes," commissioned by Reporters Without Borders (www.RSF.org).

be readily inspected, raises political concerns.[16] The substance of democracy in the modern information society extends beyond formal democratic processes of political representation to the informal interactions of civil society and culture.[17] Technical standards that empower, rather than restrict society's ability to interact, produce, and inquire within their community's cultural horizon may significantly impact conditions of democracy. Open technical standards such as HTML and TCP/IP, which have been freely available to access and use, have provided citizens with the tools to contribute to discursive political spheres and to promote creative innovation and collaboration.

In these cases, the question of who sets technical standards is highly relevant. Power over these standards is not restricted to market power but the ability to make decisions directly impacting the citizens who use technologies. This form of public policy is not established by democratic, representative government mechanisms or with public input, but by private actors. The more open a standard's development process, the greater the legitimacy of its effects on the public interest. One problem is that the technical nature of ICT standards usually precludes significant public participation, but processes that include participatory openness, information transparency, and public-document availability enable some public-interest input into the setting of standards. Organizations with open-membership policies at least allow for the possibility of public participation. Transparency in proceedings and mailing lists provides some accountability for decisions. A publicly available standard provides a level playing field for competition that can maximize user product choices and the ability of citizens to create innovative products based on the standard.

Openness is not a given, but a matter of choice. In reality, the degree of openness in standards-setting institutions varies considerably by organization. Maximal openness is important, because it increases the ability of the public to affect decisions on standards that determine civil liberties and the ability of citizens to engage in other democratic processes.

STANDARDS AND DEVELOPING COUNTRIES

Information and communication standards are a basic requirement for the diffusion of information and communication technologies in the developing world, as well as in the so-called developed world.[18] However, the underlying intellectual property arrangements of ICT standards do not necessarily reflect the political and economic interests of developing countries. Developing countries are both producers and users of information and communication technologies, but encounter unique standards-related intellectual property rights and institutional barriers.

The Internet's protocols have historically been openly available for use with minimal intellectual property restrictions, a characteristic that has contributed to the Internet's rapid innovation and global growth. However, the Internet's intellectual property rights environment has become increasingly complex. The number of standards related to access and the creation of information has increased over time as the Internet has expanded from text to multimedia applications (video, audio, images) and as the types of devices connected to the Internet (cell phones, consumer electronics, laptops) have expanded. Furthermore, an average Internet-connected device, such as a mobile phone, provides multiple functions such as voice, Internet browsing, text messaging, digital imaging, and video recording and can include literally hundreds of embedded technical standards. Most of these standards are established within institutions dominated by private corporations with significant economic interest in the selection of standards. The building blocks of information exchange, such as image formats, video encoding standards, audio compression standards, network protocols, and office application formats, are standards with deeply embedded intellectual property rights.

These complexities have pronounced effects on developing countries, which are typically late entrants in both the development of standards and the manufacturing of ICT products based on these standards. Unlike large Western companies, entrepreneurs in developing countries and other new entrants are not well versed in the traditions of standards-setting institutions. In addition, they do not necessarily have the large legal departments to deal with intellectual property rights, do not have existing patent portfolios, and do not have a history of engaging in cross-licensing agreements with other companies.[19] Andy Updegrove describes the problem as "standards based neocolonialism": "Royalty bearing patent claims are embedded in the standards for products such as DVD players and cellular phones. If the royalties are high enough, the patent owners can have such products built in emerging countries using cheap local labor, and sell them there and globally under their own brands. Meanwhile, emerging company manufacturers can't afford to build similar products at all."[20]

Developing countries, as later entrants in the information society, bear a greater burden in paying royalty payments to participate in its cultural and economic benefits and to deal with the increasing complexity of intellectual property rights. Furthermore, lack of disclosure of standards-based intellectual property rights presents a disincentive to innovate based on standards, only to face a patent-infringement lawsuit at some later date. The implication is that the increasing complexity of intellectual property rights discourages developing countries and other later ICT entrants from innovating and competing in product markets based on the standards necessary for participating in global information infrastructures.

Even when developing countries are able to become involved in standards setting, institutional and cultural barriers exist. There are hundreds of standards-setting organizations, but few norms about who can participate and how standards are set. The legitimacy of the development of Internet standards has always derived in part from its participatory and open institutional approach, but some organizations that set essential standards have closed membership policies and scant information transparency. Barriers can exist even in organizations such as IETF that are completely open to public participation. Participation requires money. Individuals involved in standards setting are funded by their employers. Smaller companies or individual citizens do not necessarily have the resources necessary to fund the time commitment required to participate in these activities or to fund travel to in-person meetings. The technological expertise required to participate in working groups also creates challenges for those joining the process as later entrants.

Relatively closed standards also have pronounced effects on developing countries as users of information and communication technologies. Costs and restrictions produced by standards-related intellectual property rights are ultimately passed to users. This has disproportional effects on countries lacking the ICT infrastructure inherent in the West. Countries with an installed base of fiber-optic and copper backbone infrastructures view broadband wireless technologies such as GSM (global system for mobile communications), Wi-Fi, and WiMax (worldwide interoperability for microwave access, a wireless communications technology) as "access technologies" that connect users to the backbone. In countries without backbone infrastructures, these technologies potentially serve as both backbone and access mechanisms. GSM, Wi-Fi, and WiMax are standards with significant embedded intellectual property rights. For example, WiMax is an emerging standard for metropolitan broadband wireless Internet access. Implementing this standard in a new product requires licensing numerous patents owned by various technology companies. The burden of the escalation in standards-based intellectual property rights is borne to a greater extent by those without an installed infrastructure base.

Even if products are manufactured in a developing country, that country often must purchase these same products from Western companies. If the standards embedded in these products are proprietary, this can create vendor lock-in, minimizing user product choice, competition, and associated cost reductions. Countries that want global interoperability sometimes have no choice but to use prevailing standards, which have a unique tenaciousness once adopted because of network effects, institutional commitments, and vendor investments. For example, there are not many alternatives except to use the Wi-Fi standard for wireless local area network access and the MP3 and related standards for audio compression.

The rise of intellectual property rights-laden ICT standards, for example, has resulted in large emerging countries such as China developing some national standards, rather than using globally interoperable standards. The World Trade Organization's Agreement on Technical Barriers to Trade (TBT) asserts that standards should not create unnecessary obstacles to trade, but ICT standards sometimes serve as nontariff barriers in global markets.[21] A response from China has been the development of Chinese national standards such as WAPI (Wireless Local Area Network Authentication and Privacy Infrastructure), the Chinese national standard for wireless local area network encryption; and UOF (Unified Office Format), China's document format for office applications.

The result of such standards lock-ups is that the poorest countries cannot move up the value chain and compete as ICT manufacturers, and the more technologically sophisticated and large developing countries have incentives to develop national standards, potentially creating global incompatibility or national proprietary infrastructures that can potentially isolate citizens from the larger global information society.

The many issues and concerns described in the previous sections have led to calls for greater promotion of open standards. However, the definition of "open standards" is contestable. Many definitions exist,[22] and various stakeholders are likely to arrive at different, contextually specific definitions. Nevertheless, most definitions have some common denominators.

First, to promote values of transparency and accountability and the possibility of multiple stakeholder participation, the standards-development process should be open to any interested party, should involve well-defined procedures, and should make deliberations and working-group discussions publicly available. As mentioned, the membership requirements, openness, and transparency of standards-setting institutions vary considerably, with traditional Internet standards-setting institutions such as the IETF providing the greatest openness and with small consortia involving a handful of private companies typically the most closed.[23] Participatory openness and procedural and information transparency provide a degree of legitimacy to organizations that establish policies through technical standards.

Second, the standards specification, the actual knowledge-embedded tool necessary to develop products, is usually considered open if it is made publicly available. Some open-standards definitions stipulate that there should be no fee to access the standard, while others acknowledge that a fee may be necessary to recoup some costs of developing and publishing the standard. An unpublished standard is truly a proprietary standard that precludes the possibility of innovation based on the standard.

Third, a standard is considered open if it is available on either a reasonable and nondiscriminatory (RAND) basis or on a royalty-free basis. The IETF working groups have always given preference to technologies with no known intellectual property rights claims or, if the technology has a claim, with royalty-free licensing. The World Wide Web Consortium (W3C),[24] citing the objective of promoting ubiquitous adoption of Web standards, has established a policy of issuing recommendations only if they can be implemented on a royalty-free basis, although there is a mechanism for allowing exceptions.[25] As mentioned, the Internet's underlying protocols have historically been made available on a predominantly royalty-free basis, a characteristic often cited as contributive to the Internet's rapid growth, product innovations, and democratic participation.

However, policies that mandate royalty-free standards can produce inadvertent consequences, such as eliminating the possibility of using popular royalty-bearing standards such as Wi-Fi and GSM. Conversely, RAND licensing policies are problematic because of definitional ambiguities over the meaning of "reasonable" and "nondiscriminatory." RAND is a term commonly given by standards-setting institutions to define their intellectual property rights policies, but, as Mark Lemley has noted, most of these policies do not actually define what is meant by RAND. This creates questions for companies interested in licensing a standard: Will intellectual property rights holders license universally or merely to other members of the standards-setting institutions? And what actually constitutes a "reasonable" royalty fee? RAND licensing lacks a clear and consistent definition.

Other definitions of open standards provide details about how well the standard can be implemented by those developing products based on the standard, whether the standard results in multiple competing implementations, and whether the interests of a single corporation have dominated the process. For example, the International Telecommunication Union's definition of openness says that the process should not be dominated by any one interest and that it should be written in sufficient detail to facilitate developing of diverse, competing products that implement the standard.[26]

The definitions of an open standard vary depending on the stakeholders and the cultural and political contexts involved, but definitions of openness share the common framework of providing openness in development (membership, transparency, process), openness in implementation (public document availability, RAND or royalty-free licensing), and openness in implementation (multiple competing implementations, user choice).

THE DYNAMIC COALITION ON OPEN STANDARDS

One multistakeholder response has been the formation of the Dynamic Coalition on Open Standards (DCOS) at the inaugural Internet Governance Forum (IGF) held in Athens, Greece, in November 2006. The purpose of the IGF was to create a formal space for multistakeholder policy dialogue to address public-policy issues related to Internet governance, facilitate discourse among international Internet governance institutions, and promote the engagement of stakeholders, particularly developing countries, in Internet governance mechanisms.[27] The formation of the IGF was in part a response to international concerns and perceptions of U.S. control over ICANN. Therefore, a contentious topic at the IGF has been control over ICANN domain-name functions, although ICANN members have not attended the IGF to any great extent. IGF participants primarily have included governments, nongovernmental organizations, academic institutions, corporations, and activists.

One tangible development at the inaugural IGF meeting in Athens was the creation of a number of mulitstakeholder "dynamic coalitions." One of the first groups to coalesce at the IGF was the Dynamic Coalition on Open Standards. While there is no formal definition of the IGF dynamic coalitions, they are multistakeholder groups that self-form under the auspices of the IGF around Internet governance topics. In addition to the open-standards coalition, other dynamic coalitions address issues such as privacy, freedom of expression and freedom of the media on the Internet, spam, an "Internet Bill of Rights," and A2K generally. The mission statement of DCOS declares: "Our mission is to provide government policy makers and other stakeholders with useful tools to make informed decisions to preserve the current open architecture of the Internet and the World Wide Web, which together provide a knowledge ecosystem that has profoundly shaped the multiplier effect of global public goods and improved economic and social welfare."[28]

The purpose of DCOS has been to frame and define the most pressing global problems related to open technical standards, to offer recommendations for promoting principles of openness and interoperability, and particularly to address best practices in government policy and procurement practices related to the promotion of open Internet standards. DCOS was formed by stakeholders representing governments (for example, Sri Lanka and Republic of South Africa), standards organizations (for example, the World Wide Web Consortium), industry (for example, Sun Microsystems, Free Software Foundation Europe), civil society (for example, Knowledge Ecology International, IP Justice, the South Centre), and academia and learning institutions (for example, the Yale Information Society Project, Bibliotheca Alexandrina, and United Nations University).

One outcome of the dynamic coalition's formation appears to have been to advance the dialogue on open standards, especially among stakeholders in the Global South and within the IGF discourse generally. In the Internet governance community, perceptions about "open standards" have progressed from viewing it as an arcane technical issue to accepting it as a mainstream issue of significant global policy importance. This has been the case even in the IGF community. At the second annual IGF, held in 2007 in Rio de Janeiro, Brazil, the issue of open standards was a pervasive theme, beginning with the opening session and continuing with three open-standards workshops, including one addressing government policies on open standards, and a broad "openness" session which, among other things, highlighted the development dimension of open standards

GOVERNMENT OPEN-STANDARDS POLICIES

One of the objectives of the Dynamic Coalition on Open Standards has been to promote open Internet standards in government policy and procurement practices. As a response to technical, economic, and political concerns about the degree of openness in global ICT standards, governments have increasingly established policies addressing these issues. Governments have several potential avenues for intervention, including development, regulation, and procurement. They can become directly involved in the development of standards by participating in standards-setting processes, mandating the development of certain standards, or funding companies to design specific standards. Governments also have the option of directly regulating which standards, and how standards, must be used within products used by publics. Finally, governments, as large users of ICT products, can exert market influence through procurement policies.

Of the three options, this procurement mechanism provides the lowest level of government intervention and therefore has become the favored option of governments seeking to promote open standards. Many governments believe that the private sector, rather than government officials, most efficiently produces ICT standards. Many also hesitate to mandate that the private sector and citizens adopt specific standards.

There have been numerous instances of governments establishing policies to procure ICT products that adhere to principles of openness and interoperability. Often called "government interoperability frameworks" (GIFs),[29] open-standards policies have been introduced by countries such as Australia, Bangladesh, Belgium, Brazil, China, Croatia, Denmark, France, Hong Kong, India, Italy, Malaysia, New Zealand, Sri Lanka, and Thailand.

As is evident from this list, open-standards policies are a phenomenon in both the developed and the developing world. While definitions of what constitutes an open standard differ, the overarching purpose of these government policies is interoperability, the ability of government agencies to exchange information with each other and with citizens, with open standards cited as the primary method for achieving this interoperability.

For example, the Brazilian federal government issued an interoperability policy establishing the adoption of open standards for technology used within the executive branch of the federal government. The Brazilian model is representative of many open-standards policies. The Brazilian framework is limited to internal government communications and information exchanges with citizens and specifically states that the policies cannot be imposed on the private sector, on citizens, or on government agencies outside of the federal government, although it does request voluntary adherence to the standards. Additionally, the federal standards policies apply to new purchases and upgrades to existing systems, rather than mandating a complete changeover to new products. Like other open-standards policies, Brazil's cites a combination of technical, political, and economic justifications. Most policies express public-service rationales such as improving services to citizens and avoiding locking users into a single vendor's products, technical goals of interoperability and seamless information exchange between agencies, and economic goals of lowering costs, promoting economic competition and innovation, and competing in global markets and exchanging information with global trading partners.

Brazil's definition of interoperability primarily addresses a standard's *effects*: enabling multiple, competing technologies, creating the ability to exchange information among heterogeneous ICT environments, providing users with a product choice, and preventing single-vendor lock-in. Principles of openness, choice, and heterogeneity thus underlie this interoperability definition.[30] The Brazilian framework also seems to embrace principles of transparency and participatory openness by stating that interoperability policy documentation should be open to public review with mechanisms for feedback. The underlying policy specifies a preference for the adoption of open standards within the federal government.[31]

Brazil's interoperability framework makes specific recommendations for standards that meet its requirements for openness and interoperability. Among the numerous pages of recommended standards are well-established Internet protocols such as HTTP/1.1, SMTP/MIME, and TCP and UDP, popular document formats such as RTF and PDF, and also newer and less entrenched open standards such as the Open Document Format (ODF).

Other government interoperability frameworks have similar elements. For example, the objectives of Malaysia's interoperability framework are a response to the need for different government systems (both hardware and software) to be able to exchange information more expeditiously and cost effectively using open standards "that are vendor and product neutral" than is possible using "proprietary alternatives."[32] Where the interoperability frameworks differ is in their definitions of open standards. The European Union's European Interoperability Framework for Pan-European eGovernment Services was established to promote pan-European electronic interoperability between technologies used by public administrators, citizens, and corporations. The definition of open standards in the European framework describes an open standard as meeting the following minimum requirements: It must be developed in an open decision-making process; the standard must be published and available either freely or at minimal cost; and the intellectual property (e.g. patents) "of (parts of) the standard is made irrevocably available on a royalty-free basis."[33] Many other open-standards policies do not require that standards-based intellectual property be made irrevocably available on a royalty-free basis, but may give preference to royalty-free standards where possible.

The emergence of government procurement policies based on open standards, as well as multistakeholder open-standards advocacy, are recent phenomena, so the results remain to be seen, but these nascent government policies, influenced by increasing multistakeholder activism in promoting technical openness, affirm the increasing recognition of technical standards as a critical area for A2K and reinforce the notion that standards have economic, legal, and political implications beyond the merely technical. These policies are also one potential area of intervention for developing countries seeking to ameliorate some of the pronounced challenges that restricted and proprietary standards present. Control over standards and the ability to use standards as knowledge-embedded tools is about politics, innovation, and creativity and ultimately about the ability to participate effectively and equitably in the knowledge economy. For the access to knowledge movement, open technical standards are a necessary precursor to achieving the goals of distributive justice, economic development, and democratic participation in the information society.

NOTES

1   See Bob Sutor's definition of a standard as a blueprint in *Open Standards v. Open Source: How to Think about Software, Standards, and Service Oriented Architecture at the Beginning of the 21st Century* (2006), available on-line at http://www.sutor.com/newsite/blog-open/?p=130 (last accessed December 27, 2009).

2   The IEEE 802.11 wireless LAN standards are collectively referred to as "Wi-Fi."

3   MPEG Audio Layer 3.

4   Hypertext Transfer Protocol.

5   Intellectual property scholar Mark Lemley generally defines a standard as "any set of technical specifications that either provides or is intended to provide a common design for a product or process." See Mark Lemley, "Intellectual Property Rights and Standard-Setting Organizations," *Boalt Working Papers in Public Law*, Paper 24 (2002): pp. 7–8, available on-line at http://www.escholarship.org/uc/item/776358p2 (last accessed December 27, 2009).

6   Economist Rishab Ghosh suggests that open standards can be defined so as to promote full competition and therefore innovation among vendors developing products based on these open specifications. See Rishab Ghosh, *An Economic Basis for Open Standards* (December 2005), available on-line at http://flosspols.org/deliverables/FLOSSPOLS-D04-openstandards-v6.pdf (last accessed December 27, 2009).

7   See also Laura DeNardis and Eric Tam, "Open Documents and Democracy: A Political Basis for Open Document Standards," *Yale Information Society Project White Paper*, available on-line at http://papers.ssrn.com/sol3/papers.cfm?abstract_id=1028073 (last accessed December 27, 2009).

8   Yochai Benkler, *The Wealth of Networks: How Social Production Transforms Markets and Freedom* (New Haven, CT: Yale University Press, 2006).

9   See Berkman Center for Internet and Society, *Open ePolicy Group's Roadmap for Open ICT Ecosystems* (September 2005), available on-line at http://cyber.law.harvard.edu/epolicy/roadmap.pdf (last accessed December 27, 2009).

10 *Ibid.*

11 Jack Balkin's opening remarks in the Plenary Session of the Access to Knowledge (A2K) Conference at Yale Law School, April 21, 2006, available on-line at http://balkin.blogspot.com/2006/04/what-is-access-to-knowledge.html (last accessed December 27, 2009).

12 Internet Protocol Version 4.

13 A 32-bit number is a number with a combination of 32 0s and 1s such as 01101001001010010110 00111110100. Jon Postel (ed.), *DOD Standard Internet Protocol*, RFC 760 (January 1980), available on-line at http://www.faqs.org/rfcs/rfc760.html, documents the original Internet Protocol specification. See also Jon Postel, *Internet Protocol, DARPA Internet Program Protocol Specification Prepared for the Defense Advanced Research Projects Agency*, RFC 791 (September, 1981), available on-line at http://www.ietf.org/rfc/rfc791.txt (both last accessed December 27, 2009).

14 Internet Protocol Version 6.

15 See Alan Davidson, John Morris, and Robert Courtney, *Strangers in a Strange Land: Public Interest Advocacy and Internet Standards*, Telecommunications Policy Research Conference, Alexandria, VA, 2002, available on-line at http://www.cdt.org/publications/piais.pdf (last accessed December 27, 2009).

16 See, for example, Rebecca Bolin and Eddan Katz, "Electronic Voting Machines and the Standards-Setting Process," *Journal of Internet Law* 8 (August 2004), available on-line at http://ssrn.com/abstract=945288. Also see Jason Kitcat, "Government and ICT Standards: An Electronic Voting Case Study," *Information, Communication, and Ethics in Society* 2, no. 1 (2004), available on-line at http://www.j-dom.org/files/Kitcat-evoting_case.pdf (last accessed December 27, 2009).

17 See, for example, Jack Balkin, "The Constitution of Status," *Yale Law Journal* 106, no. 8 (June 1997).

18 See, for example, para. 44 of the World Summit on the Information Society (WSIS), *Declaration of Principles. Building the Information Society: A Global Challenge in the New Millennium* (Geneva, December 12, 2003), Document WSIS-03/GENEVA/DOC/4-Em, available on-line at http://www.itu.int/wsis/docs/geneva/official/dop.html (last accessed June 4, 2010): "We are resolute to empower the poor, particularly those living in remote, rural and marginalized urban areas, to access information and to use ICTs as a tool to support their efforts to lift themselves out of poverty."

19 See Andrew Updegrove, "It's Time for IPR Equal Opportunity in International Standards Setting," *Consortium Standards Bulletin* 6, no. 7 (August–September 2007) available on-line at http://www.consortiuminfo.org/bulletins/aug07.php#editorial (last accessed December 27, 2009).

20 Andrew Updegrove, "Government Policy and 'Standards-Based Neocolonialism,'" *Consortium Standards Bulletin* 6, no. 7 (August–September 2007), available on-line at http://www.consortiuminfo.org/bulletins/aug07.php#feature (last accessed December 27, 2009).

21 See, for example, "Intellectual Property Right (IPR) Issues In Standardization. Communication from the People's Republic of China. Addendum," Background Paper for Chinese Submission to WTO on Intellectual Property Right Issues in Standardization (G/TBT/W/251/Add.1) (November 9, 2006), available on-line at chinawto.mofcom.gov.cn/accessory/200702/1171346578955.doc (last accessed March 6, 2010).

22 See for example, the International Telecommunication Union's definition of open standards, available on-line at http://www.itu.int/ITU-T/othergroups/ipr-adhoc/openstandards.html;

Ken Krechmer's "Open Standards Requirements" (2005), available on-line at http://www.
csrstds.com/openstds.pdf; Bruce Perens's definition in "Open Standards: Principles and Practice," available on-line at http://perens.com/OpenStandards/Definition.html, and Ghosh,
An Economic Basis for Open Standards (all last accessed December 27, 2009).

23 Gary Malkin, "The Tao of IETF: A Guide for New Attendees of the Internet Engineering Task
Force," RFC 1718 (November, 1994), available on-line at http://tools.ietf.org/html/rfc1718 (last
accessed March 22, 2010).

24 The W3C sets standards for the World Wide Web.

25 See the W3C's patent policy at http://www.w3.org/Consortium/Patent-Policy-20040205 (last
accessed December 27, 2009).

26 See http://www.itu.int/ITU-T/othergroups/ipr-adhoc/openstandards.html.

27 Second Phase of the WSIS (November 16–18, 2005, Tunis), Tunis Agenda for the Information Society WSIS-05/TUNIS/DOC/6 (rev. 1), available on-line at http://www.itu.int/wsis/
documents/doc_multi.asp?lang= en&id=2267|0 (last accessed December 27, 2009).

28 The mission statement is available at http://dcos-igf.org (last accessed December 27, 2009).

29 See United Nations Development Programme, "Government Interoperability Frameworks in
an Open Standards Environment: A Comparative Review," APDIP e-Note 20 (2007), available
on-line at http://www.apdip.net/apdipenote/20.pdf (last accessed December 27, 2009).

30 Other policies in Brazil's interoperability framework include compliance with dominant Internet standards, compliance with XML (eXtensible Markup Language), scalability, and compliance with internationally accepted standards.

31 Brazilian Government Executive Committee on Electronic Government, *e-PING Standards
of Interoperability for Electronic Government*, Reference Document Version 2.0.1 (December
5, 2006) at 9 (translated by the Brazilian government), available on-line at https://www.
governoeletronico.gov.br/acoes-e-projetos/anexos/E15_677e-PING_v2.0.1_05_12_06_eng-
lish.pdf (last accessed December 27, 2009).

32 Malaysian Government Interoperability Framework (MyGIF) Version 1.0 (August, 2003), p. 7,
available on-line at http://www.mampu.gov.my/mampu/bm/program/ICT/ISPlan/ispdoc/
Interoperability%20 Framework.pdf (last accessed December 27, 2009).

33 Interchange of Data between Administrations, "European Interoperability Framework
for Pan-European eGovernment Services," Version 4.2 (January 2004), available on-line at
http://ec.europa.eu/idabc/servlets/Doc?id=1674 (last accessed December 27, 2009).

Image produced by the Royal
National Institute of Blind People
and used worldwide to support
the Right to Read campaign.

Back to Balance:
Limitations and Exceptions to Copyright

Vera Franz

Imagine that you are a visually impaired person who cannot read a traditional book. You have identified a book you need to read for your business class and hence will need to spend three hours scanning each page of the book to produce an imperfect optical character recognition (OCR) file that can be read aloud by text-to-speech software. You will not only produce an error-filled copy of the book you seek to read, but will also have no legal way of sharing that OCR file with proofreaders or with other disabled people who want to read the same book.[1] To help solve this problem, in 2002, Benetech launched bookshare.org, which allows print-disabled readers to share scanned books and eliminate these barriers. Bookshare.org is now the world's largest on-line digital library for qualified print-disabled readers. However, there is one serious drawback: Only print-disabled readers based in the United States can currently access Bookshare.org's full catalogue of more than sixty thousand books. An exception in the U.S. copyright law makes this possible. As Jim Fruchterman, founder and CEO of Benetech points out, "Outside the U.S., we stop relying on copyright exemptions. Today's piecemeal national approach leads to a few countries with incompatible exemptions for the disabled without an effective approach to cross-border sharing, forcing us to go back to asking permission of the publisher or author to serve anybody outside our country."[2] So while new technologies make it possible to imagine a world where visually impaired persons have access to a broad variety of knowledge, the out-of-date legal environment is a serious barrier.

It is this, as well as many other problems of access to knowledge (A2K), that the A2K movement set out to solve when it first came together several years ago. In the following, I will focus on one of the major reform efforts currently pursued by the A2K movement: the rebalancing of our copyright regime through the reform of limitations and exceptions.[3] I will highlight some of the major problems with the current regime, then sketch the contours of the principal vision for reform—a proposed new international instrument dealing with minimum mandatory limitations and exceptions to copyright—and I will elaborate on how the A2K movement is working to turn this vision into reality. I will also shed some light on the diverging interests and heated debates characterizing this international reform effort.

LIMITATIONS AND EXCEPTIONS TO COPYRIGHT: THE FORGOTTEN "OTHER HALF" OF THE SOCIAL CONTRACT

As has been pointed out repeatedly, over the past two decades, and in particular since the World Trade Organization's Trade-Related Aspects of Intellectual Property Rights Agreement (the TRIPS Agreement) in 1994, we have witnessed a rapid global expansion and upward harmonization of rights holders' rights. Yet we have not seen a comparable development of the rights of users and in particular of limitations and exceptions to copyright. "As a result, the international copyright regime is less balanced than it has been at any one point in the past," concludes Gwen Hinze from the Electronic Frontier Foundation.[4]

This "explosion of intellectual property rights"[5] means, for example, that in many countries, the term of copyright protection has expanded from fifty to seventy years after the death of the author—and even longer in some countries. Protection is awarded to new subject matter such as computer programs and non-original databases. Rights holders can use technological protection measures to control access and use by users, often overriding the limitations and exceptions set out in the law. In addition, some countries—in particular, developing countries—have not implemented the full range of limitations and exceptions available to them under international law. As Consumers International has found in a study in 2006,[6] ten of eleven developing countries surveyed in Asia Pacific had not, for example, incorporated teaching exceptions to the extent allowed by the international copyright treaties. And none of the eleven developing countries had taken advantage of all the limitations and exceptions available to them under international law. This probably is at least in part due to the fact that developing countries, when drafting national copyright laws, were often not fully aware of the benefits of those exceptions to copyright. And while rights holders lobbied for strong copyright protections, nobody was there to lobby for copyright flexibilities.

As I think has quickly become clear, the current regime of limitations and exceptions is an uneven patchwork, at best, and it is far from providing a coherent global framework with the ability to match the increasingly globalized nature of our digital communications networks. The social, cultural, and political, but also economic costs are substantial. I began by mentioning the problems confronting blind and visually impaired persons. Libraries are another constituency facing

serious challenges as they strive to preserve and make available our cultural and scientific heritage. Appropriate exceptions for preservation, for example, are critical and become all the more important in an age when formats and technologies are constantly evolving, when the amount of content to preserve is growing exponentially, and when the original medium becomes increasingly fragile. For example, the British Library has a 1964 speech by Nelson Mandela, recorded on Dictabelt, for which the playback equipment is no longer available. Because in the UK the exception for preservation under current law does not extend to sound recordings, the British Library cannot make this speech available to the public.[7] Also, libraries cooperate in major cross-border digitization projects such as the i2010 European Commission Digital Libraries Initiative and Europeana, a multilingual on-line collection of millions of digitized items from European museums, libraries, archives, and multimedia collections. However, if preservation exceptions are different in each country, libraries may not be able to participate equally in digitization projects, placing the burden of digitization on those with more liberal regimes. As a library representative pointed out, "for financial and political reasons, this probably means that the projects might fail. As a result, the material will effectively disappear and the uniform access principle (UAP) cannot be upheld."[8]

Businesses are running into similar problems. Think of the practice of image search, a practice on which search engines heavily are relying. In the United States, it is covered by fair use. In the European Union, it legal in some countries, yet courts in other countries have declared that it constituted copyright infringement. A harmonized regime of copyright exceptions would put an end to these inconsistencies and encourage the development of these and other innovative services made accessible to global audiences.

Given these challenges, critiques as well as reform proposals have been advanced by many quarters over the last years. These include a proposal for "a balanced interpretation of the three-step test," a clause in the Berne Convention for the Protection of Literary and Artistic Works (1967) that imposes constraints on the possible copyright limitations and exceptions under national law.[9] Other experts are calling for a "reverse notice and take down" regime that would allow users to give copyright owners notice of their desire to make public-interest uses of protected works and require rights holders to "take down" technological protection measures on these works.[10] Yet the principal vision driving the A2K reform efforts at the moment is the idea of a new international instrument that would mandate a minimum set of limitations and exceptions to copyright globally.

PROPOSED NEW INTERNATIONAL LIMITATIONS AND EXCEPTIONS TO COPYRIGHT

The idea behind a new global regime for copyright limitations and exceptions is to match the minimum mandatory right of rights holders with a set of minimum mandatory rights of users and hence restore the balance originally enshrined in the social contract called copyright law. However, instituting a global minimum mandatory regime for copyright exceptions raises complex legal questions. For example, can current international law ever accommodate such a regime? A timely and helpful study by P. Bernt Hugenholtz and Ruth L. Okediji concludes:

> despite an unmistakable "ratcheting up" of levels of copyright protection at the international, regional and bilateral levels, enough "wiggle room" appears to be left to the parties to the main copyright Conventions to make framing an international instrument on L&Es [limitations and exceptions] within the confines of the international acquis [the total body of international copyright law accumulated thus far] a worthwhile exercise. Despite over a century of international norm setting in the field of copyright, limitations and exceptions have largely remained "unregulated space." This is not to say, of course, that the international acquis is inherently balanced. What it does mean is that there is ample scope for rebalancing without having to deviate from the current acquis.[11]

Hugenholtz and Okediji, who worked on this study with eight other leading copyright scholars, provide some further thoughts on the shape of such an instrument, suggesting that it could take the form of a special agreement listing in an exhaustive or enumerative manner those copyright limitations that are permitted within the confines of the three-step test.[12] They propose a special agreement with provisions addressing three main issues: exclusion from protection; limits to economic rights; and limitations and exceptions proper, including both mandatory and optional exceptions. Even though Hugenholtz and Okediji suggest including both mandatory and optional exceptions, they are very clear about the disadvantages of the optional approach, pointing to the European Union Information Society Directive of 2001: "The Directive's chapter on limitations and exceptions is...proof of the draw-back of an *optional* approach towards limitations and exceptions. Of the 27 Member States of the European Union, not a single one has seen fit to implement *all* the limitations and exceptions permitted under the Directive. In fact, the actual harmonizing effect of the Directive has remained quite limited." And they conclude that this situation is detrimental to the internal market that the European Union is seeking to strengthen.[13]

Hugenholtz and Okediji also put forward suggestions with regard to the actual content of the new international instrument, proposing that it could include the following clusters of limitations and exceptions: a cluster to address the needs of vulnerable members of society, such as the visually impaired; a cluster to facilitate the exercise of fundamental freedoms, such as free speech; a cluster of limitations and exceptions that would safeguard the role of institutions charged with the provision of public goods, such as educational institutions, libraries, and archives; and last, but not least, a cluster that would address limitations and exceptions necessary to promote innovation, such as reverse engineering.[14]

When developing this new international instrument, it seems clear that we need to be ready to think outside the box. For example, self-expression on the Internet often means "remixing." In other words, while the current copyright regime creates problems of access—in particular, when access is provided across borders—it also creates problems for creators who, on the Internet, build on the works of others and "remix." As Lawrence Lessig has pointed out, "It is now anybody with access to a $1,500 computer that can take sounds and images and use them to say things differently. These tools of creativity have become tools of speech. And it is the literacy for the new generation. It is how our kids speak. It is how our kids think. It is what our kids are."[15] How can we ensure that a dominant social practice driving today's creativity is not being criminalized and is instead declared legal and even encouraged? How can we restore the legitimacy of copyright law, in particular among the young? As the European Patent Office *Scenarios for the Future* points out, one of the biggest current challenges for our intellectual property regime is a serious crisis of legitimacy.[16]

A new copyright exception for "transformative use" is one of the ideas put forward in this context. In fact, in 2006, the *Gowers Review of Intellectual Property* introduced the idea of a new exception for transformative use, recommending that the UK government call for the "Directive 2001/29/EC to be amended to allow for an exception for creative, transformative or derivative works, within the parameters of the Berne three-step-test."[17] The idea of such a new exception may be useful to remedy the lack of flexibility of the regimes of copyright exceptions and limitations based on an exhaustive list of strictly limited exceptions, such as the European regime. And it would benefit on-line creativity, as well as the development of innovative on-line services. Some, for example, argue that a transformative-use exception should also cover the activity of automatically transforming or repurposing works with the sole purpose of allowing or improving user on-line accessibility to these works on any platform.

TURNING THE VISION INTO PRACTICE: THE TREATY FOR THE VISUALLY IMPAIRED AND BEYOND

Aware of the A2K barriers caused by weak and incompatible limitations and exceptions and armed with important insights and proposals provided by the academic community, the A2K movement set out to reform international copyright law. In what follows, I will outline how this movement worked to turn a vision into reality, highlighting both progress and challenges.

As detailed elsewhere in this book, in 2004, the A2K movement and developing-country governments initiated a major effort to reform WIPO.[18] This spirit of change blowing through the halls of WIPO has inspired the government of Chile to propose, in November 2004, that WIPO include the subject of limitations and exceptions to copyright on the agenda of one of its norm-setting committees, the Standing Committee on Copyright and Related Rights (SCCR).[19] A year later, Chile detailed its proposal suggesting three types of work that the SCCR could undertake in this area:

1. Identification . . . of national models and practices concerning exceptions and limitations.
2. Analysis of the exceptions and limitations needed to promote creation and innovation and the dissemination of developments stemming therefrom.
3. Establishment of agreement on exceptions and limitations for purposes of public interest that must be envisaged as a minimum in all national legislation for the benefit of the community; especially to give access to the most vulnerable . . . sectors.[20]

With these proposals, the government of Chile not only provided unique leadership on the issue at WIPO, but also brought about a paradigm shift at this UN agency. By putting limitations and exceptions on the agenda of a norm-setting committee and calling for action, it planted a seed that I hope eventually will put an end to the approach and even blind belief that WIPO had followed for so many years: that more intellectual property protection is always better. Chile dared to suggest that sometimes less can be more.

Yet for the following three years, the limitations and exceptions item did not much more than linger on the agenda of the SCCR. The SCCR was entirely focused on the so-called Broadcast Treaty, a treaty that would provide broadcasters, cablecasters, and Webcasters with exclusive, copyrightlike privileges, creating an additional layer of "broadcast rights" on top of existing copyright laws. Interestingly, the discussions on the Broadcast Treaty provided a unique opportunity for the A2K movement to build strategic alliances with industry. A2K groups worked with U.S. information technology businesses such as AT&T, Verizon Communications, Intel, Hewlett-Packard, and others to oppose the proposed terms of the Broadcast Treaty.[21] The view that this alliance between industry and public-interest organizations stopped the Broadcast Treaty is widely shared. The reason this alliance is possible is that, simply put, both parties share the vision of the Internet as a competitive and innovative communications space where bits can flow as freely as possible.

This type of coalition is important strategically: First, the A2K movement is aware that alliances with industry can help move other agenda items, including the one on limitations and exceptions to copyright. Second, these alliances reinforce the movement's credibility, neutralizing allegations often made by rights holders that it is "anticapitalist" or "Communist." And third, these coalitions demonstrate that industry is not a monolithic bloc. And policy makers sitting in Washington or Brussels will come to understand that they serve their countries' GDPs by promoting a more balanced approach to copyright.

In June 2007, negotiations on the Broadcast Treaty broke down, and limitations and exceptions to copyright, the next item on the agenda, finally moved up and into the spotlight. At the SCCR meeting in March 2008, the governments of Brazil, Chile, Nicaragua, and Uruguay presented a joint proposal that elaborated on the Chilean proposal from 2005, calling upon WIPO to adopt a work plan to implement the 2005 proposal. The reactions to the proposal ran the gamut from hot to cold, as one observer put it.[22] While many countries expressed general support, the United States, on behalf of the developed countries at WIPO (Group B), opposed both the analysis of limitations and exceptions needed to promote creativity and innovation (paragraph 2 of the proposal) and any WIPO norm-setting activity in the area (paragraph 3). This position seemed paradoxical to some observers, given the fact that domestically, the United States provides its users with relatively strong intellectual property flexibilities or user rights.

When seeking consensus on the SCCR meeting's conclusions, one of several dramatic moments unfolded: The United States took the floor to object to language that said that the SCCR as a whole "underlined the need for speedy action to improve access of visually impaired persons to protected works." It insisted that the report say only that some or several delegations supported this. As an observer noted, the effect of this objection was to deny an implied endorsement of the committee to take action to assist the blind.[23] As soon as the official meeting had ended, Chris Friend from the World Blind Union approached the U.S. delegation, evidently not happy with what had just happened. Directly confronted by a representative of the community of the blind, the United States could maintain its position only with difficulty. What transpired after an informal conversation was that the U.S. delegation declared that it would need more evidence for the need of an international instrument and that at this point it would neither support nor object to it.

In October 2008, the World Blind Union (WBU) increased pressure on governments to act by publishing an actual draft text for a Treaty for Blind, Visually Impaired and Other Reading Disabled Persons (TVI). The WBU estimates that around one hundred and twenty countries don't currently have limitations and exceptions for the blind in their national laws.[24] Also of particular concern is the fact that the current regimes of limitations and exceptions do not permit the import and export of accessible works. Accordingly, the two main features of the proposed treaty are "(1) to provide a minimum standard for limitations and exceptions for the blind and visually impaired, and (2) to allow and encourage the import and export of works in accessible formats." [25] The basic structure of the proposal is a two-tiered set of limitations and exceptions to the rights of copyright owners.

> Non-profit institutions would have the right to publish and distribute works in accessible formats if four conditions were met: (a) The person or organization wishing to undertake any activity under this provision has lawful access to that work or a copy of that work; (b) the work is converted to an accessible format, which may include any means needed to navigate information in the accessible format, but does not introduce changes other than those needed to make the work accessible to a visually impaired person; (c) copies of the work are supplied exclusively to be used by visually impaired persons; (d) and the activity is undertaken on a non-profit basis. The Treaty proposal also provides for more limited exceptions for commercial publishers.[26]

While many developing countries are supportive, the European Union remains opposed to the treaty at the time of writing. The United States, however, has softened its resistance. In fact, at the SCCR meeting in December 2009, the U.S. delegation delivered what some call a historic intervention,[27] abandoning its push for an unbalanced intellectual property regime: "We [the United States] recognize that some in the international copyright community believe that any international consensus on substantive limitations and exceptions to copyright law would weaken international copyright law. The United States does not share that point of view. The United States is committed to both better exceptions in copyright law and better enforcement of copyright law."[28] This statement gives hope that progress is possible. But as is often the case in international negotiations, all is in flux, and the outcome at this point is unclear.

One of the critical open questions is whether any new international "consensus" on strengthening limitations and exceptions will actually take the form of a *binding* instrument. The U.S. statement leaves this highly important point open. Yet any solution that promotes a *voluntary* approach will not solve one of the key problems that this reform effort set out to solve—the need to make limitations and exceptions compatible throughout the world. The European Union, with its menu of thirteen optional exceptions that European Union member states can choose to adopt—serves as an example that WIPO should *not* follow. As Okediji and Hugenholtz point out, the optional approach did little to harmonize the European regime and "has left Europe with a patchwork of incompatible limitations and exceptions, causing legal uncertainty to the detriment of commercial providers of cross-border services, such as online music stores, and of cultural institutions, such as libraries, archives, public broadcasters, offering content across Europe."[29]

Also, rights holders argue that voluntary licenses rather than statutory rights enshrined in law will help to rebalance the copyright regime. For example, publishers made a counterproposal to the Treaty for Blind, Visually Impaired and Other Reading Disabled Persons by proposing a so-called WIPO Stakeholder Platform.[30] The platform would help to solve the access crisis on a voluntary basis by transferring licensed material in accessible formats across jurisdictions. The platform also is supposed to develop more accessible publishing processes.[31] While this is a welcome initiative, it is important to understand that it is *not* a replacement for a treaty such as the Treaty for Blind, Visually Impaired and Other Reading Disabled Persons. As Knowledge Ecology International points out, nobody can honestly argue that all books will be provided by all rights holders to all reading-disabled people on a voluntary basis in the foreseeable future. For the many instances where the rights holder's files cannot be obtained, national and international law needs to provide for organizations of reading-disabled people to make and share accessible copies. This provision is not currently in place.[32] Voluntary licensing is also put forward by the European Commission in a recent Green Paper as an option for addressing the shortcomings of the current regime of limitations and exceptions to copyright.[33] Again, the problem with any voluntary licensing regime is that the terms of the license are determined by the copyright owner, and the copyright owner always has the option of not providing a license and hence denying access. A voluntary license is very different from a statutory right guaranteed to users by law. Also, in several cases, rights holders may be difficult to identify or to locate, and hence obtaining a voluntary license will be impossible.

Finally, can the vision of an international minimum mandatory regime be defended and moved forward as an omnibus solution? Or does one "cut the vision up" and move the different pieces—such as the Treaty for the Visually Impaired—forward when and if doing so is politically feasible? As every activist knows, the world looks different to the visionary than to the activist in the trenches. And a vision doesn't tell you much about the appropriate strategy and tactics for making this vision a reality. But knowing the A2K movement, I am confident it will play its cards right, keeping the vision alive while focusing on possible wins that will provide concrete solutions to real problems.

CONCLUSION

A regime of strengthened, harmonized limitations and exceptions to copyright is critically important for several reasons. It would make our intellectual property system fit for a global information society and economy, because it would foster cultural, educational, and economic activity across borders. It could also help to clarify and harmonize the interpretation of individual limitations and exceptions. Importantly for developing countries, it would alleviate the institutional weakness of states that need the diffusion of access to knowledge most. And from a strategic point of view, it would be able to rebalance our current copyright regime, in which the powers at the negotiating table are most unequal, as in bilateral trade negotiations. A minimum mandatory regime of limitations and exceptions cannot be negotiated away. Also, such a regime would be a badly needed response to the aggressive intellectual property enforcement agenda currently being negotiated, for example in the form of the Anti-Counterfeiting Trade Agreement. One could even argue that a strong regime of limitations and exceptions is a sine qua non for any new intellectual property enforcement regime and would in turn make any intellectual property enforcement agenda more acceptable.

Whether WIPO or any other policy forum will be able to deliver and actively rebalance our copyright regime remains to be seen. WIPO certainly is under pressure to deliver and it should seize the moment. After many years of failed negotiations, it could prove its renewed relevance in the space of international intellectual property norm setting by moving the copyright flexibilities agenda forward.

But the challenges are huge. Rights holders are fighting back, and the fact that some of them have deep pockets and will command armies of lobbyists to descend upon WIPO makes this a risky endeavor. The larger the vision, the more is at stake and the more one can lose. But not fighting this fight might be the biggest risk of all.

In fact, in this struggle over the future of our copyright regime much more than the rights of the blind, or the library user, or the remix artist is at stake. This is a fight about the future of our digital communications networks — the networks that we use to learn, speak, advocate, discover, create, and love. There is some truth to what James Boyle said back in 2005: "We probably would not create the World Wide Web, or any technology like it, today. In fact, we would be more likely to cripple it, or declare it illegal... What would a web designed by the World Intellectual Property Organisation or the Disney Corporation have looked liked? It would have looked more like pay-television."[34] The A2K movement is here to avert this future.

NOTES

1   Unpublished briefing paper by Benetech, July 2008. The nonprofit organization Benetech promises "to create new technology solutions that serve humanity and empower people to improve their lives." See http://www.benetech.org/about for its products (last accessed January 5, 2010).

2   Private e-mail conversation, October 2008.

3   Limitations and exceptions to copyright refer to "situations in which the exclusive rights granted to authors...do not apply. Some prefer to regard 'limitations and exceptions' as 'user rights'; that is, rather than cutting down or modifying some idealized form of copyright, user rights provide an essential balance to the rights of copyright." Wikipedia, s.v. "Limitations and Exceptions to Copyright," http://en.wikipedia.org/wiki/Limitations_and_exceptions_to_copyright (last accessed January 5, 2010).

4   Gwen Hinze, "Action Needed to Expand Exceptions and Limitations to Copyright Law," Third World Network Briefing Paper, no. 49 (July 2008), available on-line at http://www.twnside.org.sg/title2/briefing_papers/No49.pdf (last accessed January 5, 2010).

5   James Boyle, "Supersize My Rights," *Financial Times*, April 19, 2006, available on-line at http://us.ft.com/ftgateway/superpage.ft?news_id=ft004192006081920491 4 (last accessed January 5, 2009).

6   Consumers International, *Copyright and Access to Knowledge: Policy Recommendations on Flexibilities in Copyright Law* (Kuala Lumpur: Consumers International, 2006), pp. 29–31, available on-line at http://www.consumersinternational.org/Shared_ASP_Files/UploadedFiles/C50257F3-A4A3-4C41-86D9-74CABA4CBCB1_COPYRIGHTFinal16.02.06.pdf (last accessed

January 5, 2010).

7   See Kenneth Crews, *WIPO Study on Copyright Limitations and Exceptions for Libraries and Archives*, World Intellectual Property Organization doc. no. SCCR/17/2 (August 26, 2008), p. 52 (available on-line at http://www.wipo.int/edocs/mdocs/copyright/en/sccr_17/sccr_17_2.pdf (last accessed January 5, 2010).

8   Private e-mail conversation, October 2008.

9   The three-step test allows exceptions to copyright in special cases that result in no conflict with the normal exploitation of a work and don't unreasonably prejudice the legitimate interests of the rights holder. For the recently advanced proposal concerning the three-step test, see the Max Planck Institute for Intellectual Property, Competition, and Tax Law, "Declaration: A Balanced Interpretation of the 'Three-Step Test' In Copyright Law," available on-line at http://www.ip.mpg.de/shared/data/pdf/declaration_three_step_test_final_english.pdf (last accessed March 8, 2010).

10 *Ibid*. See also Jerome H. Reichman, Graeme B. Dinwoodie, and Pamela Samuelson, "A Reverse Notice and Takedown Regime to Enable Public Interest Uses of Technically Protected Copyrighted Works," *Duke Law Faculty Scholarship* Paper 1861, available on-line at http://papers.ssrn.com/sol3/papers.cfm?abstract_id=1007817 (both last accessed January 5, 2010).

11 P. Bernt Hugenholtz and Ruth L. Okediji, "Conceiving an International Instrument on Limitations and Exceptions to Copyright" (March 6, 2008), p. 26, available on-line at http://www.ivir.nl/publicaties/hugenholtz/finalreport2008.pdf (last accessed January 5, 2010).

12 With regard to the three-step test in particular, they conclude: "Read in a constructive and dynamic fashion, the three-step-test becomes a clause not merely limiting limitations, but empowering contracting States to enact them." *Ibid*., p. 25.

13 *Ibid*., p. 27.

14 *Ibid*., p. 43.

15 Lawrence Lessig, "Larry Lessig Says the Law is Strangling Creativity," transcript, TED: Ideas Worth Spreading, available on-line at http://dotsub.com/view/d3509948-261f-4fbb-9b7c-c63110f13451/viewTranscript/eng (last accessed January 5, 2010). See also Lawrence Lessig, *Remix: Making Art and Commerce Thrive in the Hybrid Economy* (New York: Penguin, 2008).

16 European Patent Office, *Scenarios for the Future: How Might IP Regimes Evolve by 2025? What Global Legitimacy Might Such Regimes Have?*, available on-line at http://documents.epo.org/projects/babylon/eponet.nsf/0/63a726d28b589b5bc12572db00597683/$file/epo_scenarios_bookmarked.pdf (last accessed January 5, 2010).

17 *Gowers Review of Intellectual Property* (December 2006), available on-line at http://www.hm-treasury.gov.uk/media/6/e/pbr06_gowers_report_755.pdf (last accessed January 5, 2010).

18 See, especially, Ahmed Abdel Latif's essay, "The Emergence of the A2K Movement: Reminiscences and Reflections of a Developing-Country Delegate."

19 See World Intellectual Property Organization, "Proposal by Chile on the Subject 'Exceptions And Limitations To Copyright And Related Rights,'" WIPO doc. no. SCCR/12/3 (November 2, 2004), available on-line at www.wipo.int/edocs/mdocs/copyright/en/sccr_12/sccr_12_3.doc. See also Sam Ricketson's seminal paper, "Limitations and Exceptions of Copyright and Related Rights in the Digital Environment" (2003), available on-line at http://www.wipo.int/edocs/mdocs/copyright/en/sccr_9/sccr_9_7.pdf (both last accessed January 5, 2010).

20 World Intellectual Property Organization, "Proposal by Chile on the Analysis of Exceptions and Limitations," WIPO doc. no. SCCR/13/5, available on-line at http://www.wipo.int/edocs/mdocs/copyright/en/sccr_13/sccr_13_5.pdf (last accessed January 5, 2010).

21 See CPTech et al., "Statement by NGOs Concerned with the Protection of Broadcasts and Broadcasting Organizations," available on-line at http://www.eff.org/files/filenode/broadcasting_treaty/20040608_Draft_Joint_Position_v1.3.pdf, and "Statement Concerning the WIPO Broadcast Treaty Provided by Certain Information Technology, Consumer Electronics and Telecommunications Industry Representatives, Public Interest Organizations, and Performers' Representatives," available on-line at http://www.eff.org/files/filenode/broadcasting_treaty/Stmt-for-USPTO-forum-final-20060905.pdf (both last accessed January 5, 2010).

22 Thiru Balasubramaniam, "Of Limitations, Exceptions and Verse (WIPO Copyright Committee)" (April 2008), available on-line at http://www.keionline.org/node/91 (last accessed January 5, 2010).

23 See James Love, "[A2K] at the SCCR, It's Over for Now, the Actual Text of the Decision Won't Be Available Right Away," available on-line at http://lists.essential.org/pipermail/a2k/2008-March/003058.html (last accessed January 5, 2010).

24 Briefing note, "WBU Launches Global Right to Read Campaign," available on-line at http://www.freelists.org/post/bcab/WBU-launches-Global-Right-to-Read-Campaign (last accessed January 5, 2010).

25 KEI Briefing Note, "World Blind Union Proposal for a WIPO Treaty for Blind, Visually Impaired and Other Reading Disabled Persons," October 14, 2008, available on-line at http://www.keionline.org/misc-docs/tvi/tvi_memo_en_pdf.pdf (last accessed, January 5, 2010).

26 Ibid.

27 "Bravo the United States," Moral Panics and the Copyright Wars: A Blog about Copyright Discourse, http://moralpanicsandthecopyrightwars.blogspot.com/2009/12/bravo-united-states.html (last accessed March 8, 2010).

28 World Intellectual Property Organization Standing Committee on Copyright and Related Rights, "Statement on Copyright Exceptions and Limitations for Persons with Print Disabilities" (December 2009), available on-line at http://keionline.org/sites/default/files/SCCR%2019%20US%20Intrvntn%20PD.pdf (last accessed March 8, 2010).

29 Hugenholtz and Okediji, "Conceiving an International Instrument on Limitations and Exceptions to Copyright," p. 27.

30 World Intellectual Property Association Standing Committee on Copyright and Related Rights, "Stakeholders' Platform: Interim Report," WIPO doc. no. SSCR/18/4 (May 11, 2009), available on-line at http://www.wipo.int/edocs/mdocs/copyright/en/sccr_18/sccr_18_4.pdf (last accessed March 10, 2010).

31 Knowledge Ecology International, "Six Myths about the Treaty for People with Disabilities that Should Be Debunked Next Week?" available on-line at http://keionline.org/node/795 (last accessed March 8, 2010).

32 Ibid.

33 World Intellectual Property Association Standing Committee on Copyright and Related Rights, "Stakeholders' Platform: Interim Report."

34 James Boyle, "Web's Never-to-Be-Repeated Revolution," Financial Times, November 2, 2005, available on-line at http://www.ft.com/cms/s/2/f3fe9c4a-4bd1-11da-997b-0000779e2340.html (last accessed January 5, 2010).

This image was created by James Love's Knowledge Ecology International with Wordle. It was used on a T-shirt during a meeting called "the Paris Accord" that took place in Paris in October 2009 and was organized by the Trans Atlantic Consumer Dialogue (TACD).

New Medicines and Vaccines: Access, Incentives to Investment, and Freedom to Innovate

Spring Gombe and James Love

Since 2002, a series of proposals to make radical changes in the approaches to financing the development of new medicines and vaccines have been advanced in various arenas by various actors, including NGOs and activists, country representatives, and academics, all in some ways participating in the A2K dynamic. The proposals feature deep reforms of the system of incentives to reward successful drug development, expand access to knowledge, and allow greater freedom to undertake follow-on research.

As with other advocacy movements, the advancement of proposals proceeded concurrently on two tracks and in several arenas, which we will here abbreviate as the political and technical tracks, though needless to say, these were necessarily tightly interwoven. On the political track, policy discussions have evolved from "We have an intractable crisis in health R&D for poor people in poor countries" to "Actually, there is a way out—how can we implement radically new solutions?" On the technical track, scenarios and models for these radical reforms have been introduced, teased out, and developed into specific proposals. In particular, these proposals have involved incentives in the form of prizes for research and development in the area of medicines needed in and by the developing world. This approach promises to achieve the goal of breaking the link in the pharmaceutical industry between incentives to research and development and product prices, one of the principle barriers to access to medicines worldwide. In what follows, we lay out the timelines along which the movement for access to medicines has made progress along these two tracks and then discuss a range of alternative options for delinking R&D and drug prices, along with the arguments in favor of these options as efficient strategies for the A2K movement to promote.

Getting access to medicines on the agenda at all was the principal task under-taken by those who worked along the political track. Strategies and ideas to pro-mote alternatives to the current system to finance the research and development of medical innovations began to take shape in 2003, following several years of intense, behind-the-scenes discussions by experts brought together by the global NGO health-policy community keen to act to resolve the crisis in access to medi-cines. These discussions triggered a partial relocation of the debate from the World Trade Organization (WTO) to the World Heath Organization (WHO).[1] Until 2002, there had been both political and technical resistance to discussion of intellec-tual property issues related to health at the WHO. A curious feature of the debate up to that point, and one that would continue to accompany the discussions for many years to come, was that it was somehow perfectly legitimate for policy mak-ers in the areas of trade and intellectual property to bring health-related issues to the center of their work, but totally illegitimate for the WHO, which holds the global mandate for public health, to discuss the effects of intellectual property on health worldwide.

A sea change occurred in 2003. In May that year, the Fifty-Sixth World Health Assembly passed Resolution 56.27 establishing the Commission on Intellectual Property Rights, Innovation, and Public Health (CIPIH) to examine the links between those three issues and particularly to assess the effect of intellectual property regimes on access to health care in poor countries and to make recom-mendations for action. The establishment of the CIPIH was in turn the result of a previous commission, the Commission on Intellectual Property Rights (CIPR), which had been established by the then UK Secretary for Development Clare Short to examine the links between the enforcement of intellectual property rights and access to health care in developing countries. The report issued by this commission was seminal and highly influential in bringing the debate on intellectual property in issues affecting health into the global health-policy arena.

A key issue that the CIPR brought to the attention of the global health-policy community, and one that access-to-knowledge advocates had been attempting to raise without much resonance until then, was that not only does the manner of intellectual property rights enforcement hamper access to medicines and in so doing undermine the right to health, but by placing severe restraints on the cir-culation of knowledge, it severely constrains the capacity of research scientists to advance the development of badly needed remedies for all diseases. This is both vexing in itself and particularly aggravating for diseases affecting the poor. The CIPR report played an important role in bringing together the movements for access to medicines and access to knowledge. It importantly also played a role in mobilizing the research-science community to act as advocates in the area of global health policy.

In February 2005, a group of 162 scientists, economists, doctors, public-health experts, members of Parliament, and other experts sent a letter to the WHO Executive Board and to the CIPIH requesting an evaluation of proposals for a medical research-and-development treaty. In their letter, the experts laid out the arguments for a paradigm change:

> The current global framework for supporting medical R&D suffers from profound flaws. A growing web of multilateral, regional, bilateral and unilateral trade agreements and policies focus nearly exclusively on measures that expand the scope and power of intellectual property rights, or reduce the effectiveness of price negotiations or controls.
>
> These mechanisms are plainly designed to increase drug prices, as the sole mechanism to increase investments in R&D. Stronger intellectual property rights and high drug prices do create incentives to invest in medical innovation, but also impose costs, including:
> 1. problems of rationing and access to medicine,
> 2. costly, misleading and excessive marketing of products,
> 3. barriers to follow-on research,
> 4. skewing of investment toward products that offer little or no therapeutic advance over existing treatments, and
> 5. scant investment in treatments for the poor, basic research or public goods.

A trade framework that only relies upon high prices to bolster medical R&D investments anticipates and accepts the rationing of new medical innovations, does nothing to address the global need for public sector R&D investments, is ineffective at driving investments into important priority research projects, and when taken to extremes, is subject to a number of well-known anticompetitive practices and abuses. Policy makers need a new framework that has the flexibility to promote both innovation and access, and which is consistent with efforts to protect consumers and control costs.

To this end, a number of experts and stakeholders have proposed a new global treaty to support medical R&D. This effort has produced a working draft (http://www.cptech.org/workingdrafts/rndtreaty4.pdf) that illustrates a particular approach for such a treaty—one that seeks to provide the flexibility to reconcile different policy objectives, including the promotion of both innovation and access, consistent with human rights and the promotion of science in the public interest. The draft treaty provides new obligations and economic incentives to invest in priority research projects, and addresses several other important topics.[2]

In November 2005, the Republic of Kenya submitted a draft resolution to the One Hundred and Seventeenth Session of the WHO Executive Board for the establishment of a working group to examine global health priorities for medical research and development. Then, in January, 2006, at the WHO Executive Board, Kenya submitted a background document for its proposed resolution, and 285 scientists from fifty-one countries, including five Nobel Prize winners, sent a letter to the WHO Executive Board supporting the Kenyan resolution. In their letter, the scientists said:

> Although we have very varied scientific backgrounds, from basic research to specific clinical research, we are all deeply concerned with deficiencies in the way that biomedical research science is supported and translated into treatments that improve health outcomes around the world. . . . We have all felt the impact and promise of the free availability of genome sequence data, notably from the human genome project. At the same time we see research activities increasingly complicated by legal restrictions, such as intellectual property rights, which can interfere with free data exchange and can limit biomedical research progress. We do not see a good balance between medical need and resource allocation in the existing system to support R&D.[3]

On January 24, Brazil joined Kenya in submitting a revised draft resolution proposing a WHO initiative to create a bold new global framework for essential health R&D. In April, the CIPIH issued its report, which, despite having to be a compromise document between stakeholders with very different interests, reached important conclusions about the adverse direct effect of the current intellectual property regime on both research and development of and access to medicines for poor countries.

In May 2006, the Fifty-Ninth World Health Assembly passed Kenya/Brazil Resolution 59.24, calling for the establishment of an Intergovernmental Working Group on Public Health, Innovation and Intellectual Property (IGWG) to draw up a global strategy and plan of action based on the recommendations of the CIPIH, including researching a new framework to support sustainable, needs-driven, essential R&D work on diseases that disproportionately affect developing countries.

In 2007, at the Sixtieth World Health Assembly, Brazil introduced a resolution on public health, innovation, and intellectual property that was passed as Resolution WHA 60.30. This document formed the basis for exploration of the proposals being introduced for alternative approaches for financing medical R&D, giving clear instructions for the development of a plan of action and a strategy for action. The second meeting of the IGWG was held later that year in Geneva, which produced drafts of these, but which did not fully respond to its mandate for providing, for example, cost estimates for the strategy and plan of action. The political process allowed for public commentary on the negotiations, to which there was a good response, though there was some dissatisfaction about the scope and process of the hearings.

In 2008, the discussions on the plan of action presented in the previous year continued, with dissatisfaction being expressed by many countries about the technical documents presented by the WHO Secretariat. At this juncture, in March, it looked as if talks could fail, but a series of white papers outlining both process and content were eventually agreed to, and these formed the basis for final negotiations in late April and early May. On May 24, 2008, a global strategy and plan of action was approved by the Sixty-First World Health Assembly in Resolution WHA 61.21.

THE TECHNICAL TIMELINE

Concurrent with developments along this political track and intrinsically related to them, the developments along a more technical track progressed, as well. Experts in health, pharmaceuticals, and economics began to discuss proposals for radical reform of the financing of medical R&D. (Many of the actors on this track were of course also actors in the political process.) These proposals increasingly focused on the use of prizes as incentives to undertake the research and development needed to produce medicines necessary to improve the prospects of health care for populations in the developing countries.

In 1999, Michael Kremer and others proposed creating large rewards for investments in vaccines for malaria and other tropical diseases.[4] In 2001, the pharmaceutical company Eli Lilly created the firm InnoCentive to administer a series of commercially sponsored prizes to solve problems in the area of life sciences. Later, a number of philanthropic organizations sponsored prizes for medical innovations, including, but not limited to the X-Prize Foundation, the Prize4Life Foundation, and the Gotham Prize.

In 2002, the pharmaceutical company Aventis held discussions on possible future pharmaceutical scenarios, including a scenario proposed by Tim Hubbard and James Love that featured prizes and the elimination of monopolies on all new medicines. This scenario was presented at a number of academic and policy workshops in 2003 and 2004. Separately, in August 2003, the economist Burton Weisbrod published an editorial in the *Washington Post*, "Solving the Drug Dilemma," that called for "two prices—one for the R&D, another for the resulting pills." Weisbrod noted "this solution is not painless, but neither is the course that public policy is now on."[5]

In 2005, U.S. Representative Bernard Sanders introduced to the U.S. Congress the first Medical Innovation Prize Fund proposal (HR 417, One Hundred and Ninth Congress), setting out a particular implementation of the new approach, stimulating additional interest among academics. In 2006, Joseph Stiglitz began publishing a number of widely read articles calling for the use of prizes to reward drug development.

In May 2006, the World Health Assembly passed resolution WHA60.30, which, among other things and in the context of addressing the unmet health needs of developing countries, called upon the director general of the WHO to "encourage the development of . . . incentive mechanisms . . . addressing the linkage of the cost of research and development and the price of medicines, vaccines, diagnostic kits and other health-care products."[6]

In 2007, Sanders, now a senator, reintroduced the Medical Innovation Prize Fund as S.2210. Also in 2007, U.S. senator and presidential candidate John Edwards called for prizes, rather than monopolies to stimulate drug development.[7]

Other notable contributions to the technical track from 2004 to 2008 were papers and articles by academics such as Joe DiMasi and Henry G. Grabowski, Aidan Hollis, Thomas Pogge, Marylynn Wei, Kevin Outterton, Carl Nathan, and Stan Finkelstein and Peter Temin, NGO experts and consultants such as Dean Baker and Ron Marchant, and journalists such as Scott Wolley.[8]

In April 2008, in a WHO negotiation over new approaches to stimulating medical R&D, the governments of Barbados and Bolivia made five separate proposals to use prizes for medical innovation.[9] In May 2008, the World Health Assembly adopted a Global Strategy and Plan of Action, which, among other items, agreed to "explore and, where appropriate, promote a range of incentive schemes for research and development including addressing, where appropriate, the de-linkage of the costs of research and development and the price of health products, for example through the award of prizes, with the objective of addressing diseases which disproportionately affect developing countries."[10]

THE STRATEGY

Advocates for the reform of essential R&D by offering incentives in the form of prizes and awards were aware that they would have to persuade a number of different constituencies of both the basis for and the viability of such an alternative approach to financing R&D. It was important in order to advance new models, that the intellectual property policy community addressing global health issues (at the time, largely centered on the WTO), the global health-policy community centered on the WHO, but also the policy and political community of regional and national governments be persuaded that on balance, the new proposals were advantageous for the constituencies they represented. It was also important to ensure that acknowledged experts in these policy areas were seen to be part of the movement. The shared conviction of research scientists, health economists, intellectual property specialists, and public-health specialists that the need for change was urgent proved to be instrumental in advancing the technical and political movement for access to knowledge and medical technologies.

The central strategy employed by the movement was to present a persuasive case to policy makers and other stakeholders in the research and development of medical technologies that it is not only feasible, but also imperative that new mechanisms be invoked that separate the market for innovation from the market for innovative products. Key to the success of the approach would be to help policy makers understand that this would allow generic competition for all products as soon as they enter the market, driving prices down, and also would mean that the developers of new medicines and vaccines would be rewarded directly by prizes. The prizes would be linked to the effect of innovations on health-care outcomes, regardless of who actually delivered the product to patients. The new approach would be beneficial for both the producers of advancements in science and the recipients of these advances, protecting the rights of both as enshrined in human rights conventions.

Ample justifications existed for the reform of the current system. Globally, access to new medicines was and continues to be constrained by high prices that are a consequence of patent-enforced monopolies. Products that cost pennies to manufacture are subject to astronomical mark-ups, leading to hardships and barriers to access.

High prices are not the only shortcoming of the current system, either. The prices and rewards for new drugs are loosely based upon the prices charged for comparable medicines, leading to wasteful investment in products that are similar to each other in terms of mechanisms and therapeutic results, offering few, if any advantages over products already on the market. Inventions that address the needs of low-income and uninsured persons receive scant attention by investors. Even for persons who can best afford new medicines, the current system seems to offer low and decreasing productivity, because sharp increases in investments in R&D have not yielded much in terms of useful innovations.

Early specifications of proposals were developed in a scenario-planning exercise with the pharmaceutical company Aventis and were later refined by contributions from a number of persons. In terms of the degree of change from the current system, the proposals can be considered in phases, each building upon each other and departing further from the status quo.

First, separate drug development incentives from the prices of the products by replacing the exclusive rights to make or sell products with large cash prizes that are linked to the effect of the product on health-care outcomes. The delinking of incentives from product prices makes it possible to eliminate most price-based formularies and dramatically expand access to new medicines and vaccines. The new approach also creates an opportunity to focus R&D investments on the most important medical needs and eliminate wasteful expenditures on marketing activities.

Second, encourage the open sharing of knowledge, materials, and technologies. Prize rewards for development are shared through a set of "open-source dividends" that are awarded to entities that make it possible for others to develop products.

Next, create a new set of competitive financial intermediaries that fund basic science and translational or early phases of development, which in turn generate rewards on the basis of their measurable and objective contributions to products that actually succeed.

Finally, remove the exclusive right to use inventions in favor of a system that gives the freedom to use inventions when patent owners receive remuneration, creating more competition for follow-on innovations.

These proposals were not designed as or considered to be a replacement of the important and significant role of governments and donors in direct funding of research through grants. The access-to-medicines movement remains aware that traditional grant programs are necessary to promote and sustain research programs and will always play an important role in the advancement of science and the development of new products. The new mechanisms were designed as complements to the traditional "push" mechanisms and as a much-needed alternative to the "pull" incentives now implemented as legal monopolies to make, use, or sell products.

IMPLEMENTING PRIZE FUNDS — FOUR OPTIONS

Over time, several options were presented for the implementation of prize funds in both national and global permutations. The first legislative proposals were introduced in the United States as HR417 (One Hundred and Ninth Congress), the Medical Innovation Prize Fund. Building on this proposal and modifying it for global application, proposals were made at the WHO by Bolivia and Barbados. The common underlying principle is that the incentives now associated with expected monopoly profits for products would be replaced with large cash rewards for successful development efforts. It is thus that there would be a delinking of R&D incentives from product prices.

The model proposals advanced four options elaborated from the academic, policy, and legislative discussions around prizes. Each carries distinct opportunities and challenges. The proposals that have been made to date fall into four broad groups, presented as options for implementation.

OPTION 1 Replace the exclusive rights to make or sell a product, following drug regulatory approval, with large cash prizes that are linked to the effect of the product on health-care outcomes. This option tackles the most important reform in the system of incentives for drug development: the elimination of the set of exclusive rights that are now offered to induce the development of new drugs. These rights are associated most importantly with patents on pharmaceutical inventions, but also include a growing set of nonpatent mechanisms to bar competition, including, for example, market exclusivity associated with pediatric drug testing as a reward for the development of "orphan" drugs and biologics and to prevent unauthorized competitors from relying upon clinical trial data from clinical trials to register new products.[11] Taken together, these measures are explicitly designed to grant legal monopolies on new medicines with the intention that the monopoly profits will stimulate useful R&D. The shortcomings of such a system are many, including not only the hardships associated with high prices (a barrier to access and a burden for consumers, employers, and society at large), but also inefficient investment in the development of medically unimportant products, as well as wasteful and often harmful marketing activities.

This option constitutes a key change in the business model for rewarding developers of new medicines. By delinking R&D incentives from product prices and allowing for competition and low generic prices for the products themselves, utilization of newer products would no longer be discouraged simply because of the high prices now associated with patented inventions.

The two U.S. Medical Innovation Prize Fund (MIPF) bills (HR 417, One Hundred and Ninth Congress and S. 2210, One Hundred and Tenth Congress) presented the first fully specified possible implementation of prizes in the context of a major market for medicines. The basic approach followed one of the scenarios outlined in the 2002 Aventis exercise. The drafters of the MIPF sought to address several important technical questions relating to the allocation of prizes:

1. The prize fund will have an annual contribution based upon a fraction of GDP.[12]
2. All of the annual funding is to be spent every year on qualifying products and processes.
3. The prize fund of a fixed size is to be divided among qualifying products in a zero-sum competition. The more that is given to one product, the less is available to competitors.
4. Every new product "wins" something, but products that have a greater effect on health outcomes receive more.

5. New products (and processes) participate in the fund for ten years. The reward in any given year is independent of future rewards and is based upon the best available evidence (at that time) of the effect of the product on health outcomes.

6. The effect on health outcomes is benchmarked against available technologies, rather than against placebos.

7. Products that are registered with the FDA at roughly the same time are compared with products that were not recently developed.

Three other features of this implementation deserve discussion. First, the MIPF provided that the amount of the prize money that any one product can receive in a given year is limited to 5 percent of the annual prize fund payments.[13] Second, in cases where there is a follow-on product, that MIPF would continue to make payments to the original product, even when its market share falls to zero, to "the degree that the new... product, or manufacturing process was based on or benefited from the development of the existing... product, or manufacturing process." Third, the MIPF provides specific set-asides for areas of public health priority, including current and emerging global infectious diseases, severe illnesses with small client populations, and neglected diseases that primarily afflict the poor in developing countries.

While the MIPF eliminates market exclusivity for products, it does not eliminate patents. Patents will continue to be available for new medicines and will continue to be valuable assets, not as legal monopolies on products, but as ways to make claims on the prize fund money. Strictly speaking, the Sanders bills did not change the system of exclusive rights for drug, biological, or vaccine development until after a product received FDA approval. Drug developers would have to litigate or negotiate with patent owners to obtain the necessary rights to register products, as they do today. In practice, however, the elimination of the postmarketing approval of exclusive rights would create a very new dynamic for patent owners. The total reward for drug development would be fixed by the size of the prize fund. A patent system that created too many barriers to product development then would be easier to reform, because changes would not change the overall system of sustainable rewards.

OPTION 2 Create open-source dividends to reward the sharing of knowledge, data, and technology. One explanation of the low productivity in drug development is that scientists and firms involved in medical research and development are not sufficiently open in terms of access to knowledge and that restrictive licensing practices discourage research in areas where patents exist. New development models in the field of software and innovative information services have enhanced interest in approaches that promote more access to knowledge and greater sharing and freedom to use and improve upon innovations pioneered by others. If prizes were used to reward innovations, it would be possible to expand access and redesign R&D incentives to stimulate more efficiently investments that improve health outcomes.

The issue of secrecy, openness, and prizes was discussed at length in a 2008 workshop on medical-innovation prizes at the United Nations University at Maastricht, the Netherlands, and in a Médecins Sans Frontières (MSF) workshop examining the possible use of prizes to stimulate the development in a rapid, low-cost point-of-delivery test for tuberculosis.[14] In the spring of 2008, the governments of Barbados and Bolivia submitted to the WHO five proposals to use prizes to stimulate medical R&D. Several of these prizes formally introduced the notion of "open-source" dividends to encourage greater openness. Among the specific proposals were the notion of sharing the final-product prizes with individuals, firms, and communities that share knowledge, materials, and technologies in a nondiscriminatory and royalty-free manner. There was even a proposal to share the open-source dividends with journals that would publish research in full text without subscription fees, creating a new incentive for journals to share research findings more openly. The Barbados/Bolivia prize proposals also created systems of rewards for interim research results that were available only to entities that would offer royalty-free open licenses for inventions, data, materials, and know-how. They would reserve 10 percent of prize money for scientists and engineers who are unaffiliated with and uncompensated by the winning entrant, but who have published and shared research, data materials, and technology, an award to be determined on the basis of who provided the most useful external contributions to achieving the end result. This would include research, data, materials, and technology that are either placed in the public domain or subject to open, nonremunerated licenses. In addition, biannual "best contributions" prizes would be available only to technologies that were placed in the public domain or licensed to a patent pool for open, royalty-free use.

To qualify for the "best contributions" prize, published research findings would have to be freely available on the Internet in full text. As an incentive to journals to make articles available to the public for free, 10 percent of the "best contributions" prize given for a published article would be available to a peer-reviewed journal that published the article, on the condition that the journal make the article available for free immediately upon publication.

OPTION 3 Add prizes for interim results, such as reaching interim benchmarks and solving discrete technical problems and for research undertaken by competitive intermediaries. Prizes that reward successful outcomes could be implemented as an alternative to a set of exclusive rights. Unfortunately, it is much more difficult to evaluate the value of interim benchmarks than it is final products that have concrete and observable utility in terms of influencing health outcomes. In any centrally managed prize program for medical research, the criteria to reward interim outcomes would be controversial and difficult to evaluate and would suffer from many of the limitations that now exist with systems of grants.

Rather than search for consensus on one or more ingenious systems for valuing interim results, policy makers could create an environment where decentralized institutions make such valuations. In this approach, there would be no need to legitimize the valuation criteria. Instead, it would be necessary only to legitimize the actors that make such valuations. Multiple intermediaries would be resourced to award prizes for interim results, using their own methods. The legitimacy of the intermediaries would be based upon the competition to obtain funds.

The notion of competitive intermediaries was recently proposed by Bolivia and Barbados in connection with a proposal for priority medicines and vaccines through which 20 percent of the proposed Priority Medicines and Vaccines Prize Fund money would be allocated to three or more institutions that run prize competitions to reward earlier stages of product development. These would include smaller technical challenges and also rewards for the successful development of early benchmarks in drug development, such as the completion of phase I/II clinical trials.

OPTION 4 Create a system of compensatory liability to reduce the problem of patent thickets in upstream research, a system of liability rules, rather than of exclusive rights. Patent rights are often implemented as a set of exclusive rights, subject to some limited permitted uses, and the understanding that abuses of exclusive rights are sanctionable by governments or courts.[15] But patents can also be implemented so that everyone has the freedom to use the invention, subject only to an obligation to pay remuneration.

Compulsory licenses or a system of prizes such as the one envisioned by S.2210 are types of liability rules. There are many possibilities for liability rules, including cases where exclusive rights have some role, but where the threshold for obtaining compulsory licenses is low and one can realistically anticipate obtaining nonvoluntary authorizations in the event that voluntary negotiations fail. The range of possibilities is large along a continuum that begins with automatic rights to use inventions and that ends with no rights to use inventions outside of voluntary authorizations from patent owners.

Patent owners often express concern that compulsory licensing would lead to a bias in favor of low rates of compensation and insufficient rewards for innovation. They also question the assumption that patents inhibit innovation, since in a voluntary context, patent owners always have the option of licensing welfare-enhancing innovations, rather than blocking them. In practice, however, both patent owners and users have incomplete and asymmetrical information about the value of patents in a particular use, and each has incentives to act strategically, factors that contribute in practice to an underutilization of patented inventions.

The disputes about compulsory licensing or other types of liability rules are particularly heated in the context of pharmaceutical inventions, which are often very inexpensive to copy. In a world where R&D incentives are linked to drug prices, any relaxation of exclusive rights could lead to competition that reduces monopoly rents and undermines incentives to invest in R&D. For this reason, patent offices and courts have often allowed patent doctrines to stray far from their putative purpose of rewarding invention. The expanding subject matter for patents and the low standards for what is known as "inventive step," a criterion for identifying a genuine innovation, could be thought of indirect efforts to use patents to protect investment.

The importance of patents for protecting investments is greater for pharmaceutical inventions than it is for many other technology fields. One observes very different perspectives on the patent system in other industry sectors. For example, in the software, telecommunications, and computing sector, a growing number of firms favor the abolition or weakening of software patents, and patent reform has been framed as a call for higher standards for patentability and for a relaxation of exclusive rights in favor of incremental steps toward liability rules.

With the introduction of a system of prizes to reward drug development, breaking the link between R&D incentives and product prices, the pharmaceutical industry's interest in patent reform would change dramatically. The relevant factor in determining the overall R&D industry revenues will no longer be the potential revenues from drug sales, which are influenced in part by the ability to exercise exclusive rights over products, but rather the size of the prize fund. A patent system that requires costly litigation, creates long bargaining delays, and blocks innovation will be seen as a negative, as will a system that rewards inventions excessively at the expense of investments.

CONCLUSION: PRIZE FUNDS AND THE INTELLECTUAL PROPERTY REGIME

Proposals for prize systems, in their most ambitious form, entail the modification of current intellectual property rules. This includes an elimination of all legal barriers to the competitive supply of the products. Patents would be used not as monopolies, but as a mechanism to stake claims on the prize money. Prizes would also reward unpatented innovations and investments.

What the movement has advanced is the premise that the use of cash prizes to eliminate legal monopolies for products would provide a powerful opportunity to address several flaws that plague the current system. In particular, policy makers would have far more freedom to design incentives efficiently. For example, rewards could be directly linked to improvements in health outcomes when benchmarked against existing treatments, rather than being linked to the replication of benefits already available from existing products, as they often now are. This would have the benefit of driving investment toward treatments that address unmet needs. The elimination of product monopolies and enabling generic competition would also lower prices, reducing treatment costs and personal hardships while expanding access.

Prize systems could further be designed to reward and encourage collaboration and the sharing of knowledge, materials, and technologies. Prizes encourage investments in translational research that has low prospects for commercial success, but that is of significant value in terms of advancing scientific knowledge.

The options discussed above, while not constituting an exhaustive list, provide a starting point for potential pilot programs that can be implemented immediately, either under the WHO Global Strategy and Plan of Action or by national governments or regional groupings.

This paper has briefly summarized a number of developments in the area of access to medicines and a number of ideas and arguments that were developed by various actors committed to introduce a shift in the paradigm regarding access to knowledge and intellectual property. In the past fifteen years many actors, from activists to academics, have pointed out the limits and barriers set by the current system to finance research and development of medical innovation, a system mostly based on product monopolies enforced by intellectual property regimes. But the involvement of these actors does not end in identifying and denouncing the barriers and injustices. They represent an active and creative force bringing specific options and solutions to the task of making access to knowledge a reality and thus to creating an environment allowing innovations to respond to the needs of all and to be accessible to all.

Today, most R&D spending is wasted on products that offer almost no realistic chance of offering significant health benefits over existing products. By eliminating product monopolies in favor of rewards from prize funds, it is possible to reduce incentives dramatically for investments in medically unimportant products while making it profitable to invest in products that truly improve health-care outcomes.

Pharmaceutical marketing practices today are rational responses to our system of R&D incentives. With a legal monopoly to make and to sell medicines and with returns based upon the number of units sold at monopoly prices, companies have enormous incentives to invest in the marketing of products to doctors and consumers, even for uses where the drug is of marginal use or even dangerous.[16] If prizes are implemented as rewards for improvements in health outcomes, irrational uses of medicines become a negative, rather than a positive incentive.

The current system for supporting innovation in the area of medical technologies is costly, inefficient, and leads to both underutilization of new inventions and unequal access. It is possible to do a better job of managing the current system, but it is also possible to refashion the approach radically in order to provide for more innovation and more access at less cost. Unlike other reforms in the health-care sector, which rely upon rationing of access to control costs, the use of prizes to reward innovation would expand access and increase investments in areas where innovation is most important. It is not easy to change existing systems of innovation, but neither will it be easy to not change. It is difficult to imagine a more expensive system of innovation that produces so little in terms of new medicines and vaccines. A reform of the reward system for new medicines has enormous potential to enhance innovation and access, not only in the United States, but everywhere. Those who object to change should have the burden of justifying the costly system of monopolies that we struggle with today.

NOTES

1 On the strategy of forum shifting, see Gaëlle Krikorian, "Free-Trade Agreements and Neoliberalism: How to Derail the Political Rationales that Impose Strong IP Protection," and Susan K. Sell, "A Comparison of A2K Movements: From Medicines to Farmers," in this volume.

2 "A Letter to the WHO Proposing a Medical R&D Treaty," available on-line at http://tacd.org/index2.php?option=com_docman&task=doc_view&gid=17&Itemid= (last accessed January 8, 2010).

3 "The Scientists' Letter: Global Framework on Essential Health Research and Development," January 25, 2006, in Médecins Sans Frontières, *Access News*, no. 13 (May 2006), available on-line at http://www.cptech.org/ip/health/who/59wha/msf-accessnews052006.pdf (last accessed January 8, 2010).

4 Michael Kremer, "Creating Markets for New Vaccines Part I: Rationale," NBER Working Paper No. W7716 (May 2000), available on-line at http://www.economics.harvard.edu/faculty/kremer/files/vaccine1.pdf; Michael Kremer, "Creating Markets for New Vaccines Part II: Design Issues," NBER Working Paper No. W7717 (May 2000), available on-line at http://www.economics.harvard.edu/faculty/kremer/files/vaccine2.pdf (both last accessed January 8, 2010). Unlike the approach recommended here, Kremer and colleagues focused on subsidies to the prices paid by consumers, rather than on prize rewards operating independently of prices.

5   Burton Weisbrod, "Solving the Drug Dilemma," Op-Ed, *Washington Post*, August 22, 2003, p. A21.

6   World Health Organization, "Public Health, Innovation, and Intellectual Property," WHO doc. no. WHA60.30 (May 24, 2007), available on-line at http://apps.who.int/gb/ebwha/pdf_files/WHA60/A60_R30-en.pdf (last accessed January 8, 2010).

7   Sarah Rubenstein, "Edwards Pushes Prizes Over Patents for Drugs," WSJ Health Blog (November 14, 2007), available on-line at http://blogs.wsj.com/health/2007/11/14/edwards-pushes-prizes-over-patents-for-drugs (last accessed January 8, 2010).

8   For detailed citations to the literature, see "An Annotated Bibliography of Scholarly and Technical Articles and Books on Innovation Prizes," KEI Research Note 2008:2, available on-line at http://www.keionline.org/content/view/82/1 (last accessed January 8, 2010).

9   The proposals were: a prize fund for development of a low-cost, rapid diagnostic test for tuberculosis, a prize for the development of new treatments for chagas disease, a priority medicines and vaccines prize fund, a prize for cancer medicines and vaccines in developing countries, and a licensed products prize fund for donors.

10 World Health Organization, "Global Strategy and Plan of Action on Public Health, Innovation and Intellectual Property" (May 24, 2008), doc. no. WHA61.21 (last accessed January 8, 2010).

11 In the United States, the period of exclusivity for pharmaceutical drugs is five years for new chemical entities and three years for a new indication for a drug. In Europe, the period of exclusivity for pharmaceutical test data is eleven years. Manufacturers of biological products are asking for twelve years of exclusivity in the U.S. market as a condition to the introduction of a system of biosimilars to promote entry by generic products. By requiring generic suppliers to replicate experiments on humans, these intellectual property regimes routinely violate the Declaration of Helsinki on Ethical Principles for Medical Research Involving Human Subjects, originally adopted in 1964 and most recently amended in 2004, available on-line at http://ohsr.od.nih.gov/guidelines/Helsinki.html (last accessed January 8, 2010). See, for example para. 8 of the declaration, which states: "Physicians should abstain from engaging in research projects involving human subjects unless they are confident that the risks involved have been adequately assessed and can be satisfactorily managed. Physicians should cease any investigation if the risks are found to outweigh the potential benefits or if there is conclusive proof of positive and beneficial results."

12 The fraction would be fifty basis points in the 2005 bill and sixty basis points in the 2007 bill.

13 At present levels of proposed funding, this would be roughly $4 billion per year, or $40 billion (plus the rate of growth in the GDP) over a ten-year period.

14 James Love, "The Role of Prizes in Developing Low-Cost, Point-of-Care Rapid Diagnostic Tests and Better Drugs for Tuberculosis," Médecins Sans Frontières expert meeting on IGWG and R&D for tuberculosis, April 11, 2008, available on-line at http://www.keionline.org/misc-docs/Prizes/prize_tb_msf_expert_meeting.pdf (last accessed January 8, 2010).

15 This also applies to some degree with other intellectual property rights regimes, including copyright, and several sui generis regimes, including those involving drug-registration data or plant-breeder rights.

16 This is illustrated, for example, by the extensive marketing of Vioxx to patients who could have used equally effective and safer medicines.

Virtual Roundtable on A2K Strategies: Interventions and Dilemmas

Amy Kapczynski and Gaëlle Krikorian

with Harini Amarasuriya, Vera Franz, Heeseob Nam,

Carolina Rossini, and Dileepa Witharana

A2K could be described as an emerging social movement or as a fundamentally disjointed and dynamic coalition. Perhaps more precisely, the A2K movement might be understood as characterized by the series of actions and arguments that are taking shape as groups develop overlapping agendas and coordinated actions in the conceptual space previously mapped using the concept of "intellectual property."

As the A2K movement has evolved, it has also increasingly gained recognition for its creativity and productivity. The movement has made connections between groups from very different contexts that occupy different places on conventional ideological spectrums. It works with and against governments, intergovernmental organizations, and industry. It mobilizes as well as troubles the rhetorics of law and markets. Those involved in the A2K movement also regularly make choices about organizational, conceptual, and strategic issues, with different groups taking sometimes substantially different approaches.

The following "virtual roundtable" was put together via e-mail to create space for a conversation on these issues and to provide a further sense of the stakes and divergences within the A2K movement. (See the "Virtual Roundtable on A2K Politics" in this volume.) Questionnaires were sent to five individuals who represent different approaches and involvements in the movement. Their responses were gathered and organized in a single document to facilitate the perception of the variety of opinions. As with the "Virtual Roundtable on A2K Politics," what follows does not intend to give an exhaustive account of the variety of positions and thoughts that exist in the movement, but give a sample of different perspectives that go through it.

The four perspectives represented here are those of Vera Franz, senior program manager at the Open Society Institute's Information Program, where she launched and heads the Program on Intellectual Property Rights Reform; Heeseob Nam, a patent attorney with the organization Intellectual Property Left; Carolina Rossini, an intellectual property attorney and innovation policy scholar, currently a fellow at the Berkman Center at Harvard University coordinating the Industrial Cooperation Project and the coordinator and main author of the Brazilian Open Educational Resources Project: Challenges and Perspectives, funded by the Open Society institute; and Dileepa Witharana, a senior lecturer at the Department of Mathematics and Philosophy of Engineering of the Open University of Sri Lanka who is currently involved with an interfaculty study of intellectual property rights and access-to-knowledge initiatives in Sri Lanka with Harini Amarasuriya, a doctoral student in Social Anthropology in a joint PhD programme with the University of Edinburgh and Queen Margaret University.

QUESTION 1 How would you describe the strategies and tactics that you have adopted regarding intellectual property and/or A2K issues? Has the A2K movement developed strategies that are particular to this coalition, or to activism at this point in history, from which other groups might learn? Do you see an evolution in the modus operandi and tactics of the A2K movement?

QUESTION 2 Where did the A2K movement learn its strategies? From where or whom could it learn?

QUESTION 3 The architecture of the A2K movement makes law central, because the rubric of the coalition is arguably "intellectual property"—a legal construct. Legal institutions (the World Trade Organization [WTO], the World Intellectual Property Organization [WIPO]) and legal mechanisms (open licenses, creative-commons licenses, compulsory licenses) are key locations for A2K interventions. The A2K movement also operates very often in the register of legal argumentation—that is, treating as its main opponent a theory of intellectual property that is commonly advanced in legal circles, a theory of intellectual property as an entitlement for creative labor or as an "incentive for innovation." A2K groups must thus regularly reckon with the influence of law and legality and develop an ethic or set of strategies appropriate to the challenge of operating in a legal context. How do you reckon with the constraints and/or attractions of law? Do you encounter tensions between mobilization through law and mobilization against law, and if so, how do you deal with these tensions? Should the A2K movement be understood as responding to a set of commitments, principles, or ethics that exist beyond the text and interpretation of law, that should guide the writing and interpretation of law? If so, what are these, and where do they come from?

QUESTION 4 What is your position toward institutions of governance and, more generally, toward those who govern?

QUESTION 5 Do you have a specific attitude toward the private sector? Can governments and industry be considered part of the A2K coalition? Should they be?

QUESTION 6 How would you describe A2K—as a movement, a coalition, an aspiration, or what? Does the A2K movement represent a simple pooling of resources or a coalescence of actions and initiatives that are facilitating the creation of an original form of action? Do coordinated or aligned actions with other "A2K" groups represent an important aspect of your work? How should A2K groups relate to one another?

QUESTION 7 Do you draw a distinction between short-term and long-term strategies? How does or should the A2K movement mediate between the pressing needs of the short term and the need for a vision and strategy of long-term change?

QUESTION 8 Do you see the strategies of A2K actors merely as acts of resistance to the agenda and profound successes of the movement for greater intellectual property restrictions? Is the A2K movement predominantly an attempt to "domesticate" or "balance" the existing intellectual property system, or does the movement advocate for more thoroughgoing alternatives to this system?

QUESTION 9 Do you see tensions or struggles for power between the A2K movement's various components? If so, how do you interpret these in terms of the shape or governance of the movement and the prospects for the future?

QUESTION 10 Some characterize A2K as an elite movement, shaped primarily by NGOs, academics, and government actors and by agendas set predominantly in the Global North. Are these perceptions accurate? Does the A2K movement have or aim for a broad base of nonexpert participants? How are people who are not professional A2K actors to participate in priority setting and campaigning?

VERA FRANZ

QUESTION 1 The much-quoted analogy to the environmental movement that's used to describe the strategies and tactics of the A2K movement I think still holds true. The main achievement of the A2K movement is that it has made people aware of a common resource—the knowledge commons—that needs to be protected by keeping it "common" and free. The main task of the A2K movement is to monitor and promote the health of our knowledge ecology, that is, to watch the precarious balance between private and common knowledge resources. This means mainly two things: Monitor and push back intellectual property rules that endanger the knowledge commons and promote the knowledge commons by planting seeds that everyone out there can grow. Yochai Benkler gives a wonderful account of the knowledge commons in *The Wealth of Networks*.[1]

In terms of strategy, two main things come to mind: First, it was critical that the A2K movement was able to unite very different stakeholders under one umbrella: librarians, free-software developers, video activists, access-to-medicines campaigners, digital-rights advocates, consumer groups, teachers and students, the visually impaired, and many others. "A2K" as a term was broad and vague enough to bring all these people together, and at the same time, the term captured a concern that everybody shared: the quest for a balance between private and common knowledge resources.

Second, I think the alliances and coalitions that the movement built with the private sector were critical. Claims about human rights and the public interest get you only so far. Once companies such as IBM, Google, Sun, AT&T, HP, and others joined the chorus, governments started to listen more carefully. It started to dawn on policy makers that those "A2K businesses" substantially contribute to U.S. competitiveness and GDP and would have to shut down in a world of maximalist enforcement of intellectual property rights.

In terms of changes over time, if anything, the A2K movement has become more mature and sophisticated in its work. The strategies haven't fundamentally changed, but the demands have gotten more concrete. For example, while in the beginning the quest was for a more balanced copyright regime and stronger limitations and exceptions to copyright, at this point, the movement is asking the WIPO to adopt a Treaty for the Visually Impaired within the coming two years.

QUESTION 2 The movement from the very beginning was joined and led by experienced campaigners, originally active in other fields, and they brought with them effective strategies. This ensured campaigns were of highest quality. And those experienced campaigners taught young A2K activists important skills. I was able to witness how several young people grew into professional activists over time—an absolutely wonderful thing to see!

QUESTION 3 The A2K movement is eclectic in many ways, and that's true of how it views the role and centrality of the law. The part of the movement that is focused on, for example, the reform of the WIPO takes the rule of law as its starting and end point. By changing legal norms, it tries, among many other things, to improve access to medicines for the sick and access to knowledge for the visually impaired. The law in that sense is the main tool in the toolbox. Very importantly, these legal changes, in addition to better access, bring about new more open models for the creation and remuneration of knowledge-based goods. Think GNU/GPL, think Creative Commons, think the Medical Innovation Prize Fund, or even the idea of an alternative compensation system for musicians. And it's not better access that is the revolutionary development here. It is those new and more open

NGO-in-a-Box, AV edition, Tactical Technology Collective, 2009 (www.tacticaltech.org).

models that incentivize creativity and innovation, excite so many, and will drive the change.

All of this said, there are important forces in the A2K movement that question the centrality of law as such. Reality is shaped by other forces than the law, they say. And I think those voices in the movement are equally important to the overall change we want to bring about in the coming years. These people take social practice as the starting and end point and argue that, in times of great technological progress, it might well be that we will improve the condition of humanity more if guided by social practice rather than by the existing law. For example, if the world is looked at through the lens of WIPO, "piracy" is classified as an "illegal activity" that needs to be extinguished at all cost. But looking at matters through the lens of human development and economic progress, for example in developing countries, one might not always come to the same conclusion. Further, some important businesses would never have developed by always adhering to the narrowest interpretation of the current legal framework. In summary, parts of the A2K movement challenge the centrality of the law and argue that social practice, rather than the law, will best guide us toward a better future.

QUESTION 4 Working for the Open Society Institute, I obviously believe in the rule of law, and I hence respect the institutions that shape and enforce the law. That said, if a Soviet dissident would have followed the law, he would never have known the "freedom" he was so desperately seeking. This is to say that we should always and at every moment question institutions of governance and those who govern. So no, I don't trust them. I am not saying anything new here, but basically, let's build on our values and use common sense when engaging in public-policy debates.

QUESTION 5 Can governments and industry be considered part of the A2K coalition? While governments and industry are not part of the A2K movement as such, many of them are very close allies. Everyone who rejects a "maximalist IP" regime and engages in this struggle is an ally. In fact, we have many examples of successful issue coalitions between civil society, governments, and businesses that prove this point. Take the campaign against an overbroad WIPO Broadcast Treaty or the effort to strengthen limitations and exceptions to copyright. To be sure, these alliances between the different sectors are not always easy and might even cause temporary friction. But I firmly believe that we can bring about the change we are seeking only by working closely with everyone who shares the same vision and goals.

QUESTION 6 A2K is probably all of it—an aspiration, a movement, a coalition. One can best think of it as consisting of concentric circles. The outer, largest circle unites all those who share the aspiration or vision of a healthy knowledge ecology, one where the commons thrives and occupies an important place. The next circle closer to the center gathers the actors that came together, initially around the A2K Treaty, which are committed to making it, or a version of it, a reality. Some people were instrumental in getting this movement off the ground, but it is not led by any one entity or person at this point, and new actors join as they feel this work is relevant to them and they want to make a contribution. Finally, within the innermost circle, concrete progress is being achieved in the form of issue coalitions, which form around concrete threats and opportunities. They spring up organically and may include NGOs, academics, governments, and industry. They share a concrete goal and get on with the work. Some are more public than others. All of them wage a concrete, modest, yet important fight in terms of the bigger picture. In a way, this inner concentric circle is driven by networked actors that jump into action when and where needed.

QUESTION 7 Very clearly, one always needs to think about short-term and long-term strategy and vision. I would argue that the A2K movement is better at short-term engagement, responding to immediate threats and opportunities. Some players have identified a longer-term vision, but it is not always clear what the strategy for implementing this vision is. I would argue that one of the weaknesses of the A2K movement is the vagueness of a longer-term strategy. Because time is a luxury, the longer term is forgotten in the daily heat of smaller fights. In a world of radical uncertainty, it is not easy to come up with a long-term strategy. But again, we need to improve and make time for longer-term thinking. The development of the European Patent Office's Scenarios for the Future was a good step in the right direction.[2] Following up on the EPO's work, we used A2K3, the Third Access to Knowledge Conference, in Geneva in 2008, for a retreat type brainstorming exercise trying to clarify the vision and strategy for where we want to go. This is only the beginning. More work needs to be done.

QUESTION 8 The A2K movement is about both resistance to overbroad intellectual property rules and alternatives to them. Resistance is born out of necessity. It's the response to a crisis. But also and more importantly, A2K is about radical alternatives to the maximalist intellectual property regime. It's simple — we can't keep fighting all the small fights forever. Civil society has limited resources and time. And unlike human rights, I think this is an issue that can be fixed. Once we are able to push through fundamental reforms such as, for example, the Medical Innovation Prize Fund and R&D Treaty, we won't need to fight against extended data exclusivity, patent-term extension, and more generally, dangerous intellectual property chapters in free-trade agreements, where our ability to influence the course of action is minimal.

QUESTION 9 Tensions within the A2K movement? In the early days, I was amazed by how excellent the collaboration between groups in the A2K movement was. With the maturing of the movement — and of the proposals put forward — differences in opinion and strategy have emerged. I guess anything else would be a surprise. This is an eclectic movement, and debates and disagreements will be an integral part of it.

I am not sure one can actually speak of "governance" of the movement in the traditional sense. For example, nobody has ever dared to trademark the term "A2K." And there is no need to enforce its use. It's a vision that you join in because you share it. "Organized chaos" would probably be one way to describe the movement's modus operandi: It is a distributed network, and every node can take action and focus on what it thinks is the highest priority at any given moment. The more urgent the cause or compelling the idea, the more nodes in the network will join and amplify the movement's voice.

QUESTION 10 I agree that A2K is an elite movement shaped to a large extent by civil society and industry in the Global North. In the beginning, I struggled a lot with that notion, because I thought this movement needs to be more global and bottom-up. But I came to realize that this movement is different from others, yet I think this is changing as we speak.

Some thoughts on why the movement started out as it did: When the movement started out, the vision was broad, general, and sometimes abstract. And the issues were new to many of us. It took some time to zoom in and identify the concrete problems one could mobilize around, in all parts of the world. I think a first important success is the fact that the global community of the blind has now joined the movement, and we are together trying to solve the concrete problem of access to knowledge for the visually impaired through a WIPO Treaty that will mandate strong limitations and exceptions for the visually impaired.

Also, it is always easier to mobilize successfully around a current, rather than a possible future problem. And the A2K movement is very often about preventing a bad future or creating a new future that is not always easy to imagine for a nonexpert. The fight against the WIPO Treaty for Broadcasters, Cablecasters, and Webcasters is a good case in point. Related to this, piracy alleviated many of the problems caused by a restrictive copyright regime in developing countries. And important parts of the developing world were not on-line yet, and yet so many of the A2K fights relate to the Internet. Both of the latter two factors are changing. Antipiracy enforcement is getting more serious, and the developing countries are moving on-line. Hence, issues with which the Global North was concerned, such as for example Internet service providers liability, are fast becoming an issue in the South.

On a more general level, I recently participated in an interesting discussion with developing country participants about open-access journals. Northern groups have for years successfully advocated for things like the National Institutes of Health Open Access Policy.[3] Many of us thought that the South would follow suit. However, Southern representatives have concerns around possible exploitation of Southern knowledge by Northern commercial companies, and they view open-access journals as one potential means for this type of exploitation. This is to say that the A2K movement has much more work to do here. The world is complex, and it is a big mistake to assume that the South will blindly follow a Northern logic.

All of this means that the A2K movement has absolutely to focus on getting more Southern players involved. The time is right for that more intensive engagement with and by the South. Whether A2K will ever become a fully fledged grassroots movement, I don't know. It will in certain instances, with certain campaigns. But I also think that to win important fights, you do not necessarily always need the masses on the streets. You need highly committed people with deep technical and strategic knowledge, from all parts of the world, who can effectively work the corridors of power and, where appropriate and possible, mobilize the masses.

HEESEOB NAM

QUESTION 1 My group—IPLeft—was established in 1999 by members of social-movement organizations who were concerned about the ever increasing monopolistic control of private enterprises over information and knowledge. We believed that a monopoly on social assets such as information and knowledge had been institutionalized and systemized by the intellectual property system, following a trend rooted in global capitalism. The intellectual property system plays a central role in information capitalism by making knowledge and information private property and commodities that can be produced and exchanged in the market, thereby causing problems such as the digital divide, conflicts with basic human rights, growing gaps in the distribution of wealth between rich and poor nations, and the nondemocratic governance of social assets.

In terms of strategies and tactics, our aims thus are not simply to criticize the current intellectual property system. We also strive for a society where knowledge and information are free, not commodities, and can be freely created and used as social assets to be shared universally among people.

Among the challenges and obstacles we have faced is the widespread assumption framing the issue that the protection of intellectual property is beneficial for our nation's competitiveness. Actually, this assumption lacks theoretical and practical justification in local contexts such as Korea, which has long been a net importer of intellectual products. However, it has prevailed in the administrative branches that are in charge of intellectual property matters and that have governed the intellectual property agenda. In addition, intellectual property issues are too arcane to be grasped by the public and even by legislative and judicial policy makers. Those in other administrative branches and in the National Assembly neither fully understand the meaning and implications of the intellectual property issues nor have sufficient resources to take part in decision-making processes. This is why the internal governance of those branches of the government concerned with intellectual property matters remains in the hands of a narrow group of experts in intellectual property issues who subscribe to the assumption that intellectual property protections are beneficial.

The notion that stronger intellectual property protections are good for our society is reinforced by global intellectual property rules and the influence of ideas promoted by Korea's major trade partners, the United States, the European Union, and Japan. Most domestic intellectual property norms have been discussed and determined in the context of international trade. In this process, the primary focus of policy makers is on increasing our nation's competitiveness in the global economic wars. But this concern with competitiveness is a narrow concept confined to issues of economic growth, or, more accurately, to the international competitiveness of domestic corporations. No other values are considered or seriously taken into account by policy makers. Meanwhile, they view maximalist intellectual property norms as international standards that were successfully implemented by our industrialized trade partners in the globalized economy and therefore as something on which they can rely, as well.

Under these circumstances, our tactics are twofold: within the government, debating with policy makers (mainly in the administrative branch) and lobbying the legislative branch and in society at large, mobilizing social and cultural movements in favor of workable models of knowledge production and dissemination that don't rely on the intellectual property protection. For the purposes of the first tactic, it is necessary to frame the issue differently, showing that contrary to the reigning assumption, enhancing access to knowledge without relying on the stronger protection of intellectual property rights also serves the progress of the national economy and that the narrow concept of corporate competitiveness may lead to unintended consequences that are harmful to the proper production of socially needed knowledge.

The A2K movement is helpful in pursuing these two tactics in that it changes the horizon of global discussion, especially at the WIPO, which is a central forum for the global intellectual property norms. International movements that can be categorized under the rubric of A2K have evolved to provide a counterframe that covers overall intellectual property norms and is not specific only to some particular intellectual property issues.

But I would like to point out a possible limitation for the A2K movement that is related to an element crucial for its success: the localization of the global movement. The framing of A2K discussions as a debate between the Global North and the South can raise obstacles to A2K mobilizations in middle-income countries that occupy the ground between the wealthy, powerful countries and the poor and less developed ones. For instance, some delegations from middle-income countries downplay A2K discussions because they believe they are driven by the South for the purposes of getting more money from the North. These delegations regard the issues raised by the A2K movement as having little relevance to their nations. This may marginalize nations that could play a leveraging role in the global discussion and may make it more difficult to move A2K issues higher on the agenda in marginalized regions.

QUESTION 2 In terms or sources for strategies for the A2K movement, I think that resisting the TRIPS Agreement has been decisive in the evolution of A2K. Farmers' groups and access-to-medicine campaigners have long struggled against intellectual property maximalism. They've achieved something great. For instance, the Doha Declaration is a great success. The activists mobilized during the struggles against intellectual property maximalism came to understand that they have interests in common with other groups fighting against the intellectual property maximalism in sectors such as education, libraries, software, privacy, freedom of expression, and so on. The A2K movement is a result of efforts to bring together diverse groups against a common enemy. However, achievements such as the Doha Declaration solved only some of the problems: It redressed some of the imbalances of practices and intellectual property norms between the North and the South. It addressed problems in a specific sector, but it did not tackle the underlying mechanism of global intellectual property systems.

QUESTION 3 In general, an approach that is centered on the law entails both opportunities and constraints. Social reform through the law can be a useful strategy in that the legitimacy of the reform can be easily established and the reform can be sustained by the regulatory nature of the law. Considering the fact that intellectual property norms have been extended to cover a wide range of fields and have had important economic, social, and cultural consequences, it is partly inevitable that the movement would concentrate on the legal aspects of intellectual property issues for the purpose of short-term reform. Moreover, in consideration of the fact that policy options at the national level are highly constrained by the norms created in international legal institutions such as the WTO and WIPO, it is a necessity to concentrate on legal issues.

However, the strategy of pursuing reforms through law has limitations, because the law is an expression or a reflection of power relations in a given society, indirectly, if not directly. Therefore, the A2K movement can hardly change the essential elements of intellectual property without including broader cultural, political, and social movements. Individuals, as social entities, can change the existing social structure. However, this is possible only when they identify and use the political opportunities open to them and turn their cultural constraints into opportunities. The recent shift of U.S. courts toward the weaker protection of patent rights and the advent of the Obama administration, for example, may offer an opportunity to alter U.S. foreign policy on intellectual property matters. Collective actions not directed at changes in the laws should be encouraged as cultural movements. The Creative Commons and open access movements are good examples of such mobilizations. A model for the production of knowledge on the basis of democratic and socially planned systems such as the democratic planned participatory socialism (DPPS) model proposed by David Kotz is an even better example.[4]

QUESTION 4 My attitude toward institutions of governance and the governing class is that I think the governing power is owned by the capitalistic class—the transnational corporations. Although the nation-state and international or intergovernmental organizations play key roles, the internal actors are transnational corporations.

QUESTION 5 We need coalitions with both the private sector and governments. I believe that unless governments and industries are members of the A2K coalition, we cannot accomplish what we are trying to do. The structures imposed on us are too strong and powerful, and unless or until the A2K movement overturns capitalism or fundamentally changes the capitalist world, we need the support of sympathetic voices from mainstream groups. Moreover, reform of the intellectual property laws is impossible without the participation of those in government.

QUESTION 6 How we describe the A2K movement may depend on how we move forward. I believe it should be a form of social movement. To bring together diverse groups under the umbrella of A2K, we may need to design a more general ideational frame. I don't know if "access to knowledge" can be the one we need. It appears to me that access to knowledge is more a means than an end. Unlike the environmental movement, in which the catchwords of environment conservation can easily be linked to the end—our survival—A2K seems to need discursive link to an end. Why do we need access to knowledge? For innovation or culture? For empowerment? Or for ensuring human rights?

QUESTION 7 We already have short-term and long-term strategies. Changing the mandate of WIPO through the Development Agenda, harmonizing copyright limitations and exceptions, patent-system reforms, and establishing global strategy and action plans for the development of drugs for neglected diseases may be categorized as short-term strategies. The proposed Medical Research and Development Treaty is a candidate for a long-term model, as long as it changes the existing price-incentive mechanism to a DPPS system where all of the interest groups take part in every cycle of innovation.

As Susan Sell suggested in *Private Power, Public Law*, structural opportunities and constraints need to be exploited in both short-term and long-term strategies.[5] Existing structures offer either constraints or opportunities to those who are seeking to change the structure in some way. In this regard, structural changes in the global economy and institutional changes, for instance, altering the attitude of the

U.S. Supreme Court on patent issues, need to be analyzed and pursued in more systematic ways.

QUESTION 8 I think the A2K movement already advocates thoroughgoing alternatives to the current intellectual property system and is on the right track. The World Intellectual Property Organization Development Agenda and World Health Organization Medical Research and Development Treaty are good examples. Seeking radical changes in intergovernmental institutions is not a good strategy. Gradual changes can lead to a radical reform in the long run.

QUESTIONS 9 AND 10 I agree that the agendas of the A2K movement are set predominantly by NGOs and academics in the Global North, but I think that the issue of "for what" they're set matters more than "by whom" they're set. For more participation by nonexperts and more importantly by grassroots individuals and organizations, we may need a larger conceptual umbrella—for instance, human rights perspectives on intellectual property.

Many scholars and commentators have identified a conflict between intellectual property rights and human rights. A comprehensive human rights framework still requires further elaboration, and coherent solutions to address the conflict between the two norms need to be proposed and examined. In order to develop a comprehensive and coherent framework for human rights approaches in global intellectual property policies, I would suggest establishing a link between international human rights organizations and international intellectual property institutions such as WIPO and the WTO/TRIPS Council. For instance, WIPO and the TRIPS Council could establish special standing bodies mandated to assess and monitor their activities and their operating treaties and to evaluate what are the possible or actual effects on human rights. Such assessments and monitoring could be carried out by consulting with other international human rights organizations and made public regularly.

CAROLINA ROSSINI

QUESTION 1 I have primarily adopted strategies and tactics that are familiar from the worlds of political economy and law—I do a lot of research on the foundations of the issues, and I have worked on legal tools (contracts and policy analysis) that achieve some of the goals of the A2K movement, such as Creative Commons licensing and open-innovation legal frameworks. I have also used some traditional pedagogical tools as a law professor, including regular e-learning courses, class syllabus and content creation, and the hosting of conferences and workshops around topics central to A2K issues such as copyright, open access and open educational

Public funeral for digital rights management (DRM) organized by the Harvard College Free Culture group, May 24, 2008, http://www.hcs.harvard.edu/~freeculture/blog/ (Christina Xu, http://creativecommons.org/licenses/by-sa/2.0/deed.en).

resources, open innovation, and distributive innovation and its effects on patent law and policy and cooperation and technology transfer agreements. These are strategies that are not specific to the A2K coalition, but rather more traditional strategies related to capacity building and awareness raising.

I think the A2K movement has clearly learned some lessons from several other movements, however. The environmental movement—and actually, the human rights movement—bring aspects of both political economy and direct protest action to bear on global public opinion, and that dual-aspect approach can be seen reflected in the legal work of James Boyle, Yochai Benkler, and Lawrence Lessig, as well as at FreeCulture.org and other more activist groups.

My hope would be that we could follow Boyle's call of an environmentalism of the public domain and learn to work together a little more, though not to formalize that working together through defined governance structure. I mean more a common agreement on philosophy and an agreement to spend more time on common problems than on intermovement disagreements. There is some internal personal conflict between people and organizations in the movement, which is not so unusual in a broad and diverse coalition of stakeholders, but it would be better if we could all agree we are working toward a common goal of more access to knowledge and focus our energy on those who oppose us, rather than on those of us within the movement with whom we disagree on matters of degree, rather than matters of importance. We also could learn from the growth of on-line communities and their associated governance norms and dispute-resolution alternatives, on which the recent work of Yochai Benkler and others at the Berkman Center focus. Transparency and cooperation are core values for the success of the movement.

QUESTION 2 The current organizations that are seen as part of the A2K movement have given it its character. You have elements from civil society that come from the AIDS/HIV movement of the 1990s turning into access-to-medicine discourse and student movements such as FreeCulture doing cause-based activism and bringing a public-interest conscience to courses of law, communications, media, and medicine. You also have lawyers who moved from the entertainment, telecommunications, and biotech industries to develop A2K strategies based on lobbying strategies, as the WIPO Development Agenda and also the open-access movement may show. In this sense, the movement can be considered a result of a convergence of old movements and new movements that found a common point for discussion and discourse building.

QUESTION 3 As a lawyer, I am perhaps the wrong person to ask about whether the centrality of legal struggles in the A2K movement is a good idea! However, it is true that we too often battle in the world of the law, and the law is frequently designed in a way that favors those already in positions of power, such as the content owners. Additionally, policy makers and lawmakers look abroad to replicate the solutions they find there. The recent international emulation of the Bayh-Dole University and Small Business Patent Procedures Act, which gave universities, small businesses, and nonprofits intellectual property control over the results of government-funded research, is an example. Thus, law can be an answer, but also a problem.

In regard to the A2K movement, there are perhaps two segments. The first would see the goal as an almost complete change in the system. This could include both the A2K Treaty, which does involve a lot of work with lawyers, WIPO, and international governments, but also the direct activism of the Defective by Design and FreeCulture groups. This group could also include those who want to reform the international conventions and treaties around copyright and who challenge the national legal implementations of those treaties and conventions. We could attribute to this group the recent efforts on exceptions and limitations to copyright and patent rights currently under discussion at WIPO.

The second group would be those who are working to make the effect of the current system less bad, using the system we have to achieve open goals today even before international legal change is achieved. This would be groups such as Creative Commons, the Free Software Foundation, and Cambia in Australia. This set of groups is definitely based on using legal tools as a practical method to make life better for people during the long struggles to change the laws and treaties. These and others try to provide standards that will make things more interoperable and transparent and try to foster cooperation in different sectors. They attempt to make the right to participate in culture and science a reality.

However, as the A2K debate evolves, the line between the two segments is becoming less and less clear. The increasing collaboration between groups and their openness to a diversity of voices, including the participation of the Global South, has brought an incredible maturity to the movement, which is reflected in much more complex and sustainable proposals.

QUESTION 4 My attitude toward institutions of governance and those who govern has been shaped by the fact that the majority of my life has been lived in countries with established or new democracies. In all cases, I believe that government plays a crucial role in balancing power within the society through mechanisms of representation. I also see the use of political instruments that may be able to open the opportunity for a more concrete societal participation in the decision-making process. This is true internationally, as well, via multilateral forums where governments, through decision-making processes within international organizations, have allowed increasing participation by civil-society organizations in multilateral arenas.

Other institutions of governance, such as international organizations, are also necessary to provide multilateral and possibly transparent forums for negotiation and policy making. Thus, in general, I think they are necessary. However, as in any institution of governance, mechanisms of checks and balances are crucial. These organizations can be and often are captured, which means that their benefits are very dependent on the willingness of our movement to engage within them, observe them, and so on.

QUESTION 5 I think a great part of the private sector has tended to be antagonistic to the A2K coalition, because the majority of business models that relate to A2K are based on control, and not on sharing. Business is still learning how to be open and how to make money for its investors. But there are clearly some business models that would allow companies to be part of the A2K coalition. Open-access publishers in science that use the Creative Commons attribution license, known by its acronym CC-BY, would probably meet that definition, and in 2008, the top two corporate open-access publishers created more than $15,000,000 U.S. in revenues without relying on control. We also observe these changes toward open business models in the music and textbook industries, with the innovative company Flat World Knowledge, but I think most of the private industrial models would have to be service models, and not control models, to qualify.

Governments are a hard question, because "government" is a complex entity that is very reliant on local factors, and thus it is hard to make general statements about it. Australia's government has endorsed open licensing and membership in the global digital commons for its publicly funded content. Brazil is one of the leaders of the WIPO Development Agenda and has recently proposed discussions on exceptions and limitations to patents at WIPO. The Brazilian government is also playing an important part in fostering the use of open-licensing schemes for content (open-access and open educational resources) and software. But does that make them part of the coalition? I don't know. It definitely makes Australia and Brazil friendly places for the movement and for coalition work, but it is also something that can change in a single election or with the change of leaders at the ministries.

QUESTION 6 Is A2K a movement, a coalition, an aspiration? For now, I think A2K still is a movement. There is still no clear ecosystem in place. Organizations still do not recognize other organizations that are clearly adopters of the same agenda, and you do not meet all people you would expect to meet at the annual conferences. Maybe this is a matter of dispute for funding, maybe this is a matter of a top-down approach, or maybe it is just because the movement is in its youth.

I consider myself a network maker, in addition to my role and professional focus in academia, capacity building and advocacy. By this I mean that I believe in people working together, and I believe that alliances and the recognition of diversity inside the movement is a plus, and not a minus. In all steps I take, I try to recognize common efforts and agendas and to integrate people and projects with my projects or point out possible collaborations and opportunities among projects I see within the movement. I also try to connect pieces of my professional and personal networks that represent different segments of the A2K movement to each other, such as the open-access movement to the open-education movement or those engaged in socially responsible licensing work to the patent reformers or the access-to-medicines groups. Personal connections between different groups who follow the same general philosophy can be a powerful way to coordinate us that does not require the strong powers of governance and that also therefore cannot be captured. This, however, also has its downside, because the strength of personal networks can be used to exclude divergent or new opinions.

QUESTION 7 Short-term strategies do not have to diverge from long-term strategies. Specific tactics may vary, but not the final goal. We have tools that allow us to opt out of the current system and to build capacity outside the imbalances of copyright and closed access and closed sources, and those are good short-term tools and strategies. It is immoral to have a strategy that says we should not use tools that help people gain immediate access to health information, cultural information, and scientific innovation. So we have to think about the actual needs of people as a key part of our strategy, short-term and long-term, and the reality is that A2K issues as such are not a part of lots of people's lives, but having clinics that have recent clinical documents about AIDS or access to genetic testing or generic medicines is a part of people's lives. Thus, strategies can help in this context to ease immediate pressures and problems.

But in the long term, we have to maintain pressure on the international conventions and treaties and on the effect of those in national policy and law. We can do that successfully if we stay together as a movement. Again, I'm basing my thoughts on Boyle's metaphor of the environmental movement. The short-term strategies of recycling locally or composting organics are similar to the licensing strategies of Creative Commons and the Free Software Foundation, but the strategy of thinking globally about carbon emissions, the Kyoto protocols and the negotiations on climate change and how these relate to innovation and access to knowledge and technology is more similar to treaty discussions. We need a big enough strategy to integrate both of these approaches. Otherwise, we risk not only dividing ourselves, but letting those who oppose the coalition generally find ways to attack both short-term and long-term strategies by pointing only at our divisions and thus not responding to the realities of our arguments.

Parody of iTunes billboard directed at the digital rights management (DRM) restrictions used by Apple among others (http://n3wjack.net/).

QUESTION 8 The majority of current strategies of A2K actors—even some that may be seen as positive—are, in the end, primarily resistant. This is not a terrible thing in some ways, but it also reflects the power of the movement for greater intellectual property restrictions in controlling the debate. Arguing against an intellectual property maximalist approach is probably more a reaction to strong intellectual property restrictions than it is a uniquely derived strategy. It is hard to separate the two sides of this same coin.

Each organization that can be considered part of the A2K movement has its own history, but even some that took positive actions for generating something new, such as the Creative Commons or recent efforts on innovation metrics not based just in patents, appeared as reactions to the nearly permanent extensions of copyright duration or the centrality of patents. Also, I have not yet seen true and consistent proposals of more thoroughgoing alternatives to this system. Some proposals on alternatives to patent systems, such as the creation of a prize system, are thoughtful options. However, I do not see how any of them can be applied in the context of poor countries, for example.

Finally, civil society and academe operate at a different pace in comparison with government and business. This may be a result from a lack of consistent funding or from the nature of business models, but it also reflects a gap in capacity in terms of training, research, and organization. This implies that we will always be behind and reacting.

There are some technically oriented approaches that have the potential to transcend the intellectual property debate advanced by people both inside the A2K coalition and outside, especially in the area of semantic computing. This represents ideas and knowledge in formats that are less likely to be creative expressions and thus less likely to be considered in the framework of intellectual property rights. If this approach is successful, we can have a debate about knowledge that is less situated inside the intellectual property discourses framed by Disney, for example, because Disney cares more about controlling its properties than about semantic computing, and thus there is more rhetorical space to develop positive agendas.

But because the movement was indeed created inside the framework of intellectual property rights, these kinds of approaches can actually be hard to get endorsed inside the coalition, because they do not fit the coalition's own preferred frames of rhetoric. Similarly, we tend to see problems with intellectual property rights everywhere and thus reach for intellectual property rights solutions, when if we could instead see technical solutions, or business solutions, or capacity-building solutions, we might find better answers. We should be more interdisciplinary in our conception of both the problems and the solutions, and we should look for ways to escape the old structures of intellectual property rights debates that force us to fight powerful forces advocating increases in intellectual property rights. This could be part of the movement that would be really valuable, to help get out of the intellectual property rights position and create interdisciplinary knowledge sharing inside the coalition.

QUESTION 9 I see tensions in the movement, and as I said before, these may be caused by disputes over scarce resources, as well as by the desire to establish oneself as a novel thinker, and being a "regular part of the movement" is not sufficient to do that. However, tensions are not necessarily bad things, since we need to allow the emergence of new organizations, groups, and positions that can bring diversity and creativity to the movement. We should be avoiding the consolidation of an A2K "aristocracy." The minute we begin to talk about movement governance, it stops being a grassroots movement and begins to entrench power inside structures.

QUESTION 10 I agree that A2K is an elite movement mostly shaped by Northern players. However, if you listen closely to some of these players, mainly research institutes and NGOs, you will find a Southern voice in almost all of them. Also, in recent years, organizations from the Global South have gained a voice in specific issues, such as access to medicines. However, these may still be considered the voices of an "elite" from the South. This might be due to the fact that some core strategies of the movement still ask for skills that are not well distributed in society. Taking this into consideration, core players of the A2K movement have a moral duty to include, in any action, an element of capacity building and network building. For this to happen, the movement needs to be able to recognize the vocabularies of the human rights movement, of the environmental movement, of the civil rights movement, and be able to build links that can guarantee a broader social legitimacy.

DILEEPA WITHARANA AND HARINI AMARASURIYA

QUESTION 1 In order to describe the strategies and tactics we have adopted regarding intellectual property and A2K issues in Sri Lanka, I need to begin with a brief survey of the A2K movement there. In Sri Lanka, there are several organizations, groups, and individuals who engage with what could be described as A2K-related issues concerning intellectual property rights. However, the term "A2K" is not established as yet, and those who are involved in this type of work do not necessarily recognize or realize that they are part of a common movement to ensure access to knowledge. The Open University of Sri Lanka, which is conducting a study on the issue, recognizes both "formal" and "traditional/informal" knowledge as knowledge that is under threat by the intellectual property rights regime and identifies a range of organizations, groups, and individuals who are involved in different ways in ensuring A2K. Most prominent are the existing or emerging initiatives one could call a "free and open-source software (FOSS) movement," an "antiglobalization movement," and a "seeds movement."

The FOSS movement in Sri Lanka mostly consists of experts in software who contribute to the development and spread of free and open-source software, although it can be argued that the potential membership is much larger than this. With the passing of the Intellectual Property Act in 2003 and the gradual implementation of that law since 2006–2007, there is a growing realization among academics, students, researchers, the media, and the general public who use PCs that they will not be able to afford software under the existing intellectual property rights regime. As a result, several universities, schools, state institutions, and NGOs are converting to FOSS.

The Sri Lankan antiglobalization movement, however, is not a movement in the true sense, but consists of several Colombo-based NGOs and their network of community based organizations (CBOs) spread across the country. They engage with A2K issues through their strong critique of free-trade mechanisms and the privatization of and granting of monopoly rights over intellectual contributions. The antiglobalization movement has yet to incorporate the term "A2K" into its discourse or to identify their role in protecting A2K per se.

The seeds movement is a movement whose work in safeguarding access for farmers' rights to use seeds as a public good would lead to safeguarding access to traditional knowledge (A2TK) in agriculture. Intellectual property rights are not the first or the only threat to accessing traditional knowledge in agriculture. The so-called "Green Revolution" has played a major role in the disappearance of traditional seeds and hence of the traditional knowledge that was protected and embodied in traditional seeds.[6] Intellectual property rights, however, pose the next serious threat to A2TK in agriculture. A strong lobbying group consisting of several influential and prominent lawyers, activists, and a few civil-society organizations works in this area, defending A2TK in agriculture from already implemented laws. When the Intellectual Property Bill was introduced in 2002–2003, it was challenged by these groups before the Supreme Court as being unconstitutional. The challenges resulted in the exclusion of microorganisms from patenting except for transgenic organisms and the inclusion of flexibilities, including parallel imports and compulsory licensing, in the current patent law. This in fact has led to the strange scenario of the section on patents in the Intellectual Property Act of 2003 being much more progressive than the section on copyrights from an access point of view. Education about the implications of intellectual property rights on A2TK in agriculture is spreading in Sri Lanka, while a movement to save and share traditional seeds initiated among several NGOs and farmer groups has been growing. More and more farmers in villages realize and are worried about the loss of traditional seeds in the face of the use of commercial seeds that have flooded the market since the Green Revolution.

In addition to these three trends, access to traditional knowledge in health can also be identified as a component of an A2K movement. A2K in health in Sri Lanka is mainly expressed as resistance to gene piracy in local plant varieties—the smuggling out of the country and patenting of local plant varieties, mainly in the United States, Japan, and Europe—that are historically used in treating patients. Work is also being done by a few individuals in the medical field and their organizations in relation to intellectual property rights and the health industry with regard to drugs, patents, and the pharmaceutical industry. However, the work is conducted mainly in the English language and targets special and limited audiences.

There are as yet no self-conscious links between the FOSS movement, the anti-globalization movement, and the seeds movement, reflecting the fact that these issues have not been conceptualized as a part of the broader issue of monopoly rights provided by the intellectual property rights regime over intellectual contributions.

The study and advocacy campaign conducted by the Open University can be considered one of the first attempts to conceptualize all these mobilizations under the overall theme of A2K in Sri Lanka and an opportunity to watch a local A2K movement in the making.

That said, players who can be considered part of the Sri Lankan A2K mobilization can be seen using a range of strategies. From our own experience at the Open University, defining the A2K movement in the broadest possible sense is itself a useful strategy. We define the A2K movement as a loose alliance of hundreds of actors, each playing his or her own role in distinct areas. This approach, we believe, will inject a strong dose of confidence that will help participants play their respective roles more effectively. But it also creates a need, or at least the space, for a collective entity to play a role as coordinator and to highlight the A2K aspects of each group. From a Sri Lankan perspective, it is too early to comment about an evolution of modus operandi.

QUESTION 2 In terms of shared strategies, the FOSS initiative in Sri Lanka is very much a part of the global FOSS movement. The Free and Open Software Community in Sri Lanka (http://foss.lk), the Lanka Linux User Group (http://www.lug.lk), TLC, the Linux Centre (http://thelinuxcenter.lk) and LSF, the Lanka Software Foundation (http://www.opensource.lk) can be identified as the main actors of the Sri Lankan FOSS movement. The small, Colombo-based FOSS community is spreading slowly within the country. It is very much linked with the global FOSS movement and in fact has made significant contributions for the development of FOSS. The introduction of transaction protocols, disaster-management software initiated after the tsunami, reliable messaging protocols, operating systems to address the issue of lack of support for the Sinhala language unicode system, and applications bundled with Linux distributions are some of the contributions.

The antiglobalization movement also is not a Sri Lankan product in its general content and strategies. Its themes, activities, focus, and sources of information are very much dependent on international resources. The members of the movement consist mainly of NGOs and depend heavily on funding from international sources. They operate as members of regional and global networks and hence are guided by the themes and work plans established at the international level. Some of the campaigns of the antiglobalization movement go unnoticed as a result of a lack of awareness by the general public of the themes based on which these campaigns are organized. Expertise on issues is generally developed by participation in regional and international forums, and what we see in Sri Lanka is the application of those on the ground in building Sri Lankan cases. The work, however, in some cases has resulted in sustainable initiatives. Some of the academic research conducted originally was initiated by NGOs and was facilitated by NGO funding. Some of the political parties incorporated action against adverse effects

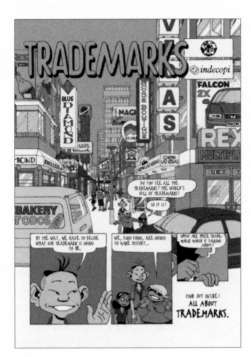

World Intellectual Property Organization comic on trademarks (http://www.wipo.int/freepublications/en/).

of globalization as a result of the initial work done by NGOs. There is, however, a long way to go to see an antiglobalization movement that is sustainable and that operates on an agenda originated from local concerns.

The seeds movement can be seen as a mixture of international and local influences. Surprisingly, the practice of saving traditional seeds, which one would expect to be a natural concern of Sri Lankan farmers, needed to be reintroduced to farmers with the facilitation of external funding channeled through local NGOs. The mind-set created by the Green Revolution that considers high-yielding varieties to be profit makers made the exercise of reintroducing the saving and cultivating of traditional seeds difficult. In contrast, protests against biotheft and biopiracy were spontaneous and considered by the Sri Lankan public to be national issues to which they could relate to easily. The public memory of mass-scale theft during the 450 years of colonial occupation may have played a role with regard to this.

The strategies of intervention by some of the activists and groups working on intellectual property rights and seeds seems to follow a strategy different from that of the general seeds movement consisting of many of the foreign-funded NGOs. While NGOs prefer to maintain minimum engagement with state mechanisms and political parties, these activists and groups have not been averse to working with such mechanisms and parties. The links they maintain with the state and with political parties have allowed them access to inside information and influence in the formulation of policies and legal texts.

One important reason they have this access is their ability to isolate the issue in question and to focus only on it when engaging with the government and state structures. For example, the stand of the government or political parties on achieving peace in Sri Lanka was not taken to be a barrier to dealing with it on other issues, such as intellectual property rights and seeds. The minimum engagement by NGOs with the state, and sometimes their hostility to it, on the other hand, and their disregard of political parties as influential actors can be traced back to their ideology of identifying "civil society" as a distinct and an exclusive category and a model of righteousness and justice and hence to their claim to occupy the moral high ground in their interventions. NGO involvement is also confined to the boundaries of the mandated projects for which money is received. The long-term commitment that is required to form a movement is not provided by project funding that has short-term goals. Funding has also created a situation where work is not done after the evaporation of funding, leaving many of the initiatives abandoned at the end of the project term. Voluntarism, which should be an essential item in forming a movement, is hence hard to find.

QUESTION 3 On the issue of the A2K movement and the law, there is a need to establish A2K as a right irrespective of the existence of the intellectual property rights regime. In the Sri Lankan context, there are many other important barriers that restrict A2K other than intellectual property rights. Two such important barriers include a serious lack of infrastructure facilities that have no significant link to intellectual property rights issues (libraries, electricity, computers, and so on) and language barriers that result from the fact that knowledge is comprehensively available only in the colonial language of English. Therefore, A2K needs to be a movement that reaches beyond a legalistic focus and beyond intellectual property rights.

Moreover, laws are made by people, and hence they can be changed by people. A2K should be understood as a movement that exists beyond the texts and interpretations of the law. A dialogue on A2K should be based on principles and values, rather than functioning as a mere response to existing intellectual property rights law. The values of equity and justice provide a solid ground for the A2K movement to stand on its own, rather than becoming a mere response to intellectual property rights issues. UN conventions and chapters on fundamental rights in the constitutions of individual countries provide legal, ethical, and moral grounds to expand the scope of the A2K movement to address a whole range of barriers—barriers that also include intellectual property rights.

QUESTION 4 At a fundamental level, the objectives of A2K and the objectives of the private sector are not in harmony with each other and in fact conflict in many occasions. A2K is about rights, while the main and the only drive behind private sector is profit. In fact, the intellectual property rights regime we see today is not a result of hard-fought campaigns by the public for the protection of their intellectual contributions, but mainly a result of advocacy by multinational software and pharmaceutical companies to enhance their profits. The origin of the A2K movement can be traced back to the date when intellectual contributions were made private property, thereby opening space for profit making.

Governments, on the contrary, as the institution responsible for formulating laws and policies and implementing plans, have a strong role to play in A2K and should be coalition partner of the A2K movement. As the mechanism that has the primary responsibility for ensuring basic rights for all citizens, there is nothing theoretically to prevent a government becoming a coalition partners of A2K. However, the fact that most governments do not work in the interests of the public needs also to be noted. Thus, it is also vitally important for the public to be aware of and motivated to protect and fight for their interests.

QUESTION 6 A2K could be a movement, a coalition, an aspiration and also a right. Bringing all groups involved with A2K issues under one A2K umbrella and getting them to communicate with each other is important because that will make them realize that A2K as an essential right of people that is being violated in many ways and on many fronts and that mobilization for A2K should not be limited to countering the monopoly power of, for example, Microsoft in software. It would also make them realize that there are many others involved in many ways with the movement and that it is necessary for these groups as a collective to overcome their feelings of disempowerment.

The seeds movement in Sri Lanka provides a good example of coordinated action. Activists on the ideological, legal, and farmer fronts complement each others' work, mobilizing people, advocating, safeguarding legal spaces in court battles, and saving and sharing local varieties of seeds.

QUESTIONS 7 AND 8 It is too early to comment on future strategies within the Sri Lankan context because there is no A2K movement as such, but movements that have A2K components within their overall aims as described above.

QUESTIONS 9 AND 10 The question of the governance of the A2K movement is bound up with the status of knowledge itself. If one investigates the history of knowledge generation and how knowledge is used as a tool of power by some over others, knowledge itself could be identified as a tool of elite domination. The questions of what knowledge is, which forms or types of knowledge are superior, and the issue of the domination of a particular type of knowledge over others have pervaded human history. Categorization of some forms of knowledge as "formal" and some other forms as "informal" or "'traditional" reflects this reality. Treating Western medicine as formal knowledge and treating the science of local medical practice as traditional knowledge is a good example. This categorization of formal and traditional knowledge has brought the practitioners of Western medicine to the center while leaving local medical practitioners at the periphery.

However, the legitimacy of a knowledge system does not necessarily depend on its ability to explain things or its use value. Attaining formal status is a process that involves power. The science of local medical treatment, which can be considered to be holistic in approach and far superior in logic, with minimal side effects, is considered "traditional" as a result of global power dynamics. The practice of the intellectual property rights regime in considering published knowledge as valid prior knowledge and in offering intellectual property regime flexibilities only for "scientific research" (as is the case in Sri Lankan patent law) makes holders of formal knowledge elites and denigrates the holders of traditional knowledge. Within such a context, it is important that when we refer to the term "A2K," we ask "Access to *what* knowledge?" And access for whom?

In fact, A2K itself can be understood as an elite movement, depending upon how we define knowledge and how we identify modes and approaches through which knowledge is formulated and accessed. For example, computers and the Internet still do not play a significant role in Sri Lanka in accessing formal knowledge because of the simple fact that a majority of the population do not have access to computers, and even when computers and Internet access are available, many find it difficult to use them productively because of the unavailability of material in local languages and because they still find it culturally alien, even when language is not a problem. For the majority of people who work with nature and who access knowledge through reading newspapers, chatting with fellow villagers, listening to the radio, and attending various social forums, sitting in front of a computer and accessing information in an unfamiliar language is rather a strange and complicated exercise. The use of computers and the Internet by school teachers and university academics still remains low due to this, as well. Narrowing the scope of A2K to FOSS itself can be understood as a way of rendering it the movement of an elite, from that point of view.

The theory of A2K must be constructed on the recognition that the separation between "formal" and "traditional" knowledge is a manifestation of power relations more than anything else and that informal and traditional knowledge still plays a significant role in the lives of everyone, and not just in Sri Lanka. Traditional knowledge in agriculture, health care, human-resource management, communications, water management, stress release, and so on are examples of knowledges still in use, some significantly and some to a lesser degree.

These traditional knowledges, however, are under threat for a range of reasons. Traditional seeds provide a good example. Traditional knowledge in agriculture is held by farmers. This knowledge is mainly saved in traditional and local seeds and planting material passed from generation to generation between farmers. When this transmission of seeds and planting material is threatened, access to traditional knowledge in agriculture is also threatened. That is the case today. However, it is also important to recognize that safeguarding traditional and informal knowledge does not imply bringing the process of disseminating traditional knowledge into formal modes and mechanisms. Asking farmers to publish their knowledge of agriculture so that the exceptions to intellectual property rights available for "prior art" can be obtained, for example, is in fact equivalent to a proposal for removing farmers from their cultural context. Otherwise, even traditional and informal knowledge becomes an elite product. Sometimes the very process of identifying and classifying traditional and informal knowledge can alienate it from the communities where it was located and accessed.

NOTES

1   Yochai Benkler, *The Wealth of Networks: How Social Production Transforms Markets and Freedom* (New Haven, CT: Yale University Press, 2006).

2   "The project was meant to consider what the patent system might look like in 20 years time, in order to gain greater clarity about the threats and opportunities the future might present. The purpose of scenarios is to examine possible uncertainties that might arise in a complex and turbulent environment. By deploying this methodology, a wider view can be taken and more relevant questions can be asked." European Patent Office, "Background to the Project," available on-line at http://www.epo.org/topics/patent-system/scenarios-for-the-future/background.html (last accessed January 12, 2010).

3   For more on this policy, see National Institutes of Health, "NIH Public Access Policy Details," available on-line at http://publicaccess.nih.gov/policy.htm (last accessed January 12, 2010).

4   See David M. Kotz, "Building Socialism Theoretically: Alternatives to Capitalism and the Invisible Hand," *Science and Society* 66, no. 1 (Spring 2002): pp. 94–108, available on-line as "Socialism and Innovation" at http://people.umass.edu/dmkotz/Soc_and_Innovation_02.pdf (last accessed February 10, 2010).

5   Susan K. Sell, *Private Power, Public Law: The Globalization of Intellectual Property Rights* (Cambridge: Cambridge University Press, 2003).

6   See Roberto Verzola's essay "Undermining Abundance: Counterproductive Uses of Technology and Law in Nature, Agriculture, and the Information Sector" in this volume.

Interview with Yann Moulier Boutang

Gaëlle Krikorian

GK: Until recently, most countries, despite enforcing their own laws protecting intellectual property, were not held to the same standards that are in force in the wealthiest countries. Now, however, every country belonging to the World Trade Organization must adhere to WTO standards, which has led to a considerable toughening of protective regulations in many countries. The question of how to apply these rules is currently at issue in most developing countries, but answers are not always evident. For intellectual property holders, the stakes are high: If the new laws are not respected, the property they hold loses all value. Yet you have put forward that the emergence of the notion of cognitive capitalism, the accumulation of immaterial assets, implies, by definition, a condemnation of the current frameworks of intellectual property law and of the tendency to impose increasingly restrictive protective measures. Should we then view this aggressive push to apply extreme, increasingly drastic policies to safeguard intellectual property, for example in the context of bilateral negotiations or international relations such as the Anti-Counterfeiting Trade Agreement, as the death throes of a dying watchdog?

YMB: Yes, I believe so. This is a classic, widely observed phenomenon. The tightening of legislation, contrary to what we might think, does not equate to a preventive implementation by the state or the authorities of a more repressive system than the one that preceded it. On the contrary, it is a proof of the inverse: that there are transgressions. It is the extent of these transgressions and the challenge to existing economic models that they express that explain the explosion of enforcement measures. It was because the poor were poaching, chopping wood, and traveling far and wide, especially according to where legislation favored them, that the Poor Relief Act was adopted in Great Britain in 1662. Under that legislation, beneficiaries could receive the poverty allowance only if they were registered in a county and remained there, unless in possession of a fixed-term work contract. (It is interesting, by the way, to consider the parallels with what is currently happening in France and elsewhere with respect to international immigrants.) In Great Britain, the goal of the 1662 legislation was to limit the circulation of the poor. We see the same phenomenon in the repressive laws governing slaves, the use of tracking dogs, bounties, and physical punishment, such as the cutting of tendons: Each of these barbarian methods expanded at a rate directly proportional to the incidence of fugitive slaves attempting escape. In the same way, I believe, major record labels' excessive zeal for protecting intellectual property and the insistence on this issue in the United States are mainly consequences of the fact that much of this legislation cannot be applied and is constantly being violated.

With any legislation, you have to consider its application. Enforcement is expensive and requires the power to monitor, to employ lawyers, to rule on cases, to oversee the application of verdicts, to supervise payment of fines, and so on. This kind of framework cannot be coordinated unless there are very large sums involved. Beyond a certain tolerated level of transgression, the major labels will take action. As long as there were only roughly fifty thousand users per day connecting to Napster to exchange data, copyright holders were unconcerned. When the number rose to 36 million users a day, it was considered unacceptable—since then, one can only guess how these exchanges have skyrocketed with the leap in video downloading made possible by broadband.

Still, we need to realize that this push to curb access is, by nature, in absolute contradiction with the spirit and practices of cognitive capitalism. The latter cannot develop except through intensive digital usage, which in turn opens up an extraordinary potential for the diffusion of content. This dynamic began first with text, then sound, and then images. Gradually, everything becomes accessible, which presents a significant problem for copyright holders, who then try to shore up the walls, their legislative and technical protection systems, such as digital rights management (DRM). The framework protecting intellectual property was set up before digital content found a wide audience, when broadband was not yet available. It was the masses' general ingenuity, more than the "general laboriousness" of which Marx speaks, that grew with respect to technology. Because the goods in question are nonrival, that is, infinitely consumable without being exhausted, copyright holders find themselves constructing further obstacles beyond those protecting material goods. These obstacles in the domain of knowledge goods are fundamentally, absolutely artificial, whether they take the form of laws or technical tools such as DRM. The tightening of rules is proportional to the frequency of the exchanges taking place. Copyright holders are not raving, paranoid, or obsessed—they are

Cartoon on copy protection by Reto Fontana, Basel (http://www.urheberrecht.ch).

simply defending their interests, blow for blow. They are defensive lobbyists, similar to the planters who feared the effects of contamination when a free labor force developed. They will use any means available, even, for example, attempting to regulate in the name of antiterrorism, to gain influence over other countries and defend against counterattacks such as Freenet, which allows people to share files, browse the Web, chat, and establish Web sites anonymously, or the practice of encrypting exchanges in order to keep their contents private.

What we have seen amounts to a series of desperate efforts. One may recall the defeat on the users' side, when Kazaa lost its case in the U.S. Supreme Court. Radio listeners started recording broadcast content, digitizing it, and then broadcasting it again without infringing copyright. Intellectual property specialists called it "the analog gap." The issue of such recordings was brought to Congress, and certain members of Congress proposed, in all seriousness, that radio programs should be aired with a technical tattoo to make these recordings impossible. That restrictive project failed, because it was completely unrealistic, but this shows how those hoping to restrict such exchanges sometimes let their imaginations run wild.

In French corporations and universities, the replacement of the position of the director of information resources, or DRI, with that of director of information systems, with security as the primary objective, led to an absolutely insane level of compartmentalizing policies. These policies made little difference in a situation that has become familiar: You can build a stronghold, but when a pirate infiltrates it, and he always does, he will be the one turning the protective systems to his advantage, becoming invulnerable to expulsion. That is what happened with Microsoft's famous Palladium project, through which the firm hoped to create an ultratough armor for computers by constantly violating users' privacy to verify that applications met intellectual property regulations. In the same way, when Sony installed spyware programs in its video games, it was in the end at great cost to the company. It wreaked havoc on users' computers. Many lost the contents of their hard drives. The company was unable to uninstall the program and was finally forced to recall the games and compensate hundreds of thousands of customers whose computers were destroyed. In the chase between cops and robbers, the cops never get a head start. There is a delay, and their route is full of pitfalls.

The only real risk is when one is dealing with very large corporations that can collude as oligopolies or cartels. The Internet would never have become available to everyone, and we would never have witnessed the critical shift from heavy computers with simple terminals (like those in the Sun system, Oracle, and those of other computer companies) to the home computer and then the laptop, if ITT-ATT had not been dismantled in 1974 into two separate firms, or, if later, IBM had not been threatened by the prospect of a split.

When the Internet was developed, Microsoft did not believe in it at first. Netscape's Web browser quickly became the most widely used. In order to push its own, decidedly inferior browser, Internet Explorer, Microsoft deliberately programmed its Windows operating system, which is installed as a matter of course on all PCs (only Macs run without it or PCs with the GNU-Linux operating system) to make it systematically crash whenever users tried to open Netscape. This monopolistic practice (along with the obligatory installation of Windows on personal computers) came before the courts in a highly publicized trial brought by Netscape and by more than thirty of the states, including the largest ones in the country. Microsoft was able to escape being dismantled in the United States thanks to George W. Bush's election in 2000.

In Europe, however, a very similar legal action against Redmont undertaken by the European Commission ended in a conviction and a record-setting fine of $480 million, while Microsoft's appeal in the European Court of Luxemburg was rejected. I believe, therefore, that the only really dangerous enemy of consumers' individual liberties in their endeavors of cooperation and the creation of new common space for innovation is very large corporations' ability to block access to digital technology by imposing new enclosures, as Yochai Benkler, Philippe

Aigrain, Lawrence Lessig, and I have emphasized. Even the president of IBM could have prevented the democratization of computer access. When he was advised to seek a patent for his technological advances after he helped develop the computer, building on the pioneering work of Alan Turing, John von Neumann refused. As president of IBM, his view was that the computer would never sell and was destined to be used by only a small handful of wealthy customers—himself included. Monopoly, elitism, and Malthusianism all go hand in hand. We have seen how far wrong his prediction proved to be.

GK: While economic self-interest may explain the logic espoused by copyright holders in wealthy countries, we see these nations also bringing their full influence to bear on the poorest countries to force them to contribute sums that appear colossal in view of their resources to help increase protective measures and enforce regulations, even though the economic impact of these measures is relatively insignificant. What is at stake in terms of controlling knowledge and technology, however, is very significant. Is this where confrontations between wealthy and developing countries are playing out? Why would developing countries join such an unequal game?

YMB: In the history of patent legislation, it is very clear that emerging economic powers have never respected patents during their period of development—whether we consider the United States, the Soviet Union, or China—because they knew it would prove too costly. Poland, for example, was favorable to software patenting, but when it joined the European Union, Poland dropped Microsoft flat because it became clear that the cost of intellectual property rights would far exceed the benefits to be gained from fiscal advantages or a low-cost workforce and that small companies would be forced out.

The United States made its move into cognitive capitalism early. The Americans also have a legal culture that lends itself especially well to the culture of patents and brands, while countries with a non–Anglo-Saxon culture are much less well adapted to this activity. The United States is fifteen or twenty years ahead and is trying to consolidate its lead to prevent other countries from catching up. Here we have a classic trait of economic liberalism: turning liberal when you are at the top of the class, in a position of hegemony, because you no longer fear competition. It is an attempt to establish one's power and push one's advantage. The United States put in place a legal system protecting intellectual property that, in fact, allowed the country to protect itself from Japan, for example, which was an emerging rival. The eight years of Bill Clinton's presidency and then George W. Bush's presidency until 2001 were characterized as the Silicon Valley years, with the boom in California. The Americans knew that intangible goods, services, and intellectual property were

Julien Milliard (www.lagouache.com).

not covered by the commercial negotiations begun in the 1960s, with the Kennedy Round of the General Agreement on Tariffs and Trade, which led to the creation of the WTO. And because free access was finally taking off at an international level, the United States, very shrewdly for their interests, set in motion a legal offensive designed to confirm their lead over other countries. They made what Laurent Thevenot calls "investment in forms," that is, an investment in an institutional framework that, once in place, allows one to reduce the cost of transactions. When the Marrakesh Agreement was signed in 1994, not one country in the Global South came out with a positive balance in terms of intellectual property. These countries could not have simultaneously accepted free access and policies protecting intellectual property unless they had some hope, in the end, of benefiting from the system in one way or another. Instead, they were offered a Paretian balance, in which no country can improve without the other worsening. The United States gained more than the others while convincing its partners that they, too, would come out winners in comparison to their situation before the agreement.

Of course, not all the Southern countries were in the same circumstances. India, for example, had already developed cutting-edge pharmaceutical, chemical, and electronic industries. Korea had invested in patents, as had Japan before. Korea began with large-scale importation of patents and went on to produce them itself. Korea currently produces a phenomenal number of patents, far more than France. There are also many Southern countries with financial investments in Northern ones—sovereign wealth funds made investments in banks with large intangible holdings. Some countries, such as India, Thailand, and Brazil therefore stand to profit from the system as it has been adopted, but not the majority of developing countries, and this system leads to inequalities between Southern countries.

Even in countries such as India, there are greatly varying positions with respect to this issue. Vandana Shiva and others are adamantly opposed to increasing intellectual property protection measures, and they won a victory in India's parliament in 2004, when new intellectual property legislation, which was supposed to be enacted a maximum of ten years after the Marrakech Agreement of 1994, turned out to be more flexible than the United States, Europe, and Japan had hoped. Nevertheless, the more that Indian capitalism advances (take, for example, the steel producer Arcelor Mital, the computer industry centered in Bangalore, or the generic-drugs industry), the less India wants to play the role of a mere subcontractor or that of the number-two factory of the world. India is rising in the productive circuit, investing in research and development, and producing large numbers of engineers—English-speaking ones moreover—and founding research centers. India's traditional business strength now goes hand in hand with the project of capitalizing on intellectual property.

GK: These diverging perspectives and inequalities between countries play into the hands of those working to develop and homogenize intellectual property protection on an international level. But despite these deliberate efforts, it seems that the push for access is difficult to contain, and myriad cracks are spreading and threatening to burst open. Copyright holders' ability to work in a partnership, to resolve their competitive differences and join forces, allowed them to obtain the Trade-Related Aspects of Intellectual Property Rights Agreement—the TRIPS Agreement. There are, however, still permanent tensions that could grow into fractures between these groups, benefiting those who are pursuing access to knowledge and information.

YMB: The copyright holders were successful in obtaining TRIPS, but they lost at Doha over the ticklish question of generic prescription medicine.[1] The success of TRIPS opened the eyes of many Southern countries, because although India had Bangalore and managed to stay afloat, the agreement put a real stranglehold on countries in Africa and Eastern Europe. Proponents of toughened enforcement of intellectual property laws tried to regroup, but it's true that cracks had begun to spread.

The first crack appeared in the area of proprietary goods and services, under the pressure of free and open-source alternatives. Consider IBM, for example. They had begun making watches and calculators in the 1930s, then shifted to very large computers, then software, then computer tech services. In marketing those services, IBM at first focused exclusively on offering proprietary services, based on very large patent holdings. As IBM was aware, however, a real culture of free access was developing, and they later changed their approach and decided to explore free solutions. The company invested 3 billion dollars in the free approach, calculating that they stood to save at least 600 million themselves on nonproprietary systems. IBM's yearly business revenue from free services is currently 3 billion. IBM opened up a new sector to develop this culture, which did not exist inside the company, a culture based on cooperation, sharing, and the use of standards acceptable to everyone, without favoring one group at another's expense, in contrast to proprietary systems, in which the objective is imposing a standard and taking users captive.

When IBM made its move to free services, the computer boom was losing momentum. Corporations that had for years cheerfully financed everything to do with computer technology began to cut off funding. In this context, it became clear that the free system was eminently economical, but often also more efficient and reliable, in part because it is easier to identify and isolate piracy in programs with open, readable source code. This is also why Java, a free programming language, became standard, even in hybrid open-source systems. The allied forces of major computer firms working to defend proprietary solutions disintegrated. This is why IBM, which manages a portfolio of fifteen thousand patents that earn the company 2 billion dollars per year, decided that its engineers and departments needed to address its clients' culture of free access and do everything it could not to alienate them.

At the other end of the spectrum is Microsoft, whose image has become so repugnant to fans of free access that they have a permanent hostility to the firm, and Microsoft cannot even get its wiki platform to work, using the company's own standards, to break into the universe of free access. Proprietary vocabulary remains so closely associated with Microsoft that its corporate managers have even considered trying to find the company a new name.

Microsoft's overwhelming presence in the market goes back to Bill Gates's stroke of genius when he decided to grant IBM a license for his MS DOS operating system in 1985–86—at a juncture when IBM was releasing its PC patents to the public domain—on condition that the system be systematically installed on all personal computers. This clause was and still is completely shocking and rapacious in terms of nonbiased competition. Microsoft acquired a position of near monopoly, no thanks to its powers of innovation. As we have seen, though, Microsoft missed the ball when the Internet arrived. Today, the company is aware that Vista and Windows 7 are the last proprietary programs it will be able to impose on customers. In his departure speech, Bill Gates even made this explicit. The plan to corner the market on all user applications connected to the operating system had backfired. This was, in fact, what the company was censured for in the wake of antimonopoly legislation: creating mandatory compatibility between the operating system and applications. Although Microsoft slipped out of its trial in the United States, thanks to the Bush administration, in Europe, the company was ordered to provide parts of its operating system to users working with related software. In other words, Microsoft failed in its bid to capture, via Windows, control of the other user applications for music, television, video, games, and office and home automation.

The stunning success of free software in the server market foreshadowed the decline of Microsoft's business model, which was unable to integrate hardware and software into a proprietary whole with public success, as Apple did, or ride the powerful wave of free cooperation, as IBM finally did. There is no doubt that creating barriers around a strictly proprietary model contributed to a kind of inertia that, for many hackers, reflects a similarity to the dinosaurs. This split between the free and proprietary approaches was the first area of conflict between major proprietary companies.

Another crack in the united front maintained by intellectual property advocates was thanks to Google. The directors of Google are aware of the increasingly immaterial nature of production. They also realize knowledge goods are difficult to sell. They have built an economic model that meshes with this free-use era. The model works because Google offers an unbelievable range of services and access. From this point of view, Google is a "small-c" communist organization, in contrast with Microsoft, which labeled developers of free software "communists," as often happened during the McCarthy era. Google realized that interaction itself would represent an economic resource.

This is what I like to call "pollination" and "coralization." If you offer people free tools, human activity will seize them and begin to pollinate as bees do, but it will produce something much greater than honey, something called "positive externalities," in this case, networks, which in a business economy normally require an enormous investment, hundreds of billions of euros. The value of pollination if compared to the value of honey and wax product is three-hundred-fifty times to one. Interaction creates networks spontaneously. Peer-to-peer services build networks, which are very valuable from an advertising point of view.

Google understood the role that interaction can play. In this system, human interaction is the raw material. Google harnesses it in a way that advertising executives do only secondarily, by using the linguistic and communicative inventiveness of people who are constantly creating perspectives on society. Google represents a huge step forward. It forced the gaps wide open and caused a crisis, or at least an awkward predicament, for those supporting the proprietary system. This is why I believe it is strategically sound to create an alliance with Google to dismantle old, archaic models, even though I feel we simultaneously need to be ready to fight it, because Google's goal is to make money, and the company could, in any case, be bought out by Chinese pension funds or anyone else at any point, which could easily lead to problems, especially around the issue of privacy, since Google uses personal data.

Google has also developed a real dynamic of social creativity within the company itself. It is the only company in California that has kept the spirit of the golden age of start-ups so alive. The company offers both innovative and well-paid working conditions, somewhat like the multinational companies of the 1950s and 1960s that paid 35 percent more than other employers. It is an entirely different conception of creativity being applied, where on Thursdays, for example, employees have no obligations in terms of work. This is a fundamentally contrasting approach to that of companies such as Microsoft, which are run through secrecy and stress. Google understood that creativity requires special conditions to develop and that keeping employees under constant pressure, like bees stripped of their own honey to make them keep producing and feeding themselves, is a fruitless approach. People who are put in an environment where they can flourish will produce results on their own. This is why it pays to offer a guaranteed minimum wage that provides the proper conditions for creative production, somewhat like the grants given to researchers. Giving someone an order to create is absurd. This kind of performative proclamation is impossible. It's impossible to sit people down in front of a sheet of paper and tell them "You shall produce," the way workers used to be stood in front a lathe and told to produce. By the same token, ordering people to cooperate is completely ineffectual. We cooperate with people we like, with whom we share affinities. Although it is possible to coordinate by force, this works only through a material apparatus. The new immaterial nature of production has freed these processes from Fordist control.

Traditional employers have a warped, negative view of these issues. They often think we could live in a society characterized by abundance and information, a society in which the state invests and certain circumscribed areas would remain free of charge, without affecting the business sector, where it would be possible to apply a host of intellectual property regulations and reap the benefits. They believe there can be periods of expansion, like the dot-com boom, but that there is always a financial stabilization afterward, bringing us back to the baseline. But that won't happen, because there is no baseline. The new situation brings a greater degree of freedom, in comparison with the capitalism that preceded it. Contrary to their hopes, there cannot be a partition between material, industrial production and life. Universities are the factories of the twenty-first century, producing codified, marketable knowledge, and in this sense, the European project to place university funding at 3 percent is ridiculous—funding will realistically need to be in the region of 20 percent.

GK: If cognitive capitalism has ushered in certain kinds of progress that tend to call into question traditional hierarchies such as employee/boss, worker/owner, and so on and that may have the "power to liberate society," there remains, for a majority of people worldwide, the problem of access to this new world. As you have commented, intellectual capital becomes discriminatory in this context, in the distribution of social divisions, and although participation in training and education has risen around the world, for a large proportion of the population, the world of cognitive capitalism is still out of reach. This is where we encounter the difference between information and knowledge. Gaining access is one thing, but being able to evolve in this period of expansion, to benefit from what is available, is another.

YMB: It would be a mistake to think that cognitive capitalism is relevant only in Northern countries. Some people say Africa isn't wired in, and poor populations will never be able to connect, and that's not true. We have seen, for example, the explosion of cell phone use in the *favelas* in Brazil. This is because, in the context, portable phones offer an incalculable range of uses: to look for a job and be contactable at any point regarding job offers, to gain access to a boundless range of things and types of information, to protect yourself by calling the police or friends when it's necessary, and so on. Cell phones are a network for everyone. This makes them even more important than housing—they even take the place of fixed housing for many people. We live in a physical circle of personal contacts, people we know physically, but outside this circle is a much larger circle that also plays a very important role. For now, the content that people exchange on cell phones is of only relative interest—soccer scores, standardized information, and so on, but things can change. And people are certainly going to be turning more and more to free systems, as has happened with peer-to-peer, because the activity alone can make these channels economically feasible.

There is no question that, on the one hand, education and training have a role to play. On the other hand, it would be wrong to think of education as an empty bottle to fill. That's not how it works. Popular cultures produce values and knowledge. And it isn't true that just because people haven't been informed or trained, they can't use technology and engage with the transformations taking place.

We also need to realize the networks are interconnected. Cell phone users, for example, represent an extraordinary mix. In Africa, these phones are important for illiterate users, people with higher education, immigrants sent back to their own country, and so on. I do not share the views of Dominique Walton, who says the Internet gives an illusion of democracy while in reality, users are trapped by codified content, and the whole system simply follows the market's rules. The dot-com crisis in 2000 was associated with a poor understanding of the phenomenon. The market sector thought it could take hold of the Internet, and it turned out to be mistaken. The market won't sell effectively on the Internet unless it makes effective use of interaction and offers things that the conventional market system doesn't. The market cannot engage with this kind of interaction unless the Internet remains open and not locked into a strictly market-oriented arena. It is interaction that empowers users and gives them agency. This potential cannot, of course, be realized without looking at the political struggle for access, open source, and a general free-of-charge culture....

GK: In developing countries, we find a certain elite (although, as they make up 10 percent of the population, we should perhaps stop referring to them as elite) that has emerged and become a player in this immaterial world, but there is also a split. Until now, the upper classes and the intellectual and cultural elites have kept control. What is more, on a national as well as an international level, inequality is growing. In the context of productive connections formed through cognitive capitalism, should we worry about the possibility of a worldwide social structure based on two main classes, one for which cognitive capital represents an opportunity to exercise its freedom and another that is excluded from this new reality, remaining subjugated under classic forms of domination and trapped in the sterile role of consumer? What possibilities are there for these populations to take part in the creative process?

YMB: There is a split between the populations, but it doesn't divide them into monolithic units. The split is related to culture, rather than only to economic factors. In Brazil, for example, we can see how the split between those in favor of a proprietary approach and those opposing it becomes a new determinant of class distinction without following the contours of people's social attachments. It actually cuts across the closed class system and divides it on different lines. Access to digital content reorganizes things. We also need to recognize that the masses, such as poor *favela* dwellers, represent a valuable resource for cultural capitalism. In Rio, the carnival is the means to exploit the creative resource of the *favelas*, serving to boost tourism, which has become the city's main industry.

Of course, managing to harness this rich resource is one means of functioning in modernity, but harnessing people in this way is also a way of co-opting them as partisans of the proprietary system. This brings us to a basic question of access and the free-of-charge system. If we simply demand piracy, making everything free, without addressing the question of intellectual property except as a limit to be transgressed, it leaves the creator out of the equation. How will Native Americans or Aborigines be repaid or compensated when multinational corporations plunder their intellectual resources at the same time that these same corporations make vehement claims about the importance of respecting intellectual property when they are the owners in question?

The debate on open source and copyleft versus the Creative Commons is very important and is helpful in thinking about this question. There are real differences between these choices. On the one hand, you have Richard Stallman, a computer scientist, who doesn't understand that although in collaborative programs, it is more effective not to protect them from interventions by others, the same does not apply to writing a text. Mallarmé would be categorically opposed to having his texts modified, and we can understand why. Similarly, in the context of a collective text or a manifesto, for example, although it is drawn up collectively, we also understand that the text cannot be changed unilaterally, even by those who support it. Nobody would let right-wing extremists alter a manifesto for the Socialist Party, and vice versa. There must be a closure clause to protect the integrity of the project. On the other hand, proponents of the Creative Commons, with whom I identify myself, are opposed to a certain kind of diabolical, paradoxical pact. We are against the alliance between the most sophisticated forms of capitalism, based on rights of ownership, and a position we might call "terra nullius," which says that anything not codified as property is public domain and can be used—in other words, that positive externalities can be harnessed. Corporations are very skilled at this. The real question is how we can reencode what is in the public domain while also protecting its authors.

In the current system in France, authors' rights are recognized only on an individual basis. In music, cinema, theater, and so on, a huge proportion of those contributing to the production and creation of a work of art are not considered its authors, but are seen as technicians. Furthermore, the cost structure represents the bulk of distribution costs. In publishing, we find the following breakdown: The distribution network takes 60 to 70 percent of the product's final value, and once the bookstores have deducted their fees, that leaves around 10 percent to be split between the publisher and the author. The distribution mechanism of membership contributions works in such a way that small sums, which should go to the authors, in fact end up in a shared account and eventually benefit only those who make the largest contributions. Only 10 percent of authors can currently make a living from writing, but they can do it because those who make 10, 20 or even let's say 50 percent of their living are contributors to their product. In this way, lower-profile authors, as well as those who contribute to a work's production without being considered authors, have little to gain from the current system and often have to endure unstable positions. Digitizing a product can lead to lower revenue, because of copying, which then leads to a crisis in this economic model. The ones who suffer the most are people already struggling to make a living.

If we look at the alliance against copying in the cinema world, it is composed, unsurprisingly, of the major SACEM (Société des Auteurs, Compositeurs, et Editeurs de Musique) authors, but also technicians. The groups taking action against copying are, in part, those profiting from the established system, but also many others struggling to survive who know the economic model is outdated, that copying affects financial returns and consequently the investment in production on which their salaries depend. Accordingly they view the expansion of digital content as a threat. If sales are affected by digital content, then producers become anxious and try to limit costs, which affects technicians first, destabilizes production teams, and so on. This is why we must tackle the issue of authors' rights and find a new form of insurance by changing the cost structure, which we can do through increased dematerialization.

There are still vast numbers of elements that could be dematerialized. If this lowers costs, then, first of all, prices will drop, making the results more affordable, but it will also be possible to shift the dividing lines, and we will then be able to change current relationships of power. We could consider the possibility of a model in which, through digital channels, products become public and accessible while production, authors, and technical teams would be subsidized. This is how the Opera is currently run in France. When you buy a ticket for 20, 50 or even 200 euros, the actual cost is around 1000 euros. The opera is subsidized, along with the press and free access. While in the current system, distribution channels take the lion's share of funding to cover costs, materials, energy and transport, we could conceivably push digitization further along, so that instead of a CD or a DVD, for example, you would use an intangible product (a digital file) that can be distributed at very low cost. Of course, there would still be distribution expenses, such as the cost of network administration, but instead of making up 70 percent of the price, they would represent 10, 15 or 20 percent. A large part of the price would in this way be made available simply to pay those with artist roles, but also technicians and other jobs associated with the creative process.

The objective is to incorporate free access into economic models, develop channels, find finance circuits that use interaction as a resource, and change the legal structure. The point is to move from an extremely primitive copyright system based on elementary, binary notions that says you either have the right to do anything or no rights at all—the root of copyright is prohibiting copies—to a system where the author and the producer exist, where each one's contribution is recognized and, by extension, brings in a living wage, and where what counts is the way a product is used.

This brings us to the importance of *usus* in relation to *fructus* and *abusus*—in other words, it brings us back to property laws. Depending on the situation, it is possible to make the public pay, give it something free of charge, or strike a balance between the two extremes. In the system I am describing, the public gains access to products, so in this model, it is the public that trains itself, providing its own references about a product. The public is constantly tagging the product. I always thought that if Paul-Loup Sulitzer's books weren't wrapped in cellophane, if they were on-line, there would be a spontaneous backlash, and nobody would want to read this literature, which is of no interest to any part of the public.[2]

The other side of this dynamic is that a good product finds its own audience, which expands and differs according to taste and use and according to people's ability to pay. As an academic, for example, I cannot afford an access fee of 150 euros for journals and am happy to read what is available on-line. If I read an interesting book, I'll recommend it to the university library and to students. It is clear that putting media on-line does not take away buyer demand. By the same token, it is absurd to view downloads as just so many products not being purchased, especially since people in a position to buy are revealed through the actions of those who aren't and who use downloads instead, because the latter become like consultants. When kids download something and show it to their parents and friends, customers with spending money are directed toward the product. This is the economic model we should be defending. It is not about destroying all definitions of property, but about removing the effects of closure. The current system is presented as being constructed to encourage inventors and producers, when in fact it tends to breed parasitic chains. We must therefore work to enact much more sophisticated legislation that makes it possible actually to support people who are inventing and creating while taking account of people's income and not creating obstacles to access.

People need to consider what type of intellectual property protection will benefit them most. The patent system favors large corporations, because patent registration is something small start-ups cannot afford. There have also been more and more problems associated with applying patent protection. There is no worldwide patent authority, and the United States issues a staggering number of patents, many of which are questionable and rightly called into question. Some Brazilian manufacturers were telling me recently that the tangle of patent administration is becoming an increasingly powerful deterrent to people interested in doing research. This explains why the science community is often particularly cool, if not hostile, when it comes to patents. The author's rights–royalties system, as used by songwriters, for example, is more favorable to small parties, but difficult to apply: It requires frameworks for listening, attribution, allocation, and so on. If we take the example of a flat fee for downloads that has been discussed, it is clear that the task of attributing and reattributing royalties will not be easy.

However, there is also a growing interest in brand names, which are much more manageable and affordable for small companies, because their capital and maintenance costs are much lower, which also explains why brand laws tend to be put in place more quickly than other types of intellectual property laws in developing countries. Furthermore, brands seem to present us with an interesting model in terms of protecting goods based on traditional knowledge and other collective rights. Generally speaking, the advertising model offers many important advantages. It allows one to publicize and recognize authorship and thus facilitates attribution. Publicizing someone's work is not payment in itself, but it does bring professional recognition and help the author to build and promote her career. It is a symbolic wage. On some level, it becomes a form of brand. I believe we should not be fighting the notion of attribution, or the project to codify usage, or the system of economic compensation.

GK: It is in the nature of intangible goods and their means of production to favor certain kinds of collective action—sharing, exchange, collaboration, and the production of collective intelligence. Would you say that the A2K movement, which has gained momentum thanks to the arrival of cognitive capitalism, will be able to defeat the movement for increased intellectual property restrictions by simply continuing to push forward in the same direction that it has heretofore followed? In this sense, piracy might be one of the first steps to take, because it allows people to participate in free expansion and also because it would represent a form of direct resistance against control and repression that we would be better off breaking through immediately, before it has a chance to solidify.

YMB: The problem with piracy is that it can function like the drug dealers' technique of distributing free samples of the product, then cruelly cutting off access in order then to *sell* it to users who have become addicts. Microsoft uses this method. The company knows it is unable to force licenses on the entire market in China, even if it can in Shanghai, on the coast. It isn't possible in the rest of the country. The important thing is for the standard to become indispensable. So Microsoft allows piracy to take place, which helps its products become the default choice. The company's wager is that when people have money to spend, they will go buy the original products and stop using pirated ones, becoming customers. The effect of piracy in this case actually hinders the development of free access.

From a certain perspective, piracy is the tribute vice pays to virtue. Some acts of piracy, namely, viruses and bugs, are deliberately developed by businesses to stimulate customer use of help hotlines—there is proof that they do this. The aura surrounding piracy also serves to legitimize the system. It's like the story of Parmentier: To allay people's anxiety about potatoes, he had to put up a fence around the potato patch, which stimulated interest and desire.

If we take counterfeiting, the business revenue for Swiss watches comes to 5 billion per year, perfume is 5 billion, and luxury industries the same. These three industries are favorite targets of counterfeiting. There is, furthermore, an unbelievably wide range of calibers of counterfeit watches, costing from a few hundred to several tens of thousands of euros. The people who buy copies can't afford to buy real ones, but by buying copies, they are contributing to the image, providing advertising for the brand. Buyers start out with crude counterfeits and progress to closer and closer approximations of the real thing, until they can actually afford it. In China, certain counterfeit retailers offer four or five levels of quality, ranging from very crude copies to undetectable ones. Studies of customs seizures show that wherever counterfeiting spreads, not only does it fail to harm the market for branded products, it in fact boosts the market. The two complement each other, rather than substituting for one another. Counterfeiting stimulates a demand, which turns into a social norm, especially among people with money to spend. It is the same economic model at work in peer-to-peer networks. Piracy, therefore, has a positive effect on sales. It is for this reason that promoters of free access often criticize piracy as being the cause of consolidation of proprietary standards. Some economists view piracy, in terms of a rather sophisticated model, as a form of "free" goods that creates a space for valorization. Having done this, piracy creates a market based on the coralization produced by this particular type of consumer.

Certain people use piracy because they don't want to pay for something, others out of necessity, others out of defiance, and others do so even while making a public issue of defending intellectual property, because people aren't always consistent in their behavior. So piracy brings together people with widely varying motivations. They have different reasons for using piracy to gain access to knowledge, but we need to make the consequences clear to them, to explain the advantages of alternative, nonproprietary systems.

We also need to make sure all computers have free programs installed on them so that people can choose freely. Logically, along with demanding access to knowledge with as few obstacles as possible, we should also be promoting the development of basic computer know-how, which will be a sort of modern civic education and allow every computer user to get into the operating system and see it functioning, see its limits and barriers, and so on. Until now, only "hackers" and "crackers" had these kinds of skills.

GK: There is a very diverse set of figures behind the movement supporting access to knowledge, people who don't always agree about aims or what kind of philosophy to put forward, but who come together to lead an offensive. With a wide range of motivations, they manage to join forces in collective action, which gives them a certain power. How do you view this multitude, and what would you say are the alliances and demands likely to emerge from the issues they share in this context? In the fight to abolish slavery, the ultimate goal could be articulated clearly and simply. In the case of A2K, how is the movement taking shape, and how can the goal be articulated? What seems desirable?

YMB: I think people agree that it is time to condemn the arbitrary reign of corporate power. The antimonopoly effort is one of their motivations. The demands made, however, are more diverse in nature. Some people are in favor of the principle of free access and believe that the domain of learning ought to be subsidized and not represent another space occupied by merchandizing, for making profits. Another category of people is concerned about the decline and contamination of a public sector as a result of market norms and would like to protect this public sector while reigning in market expansion, which is a more sophisticated position. A third type of position claims that the system of barriers protecting property is ill-suited to the domain of knowledge and represents an impediment to innovation. There is yet another category that takes a more technical perspective and raises the issue of the standards battle—which in 1999 led to Tim O'Reilly from Open Access splitting with the Free Software Foundation's policies. They are convinced that the fundamental issue in the industrial sector, as well as in the public sector, is the choice of norms, which are used by companies to try to escape the effects of competition.

Finally, there is another view, which is roughly the one I myself hold, that says the real issue consists in finding economic models that effect a compromise involving cognitive capitalism, and a viable system for the public sector and traditional learning, and the radical democratization of society—these are the three elements to be brought together. This means it is vital to integrate some parts of capitalism, the most interesting ones, because it opens up arenas that the greatest possible number of citizens, researchers, students, and customers can use. The battle certainly isn't over yet. We have already seen it in the case of slavery: Almost a century went by between the first revolt and abolition of slavery, in 1804, and the start of its absolute abolition, in Brazil in 1889. It is in any case unsurprising that certain parties in the so-called "real," material economy are opposed to a radical evolution. In some cases, it might be people drawing pensions, but it could also be employees working to produce tangible goods who turn reactionary in the hopes of protecting their jobs, which are more important to them than society becoming authoritarian and noncreative.

In addition, there are complex alliances at odds with each other. We see this with the issue of traditional knowledge. In the lawsuit over the Amazonian Roraima Reserve in Brazil, for example, which the Supreme Court decided in favor of the defenders of the reserve, there were landless groups, wealthy settlers who wanted to grow soy, the army, prospectors working for the Brazilian national corporation Petrobras, wood manufacturers, and ecologists. Their motivations were all interlinked. Native Americans, like the Aborigines in Australia who fought against multinational corporations, actually made a tacit, technical use of property rights by demanding a sort of author's right to resources in the reserve when multinationals refused to recognize the relevance of authorship—often, in this case, taking the form of collective author's rights. The Brazilian state and army had recognized the Native Americans' right to a usufruct, but not to the transferability of property, which means, in other words, that they maintained sovereignty. There was a deep-rooted conflict between those who accepted the Native Americans' sovereignty and right to oppose the state on issues such as petroleum development and those who denied it. In this conflict, the army pointed to the necessity of national security, and the settlers and landless groups took the same line, while on the other side, an alliance between proponents of Native Americans sovereignty on Amazonian soil and ecologists who wanted to limit the clearing of forest for farmland was endorsed by the local, but not the federal government. Native Americans believe the forest cannot be touched because trees have souls, a belief that they expressed tactically in the language of property because it was the only means of making their message understandable. In the same way, Stallman called for interoperability and nonclosure of systems in bourgeois law by creating the idea of "copyleft" while preventing the ransacking of public space that was attempting to consolidate private property.

In a political battle, there is unavoidably some mixing of alliances between people with different objectives. There are two parts that make up access to learning, for example. There is access considered as an absolute, timeless virtue, and then there is a more historically determined position, which is therefore more strategic and raises the issue of rearranging the power structure. The idea of knowledge communism could represent the locus of this reorganization, the field where all the different groups can reach an agreement. People don't refer to it as "communism," but say "for common use," "new common ground," "common space." The term "communism" is, in fact, derived from "commoners," and we must not forget that in revolutionary history, it was the commoners who started things, not the "communists." I use the notion of the communism of capital because I think capitalism needs to be brought into common use by the commoners, since they are the ones producing it and generating its income. Free access to learning can provide the impetus for a political representation of society, the development of a cooperative society in which exploitation is significantly reduced and gradually eliminated as innovative forms of work and creation emerge.

NOTES

1   See the essay by Sangeeta Shashikant, "The Doha Declaration on TRIPS and Public Health: An Impetus for Access to Medicines," in this volume.

2   Paul-Loup Sulitzer is a French a novelist and neoliberal chronicler of Ayn Randian capitalist myths. He was indicted in the so-called "Angolagate affair" that enmeshed several French politicians and their connections in an arms-sales and influence-peddling scandal. See, for example, "French Power Brokers Convicted over Arms to Angola," Reuters, October 27, 2009, available on-line at http://www.reuters.com/article/idUSLR21027820091027 (last accessed January 10, 2010).

Nollywood: How it Works—
A Conversation with Charles Igwe

Achal Prabhala

> Many of the much-touted values of contemporary global capital and its prophetic
> organizational models of dispersal and discontinuity, federalism and flexibility, have
> been realised and perfected in West Africa. That is to say that Lagos is not catching
> up with us. Rather, we may be catching up with Lagos.
> —Rem Koolhaas, *Lagos: How It Works*

It was with a single experimental film that the phenomenon we now know as
Nollywood came into being. Kenneth Nnebue's 1992 production, *Living in Bondage*, wasn't very expensive to make and didn't take long to film. For this, his first
Igbo-language venture, he played it safe and culled the depths of local ritual lore
to form a basis for the plot:

> Andy learned it the hard way—selling your soul to the devil stinks. He didn't mean
> for his wife to be killed in a satanic sacrifice. He just wanted to live the high life—
> get a girlfriend, a nice house, and a car (or two—a Mercedes *and* a Pathfinder). But
> the high priest was insistent, and Andy was in over his head, and well, though her
> blood tasted terrible, things did really start looking up after Merit died. Fast women,
> fancy cars, foreign wines, and all the chicken he could eat: if Merit's ghost would
> just stop haunting him, everything would be fine.[1]

Sixteen years and thousands of films later, Nollywood is everywhere. The
movement that began almost hesitatingly has catapulted itself into one of the
world's significant film factories. (In 2005, Nigeria produced 872 films—in comparison with 1091 in India and 485 in the United States.[2]) It is certainly the most prolific
and prolifically studied national film industry on the continent of Africa.

Among Nollywood's most ardent observers are researchers of culture and

"'Osuofia In London' is a first-rate comedy with a serious theme!"

Osuofia in London

A KINGSLEY OGORO PRODUCTION

DVD cover for the film *Osuofia in London* (Kingsley Ogoro Productions).

intellectual property. Culturally speaking, the content of the films may be the subject of heated discussion, but one thing is clear: Nollywood has captivated all kinds of audiences. Indeed, cinema is now Nigeria's primary cultural export. Nigerian films can be watched on satellite television in almost every country in Africa, plus a few in Europe and the West Indies. VCDs and DVDs of the most popular films— such as Kingsley Ogoro's 2003 production, *Osuofia in London*, the country's highest-grossing film of all time—can be found everywhere from Johannesburg to London.

Odia Ofeimun, the writer and poet, provided a bracing account of the culture of Nollywood at the Nigerian National Film Festival in 2003:

> Rather than wait on the imports from Hollywood which speak to our common humanity by denying or simply being indifferent to whatever we could call our own, the home-video woke up something that was once there but had been stamped underfoot by managers of the national and sub-regional cultural economy. Not to forget, this was happening while swindlers in the political marketplace were emplacing homegrown democracy with one hand and displacing it with the other. The video arrived in the most homegrown attire that it could weave for itself in a country where the search for foreign exchange had become the defining factor in national dream-making. It turned its back on the dollar trail and reached out for the Naira without hesitation. Rather than the dollar-mania that had overtaken all comers, it sought an import-substitution aesthetic which insisted on building a comparative advantage not as a subaltern of the imported Hollywood stuff but its avid displacer.[3]

It is the complex, challenging, and often unique process of these films' production and distribution, however, that has left a host of observers wondering just how it is done. This interest is not solely academic, for Nigeria has accomplished a feat that many Western countries with far greater capital resources have tried—and failed—to do. As the producer Zeb Ejiro puts it: "You know, once upon a time the French government supported arty films in Francophone countries. But now the French Cultural Centre here in Nigeria wants to bring French moviemakers here to study *our* methods. They cannot understand how we can make a movie in seven days—and still enjoy lunchtime."[4]

Charles Igwe is the principal consultant at Big Picture, a film production and services company in Lagos. He has worked in the film industry as an executive producer, financier, and lobbyist. Prior to his present career, he worked at Citibank. He is a cofounder of the Nigerian film and television expo BOBTV (Best of the Best TV), which showcases and markets Nigerian cinema to the world. He was a panelist at the Yale Law School A2K3 meeting in Geneva in 2008.

"The A2K movement," Igwe told me before we began the conversation recorded here, "is even more important to developing countries at a time when knowledge is becoming the prime currency in global affairs. The right to access knowledge is a stronger argument in certain fields of endeavor than others. It seems to me therefore that a modular approach should be taken in these deliberations to distill a graded pattern for exploiting and protecting intellectual property material."

AP: Charles, I'd like to begin by asking how it was that Nollywood came to be. Was there a particular event that catalyzed this industry? Or was there a policy that resulted in this particular outcome?

CI: Historically, Nigerian television always had a slot for local content. In the 1980s, in the midst of a major cash crunch, state television was asked to seek funding from sources other than the state. This resulted in cost cutting and a narrow revenue focus, which in turn resulted in shutting out original local content. The policy shift occurred in 1985 and was consolidated by 1990. At around the same time, there were some smart local entrepreneurs who spotted this shift and saw it as a market gap. So they took exactly the format of local entertainment that people had been used to, exactly those television actors who people already knew, and started producing their own films on VHS. That is how this new—and revived—industry now known to the world as Nollywood began.

AP: I'd like to understand what the revenue pie looks like from the point of view of a Nigerian film producer. For instance, how is revenue split between DVD sales within Nigeria, DVD sales outside (from Africa, and beyond), and television rights?

CI: I would say that the bulk of the money—far greater than 50 percent—would come from VCD and DVD sales within Nigeria. Recently, we discussed a national distribution framework for audiovisual content that seeks to define who should deal in such products and whether and how such business is to be done, with a view to ensuring that the value chain is more like in other parts of the world...because here, it's a little different from everywhere else. People who have dominated the retail business in film so far are people who were originally trading in electronic appliances, televisions, VCRs, VCD players, and so on. Up until recently, they remained the main channel for reaching the public. DVD and VCD sales outside of Nigeria are fragmented, and it is hard to say how much revenue they bring in, since most of what is sold outside is pirated.

The cinema theatre business was disrupted in the 1970s when most foreign operations in Nigeria were nationalized. Cinema theatres have not recovered since. Now, however, this distribution window is enjoying a revival, and some local media groups (working in tandem with major American studios) are driving investment into the infrastructure for cinema theatres. However, it is evident that these developments will primarily distribute productions of the self-same American studios.

In the future, I think we are going to see a lot more cinema made on digital video, which will mean greater control on our part as to how we deliver this cinema to the viewer.

AP: How does the distribution of film actually work? Is it similar to the distribution of any other commodity, where retail returns are based on the number of units sold? Or is it that in some cases, you sell to retail middlemen—intermediaries—who (in exchange for an upfront fee) might be allowed to make their own copies for sale within a territory?

CI: Both models work here. The latter model—with retail intermediaries—tends to be a lot more secure, because what happens is that the people who actually sell these films are often the producers themselves. After spending money to produce a film, they tend to be more willing to make immediate deals that will bring immediate revenue for them. So they try to arrange a sale of reproduction rights to someone who has the capacity to make copies and distribute.

AP: In India, which is where I am from, cinema distribution tends to happen along territory lines, and national theatre chains have only recently entered the fray. This would mean, typically, that there are several intermediaries to whom one would sell a film. How does it work in Nigeria?

CI: The exercise of mapping territories is what we are trying to put in place now. Up until recently, it was ad hoc: People sold films to others, who more or less sold them on at their will. What has been put in place now is a system of licenses: We have national distribution licenses, regional distribution licenses, and others. It's a framework with a structure within which producers can decide where they want their films to be seen or bought, and that defines who they would talk to.

AP: In terms of television rights, do you negotiate these sales when the film is being made? Or has it been your experience that television channels wait to gauge the popularity of a film before deciding to acquire it? A related question I have is, which are the channels you regularly speak to? I'm aware of the MultiChoice platform within southern Africa. What others are there?

CI: Well, up until recently, as I explained, television stations within Nigeria were not exactly enthusiastic about Nigerian cinema. In 2005, MultiChoice created a channel exclusively devoted to the cinema of the region—they called it Africa Magic. It proved to be enormously popular all over the continent and even in Nigeria. In the wake of that, our local television channels realized that local cinema resonates strongly with the audience. Now there is a lot of demand for our cinema, and as a result, there is a lot more conversation between television channels in Nigeria and movie producers there. Today, there are several television channels showing Nollywood, many more than just MultiChoice. We are carried on a channel distributed by Sky TV in the UK, on several channels in the Caribbean, and practically everywhere in eastern Africa.

AP: What would you say is the revenue share that a producer can expect to earn from television?

CI: I would say that, in the current time, it is between 5 and 20 percent—because there's so much interest being shown by television channels now. But prior to recent times, I would say that it was negligible. Television channels had small budgets and could not really devote their attention to our content. A lot of players in the television market expected the content to just come to them—they had designed their businesses based on merely providing broadcast space for selling airtime.

AP: I'd like to talk about piracy and relate my question to what I have seen and heard from Mumbai, India's Hindi film capital. In India, we regularly hear producers bemoaning piracy—as perhaps is the case everywhere else in the world. We also hear, however, of individual producers, directors, and actors who acknowledge the inadvertent (and enormous) publicity and outreach that piracy provides. So the reaction to piracy tends to be mixed. The other interesting thing about piracy here is that often, actors themselves earn more money from endorsements and theatrical shows than from the films they act in. This means that to an extent, piracy results in building their reputation more significantly than legal sales of the film would accomplish.

Turning to Nigeria, Brian Larkin writes that "looking at piracy and the wider infrastructure of reproduction it is part of reveals its generative side; piracy is part of the migration of the parallel economy in Nigeria into center stage, mixing legal and illegal regimes, uniting social actors, organizing common networks."[5] What do you think of piracy in the context of the Nigerian film industry?

CI: Look, when we started making films in Nigeria, it wasn't like we had some big business plan or big budgets to drive our products. But somehow, the product caught on anyway—first within the country, then in West Africa, the rest of Africa, and even elsewhere in the world. A lot of what drove that popularity was piracy—it was the circulation of pirated films that made it possible. I remember being at a conference in Cape Town and listening to someone from Namibia complaining about how Nollywood films were coming in to his country illegally. We were aware of it then, we are aware of it now—and we made a strategic decision, along with our lawyers, to recognize that we did not have the means or the finances to pursue our copyright in a restrictive way. We also recognized that Nigerian films had become a global phenomenon not because we went out there and marketed them, but because of pirate networks. What we found is that piracy decreased our own costs, because the pirates opened up markets that we might have otherwise been unable to afford to reach.

So we considered things and realized that it all balances out. Right now, for instance, we're looking seriously at markets like Kenya, Brazil, and the West Indies, and we see that our existing popularity there is not because we went in and spent money to promote our films, but because the pirates prepared the ground for us. So it works for us, for producers and distributors. Of course, our actors enjoy tremendous (tremendous!) investment from just being so popular all over the world. In some countries, they are treated like royalty. And the reason for this attention is not because we pushed our films in those countries ourselves.

AP: As more cinema theatres get built in Nigeria, you will presumably have a greater degree of distribution control over your films. How has the growth of Nollywood contributed to the inception of cinema theatres in Nigeria?

CI: What has happened is that there has been a resurgence of cinema in Nigeria. That resurgence has resulted in a joint venture to build new multiplexes all over the country, as I explained earlier: They are starting in Lagos and expanding to ten more locations. It is an interesting move, because these people are trying to get local film habits restructured. They're trying to get them to go to see films in a theatre. Once these multiplexes are up and running, I think we can work with them to show Nigerian cinema. At the current time, however, they are being designed to distribute Hollywood films.

HOME COOKING IS KILLING THE RESTAURANT INDUSTRY

AND IT'S TASTY

Parody of a graphic used in a 1980s anti-copyright infringement campaign by the British Phonographic Industry (BPI), a British music industry trade group. The original graphic included a silhouette of a cassette tape, and the words "Home Taping Is Killing Music" at the top, with "And It's Illegal" at the bottom.

AP: One experience that comes to mind from Bangalore, where I live, is the "tent screening." It usually works like this: Some enterprising individual gets a hold of a DVD or VCD (it doesn't really matter whether the version is legitimate or pirated), organizes a screening, sells tickets, and keeps the profit. Tent screenings are a common feature in low-income neighborhoods within the city and in small towns elsewhere in the state. Does this kind of thing happen in Nigeria at all?

CI: It's exactly the same here. We don't get revenue directly from these screenings, of course. But this is where it would be wise to apply some cultural analysis. Ours is a country where people like to own stuff. With things that they like, people want to buy and own them—and this is good. What happens is that when people go to these screenings and see the films, they often want to buy a copy for themselves. They use the screening as a trial, to form an opinion of the film. If the opinion is good (and this is a challenge for our filmmakers, to deliver quality that can be appreciated), then not only do the viewers go out and buy a copy for themselves, they often go ahead and recommend the film to others.

AP: So, like the appraisal of piracy, would you say that producers in general encourage such informal screenings, since there is eventually a direct commercial benefit to the producers themselves?

CI: Well, not all producers would be encouraging of this trend. If you are a strong producer who produces really good material, then yes, you would encourage such screenings, since you are fairly confident that people will go out and buy your film after watching it once. But if you are at the bottom of the chain, let us say a producer of substandard cinema, you will not encourage such screenings—since people are unlikely to go out and buy your film after having watched it once.

AP: I'd like to touch upon an issue of intellectual property within the Nigerian film industry itself. One thing that becomes pretty clear is that when a title becomes popular—as with *Living in Bondage* or *Osuofia in London*—there is inevitably a sequel to come. Have there been any issues within the film industry or disputes, as to the control of actual characters and plot lines? Are the sequels usually produced by the producer of the original film? Or are they sometimes franchised?

On a side note, I'm currently looking at the Wikipedia entry for *Osuofia in London*, and it says that the film earned $8.9 million in the United States alone! That seems like an extraordinarily high return, does it not?

CI: That is a very large sum of money, and perhaps a bit exaggerated. We had exactly that kind of dispute with the sequel to *Osuofia in London*, but it was sorted out. It is a learning process for us, as well. Many people in the industry have no idea of the legal requirements involved with characters, plots, and so on. In terms of disputes, yes, there are a lot of them related to copyright—but not quite what you are suggesting. For instance, we have had disputes with large on-line sellers of Nigerian films—there are retailers who sell in the hundreds of thousands. But each time, we weigh such infringing action against the popularity that the on-line sellers are enabling for us.

AP: In India, a recent trend in Hindi cinema has been to court the nonresident Indian, or NRI. The NRI film has now become a staple genre of Hindi cinema, which, while made squarely within the same Mumbai commercial complex as its counterparts, is primarily aimed at foreign audiences in North America and Europe and secondarily pitched to urban, middle-class audiences within India. The overseas financial success of *Osuofia* got me wondering as to whether such a trend has emerged in Nigeria.

CI: You see, to draw a parallel, let me take the Indian community in Nigeria. They seem very well organized and articulate. So it is presumably easy to target the Indian community in Nigeria (and in other places) as a market. In contrast, Nigerian communities overseas tend to be less aggregated and more fragmented—yet no doubt they will access our material if we can provide a structured market. Perhaps by selling through ethnic food stores and the like, we can target them. We need to find better distribution within those overseas markets, especially since we know they are buying anyway, even now. In terms of making products that are specifically targeted at the overseas market . . . it has not worked for us. Remember, however, that we are already making films for overseas markets—for the West Indies, South America, the UK, and of course, for other countries in Africa. It is all just coming together now, and as the market organizes, things could change.

AP: I believe you are working on a project to document Nigerian stories, especially folklore. How did this come about, and what do you plan to do?

CI: After the A2K Conference [in Geneva, 2008], it became clear to me that the world is moving toward a knowledge economy. I realized that we did not have any kind of concerted process for capturing our own knowledge here in Nigeria. At the same time, I realized that the film industry in Nigeria has no research base to refer to in terms of cultural activities within the country. Our rich oral traditions, I felt, had not been recorded well enough for those in the film industry to use. I am trying, through the BOBTV Expo, to see if we can get together and document some of the stories we refer to in our films—and thereby create a better knowledge base for everyone. This effort has already started, and we have attracted the notice of government, which is now embarking upon a project with a similar agenda.

AP: A final question. What is the future of Nollywood? What can we expect in the years to come?

CI: Well, Nigeria has 140 million people; Africa has about 840 million people; and if you count those of African origin in the Caribbean, Europe, and North America, you have altogether more than a billion people. Nigerian films are the most significant cultural factor that links these communities in the present day.

The first thing that we have to do is build the infrastructure and the capacity to ensure that we stay in business. Once that is set, the content that we produce—and the service that we do to Nigeria, to Africa, and to various parts of the world—should put us in a place where we are one of the leading suppliers of content *globally*. After all, our stories are universal. It's just that our technology hasn't yet brought us to the point where we are globally accepted.

NOTES

1 Trenton Daniel, "Nollywood Confidential, Part 2: A Conversation with Zeb Ejiro, Ajoke Jacobs, Tunde Kelani, and Aquila Njamah," *Transition* 95, vol. 13, no. 1 (2004): p. 110.

2 UNESCO, Analysis of the UIS International Survey on Feature Film Statistics, Information Sheet No.1 (2009), available on-line at http://www.uis.unesco.org/ev.php?ID=7651_201&ID2=DO_TOPIC (last accessed February 1, 2010).

3 Odia Ofeimun, "In Defence of the Films We Have Made," *West Africa Review* 5 (2004), available on-line at http://www.westafricareview.com/issue5/ofeimun.htm (last accessed February 1, 2010).

4 Zeb Ejiro, in Daniel, "Nollywood Confidential, Part 2," p. 128.

5 Brian Larkin, "Looking at Piracy," working paper, Social Science Research Council 2003, available on-line at http://mediaresearchhub.ssrc.org/looking-at-piracy/resource_view(last accessed January 13, 2010).

A2K IN THE FUTURE—VISIONS AND SCENARIOS

In this section, the authors were asked to imagine best- and worst-case scenarios of the regulation and production of knowledge in their field of interest. Unconstrained by the imperative to describe "likely" scenarios, they offer us alternative visions that illuminate the stakes of the choices that we make today and how these choices could portend radically different futures for access to knowledge.

Josh Bonnain (http://creativecommons.org/licenses/by/2.0/deed.en).

A Copyright Thriller versus
a Vision of a Digital Renaissance

Sarah Deutsch

On April 1, 2008, a mysterious piece of new legislation appeared in the e-mail in-boxes of those who follow copyright developments. The e-mail warned of a pending bill entitled The Assuring Protections and Remedies for Intellectual Property Laws Act of 2008. Given the Old Faithful–like eruptions of new copyright bills in Congress every year, this bill seeking familiar additional "protections and remedies" appeared to be genuine. The bill contained all of the standard formatting and headers one finds in copyright legislation. It contained the typical "findings" section, established a new department of intellectual property security, touted new "enhancements" to copyright rights, new enforcement powers, and additional technological protection measures. It was only upon a slightly deeper look that one figured out that this latest bill was actually a clever and elaborate April Fools joke.

The Assuring Protections and Remedies for Intellectual Property Laws Act (the "APRIL Act") parodies many of the new rights and remedies inserted into countless copyright bills introduced over the past years. For example, the recently introduced sixty-nine-page Pro IP Act proposed creating a "White House Intellectual Property Enforcement Representative" and revamping the Department of Justice's Intellectual Property Enforcement Division. The APRIL Act instead proposes creating a Department of Intellectual Property Security (DIPs), along with a Computer Hacking and Intellectual Property Program at DOJ (CHIPs). Hence, the bill's critical Section 104 is entitled "Coordination of CHIPS and DIPS." The bill also picks up on the legislative debates surrounding the *Grokster* decision,[1] codifying new forms of secondary liability for "attempted infringement," inducement, and adding a new, even deeper and more obscure layer of "tertiary liability."

In a nod to the issue of copyright-term extension, Section 204 of the bill proposes a term of perpetuity minus one day. The U.S. Constitution requires that a copyright exist only for a "limited term," yet the Supreme Court recently refused to provide any further insight on what a "limited term" means, although it has been expanded to the life of the author plus seventy years. And in an acknowledgement of those bills that seek to expand the scope of copyright law, Section 209 extends copyright to broadcasting, databases, and fashion designs (with originality to be determined by an "Originality Czar"). Many of these noncopyrightable areas that seek protection under a copyright umbrella are already adequately protected under existing law (for example, databases are adequately protected under contract law and theft of broadcast signals under a wide array of federal and state laws), and they thrive without the need for additional copyright protection. The mock bill acknowledges new "self-help" rights for copyright holders, including the right to disable, interfere with, block, divert, or otherwise impair the unauthorized exercise of any of their exclusive rights. For example, someone could find that files have been removed from their computer or that tracking devices have been placed on their hard drive under the auspices of "self-help." The Federal Communications Commission is empowered to adopt "Dragon Shield" copyright security standards, which, among other attributes, are required to be "invincible." In reference to both the old proposed "Hollings Bill" and the Digital Millennium Copyright Act, it creates liability for any person who tampers with or undermines the Dragon Shield. The final provision repeals the Copyright Act's prohibition against the federal government owning copyrighted works and bears a bold copyright notice in the name of the United States Congress.

THE AUTHORITY: A WORST-CASE SCENARIO

This legislative parody makes sense only because, as the old adage goes, behind every joke is a grain of truth. It was appreciated by those of us familiar with the recent history of copyright law, because it hints at our fears about the future of copyright. As a group, we have been trained to imagine the worst-case scenarios, so when reading the provisions in the APRIL Act, we could easily move beyond this parody and concoct a future where copyright policy in fact has gone mad.

One could imagine a bad science-fiction thriller movie, *The Authority*, that would include a sole "Copyright Authority" who is much more powerful than any of the copyright departments suggested in recent legislation. Our imaginary Copyright Authority (known as The Authority, or perhaps just "C") would be virtually omniscient when it comes to detecting copyright violations. In the opening scene of the movie, the camera would pan on "C" sitting in an office in a high tower behind a spotless modern desk, wearing the requisite black turtleneck, gold chain with shining copyright symbol, and reflective sunglasses. Copyright violations would stream across a floating electronic billboard like the Wall Street stock ticker.

In this movie, The Authority has found it necessary to declare its first virtual copyright war and would control the strategy, aided by an army of people and robots. Just as in the movie *Minority Report*, an army of mechanical spiders would crawl the Internet, scurry through the walls into suspects' homes and computers, and search for infringements, taking whatever "self-help" measures might be necessary. Perhaps they might destroy files and equipment, automatically deduct fines and penalties from bank accounts, or publicize the faces and names of violators on huge screens like those in Times Square. The movie would not have a satisfying element of conflict and violence unless the spiders also had the power to "Terminate" (with a capital T) those that the Authority deems the worst offenders.

Of course, in our movie, due to years of "clarifications" to existing laws, whatever protections existed for individuals or third-party intermediaries long ago have disappeared. The new laws have created seamless, one-click primary, secondary, and tertiary liability. Elders reminisce about their old iPods, Tivos, and home networks as sad relics of the past. The old-timers recall that they used to be able legitimately to stream content to different rooms of the house, space shift recorded programming, and create libraries of time-shifted content. All the companies who produced the old innovative products and services would have long ago been sued out of existence. Most remaining black market "antiquities" would have been seized and destroyed by the spiders. The Copyright Authority would approve all new devices and administer all Internet services. The average citizen would have almost no societal memory of the concept of privacy or reasonable uses of copyrighted works, because rights and penalties for violations were doled out like small electrical shocks to a submissive public.

The movie, however, would need a protagonist. Perhaps the protagonist would be some deeply recessive genetic mutant who chooses to reject the rules. It would be convenient if our protagonist's great-great-grandfather might have left an undiscovered cache of equipment, devices, and communications networks as weapons to fight back against The Authority. And, of course, we would need to figure out how to insert the requisite car chase scene, with the Authority and squadron of spiders in hot pursuit.

But, this overexaggerated, nightmarish yarn would never pass muster in a real Hollywood movie script. Note that in the real movie *Be Kind, Rewind*, the protagonists dangerously create and film their own retakes of famous Hollywood movies after their entire inventory of movie rentals is accidentally erased. Their innovative derivative works become huge word-of-mouth hits in their low-income community, and their video store's business begins to thrive. But, the plot does not permit this kind of creative success to continue. The studio executives show up to enforce the protagonists' copyright violations with a steamroller, destroying all the tapes as the neighbors sadly look on. Fortunately, the protagonists of our movie use their new-found talents to create their own original copyrightable film that unites their community and celebrates their history at the same time.

It's hard to imagine that our copyright thriller would ever become anything more than frightening imaginary and unmarketable plot. But unlike *Be Kind, Rewind*, the controversial copyright issues at the heart of this script could not be salvaged by the sudden appearance of a steamroller. Fortunately (at least to date), however, most of the worst-case copyright scenarios that appear in legislation never come to pass. While copyright continues to expand through new bills introduced each year, the worst-case scenarios usually do not survive the U.S. legislative process unless they reflect the input of all stakeholders.

ONLY CONNECT: A BEST-CASE VISION OF A DIGITAL RENAISSANCE

Let us now envision a completely different future—a best-case scenario. In this figment of the imagination (a mere idea not subject to the protections of copyright), copyright policy fights have long been abandoned as futile. All industry players, including copyright owners, on-line intermediaries, device manufacturers, software companies, and other players are completely engrossed in the new task at hand: devoting every possible resource to meet the exploding demand for new content, products, and services. The copyright policy disputes of prior decades now seem irrelevant, and there is money to be made. Users are also too fully occupied creating their own content and availing themselves of an endless supply of readily available content to waste their time illegally downloading. In a nutshell, everyone is too focused on creating, distributing, and using content to see any need to infringe copyrights, sue each other over them, or seek punitive legislation.

In this scenario, in the more technologically advantaged societies, where everyone now is a copyright owner and creator, all would gain newfound respect for copyright. And in the developing world, citizens would more slowly, but increasingly gain access to technology, computers, and broadband Internet access and become increasingly interested in developing their own unique cultural content and creations. At the same time, all would recognize that increased interdependence requires flexibility to allow others to make liberal uses of copyrightable works to build upon and add value to existing works. (Our scenario, conveniently, will fast-forward through at least a decade of "unpleasantness," where the rules governing who gets paid, how much, and what uses are considered fair and reasonable get ironed out.) The notion of "user-generated content" would become an oxymoron, because the new digital renaissance would encourage everyone to create music, art, films, literature, and new forms of derivative works that defy the imagination. The marketplace supporting this digital renaissance would profit equally from the distribution of creative content in both the physical and the virtual worlds.

Broadband speeds and services would continue to grow and with them supply an endless array of interactive devices. Users would continuously send streams of content across multiple platforms and devices. Almost any item capable of containing electronics would become a new mobile device. Innovative and ubiquitous payment systems would develop and along with them interfaces to interact with our constant stream of content. Home networking would extend beyond the home, and words such as "time shifting," "space shifting," and "format shifting" would seem as obvious and antiquated as the "Information Superhighway." Although digital rights management (DRM) and enforcement options would still exist, the need for onerous DRM or expensive enforcement actions would not be nearly as great in a world where the marketplace is thriving.

Large content companies would be thriving. They would profit from strong partnership arrangements with Internet service providers, device manufacturers, and other suppliers. Consumers would earn new income streams from individual partnership arrangements with content companies and distributors.

Companies would share revenue with the public and reward the public for creating, distributing, and promoting new content in unique one-to-one communications. Notably, the ability to share content in a small community means that entire specialized on-line communities could generate content and earn income in both the physical and the virtual worlds. Groups might be devoted to a particular movie, book, or music or might sell unique versions of derivative works. Profit-sharing arrangements would benefit the creators, copyright owners, and users alike. And of course, the newfound societal approval for creating derivative works would now mean that we would laugh more (and live longer) as the best new parodies of the century were celebrated and shared.

Our technology and content sectors would also embrace the business notion of "creative disruptions." Companies would understand that they must adapt to, rather than fight changes that threaten their business models and seek innovative ways to profit from those changes. For example, in the old days of telecommunications, criminals sought to profit from schemes to steal long-distance, international, or cellular telecommunications services. As communications companies shifted to selling generous buckets of minutes at reasonable and compelling price offerings, this encouraged increased usage of their services, and the need to steal suddenly seemed irrelevant. This is, after all, the history of the telephone industry, too: In the old days, long-distance or "toll calls" were very expensive, so people used to steal phone service through black-box devices and other schemes. They no longer do that, because today, the cost of calls or the Internet is low enough that it's not worth anyone's effort to steal. Likewise, in the copyright industry, content owners would explore selling volumes of content through compelling and attractive bundles, including buckets of downloads at fast speeds, and package them with additional content extras not otherwise available. Gradually, the need to download illegally would appear outmoded and irrelevant.

Our digital renaissance would borrow from the rallying cry in a famous work of literature. In *Howard's End*, E. M. Forster begins the novel with the call "Only connect..." and carries that theme throughout the book. *Howard's End* teaches the lesson that fulfillment comes when people with different values—those focused on spiritual values and those concerned with material goods—ultimately connect and engage with each other: "Only connect! That was the whole of her sermon. Only connect the prose and the passion, and both will be exalted, and human love will be seen at its height. Live in fragments no longer."

In our scenario of a digital renaissance, the power to connect requires the need for a robust and healthy broadband market, innovative new devices and services, and the ability to generate and distribute attractive content in a robust marketplace. The "only connect" scenario might be the unifying theme linking content owners, users, distributors, device manufacturers, and others in a more harmonious future.

NOTES

The opinions offered here are solely those of the author and do not reflect the opinions or views of Verizon Communications.

1 Copyright liability applies to those who are either found primarily liable (direct infringement) or secondarily liable (contributory or vicarious infringement). The *Grokster* decision explored the concept of whether one could be found secondarily liable for "inducing another to infringe," for example by taking purposeful steps in advertising infringing conduct. The APRIL Act imagines that the chain of secondary liability should be further extended down to indirect recipients of copyrighted works in the distribution chain.

Social Mutations in the Future

Gaëlle Krikorian

In the future as I envision it, two social mutations are possible. In both, the reach of the Internet and its associated technologies, already broad at the beginning of the new millennium, spread ever more expansively, infiltrating and thus altering many domains of daily life. But intellectual property does not play the same role in both, and the outcomes are very different.

A MARGINALIZATION OF "INTELLECTUAL PROPERTY"

In the first scenario, the movement for open-source software and open access is very successful, facilitating the use of open initiatives allowing users to search the Internet and store and share information (for example, Firefox, Wikipedia, blogs, and social-networking sites such as MySpace and the travel site Dopplr) and to participate in open collaborative work, first in the software sector, but soon in many other sectors. More and more information becomes freely available on-line, and scores of scientific and academic data sources are in the public domain, accessible as long as one has access to a computer and can get on-line. The plugged-in population, in turn, grows dramatically, fueled by a number of initiatives, from municipalities, companies, and other institutions, to spread and decentralize broadband and Wi-Fi technology and access. Private-sector attempts to control the "tubes" through which information is disseminated fail. Open-access policies prevail, denying broadcasters and Webcasters the copyright they seek to secure on information they publish. Although there remain plenty of content and numerous channels, programs, services, and goods accessible only through the paying property system, users generate and share so much material that people can typically find enough of what they need to avoid that system altogether. In this context, knowledge previously defined by the principle of intellectual property as a scarce resource to which access is limited

CCcommunism (http://uncyclopedia.wikia.
com/wiki/File:Creative-commies.gif).

through ownership becomes so plentiful that it overwhelms and effectively out-
dates the notion of ownership.

In one way or another, countless Internet users get involved in the production
of "information goods." Networks of users and producers of information form flexible communities, and many of these users are members of multiple, miscellaneous
communities. Of course, networks face internal power relations and experience
governance issues, power struggles, and crises. They emerge, grow, transform, or
disappear. Some become long-term fixtures, while others experience stunning success and a speedy death. Networks become learning bodies. Users acquire skills
as network members, develop them, and pass them on when they migrate to new
communities. A noteworthy evolution takes place as users and communities elaborate models to facilitate information production that remains stable, regardless of
the birth or death of any particular network or community.

An important common aspect of these forms of production and consumption
is that money is not the sinew of war. Users seek information, friends, recognition, and advice and invest time, humor, sex, knowledge, and resources. Whatever the exchange, they deem it worthwhile and participate freely in these
creative, sometimes arduous and time-consuming, activities. Their involvement

can be passionate or topical and temporary, and of course, there are many who are not members of such networks or communities at all. Substantial portions of the planet's population still do not have access to computers or the Internet, but, in the long run, across the globe, a stable critical mass develop and continues to contribute to the survival and advancement of an open system.

Studies in Germany, Spain, and India have shown how individuals' relation to work, on the one hand, and to the idea of individual performance, on the other, change as a result of the development of these new models and their appropriation by ever larger segments of the population. Most users do not see the creation, acquisition, and communication occurring on computers after hours as "work." In fact, neuroscience studies show that the brain does not treat these activities as work, assuming instead the same resting state as in response to any other hobby. The studies demonstrate that people's perception of their own performance, usually shaped first by school and subsequently by the work environment, is readily altered by their participation in creation and communication experiences on the Internet and that the increased sense of self-efficacy encourages them to engage in activities or explore fields they would not otherwise have approached in actual— as opposed to virtual—reality. Moreover, the various stratification systems that exclude potential participants in "real life," such as hierarchy or class, have less traction in the cyberworld, though language, of course, continues to present a considerable obstacle.

This overwhelming success of the open-source system is fueled by performance and efficacy. Users no longer need the big software companies when they can update their software gratuitously and reliably. Costing little or nothing, open-source software will offer many valuable programs, as well as virtual twenty-four-hour help lines of fellow beneficiaries. Of course, this doe not happened overnight. Cognizant of the comparative cost effectiveness of open-source software development, computer companies adopt that model. When Dell subsequently starts to sell Linux-installed computers, consumers quickly catch on, first on account of politics and then simply because of the lower cost. Ultimately, all major computer corporations decide to use open-source software and base upon it the services they sell to their customers.

In some cases, the enforcement of intellectual property protection is precisely the reason users turn to open-access products: Cybercafé operators in developing countries, for example, shift to open-source code to preclude the consequences of using illegal software. Governments do so, too, on account of outlay, legality, and software compatibility issues that might otherwise create obstacles to normal governance and the successful implementation of national policies. The same is true of institutions underpinned in part by principles of public service (for example, schools, universities, libraries, and hospitals), whose interests in the public good provide sufficient incentive for this sort of change. Once some of these established institutions take the plunge, others follow. Meanwhile, a wide range of open-publishing initiatives and services are created, and millions of open-publishing acts flood the Internet.

These conditions force a drastic change in the parameters of production and distribution in many sectors. A number of industries, including entertainment, communications, and pharmaceuticals, change their business models, reinstating manufacturing and selling in industrial quantities as the main source of revenue in some cases. An open-source approach to medical research takes hold, as well, in countries in both the Global North and the Global South. Inspired by the openness of the Human Genome Project and first used by public institutions conducting research on tuberculosis, the open-source model rapidly yields positive results and is embraced by the private sector, as well—in order to access the databases fed by researchers, one needs to be a member and contribute to the general effort, a condition that constitutes a powerful incentive for companies to join the club. Quickly, this open-access approach proves much more effective at fostering innovative research than the secretive patent system. A lot of companies restructure their activities and specialize in the development and organization of clinical tests, as potential drug candidates are identified. Still, probably the most spectacular mutation takes place in education, where open-source software and open access facilitate a new type of legitimization, divorced from conventional institutions and thus positioned to jam the traditional mechanism of the reproduction of the stratified social order.

In a way, in this scenario, we can witness a popular outflanking not only of the intellectual property system, but of the traditional, stratified social structure that is anchored upon it. The logic and the exercise of open-source production show a tremendous normative effect on cultural production and appropriation, social practices and representations, and the rigid social structures characteristic of most societies.

THE ERA OF FREE EXPLOITATION

In the second scenario, the spread of the Internet and the persistent enforcement of intellectual property restrictions leads to social mutations that are much less promising: the metastasization of neoliberal logics and practices throughout the knowledge economy and the information society and the consequent exacerbation of social and economic polarities.

Awarded the Nobel Prize in Economics in 2011, George Arturpán will become the theorist of the articulation between neoliberalism and knowledge capitalism.

He developed and first used the notion of "free accumulation" (or "free exploitation"), which subsequently and rapidly was adopted as a central concept by many governments. It became the heart of the Treaty for International Development, Free Accumulation, and Trade (TIDFAT), which will be signed after eight years of negotiations under the auspices of the United Nations Conference on Trade and Development (UNCTAD).

In the knowledge society as it develops in this scenario, the slogan is "double free." Immaterial goods are not only free to access, they are also free of charge, at least for those producing their quota of creativity units (CUs).

Once a year, beside one's income, individuals have to declare the CUs they have accumulated throughout the year. The threshold they reach varies from one person to another and is calculated by the state, taking into account the individual's age, creativity quotient (a sort of IQ established during schooling through tests or assessed at the request of the administration for those who did not benefit from normal schooling), and level of education. Ratified by 204 countries, TIDFAT establishes these thresholds and modalities of calculation, both of which can be reevaluated, if necessary, during the annual General Assembly of the United Nations. The CUs can be capitalized, rolled over from one year to the next in the event of a surplus above one's predetermined threshold, or transmitted to members of the same family. Individuals who meet their CU quota will receive a personal code for Access to the Resources of Information and Communication Technologies (ARICT), allowing them to connect and access the totality of the Global Resources (GR) of the Internet—as opposed to the Open Resources (OR) accessible without an ARICT code—which belong to a planetary library of data and information owned by the public as well as by the private sector.

In addition, each person is a member of a multitude of virtual networks, professional as well as social. Information and interconnection are the two elements that allow individuals to work and produce. Each person is free to perform in the field of her or his choice, which does not have to be the same from one day to another or over the years. Private companies (in fact, there are fewer and fewer traditional firms, and companies are more and more referred to simply in terms of "label" or "brand") and public institutions publish their needs and interests on the Internet. These are circulated as job descriptions or mission profiles. Anyone can register, and thus "self-employ" oneself, so to speak, in order to fulfill the requested task. Labels or institutions generally set a maximum of possible enrollments. Once the pool is full, the jobs are no longer available. The remuneration of individuals generally originates from several sources, depending on the different missions or the projects to which they contributed over a period of time. A salary for each fulfilled task is calculated and offered by the label or institution that posted the mission. It comes with a number of CUs and a bonus in cases of particularly important findings or efficient work.

The labels are put to work financially according to the pertinence and the interest of the research they are piloting and the products they make. Their quotations on the stock exchange depend as much on the products they launch as on financial experts' estimates of their potential, based on the publication of research projects and mission profiles for which they are recruiting. The one to receive the highest marks may be the first one to launch an interesting request or the first to find the answer to it. Industrial secrecy is actually limited to a minimum, because the transparency of the research conducted is almost total. Anybody can copy anybody. It is knowing how to conduct and lead innovation—innovation guided by awareness of society's needs—that makes the difference. This system allows teeming creativity and innovation in all fields of technology.

Knowledge and information are now considered the raw material accessible to individuals. They are a common good freely shared by ARICT code owners, who develop their creativity and professional activities using computer tools and Internet potential. This work, mostly contributed in a nonsubaltern fashion, is extremely productive. Individuals enjoy much freedom regarding both the content of their work and the way it is organized in time and space, which contributes to increasing their motivation, in part by enhancing the pleasure they derive from doing their work. This, connected to the type of cooperation and interconnection existing through digital networks, allows an extreme efficiency in creation and production.

However, the United Nations shows that 40 percent of the world population, which represents around 3 billion people, still does not have an ARICT code. These numbers, of course, correlate with known statistics on literacy and level of education. Back in 2000, the United Nations Educational, Scientific and Cultural Organization estimated that around 900 million people in the world did not know how to read. With the population increase in developing countries, this number has reached 1.2 billion. That same year, the Organisation for Economic Co-operation and Development calculated that in fourteen of its twenty member states, 15 percent of the adult population had reading skills so rudimentary that they had a hard time adapting to the information era—even, in the United States, the United Kingdom, Canada, and Switzerland. Almost twenty years later, this figure has barely decreased to 12 percent.

The dawn of the neoliberal era led to the progressive disintegration of the social benefits inherited from the period of the New Deal and the welfare state, while the structural-adjustment policies of the World Bank and the International Monetary Fund provoked the demolition of the public-health and education

systems in developing countries. As a consequence, access to training and education institutions is now limited for sizable proportions of the populations in poor
countries and for the poor populations in rich countries. Inequalities continue to
increase throughout the intervening years, and the capitalization and assimilation
of knowledge, which is at the heart of the system of production of free exploitation, proves to be particularly discriminatory and leads to great exclusions.
Globally, the world society is divided into two main social classes: those who live
and work in the universe of digital knowledge, play with ideas and data, and are
both users and producers of knowledge, on the one hand, and those anchored
in postindustrial capitalism, whose performance is based on more traditional
accumulation and exploitation mechanisms, on the other. There is also an intermediary population that does not necessarily reach its CU quota at the fringe of
these two systems. From one year to the next, those belonging to this population
move from one world to another, depending on the results of their work or their
personal choices.

The old system of creation based on intellectual property is still in place in the
postindustrial world, and its enforcement contributes to marking the distinction
between the two universes. It ensures a nonporosity, so that free goods in digital
networks are not able to be transferred in the market area. Of course, trafficking in
ARICT codes and free data exists, but it is staunched and remains marginal. Sanctions on the illegal use of data, whether it be raw digital data or data transferred to
physical media, is extremely harsh. It is even easier to criminalize illegal use and
to apply sanctions, since the surveillance and detection of offenses focuses only
on a portion of the world population—those for whom access to technologies is
relatively limited. Customs authorities, police, and private companies work hand in
hand at the local as well as international levels.

Surveys conducted on disadvantaged populations show that individuals who
live within the postindustrial system globally have a limited interest in the possibility of accessing another type of life and work. They are fully absorbed in the world
they know, in which they have a job, a life, objectives, and concerns. For many
reasons, transfers from one world to the other is rather limited.

Cases such as that of Miguel Hernández, the self-educated son of a shepherd
who became one of Spain's greatest poets and playwrights of the twentieth century, is rare. Despite the social handicaps they have to deal with, some especially
gifted individuals are able to access a universe that has, since birth, treated them
as outsiders. However, even those, when they develop the resources for their integration, may in the end be reluctant to cut the links with their original anchorage
and environment. Thus, as Pablo Neruda recounted about Hernández in "Confieso
he vivido" (1974):

As he did not have enough to live, I looked around to find him a job. It was difficult for a poet to find work in Spain. Finally a viscount, high-ranking official of the Relations, got interested in him and said yes, he was okay, that he had read Miguel's verses, that he admired him, and Miguel should indicate what kind of job he was willing to take so that he could write his appointment. Full of joy, I told the poet:

"Miguel Hernández, you finally have fortune. The Viscount is hiring you. You will be a high employee. Tell me what sort of position do you want to exert so that we can process your hiring."

Miguel remained pondering. His face, with large premature lines, covered itself with a musing mist. Hours went by and we had to wait until the afternoon for him to respond to me. With the bright eyes of someone who finds the solution to his life, he told me:

"Could the Viscount trust me with a herd of goat around here, close to Madrid?"

However, the sharp distinction between the two worlds generates tensions and potential conflicts. The era of free accumulation may not end as a peaceful one. As Hernández once said, and others might now repeat:

I gather my hunger, my sorrows, and these scars
that I wear from my working with stones and axes,
to your hungers, your sorrows, and your branded flesh,
because to calm our desperation of castigated bulls
we have to come together into an oceanic roar.
We will have to see the fields fertilized with wrongful
    blood,
We will have to see the fierce crescent of the sickle
    approaching the napes,
We will have to see it all nobly impassive,
We will have to do it all suffering a little less than what we
    suffer now from hunger,
that makes us reach out our innocent animal hands
toward robbery and crime, our saviors.[1]

NOTE

1   Miguel Hernández, "Smile at Me," *The Selected Poems of Miguel Hernández*, trans. Ted Genoways (Chicago: University of Chicago Press, 2001), pp. 135–37.

The Future of Intellectual Property
and Access to Medicine

Eloan dos Santos Pinheiro

In the future, would it be possible for essential medicines to be accessible to the majority of the population in developing countries? If we examine worldwide consumption of medicines today, at the end of the first decade of the twenty-first century, we find that 80 percent of transnational companies' income comes from developed countries such as the United States, Japan, Canada, and Western Europe. According to a report by the World Health Organization (WHO), 30 percent of the global population has no access to medicines. In poor areas of some countries in Asia and Africa, the population with no access to medicines is above 50 percent. The lack of public policies and the indifference of many governments have systematically excluded certain populations from a universal human right—the right to life.

This exclusionary situation was compounded by the fact that in 1994, transnational companies were able to insert intellectual property into the global system of free trade via the TRIPS Agreement, allowing them to exert more control over the worldwide market. Until the Uruguay Round of the General Agreement on Tariffs and Trade in 1986, there was greater flexibility for governments to decide what would be best for their own societies. Over fifty countries did not grant patents in areas considered strategic, such as medicines, and they did not suffer any decline in investments because of this decision. But the system of industrial patents was globalized according to the logic of maximizing profits at any cost, including social costs. The control of the market, including guarantees of noncompetitiveness by generic industries, has constrained developing countries by subjecting them to various types of pressure and threats.

Today, developing countries are technologically dependent, even when promoting appropriate public policies such as universal access to medicines for HIV/AIDS patients. They are required to pay high fees to the owners of the patents, who continue to affirm unfairly that the fees are necessary to recoup their investments. The right to life is subjugated to the right to property.

My greatest concern for the future is that strengthening patent owners' rights will have a negative effect on developing countries' efforts to improve public health and advance technologically and economically. This is particularly worrisome with respect to the effect of patents on the rising price of medicines and the availability of sources for pharmaceutical products.

In light of the present situation of the international intellectual property system, I will present and analyze two possible future scenarios of the regulatory framework for intellectual property in the future and their consequences for access to medicines and the inalienable right to life.

FIRST SCENARIO: STRENGTHENING INTERNATIONAL INTELLECTUAL PROPERTY RULES

The first scenario is based on the current status of the intellectual property system, along with the tridimensional tendency proposed by M. F. Jorge in which the TRIPS Agreement is a minimum standard in the multilateral sphere, while in the regional, bilateral, and local spheres, there is a strengthening of rules, commonly referred to as "TRIPS-plus."[1]

The flexibilities currently provided in the TRIPS Agreement—compulsory licensing, provisions for parallel imports in the Doha Declaration, and the understanding that retaliation by member countries is not allowed—are inadequate, because they do not ensure that when governments attempt to put these flexibilities to use, they are able to do so easily. The mechanisms are not simple, and many national legislatures have bureaucratic and legal impediments to putting compulsory licensing into practice.

National laws frequently foresee the importation of medicines and/or active ingredients acquired on the international market. However, currently, they link such acquisition to approval by the title bearer. They do not institute the title bearer's obligations with regard to fully disclosing the know-how for making them or to ensuring that the licensee can acquire the ingredients at accessible prices, and in the case of compulsory licenses, they do not define the procedure for acquiring or managing such licenses.

Although the current situation is thus already highly unfavorable for access to medicines, the situation could become even more unfavorable if bilateral agreements come to constitute the norm, if the proposed international patent is established, if a move to link patents to registration with regulatory agencies is approved, and if in 2016 the countries that are currently classified as very poor

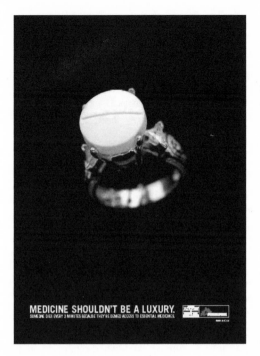

Poster from Médecins Sans Frontières's Access to Essential Medicines campaign.

lose their right not to recognize patents. Within the regulatory framework of patent concession, the situation for access to medicines will deteriorate if the criteria become more flexible in order to allow patents for second use (which is permitted, for example, in Brazilian legislation via procedures of examination of patentability) and patents for products already known to the public, but produced in new combinations such as 3TC+AZT+NVP for treating HIV/AIDS, Furosemida + Captopril for treating hypertension, and Artesunato + Mefloquina for treating malaria.

Under this scenario, the consequences will be catastrophic for access to medicines, because generic producers will not have access to the most recent and relevant technological information for processes, information necessary for the strategy of reverse engineering, which today is possible through the Bolar Exception, which holds that the rights of the patent do not apply to research activities or to the registration of the medicine for commercialization.[2]

Difficulties exist in applying compulsory licensing when availability of the product on the international market is absolutely lacking. This situation creates an absolute monopolistic power in which weapons are unnecessary for subjugating developing countries. In the future, if the decision about who will have access to products is consolidated in the hands of the holders of intellectual property rights, even when developing countries have policies that are just, democratic, and nonsubservient, even when civil-society participation is organized and coherent in order to ensure the transparency of public policy, the private rights granted by the state will overpower the rights of citizens.

SECOND SCENARIO: REFORM OF THE INTERNATIONAL SYSTEM OF INTELLECTUAL PROPERTY

The second scenario is based on the premise that the majority of United Nation member countries—developing countries—will realize that they need citizens who are healthy in order to survive as a nation and achieve development. They will come to this understanding after the dark period of the AIDS epidemic, when a significant portion of the population has died and others have had their life expectancy diminished. They will understand the necessity of establishing public policies guaranteeing treatment for their populations and that the transparency of the process will be guaranteed only through the participative representation of civil society. In essence, they will understand that human rights should prevail over private rights, guaranteeing access to essential and strategically important medicines.

They will consider that strategic investments are necessary for discovering new pharmaceuticals. Therefore, clear definitions will be established regarding what should be rewarded with prizes such as patents and how long these rewards should prevail for those discoveries that are in fact inventive and that result in a good for society.

The developed countries will argue vehemently that universal respect for intellectual property rights is a tool for advancing research and development, because exclusive market rights attract the investment of the pharmaceutical industry. Yet to which markets, precisely, does this argument refer? If we look at developing countries, we see that after the TRIPS Agreement was implemented at the end of 1994, there were no concrete results with regard to technology transfers or access to knowledge. Furthermore, issues such as the lack of an obligation for local production (a mechanism that promotes technological absorption) and for technological investments in developing countries remained merely issues under discussion, with no actions taken. For all these reasons, the first step in the future will be to delink intellectual property from the laws of commerce, returning intellectual property to its initial institutional locus, internationally, which is the World Intellectual Property Organization (WIPO).

This will break the global connection between intellectual property rights and the mechanisms for commercial sanctions between nations that do not follow World Trade Organization (WTO) rules, including TRIPS. The perverse relation

break link btw. trade & IP

embedded in TRIPS is the fact that unequals—developed nations and developing nations—are treated as equals. As Aristotle taught many years ago in the *Nicomachean Ethics*, injustice arises when unequals are treated equally.

The flexibility of the criteria for approving patents will also be changed, prohibiting the approval of patents resulting from the redeployment of already known molecules, patents for second use, patents to combine pharmaceuticals already used individually, patents for polymorphic structures, and so on. These changes will seek to valorize the inventive content of the discovery while understanding that, as the Commission on Intellectual Property Rights has noted, "Countries need to ensure that their IP protection regimes do not run counter to their public health policies and that they are consistent with and supportive of such policies."[3]

In order to stimulate growth in innovation and technology, it will be necessary to revoke patents for pharmaceutical products, allowing patents only for processes. This will allow new routes of chemical synthesis to arrive at the same product. It will stimulate new technologies and increase competition in the field of medicines.

To guarantee that the pubic interest prevails over private interests, for all diseases that are prevalent in developing countries, known as "neglected" diseases, the developed and developing countries will agree that the property rights of an invention will be acquired by the public partner, the research institution or university, that in fact originated the idea. They will agree that the private entity involved in the public-private partnership will have the right only to a nonexclusive license, which may contain clauses to provide proper rewards for its investment, since the public sector has the regulatory prerogative on the final price of the product. They will agree that the private partner can exploit the protected invention under an exclusive license only for nonpharmaceutical uses, for example, a formulation for veterinary use, that is, for a use other than to treat a life-threatening life disease. And finally, they will agree that it is the public partner that will have the right to produce, import, export, and license the protected invention to produce and commercialize drugs at low prices.

Making this scenario a reality will require a range of solutions based on an inclusive, plural vision, without power asymmetries between those who create the model and those who implement it. The UN and the WHO should establish centers of development in the least-developed countries, guaranteeing that during this period of transition, in order to deliver treatment to all those who need it, all strategic products under patent that are necessary to attend to great endemic diseases will be developed through the transfer of technology.

The participation of organized civil society also will be absolutely essential in order to strengthen this range of proposals and bring about the transition between models. Public participation and accountability will be needed to consider

collective rights as well as the relationship between the new model and the different public policies in each state, with an emphasis on social policies.

This second scenario protects the rights of the inventor and the possibility
of recuperating investments, as long as the invention of novel pharmaceuticals
involves substantial technical and scientific knowledge. In fact, it bans only what
interferes with access to medicines.

The key point for development is technological knowledge. Modifications in
the paradigms that govern the mechanisms of innovation are necessary to stimulate a favorable environment for technological development in the short term,
leading to capacity building for local industry and the strengthening of research
institutions that produce new inventions in the medium and long term.

Translated by Shanti Avirgan

NOTES

The author would like to acknowledge Fernanda Macedo for her contribution and comments
on this text.

1   M. F. Jorge, "TRIPS-Plus Provisions in Trade Agreement and Their Potential Adverse Effects
on Public Health," *Journal of Generic Medicines* 1, no. 3 (2004): pp. 99–211.

2   On the Bolar Exception, see http://www.wto.org/english/tratop_e/trips_e/factsheet_
pharm02_e.htm#bolar (last accessed January 14, 2010).

3   Commission on Intellectual Property Rights, *Integrating Intellectual Property Rights and
Development Policy* (London: Commission on Intellectual Property Rights, 2002), p. 39, available on-line at http://www.iprcommission.org/papers/pdfs/final_report/CIPRfullfinal.pdf
(last accessed January 14, 2010).

Options and Alternatives to Current Copyright Regimes and Practices

Hala Essalmawi

Under the existing regime of intellectual property rights, we already have options in international and national legal instruments, such as exceptions to copyright and limitations on it, to help improve access to knowledge. Are we exploiting these options to their full potential? And what are the alternatives to the current regime—what new initiatives, tools, and methods to disseminate knowledge are available? This essay explores possible futures by investigating two extreme scenarios in order to pose and answer these important questions: What can we expect in the next ten years in the field of copyright, particularly in developing countries? And how will the inevitable changes affect access to information and in turn education and libraries, especially research and academic libraries, and consequently access to knowledge?

WORST-CASE SCENARIO, 2018: A STRONGER, MORE STRICT INTELLECTUAL PROPERTY RIGHTS SYSTEM

In 2018, terms for the duration of copyright, even in the poorest countries, now frequently last for as long as 100 or even 150 years after a work has been published or released to the public. Some developing countries now have increased the term of copyright protection after signing free-trade agreements with the United States or the European Union. Regional harmonization processes then exported these higher standards to neighboring countries. For example, some Arab countries signed free-trade agreements with the United States during the previous decade. In 2015, Arab intellectual property laws were harmonized. This led to the increase of the term of protection in countries that had a term of life of the author plus 50 years to life of the author plus 100 years in all these countries. This has delayed indefinitely the transfer of creative works into the public domain, resulting in the weakest public domain in human history.

By 2018, the consequences of this policy have become evident in the slowing rate of innovation and the centralization of creativity in the hands of a few transnational corporations. This can be seen clearly in the fields of pharmaceuticals, information and communications technologies, and education. The price of medicine has doubled, negatively affecting public health everywhere, with poor countries being the worst affected. Education now has become a luxury, even at the elementary, compulsory level, not to mention at the university level. The high price of subscriptions to e-resources and printed materials has contributed to higher illiteracy rates in the Global South.

Even unoriginal databases are now fully protected by copyright.[1] Software that was once subject only to copyright protection now can be protected internationally by patent law. The public domain is shrinking, instead of growing and flourishing, despite the usual claims that the intellectual property system stimulates, rather than constricts the public domain.

Researchers and students in developing countries now have even less access to journals, books, articles, and new, Internet-based archives than ever before. For example, access to fifty-year-old Arabic scholarly articles in sociology is currently restricted to on-line journal subscribers. Moreover, the cost of e-journal subscriptions has soared, despite the lower costs of maintenance and distribution of on-line journals compared with those in traditional print media. A very small proportion of the population can afford the subscriptions, and most universities, with their limited budgets, now can provide access to very few internationally well-known journals. The astronomical costs associated with these subscriptions prohibit access for more than 90 percent of Arabic speakers. Furthermore, local and more specialized publications have either perished or become even more expensive.

By 2028, these trends have contributed to the decline of higher education in the Arab countries and in the Global South in general. This situation has forced more and more students and researchers to leave their home countries so that they can access the information they need to learn and work in richer countries in the Global North.

Although some groups of stakeholders have raised concerns about copyright issues, no real attention has ever been paid to building the infrastructure necessary for access. Means such as financial resources, tools, computer equipment, and Internet connections are scarce, compounding the overprotection characteristic of the current system of copyright restrictions.

Due to the increasing complexities of the existing system, in 2018, academics, researchers, librarians, content creators, and library patrons have very limited knowledge of copyright issues and of the options the system provides, including exceptions and limitations.[2] Because of the political or economic pressures that

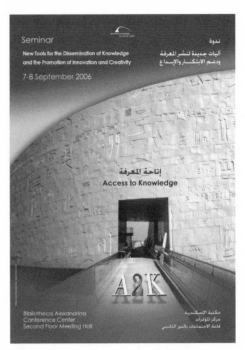

Poster for The Oil of the Twentieth-Century conference (http://oil21.org/?conference); poster for New Tools for the Dissemination of Knowledge conference, Alexandria and Cairo, September 2006.

developed countries exercise to prohibit the use of these flexibilities in developing countries, these options have become as useless as if they had never existed. Some vestiges still appear in the texts of laws, but nobody is aware of their existence, and nobody knows how to make use of them.

In 2018, content creators, users, and intermediary institutions are not aware of the new tools available to disseminate knowledge and as alternatives to facilitate access to knowledge. They are not familiar with the digital equipment needed to access the huge amount of information that has now been made available in digital form in order to promote and sustain development.

In such a world, where individuals are banned from accessing and enjoying the fruits of knowledge that have been accumulated through decades of human history, what are the options and alternatives? Either to suffer from illiteracy and poverty in a backward, unhealthy society or to seek ways to flee to a richer country. However, even with the regulations and policies current in the first decade of the twenty-first century, a citizen from a poor country could wait more than a year to be granted a visa to access a European or U.S. territory. By 2018, all these regulations have worsened the problems of illegal immigration, terrorism, and security, fostering increasingly fanatic ideologies. In 2018, the world is still not safe or fair. All suffer.

BEST-CASE SCENARIO, 2018: COPYRIGHT IN AN OPEN-ACCESS WORLD

By 2018, the unjustifiably long term of protection of copyright has led to questioning of whether the circumstances within which the Berne Convention, which was signed more than 120 years ago, are still valid today. A general consensus emerged that at the levels that prevailed at the turn of the twenty-first century, protection was in effect perpetual, if we consider the duration of protection and the average lifetime of a person. Most people would never have seen the majority of the works produced during their lifetime transferred to the public domain. Therefore, the copyright term was reexamined in the World Intellectual Property Organization (WIPO), which influenced World Trade Organization (WTO) negotiations later, and thanks to the intensive work of NGOs, experts, and academics, both developed and developing countries agreed to reach a win-win solution using determinants to these terms other than author's life.

Some countries chose to protect the work for twenty years from the date of making the work available to the public, while others chose a more sophisticated system that now allows monitoring the yearly revenues of the work. So when those revenues are below a certain amount for two years, the work falls into the public domain. Many countries also now have more than one determinant, depending on the nature of the protected work. The new terms give a real incentive to the author and give a chance for those who invest in the copyrighted materials to generate reasonable profits within a reasonable period from the time of making the work available to the public.

Moreover, by 2018, several initiatives have been launched to compile, document, and classify materials in the public domain in order to make them more accessible to as many people as possible. Systems have been developed to follow up on copyright terms and to produce freely accessible lists of works that fall into the public domain. Some of these systems also take into consideration that some works that may still be under protection in a country with a longer term of protection may already be in the public domain in a country with a shorter term of protection. In other words, proposals for the global harmonization of intellectual property laws that assume that one size fits all have failed.

By 2018, positive steps have been taken to use and activate the options already available in all the international instruments and national laws as exceptions and limitations to copyright, in order to facilitate the acquisition and utilization of protected materials. The real breakthrough was in finalizing the international treaty for exceptions and limitations that had been negotiated and discussed for over a decade. By 2018, several other alternative tools and methods to disseminate knowledge, such as free and open-source software, the GNU Free Documentation License, and Creative Commons licenses, have attained considerable success as technical and legal parallel substitutes for classically copyrighted materials and older copyright acts.

By 2018, higher awareness among stakeholders of users, creators, academics, researchers, officials, public-policy makers, and intermediary institutions working to make information available has altered the structure of copyright regimes and the approaches taken by these groups to balance the benefits that go to creative authors and those that society can claim. After long discussions and many struggles, free and open access to publicly funded research results has been guaranteed shortly after their publication. This milestone has been praised by the coalition of researchers, users, and libraries that has long called for such a step.

By 2018, governments have understood that they will boost innovation and get a better return on their investments in publicly funded research if they made research findings more widely available. Consequently, many countries, including developing ones, have adopted policies to promote and support the archiving of publications in open-access repositories after five years of publication. National and academic libraries now implement and manage these repositories. Evidence indicates that open access to research findings brings economic advantages across the world. It increases the potential benefits resulting from research and promotes scholarship in Global North and South.

The momentum is real, and figures show that in 2018, 40 percent of scholarly journals are now released under an open-access license, while free and open-source software now constitutes around 50 percent of all software in operation.

It is worth mentioning that these radical changes have not completely diminished earlier intellectual property rights systems. The intention of these developments was to modify previous systems to recognize and activate all possible solutions in favor of education, access to knowledge, and freedom of opinion and expression as essential human rights under the Universal Declaration of Human Rights, Articles 19, 26, and 27 of 1948.[3] From what can be seen today, in 2018, this goal has been largely achieved. This has also been done with ample space and tolerance for possible parallel alternatives to develop and flourish as complementary to, not superseding the previous systems.

These are two imaginary scenarios. They represent a personal vision. However, if all stakeholders do not move now to change the current situation of the intellectual property rights regime and practices by activating available options and developing new alternatives, we will be living the worst-case scenario—or in a reality that is even worse.

NOTES

1 Originality is a prerequisite for copyright. The criterion of originality is satisfied if the work in question is the work of the author and not a copy of another's work. The level of skill and labor required to satisfy the requirement of originality is generally satisfied provided that the creation of the work is not a purely mechanical exercise. The classic example of an unoriginal database is a telephone directory that is organized in alphabetical order.

2 Exceptions and limitations to copyright involve certain situations in which the exclusive rights granted to copyright holder by law do not apply.

3 The Universal Declaration of Human Rights (1948) is available on-line at http://www.un.org/en/documents/udhr (last accessed January 11, 2010).

The Golden Touch and the Miracle of the Loaves

Roberto Verzola

Dionysus...decided to reward Midas for his hospitality and granted him one wish. Midas wished that everything he touched be turned to gold. Dionysus warned him about the dangers of such a wish, but Midas was too distracted with the prospect of being surrounded by gold to listen. Dionysus gave him the gift. Initially, King Midas was thrilled with his new gift and turned everything he could to gold, including his beloved roses. His attitude changed, however, when he was unable to eat or drink since his food and wine were also changed to unappetizing gold. He even accidentally killed his daughter when he touched her, and this truly made him realize the depth of his mistake.

—Anna Baldwin, s.v. "Midas," *Encyclopedia Mythica*

The best and the worst scenarios of access to knowledge can be seen today in two opposite trends. In the genetic field, islands of proprietary genetic material are growing amid a sea of free and open-access biodiversity, while in the information field, islands of free and open-access initiatives are growing amid a sea of proprietary resources.

THE PRIVATIZATION OF GENES

In agriculture and genomics, a race to patent and thereby privately own genes continues unabated. According to a 2005 study, one-fifth of the human genome has already been patented. Patents are exclusionary devices and are therefore a form of private monopoly, in effect turning genes into private property. Genes are a natural monopoly. As Robert Cook-Deegan, the director of Duke University's Center for Genome Ethics, Law and Policy says, "You can find dozens of ways to heat a room besides the Franklin stove, but there's only one gene to make human growth hormone."[1]

The privatization of genes is, of course, a prelude to commodification. Commodified goods—or bads, for that matter, such as carbon credits representing a right to pollute—then become subject to market mechanisms and forces. If there is carbon trading, can DNA trading be far behind? If we can have commodity futures, why can't we have derivatives such as carbon futures or DNA futures?

Commodification is an all-consuming trend in economics. Commodification respects none and targets all: land, culture, knowledge, information, human beings, water, air, nature, life, genes, relationships—truly anything and everything. Driven by corporate profit seeking and gain maximization, commodification knows no end, no limits.

Like King Midas, today's corporations and other gain maximizers turn everything they touch into commodities and, subsequently, into money. Wherever they look, whatever they look at, they see a dollar sign. If we followed their lead or allowed them to continue, our entire world and everything in it as well as outside it would sooner or later be for sale or for rent. Then we would end up like the cynic who "knows the price of everything and the value of nothing."

THE FREE AND OPEN-SOURCE TREND

There is, fortunately, an opposite trend. Ideas about information freedom and sharing have percolated for some time. They are called by different names, representing subtle differences in attitudes, perspectives, and approaches toward access to information and knowledge. The earliest were ideas about the public domain and the commons. But as everything turned digital, accelerating the commodification of information, these early ideas were apparently insufficiently developed to deal with the rapidly changing nature of information. For instance, when software was simply released to the public domain, commercial interests were better positioned to take full advantage and incorporate it into their products. Also, object code (in contrast to source code) in the public domain remained largely inaccessible for modification. Thus, while many software utilities and simple programs were distributed as public domain or "freeware," no major software projects were. New approaches were also tried that relied on a license based on existing intellectual property concepts. These included the "shareware" license, the GNU Public License, the BSD License, and their variations. Out of the latter two emerged truly huge software projects such as the Linux kernel, the GNU systems-and-utilities package, the BSD operating system, software application suites such as OpenOffice, and similar software.

The idea of free and open sharing caught on and extended to other fields. The Creative Commons license extended this idea to other literary and artistic works. The Wikipedia represented another huge effort to accumulate and share human knowledge in a completely nonproprietary way.

This new social movement might be called the "free and open-source information movement." It is now being embraced in other fields and promises to become the guiding principle for access to knowledge. In the academic community, free and open on-line journals are now emerging in the spirit of this movement, challenging the entrenched publishers of printed academic journals.

In the future, this movement may merge with other "free" and "open" movements. In the educational field, a "free" schools movement—"free as in freedom"—has been simmering for some time, following the pioneering works of educators Maria Montessori in Italy, A. S. Neill of Summerhill fame in England, and John Holt in the United States. Among the ideas that contributed to the intellectual ferment and the eventual peaceful uprising of the East Europeans was the "open" society concept given impetus by George Soros. The free exchange and sharing of seeds is a freedom that farmers will defend with their lives. If a convergence happens, a truly historic shift in mindset can occur, promising a freer and more open world.

We have to divide bread to share it, but sharing knowledge multiplies it. Because knowledge is literally food for the mind, the movement to ensure free and open access to information and knowledge will turn into reality the parable in this Biblical story:

> In those days when there again was a great crowd without anything to eat, he summoned the disciples and said, "My heart is moved with pity for the crowd, because they have been with me now for three days and have nothing to eat. If I send them away hungry to their homes, they will collapse on the way, and some of them have come a great distance." His disciples answered him, "Where can anyone get enough bread to satisfy them here in this deserted place?" Still he asked them, "How many loaves do you have?" "Seven," they replied. He ordered the crowd to sit down on the ground. Then, taking the seven loaves he gave thanks, broke them, and gave them to his disciples to distribute, and they distributed them to the crowd. They also had a few fish. He said the blessing over them and ordered them distributed also. They ate and were satisfied. They picked up the fragments left over—seven baskets. There were about four thousand people.[2]

If we join the commodification race, we will all acquire the golden touch. If we adopt the free and open sharing perspective, the knowledge of some can miraculously feed all. The golden touch or the miracle of the loaves? Whichever road we take will determine whether we will enter a neofeudal period ruled by information and genetic rentiers as they increasingly privatize human knowledge and genetic material or a new flowering of human culture, thanks to free exchange of ideas, information, and knowledge.

NOTES

1 Stefan Lovgren, "One-Fifth of Human Genes Have Been Patented," *National Geographic News*, October 13, 2005, available on-line at http://news.nationalgeographic.com/news/pf/22064243.html (last accessed January 14, 2010).

2 Mark 8:1–9; see also 6:34–44.

Contributors

AHMED ABDEL LATIF, an Egyptian national, is program manager for intellectual property and sustainable development at the International Centre for Trade and Sustainable Development (ICTSD) in Geneva, an NGO whose mission is to influence the international trade and intellectual property system to advance the goal of sustainable development. As a career diplomat, he worked at the Permanent Mission of Egypt to the United Nations in Geneva from 2000 to 2004, where he was a delegate to the TRIPS Council of the World Trade Organization (WTO) and to the World Intellectual Property Organization (WIPO), as well as coordinator of the African Group at WIPO (2004). He was closely involved in efforts to launch the WIPO Development Agenda and in the emergence of the access to knowledge (A2K) movement.

HARINI AMARASURIYA is a doctoral student in social anthropology in a joint PhD program with the University of Edinburgh and Queen Margaret University. She has worked in the development and humanitarian sector for the past fifteen years as a practitioner and researcher. She has been a visiting lecturer at the University of Colombo for the Post Graduate Diploma in Counseling and Psychosocial Work and a temporary lecturer at the Open University of Sri Lanka in the Department of Social Studies.

JEFFREY ATTEBERRY currently practices civil litigation. He has a PhD in comparative literature and critical theory from the University of California, Irvine and a JD from the University of California, Berkeley, School of Law. His current research interests include postcolonial perspectives on international law and the legal character of the relationship between the state and civil society. He has previously published articles in *Modern Language Notes*, *Modern Fiction Studies*, and *Critical Horizons*.

PHILIPPE AIGRAIN is the CEO of Sopinspace, the Society for Public Information Spaces, a company specializing in free software and services for the public debate of policy issues by citizens. He was head of the software technology sector in the European Commission Information Society Directorate-General from 1998 to 2003. He is an advocate for the recognition of the information commons in international norms and is a member of the Quadrature du Net in France.

YOCHAI BENKLER is the Berkman Professor of Entrepreneurial Legal Studies at Harvard and faculty codirector of the Berkman Center for Internet and Society. In the 1990s, he played a role in characterizing the centrality of the information commons to innovation, information production, and freedom in both its autonomy and democracy senses. In the 2000s, he worked more on the sources and economic and political significance of radically decentralized individual action and collaboration in the production of information, knowledge, and culture. His work can be freely accessed at benkler.org.

YANN MOULIER BOUTANG is an activist, writer, socioeconomist, and researcher, and the editor of the French journal *Multitudes.* He is also a professor of economics at the University of Tecnologia of Compiègne, Costech, and Centre d'Economie de la Sorbonne. He is the originator of the concept of cognitive capitalism, which he developed in *Le capitalisme cognitif: La nouvelle grande transformation.*

CARLOS M. CORREA is director of the Center for Interdisciplinary Studies on Industrial Property and Economics and of the Post-Graduate Course on Intellectual Property at the Law Faculty, University of Buenos Aires. He has been a consultant to several regional and international organizations, such as United Nations Conference on Trade and Development (UNCTAD), the United Nations Industrial Development Organization (UNIDO), the United Nations Development Programme (UNDP), the World Health Organization (WHO), the Food and Agriculture Organization (FAO), the Inter-American Development Bank, the World Bank, and others. He is the author of several books and numerous articles on law and economics, particularly on investment, technology, and intellectual property.

PETER DRAHOS is a professor in the Regulatory Institutions Network at the Australian National University and Director of the Centre for the Governance of Knowledge and Development. His areas of research and publication include globalization, regulation, trade, intellectual property rights, and development. He has worked as a consultant on international intellectual property issues for a number of organizations, including the European Commission, the UK Commission on Intellectual Property Rights, the Commonwealth Secretariat, Oxfam, and the ASEAN Secretariat.

LAURA DENARDIS is the executive director of the Yale Information Society Project and a lecturer in law at Yale Law School. She is an Internet scholar with a research concentration in Internet governance, Internet protocols, and network security. She is the author of *Protocol Politics: The Globalization of Internet Governance* (MIT Press, 2009), *Information Technology in Theory*, with Pelin Aksoy (Thompson, 2007), and numerous book chapters and articles. She received a PhD in science and technology studies from Virginia Tech.

SARAH DEUTSCH is vice president and associate general counsel for Verizon Communications. Her practice covers a wide range of legal issues in the areas of global intellectual property issues, Internet policy, liability, privacy, and Internet jurisdiction. She has represented Verizon on a host of domestic and international Internet issues, ranging from Internet governance and domain-name issues and ICANN to digital copyright issues, cybercrime, and international copyright issues. She was one of five negotiators for the U.S. telecommunications industry who negotiated service-provider provisions that resulted in the passage of the Digital Millennium Copyright Act.

HALA ESSALMAWI is an attorney at law and the intellectual property rights officer at the Library of Alexandria (Bibliotheca Alexandrina, BA). She is the project leader of the BA A2K initiative and editor of its Web site, www.bibalex.org. She obtained her LLB and LLM from the Faculty of Law, Alexandria University; a postgraduate Diploma in International Law and Development from the Institute of Social Studies, The Hague; and an LLM in intellectual property from Turin University, Italy. She is the Intellectual Property Resource Person for the International Development Law Organization, IDLO, Rome, IP impact Program in Egypt, the representative of the eIFL-IP network in Egypt, and the project leader for Creative Commons in Egypt.

RICK FALKVINGE is the founder and leader of the Swedish Pirate Party, as well as the founder of the international politicized pirate movement. His leadership took the Pirate Party from nothing into the top ten parties in the last Swedish election, without a dime in the campaign chest. Rick's personal candidacy came in at rank fifteen out of over five thousand candidates for the 349 parliamentary seats. While he didn't win a seat due to threshold rules, his fight for civil liberties continues, focusing on the current copyright aggression that threatens our rights to privacy, postal secrets, whistleblower protection, and more.

SEAN FLYNN is associate director of the Program on Information Justice and Intellectual Property (PIJIP) and professorial lecturer in residence at American University's Washington College of Law. He served as a consultant to the South African

Competition Commission in its evaluation of the case of *Hazel Tau v. GlaxoSmith-Kline, et. al.* PIJIP conducts research and supports advocacy on access-to-medicines issues, including the role of competition law in promoting access to medicines. See http://www.wcl.american.edu/pijip.

VERA FRANZ has been program manager at the Open Society Institute's Information Program since 2002, heading the Program on Intellectual Property Rights Reform and Open Knowledge. She has helped to build the A2K movement through grant giving and strategic organizing. In addition to her job at OSI, she has been teaching a course on media, technologies, and globalization in the department of communications at Salzburg University, Austria. Previously, she was affiliated with the Austrian Techno-Z R&D Institute, researching information-market developments for the European Commission. She holds a master's degree in media and communications from the London School of Economics and a Mag.Phil. in political science from Salzburg University, Austria.

SPRING GOMBE, a senior health policy advisor, has fifteen years' experience in health policy analysis. She currently serves as a consultant to a number of public interest civil society organizations working in access to medicines. She has worked with Knowledge Ecology International, as a senior health policy analyst, focusing on innovation and access and the intergovernmental negotiations on public health, innovation, and intellectual property. She was a senior policy officer with Health Action International-Europe, where she worked on access-to-medicines issues. She has also worked with Médecins Sans Frontières (MSF) in Uganda and Germany, with the Institute for Urban Family Health in New York, with Oxfam International in Barcelona, and as a program coordinator with the Chicago Department of Public Health and the Chicago Health Corps, an AmeriCorps program. She holds a bachelor of science degree in biological sciences from Cornell University and completed postgraduate studies in development policy at New York University.

ANIL K. GUPTA is a professor at the Indian Institute of Management, Vastrapur, Ahmedabad. He is executive vice chair of the National Innovation Foundation and President of the Society for Research and Initiatives for Sustainable Technologies and Institutions (SRISTI), which aims at strengthening the capacity of grassroots-level innovators and inventors engaged in conserving biodiversity. He is also the editor of the newsletter of the Honeybee Network, a global network of activists, scholars, policy makers, and farmers organized around the issue of indigenous technological and institutional innovations for sustainable natural resource management and building bridges between formal and informal science.

CHARLES IGWE is principal consultant at the Big Picture, a motion picture business and services company. A banker by profession, though trained in medical sciences, he is a pioneer investor and operator in the Nigerian movie industry, now known to the world as Nollywood. He left Citibank in the early years of the Nigerian motion picture industry revival to provide the vital business infrastructure for the emerging industry. He is a cofounder of the African film and Television Programmes Expo: BOBTV, which is in its sixth year.

AMY KAPCZYNSKI is an assistant professor of law at the University of California at Berkeley Law School. Her research occupies the intersection between social movement theory and transnational law, particularly in the domain of intellectual property/access to knowledge.

KANNIKAR KIJTIWATCHAKUL is currently coordinator of the Health Consumer Protection Programme at Chulalongkom University, Bangkok, Thailand. She is a veteran journalist who has been working as a campaigner for access to essential medicines with Médecins Sans Frontières (MSF)–Belgium (Thailand) since 2006. She has written several books and articles on access-to-medicine and intellectual property issues, including "Drug Patent: Rich People's Darling" (2005) and "The Right to Life" (2007). She is also a coauthor of "Sustaining Access to Antiretroviral Therapy in Developing Countries: Lessons from Brazil and Thailand" published in *AIDS 2007*.

GAËLLE KRIKORIAN is a PhD candidate in sociology at the École des Hautes Études en Sciences Sociales in Paris, working with the Institute for Interdisciplinary Research on Social Issues (IRIS). She is focusing on the way health matters are taken into account by governments in trade negotiations and the impact of collective action. She worked for many years with Act Up–Paris on access to medicines in developing countries, then was a consultant for NGOs and organizations in France, the United States, and Morocco. She is the coeditor, with Michel Feher and Yates McKee, of *Nongovernmental Politics*, as well as the author of many articles, mostly on intellectual property, free trade, health and HIV/AIDS.

LAWRENCE LIANG is an Indian legal researcher and lawyer based in the city of Bangalore. He is a cofounder of the Alternative Law Forum, and, as of 2006, emerged as a spokesperson against the politics of intellectual property. He has been working closely with Sarai, New Delhi, on a joint research project on intellectual property and the knowledge/culture commons. He has been working on ways of translating the open-source ideas into the cultural domain. He is author of *Sex, Laws, and Videotape: The Public Is Watching* and *Guide to Open-Content Licenses*, published by the Piet Zwart Institute in 2004. He is currently working on a book on the idea of cinematic justice in India.

JIRAPORN LIMPANANONT is currently chairperson of the Foundation for Consumers (FFC) and Associate Professor at the Social Pharmacy Research Unit, Faculty of Pharmaceutical Sciences, Chulalongkorn University, Bangkok, Thailand. She is also a member of the Free Trade Agreement Watch coalition, which denounces the negative impacts of bilateral and regional trade agreements. She has been an integral member of and academic advisor to the Drug Study Group (DSG), a network of university academics. The DSG contributed to the successful campaign against the GlaxoSmithKline patent application on Combid and the three-year-long lawsuit filed against Bristol-Meyers Squibb, which resulted in the revocation of the patent on ddl (didanosine). She was also active in efforts to obtain the compulsory licensing of patents on essential medicines.

JAMES LOVE is the director of Knowledge Ecology International, a nongovernmental organization with offices in Washington, D.C. and Geneva. In 2006, KEI received the MacArthur Award for Creative and Effective Institutions. He was previously senior economist for the Frank Russell Company, a lecturer at Rutgers University, and a researcher on international finance at Princeton University. He received a master of public administration degree from Harvard University's Kennedy School of Government and a master's degree in public Affairs from the Princeton's Woodrow Wilson School of Public and International Affairs.

LEENA MENGHANEY currently works as project manager for the Campaign for Access to Essential Medicines, Médecins Sans Frontières (MSF), in India. A lawyer by training, she worked on the campaign to ensure that India's new patent law included public health safeguards to limit the impact of patents on access to affordable medicines. She then fought alongside local activists to defend these safeguards when the drug company Novartis sought to have them overturned. She continues to fight for access to a number of Indian generic medicines. She is the author of "HIV/AIDS Treatment Legal and Political Choices for India," *Journal of Creative Communications* 1, no. 2 (2006).

VIVIANA MUÑOZ TELLEZ is a program officer at the South Centre, an intergovernmental organization of developing countries based in Geneva, Switzerland. She conducts policy-oriented research and provides policy advice on innovation, access-to-knowledge and intellectual property issues, including negotiations at the World Trade Organization (WTO) and World Intellectual Property Organization (WIPO). She holds a master's degree in development management from the London School of Economics. Previously, she worked at Queen Mary Intellectual Property Research Institute, University of London, as a research assistant for an Economic and Social Research Council (ESRC) project (www.ipngos.org).

SISULE F. MUSUNGU, a Kenyan national, is president of IQsensato (www.iqsensato. org), a Geneva-based policy think tank that serves as a platform for promoting the research and thinking of developing-country researchers and experts in international policy making. His research specialization and expertise are in intellectual property law and policy, particularly the relationship between various categories of intellectual property and development; knowledge governance and access to knowledge; innovation for development; and international human rights law, especially issues related to the implementation of economic, social, and cultural rights. He also has an interest and has participated in a number of scenario-building processes. He is widely published in his areas of specialization.

HEESEOB NAM is a patent attorney in South Korea. He is also executive director of IPLeft, a social group created in 1999 to advocate for an information commons. IPLeft has been concerned about the social and digital divide, has criticized the strengthening of intellectual property rights, and has researched alternative policies against existing intellectual property rights. IPLeft launched the Korean Open Access License (KOAL), an information open-access model, in October, 2004. With others, he submitted several requests for compulsory licenses to the Commissioner of the Korean Intellectual Property Office. He also drafted Korean legislation for the granting of compulsory licenses for exporting pharmaceutical products.

CHAN PARK was formerly with the Lawyers Collective HIV/AIDS Unit in New Delhi, working with the Affordable Medicines and Treatment Campaign. He was involved in many mobilizations opposing the grant of questionable patents in Indian Patent Offices and contributes to the filing of pregrant motions opposing the patenting of many essential drugs. He is currently based in India, researching issues related to intellectual property and access to medicines. He is the author, with Arjun Jayadev, of "Access to Medicines in India: A Review of Recent Concerns" (July 20, 2009), available at SSRN: http://ssrn.com/abstract=1436732.

ELOAN DOS SANTOS PINHEIRO has a degree in chemistry with a specialization in pharmaceutical technology. She spent thirty years working in the pharmaceutical industry. She is currently working as a consultant in Brazil and other countries. She is former director of Farmanguinhos, a public pharmaceutical laboratory in Brazil that was one of the first local firms to produce generic antiretroviral medicines. She has also worked for the AIDS Medicines and Diagnostics Service (AMDS) of the World Health Organization Department of HIV/AIDS in Geneva. With Octavio Augusto Ceva Antunes Fortunak, she recently authored a survey of the syntheses of active pharmaceutical ingredients for antiretroviral drug combinations critical to access in emerging nations.

ACHAL PRABHALA is a researcher and writer based in Bangalore, India. Between 2004 and 2006, he coordinated a campaign for access to learning materials in Southern Africa, working from Johannesburg, South Africa, as part of a collective of consumer organizations and university groups. He has worked with the Alternative Law Forum on a proposed overhaul of the Indian Copyright Act and with the Lawyers' Collective HIV/AIDS Unit on patents and access to medicines in India. He is a research advisor to the African Copyright and Access to Knowledge Project at the Link Centre, University of the Witwatersrand, and serves on the Advisory Board of the Wikimedia Foundation.

ONNO PURBO is a former electrical engineer who has dedicated his life to promoting the dissemination of knowledge through information and communication technologies in Indonesia, both as a professor at the Institute of Technology, Bandung, and as the author of numerous articles and books. He has been a Research Fellow with the Information and Communication Technologies for Development Program area of the International Development Research Centre, meeting policy makers in developing countries to discuss new initiatives. Since September 2008, he has been, among many other things, active as a qualified trainer at the Wireless University, an open-access repository of educational materials, training people to provide wireless Internet access throughout the world.

MANON A. RESS is the founding editor of *Knowledge Ecology Studies*, an open, peer-reviewed policy journal. At Knowledge Ecology International, she is currently focusing on intellectual property issues, building public awareness of the debate over the value of the public interest in intellectual property rights. She is an active participant at World Intellectual Property Organization's meetings of the Standing Committee on Copyright and Related Rights and in other multilateral forums where access-to-knowledge issues intersect with intellectual property laws. She works on promoting fair-use rights, open standards, open-access publishing, and other e-commerce and Internet-related consumer-protection issues.

CAROLINA ROSSINI is a Brazilian lawyer. She currently acts as a consultant for the Vision Team of the USAID Global Development Commons Project, focused on open educational resources, intellectual property policy, and community building. She is also a research associate at the Berkman Center for Internet and Society at Harvard Law School for the Copyright for Librarians Project. She is affiliated with the Diplo Foundation as an expert on intellectual property law and coordinates the intellectual property research area. In Brazil, she was the coordinator of Legal Clinical Programs at the Getulio Vargas Foundation School of Law, where she was also a law professor and coordinated the Open Business Latin America, a project of Creative Commons Brazil.

SUSAN K. SELL is a professor of political science and international affairs and director of the Institute for Global and International Studies at the George Washington University in Washington, D.C. She is the author of *Private Power, Public Law: the Globalization of Intellectual Property Rights* (Cambridge University Press, 2003) and, with Christopher May, *Intellectual Property Rights: A Critical History* (Lynne Rienner, 2005). She is currently working on a project entitled "Who Governs the Globe?" and another project entitled "Books, Drugs, and Seeds: The Politics of Intellectual Property." She serves on the board of IP-Watch in Geneva (www. ip-watch.org.

SANGEETA SHASHIKANT is senior legal adviser to the Third World Network, as well as coordinator of the Third World Network office in Geneva. She is a regular contributor to the South-North Development Monitor, which provides information and analyses on international-development issues from the Southern perspective, with particular focus on North-South and South-South negotiations. She is the author of numerous articles on intellectual property, public health, international trade, and climate change. She recently published *The Substantive Patent Law Treaty: The Dangers of Global Patent Policy Harmonization* (Third World Network, 2009).

ELLEN 'T HOEN is an expert in intellectual property and medicines policy. She was director of policy and advocacy of the Campaign for Access to Essential Medicines of Médecins Sans Frontières (MSF) from 1999 until 2009. She is currently senior adviser for Intellectual Property and Medicines Patent Pool at UNITAID, an international facility for the purchase of drugs against HIV/AIDS, malaria, and tuberculosis. In 2005 and 2006, she was listed as one of the fifty most influential people in intellectual property in the world by the journal *Managing Intellectual Property*. She has a master's degree in law from the University of Amsterdam and in 2009 was a visiting fellow at the IS HIV/AIDS Academy of the University of Amsterdam, doing research on the implementation of the WTO Doha Declaration on TRIPS and Public Health.

ROBERTO VERZOLA is a social activist and a trained engineer who has focused on the social impact of new technologies. His current involvements include the Philippine Greens, Halalang Marangal (Network of Citizens for Honest Elections and Truthful Statistics), SRI-Pilipinas, and the Bulletin Board. He is also a part-time consultant with the Philippine Rural Reconstruction Movement. He has done policy studies as well as having been involved in political actions involving a wide range of issues, from nuclear power and clean renewables to genetic engineering and organics. He has written *Towards a Political Economy of Information* (2004).

JO WALSH is an open-source programmer and author who became involved with issues of access to geographic information during the writing of *O'Reilly Media's Mapping Hacks*. She works on metadata and structured data search projects. Her interests lie in the intersection of the semantic Web, spatial modeling, community wireless, public transport, and bots, helping build different open-source software projects to augment the semantic Web and bring knowledge representation to more people. She has worked with the Open Knowledge Foundation to transplant open-source practices into open data. She was a founding director and is now a board member of the Open Source Geospatial Foundation.

DILEEPA WITHARANA has done undergraduate and postgraduate studies in the field of electrical engineering and is currently working as a senior lecturer in the department of mathematics and philosophy of engineering of the Open University of Sri Lanka. His research interests are in the fields of energy, the philosophy of engineering, and globalization. He is currently involved with an interfaculty study conducted by the Open University of Sri Lanka on intellectual property rights and access-to-knowledge initiatives in Sri Lanka.

Series design by Julie Fry
Typesetting by Meighan Gale
Image placement and production by Julie Fry
Printed and bound by Thomson-Shore, Inc.